Basic Conversion Constants

Quantity	To convert from	To	Multiply by
Length	foot	meter	0.3048
	mile (statute)	kilometer	1.609344
	mile (nautical)	mile (statute)	1.150779448
	mile (nautical)	kilometer	1.852
	yard	meter	0.9144
Force	pound	newton	4.4482216152605
	dyne	newton	10^{-5}
Density	$slug/ft^3$	$kilogram/m^3$	515.4
	lbm/ft^3	$kilogram/m^3$	16.018463
	gm/cm^3	$kilogram/m^3$	1000
Mass	lbm	kilogram	.45359237
	slug	kilogram	14.59
	slug	$lb\text{-}sec^2/ft$	1
	lbm	slug	0.031081
Pressure	lb/ft^2	N/m^2	47.880258
	lb/in^2	N/m^2	6894.7572
	Pa	N/m^2	1
	bar	N/m^2	100,000
	atmosphere	N/m^2	101,325
	1 in Hg (0°C)	N/m^2	3386
Temperature	Celsius	kelvin	Add 273.15
	Fahrenheit	Rankine	Add 459.67
	Fahrenheit	Celsius	(T-32) × 5/9
Kinematic viscosity	ft^2/sec	m^2/s	.09290304
	stoke	m^2/s	10^{-4}
Viscosity	poise	$N\text{-}s/m^2$	0.1
Power	Btu/sec	kW	1.054350264488
	calorie/s	watt	4.184
	ft-lb/sec	watt	1.3558179
	Joule/s	watt	1
	N-m/s	watt	1
	horsepower	ft-lb/sec	550
	horsepower	kW	0.74569987

Useful constants

Acceleration of gravity $= 32.174$ ft/sec$^2 = 9.80665$ m/s^2

Specific heat of water $= c_p = 1$ kcal/kg-°C $= 4187$ N-m/kg-°C $= 1$ Btu/lbm-°R

Specific heat of air $= c_p = 1005$ joule/kg-°K $= 0.240$ Btu/lbm-°R

Gas constant for air $= R = 1715$ ft-lb/slug-°R $= 286.8$ N-m/kg-°K

Specific gravity of mercury $= 13.56$ (at 15°C)

FLUID MECHANICS

FLUID MECHANICS

James A. Liggett

Cornell University

McGraw-Hill, Inc.

New York   St. Louis   San Francisco   Auckland   Bogotá
Caracas   Lisbon   London   Madrid   Mexico City   Milan
Montreal   New Delhi   San Juan   Singapore   Sydney   Tokyo   Toronto

FLUID MECHANICS
International Editions 1994

Exclusive rights by McGraw-Hill Book Co. - Singapore for manufacture and export. This book cannot be re-exported from the country to which it is consigned by McGraw-Hill.

2 3 4 5 6 7 8 9 0 KKP FC 9 8 7 6 5

This book was set in Times Roman by Electronic Technical Publishing Services.
The editors were B. J. Clark and John M. Morriss;
the production supervisor was Paula Keller.
The cover was designed by Carla Bauer.
Project supervision was done by Electronic Technical Publishing Services.

Library of Congress Cataloging-in-Publication Data

Liggett, James A.
    Fluid mechanics / James A. Liggett.
        p.   cm.
    Includes index.
    ISBN 0-07-037805-3
    1. Fluid mechanics.   I. Title.
QA901.L54   1994
532-dc20   93-28418

When ordering this title, use ISBN 0-07-113449-2

Printed in Singapore

ABOUT THE AUTHOR

James A. Liggett is Professor of Civil and Environmental Engineering at Cornell University where he has taught since 1961 except for several sabbatical leaves. These include positions at the National Center for Atmospheric Research in Boulder, Colorado; at the Universidad del Valle and the Corporación Autónoma del Cauca in Cali, Colombia; at the University of New South Wales and the University of Adelaide, Australia; at the University of Canterbury, New Zealand; and the Instituto de Investigaciones Eléctricas in Cuernavaca, Mexico. He is the author of many publications in hydraulics, fluid mechanics, and computational fluid mechanics, and at various times in his career has worked in almost all phases of hydraulics. He has served as editor of the *Journal of Hydraulic Engineering*.

CONTENTS

Preface   xvii

Corrigenda   xxi

Notation   xxiii

1 The Basic Equations of Fluid Mechanics   1

1.1	Methods of Description	1
1.2	The Control Volume Equation	3
1.3	Conservation of Mass—The Equation of Continuity	7
1.4	Conservation of Momentum—The Equation of Motion	8
1.5	Conservation of Energy—The Energy Equation	11
1.6	Time Derivatives	14
1.7	Translation, Rotation, and Rate of Strain	16
1.8	Stress and Rate of Strain	19
1.9	Thermal and Mechanical Energies	27
1.10	Perfect Fluids	28
1.11	Tube Flow	32
1.12	The Vorticity Equations	34
1.13	Boundary Conditions	36
1.14	The Lagrangian Equations	38
	Problems	41
	Appendix	42

2 Similitude and Modeling   45

2.1	Modeling, Dimensions, and Units	45
2.2	The Pi-Theorem	47
2.3	Similitude	51
2.4	Incomplete Similitude	54
2.5	Froude and Weber Modeling	56
2.6	Approximate Similitude	57
2.7	The Reynolds Number	59
2.8	Compressible Flows	63

2.9	Cavitation Modeling	65
2.10	Distorted Models	66
2.11	A Few Dimensionless Numbers	69
2.12	Conclusion	78
	Problems	79

3 Potential Flow (Fundamentals)   **80**

3.1	Applications	80
	3.1.1 Incompressible, Inviscid Flow	80
	3.1.2 Compressible, Linearized Flow	81
	3.1.3 Flow through Porous Media	83
	3.1.4 Waves and Free Surface Flow	84
3.2	The Stream Function	85
3.3	Flow Nets	87
3.4	The Fundamental Singularities	89
	3.4.1 Parallel Flow	89
	3.4.2 Source (Sink) Flow	90
	3.4.3 Vortex Flow	91
3.5	Superposition	92
3.6	Circulation, Rotation, and Vorticity	93
	3.6.1 Circulation and Vorticity	93
	3.6.2 Vorticity	95
	3.6.3 Vortex Lines and Vortex Motion	97
	3.6.4 Vortex Sheets	99
3.7	The Doublet	101
3.8	The Method of Images	104
3.9	Axisymmetric Flows	105
	3.9.1 The Fundamental Singularities	106
	3.9.2 The Line Source	107
	3.9.3 Line Doublets	109
3.10	Surface Sources	109
3.11	Unsteady Flow	110
	3.11.1 The Moving Sphere	111
	3.11.2 Virtual Mass	112
	3.11.3 Kinetic Energy	114
3.12	Complex Notation	117
	3.12.1 Basics	117
	3.12.2 The Fundamental Singularities	119
	3.12.3 Conformal Mapping	119
	3.12.4 Cauchy and Blasius	122
	3.12.5 The Schwarz-Christoffel Theorem	125
	Problems	129

4 Potential Flow (Free Surfaces)   **133**

4.1	The Equations of Free Surface Flow	134
	4.1.1 The Kinematic Boundary Condition	134
	4.1.2 The Dynamic Condition	136
	4.1.3 Calculations	137
	4.1.4 The Steady State	138

4.2		Porous Media	139
	4.2.1	The Kinematic Condition	139
	4.2.2	The Pressure Condition	141
	4.2.3	The Seepage Surface	141
4.3		Far-field Conditions	142
4.4		The Use of Complex Variables	145
	4.4.1	Hydrodynamics	145
	4.4.2	Porous Media	150
		Problems	153

5 Inertialess Flow — 155

5.1		Tube Flow	155
	5.1.1	Couette and Poiseuille Flow	156
	5.1.2	Axisymmetric Tube Flow	157
	5.1.3	Flow in an Arbitrary Cross-section	158
5.2		Stokes' Flow	160
	5.2.1	The Equations of Slow Flow	161
	5.2.2	Properties of Slow Flow	161
	5.2.3	Slow Flow in Infinite Domains	162
5.3		Hele-Shaw Flow	165
	5.3.1	The Hele-Shaw Equations	166
	5.3.2	The Approximation	167
5.4		Lubrication	168
		Problems	170

6 Laminar Flow — 172

6.1		Stagnation Flow	173
	6.1.1	Two-dimensional Stagnation Flow	173
	6.1.2	Axially Symmetric Stagnation Flow	175
6.2		The Boundary Layer Approximation	177
6.3		Separation (Stall)	182
	6.3.1	Definition of Separation	182
	6.3.2	Some Consequences of Separation	184
	6.3.3	Unsteady and Three-dimensional Separation	187
6.4		Boundary Layer Thickness	191
6.5		Solution for the Boundary Layer on a Flat Plate	195
	6.5.1	The Blasius Equation	195
	6.5.2	Flat Plate Results	197
6.6		Similarity Solutions	199
6.7		The Kármán Momentum Equation	200
6.8		The Kármán-Pohlhausen Method	201
6.9		Integration for Two-dimensional Boundary Layers	203
	6.9.1	The Dimensionless Equations	203
	6.9.2	The Starting Solution	204
6.10		Wakes and Jets	204
	6.10.1	The Wake Behind a Plate	205
	6.10.2	A Two-dimensional Jet	207
	6.10.3	The Axisymmetric Jet	209
		Problems	210

7 Turbulent Flow   213

7.1   Stability   215

    7.1.1   Interfacial Stability—Kelvin-Helmholtz Instability   216

    7.1.2   The Orr-Sommerfeld Equation   220

    7.1.3   Comments on Stability   224

7.2   The Mean Flow Equations   226

    7.2.1   Time Averages   226

    7.2.2   The Equations of Motion   228

    7.2.3   Vortex Stretching   230

    7.2.4   Relative Size of the Reynolds Stress   232

7.3   Empirical Formulas   234

    7.3.1   Eddy Viscosity   234

    7.3.2   Prandtl's Mixing Length   235

    7.3.3   The von Kármán Similarity Theory   235

    7.3.4   Deissler's Formula   238

7.4   Jets and Wakes   239

    7.4.1   Length and Time Scales for Wakes   240

    7.4.2   Length and Time Scales for Jets   242

    7.4.3   Similarity Hypothesis in Wakes   243

    7.4.4   The Axisymmetric Wake   245

    7.4.5   The Plane Jet   246

    7.4.6   The Axisymmetric Jet   248

    7.4.7   Laminar and Turbulent Jets and Wakes   248

7.5   Wall Turbulence   248

    7.5.1   Smooth Walled Channels and Pipes   249

    7.5.2   Rough Walls   251

    7.5.3   Flat Plates and Pipes   253

    7.5.4   Boundary Layers   254

7.6   Noncircular Conduits   255

    7.6.1   Friction   255

    7.6.2   Secondary Currents   256

7.7   Turbulent Energy   260

    7.7.1   The Energy Spectrum   260

    7.7.2   The Energy Cascade   262

    Problems   263

8 Shallow Water Flow   264

8.1   The Shallow Water Equations   265

    8.1.1   Conservation Equations and Boundary Conditions   265

    8.1.2   Conservation of Mass   267

    8.1.3   Conservation of Momentum   267

    8.1.4   The Compressible Flow Analogy   271

    8.1.5   Linearization   272

8.2   The Theory of Characteristics   274

    8.2.1   Directional Derivatives   275

    8.2.2   Transformation of the Quasi-linear Equations   276

    8.2.3   Examples of Equations   277

    8.2.4   Boundary Data   280

8.3		Unsteady, One-dimensional Flow in Rectangular Channels	282
	8.3.1	The Characteristic Equations	282
	8.3.2	The Equation Without Slope and Friction	284
	8.3.3	Dam Breaking	285
8.4		Discontinuous Solutions	290
	8.4.1	The Basic Relationships in One Dimension	290
	8.4.2	The Equations of Bore Movement	291
	8.4.3	Dam Break Revisited	293
8.5		One-Dimensional Flow in Channels	298
	8.5.1	The Basic Equations	298
	8.5.2	Steady Flow	301
	8.5.3	Unsteady Flow	302
	8.5.4	Friction	303
8.6		Stability in One-Dimensional Flow	305
	8.6.1	Stability Criteria for Uniform, Steady Flow	305
	8.6.2	Channels of Various Shapes	307
	8.6.3	Distance Required for Bore Formation	311
	8.6.4	Breaking Wave Due to Rising Water	315
	8.6.5	Roll Waves	317
8.7		Two-Dimensional, Steady, Supercritical Flow	319
	8.7.1	The Equations	319
	8.7.2	The Frictionless Solution	322
	8.7.3	Simple Waves	323
8.8		Diagonal Jumps	326
8.9		Boussinesq Equations	328
		Problems	334

9 Circulation

			338
9.1		Rotation	338
	9.1.1	Particle on a Rotating Earth	339
	9.1.2	Conservation of Momentum	343
	9.1.3	The Dimensionless Equations	345
9.2		Frictionless Flows	347
	9.2.1	Geostrophic Flow	347
	9.2.2	Gradient Flow	348
9.3		Inertialess, Nearly Horizontal Flows	351
	9.3.1	Ekman Surface Current	352
	9.3.2	Ekman Bottom Current	354
	9.3.3	Surface Current in Geostrophic Flow	355
	9.3.4	Oceanic Circulation	356
9.4		Slope Current	358
9.5		The Vertically Integrated Equations	361
9.6		Homogeneous Lake Circulation	362
	9.6.1	Wind Stress	364
	9.6.2	Steady State	365
	9.6.3	Unsteady Circulation	368
9.7		Seiche	372
9.8		Stratification	374
	9.8.1	Lakes	374

	9.8.2 Oceans	377
9.9	Stratified (Baroclinic) Circulation	380
	9.9.1 Lakes	380
	9.9.2 Geostrophic Flow	382
	Problems	385

10 Waves — 389

10.1	The Linearized Series of Free Surface Equations	389
10.2	Linear Waves	393
	10.2.1 Progressive Waves	393
	10.2.2 Standing Waves	394
	10.2.3 Comparison of Linear Waves	395
	10.2.4 Particle Paths	397
	10.2.5 Pressure	398
	10.2.6 Energy	398
	10.2.7 Power	399
10.3	Waves Trains	400
10.4	Fourier Transform	404
10.5	Capillary Waves	408
10.6	Refraction	413
10.7	Diffraction	415
10.8	Internal Waves	417
	10.8.1 Discrete Layers	417
	10.8.2 Continuous Stratification	420
	Problems	422

11 Transport — 425

11.1	Equations	426
	11.1.1 Advection	426
	11.1.2 Diffusion	426
	11.1.3 Advection-Diffusion	427
11.2	Solutions for Constant Diffusion Coefficient	428
	11.2.1 One-dimensional Diffusion	429
	11.2.2 Multi-dimensional Diffusion	430
	11.2.3 Simple Advection-Diffusion	431
11.3	Dispersion	432
	11.3.1 Hydrodynamic Dispersion	432
	11.3.2 Porous Media	433
11.4	Heat	435
11.5	Turbulence	436
11.6	Solutions	437
	11.6.1 Shear Flow Solution	437
	11.6.2 Ray Method	438
	Problems	441

Appendix A   Cartesian Tensors — 443

A.1	Vectors in a Cartesian Coordinate System	443
A.2	Vector Algebra	444
A.3	The Summation Convention	446

A.4 Vector Transformation between Rectangular Coordinate Systems   448
A.5 Cartesian Tensors   449
A.6 Tensor Algebra   451
A.7 Principal Axes of a Symmetric Tensor   452
A.8 Differential Operations   455
A.9 Transformations   457
A.10 Notation   459
A.11 Summary of Differential Operations   460
   A.11.1 Rectangular Coordinates   460
   A.11.2 Cylindrical Coordinates   461
   A.11.3 Spherical Coordinates   462
   Problems   463

Postface   466

References   468

A Fluid Mechanics Bibliography   473

Index   487

PREFACE

This book is an outgrowth of class notes that were used in a course for graduates and advanced undergraduates at Cornell University. It could be used as the basis for either a first or second course in fluid mechanics. As a first course it would appeal to those interested in a mathematical presentation such as the engineering physicist. The primary enrollment at Cornell has consisted of students who have had one course in fluid mechanics. There is more material than is normally given in a one semester course, which allows the instructor to select those chapters that are most important to a particular course. Alternatively, the book could be used in a two course sequence.

In engineering a first course in fluid mechanics is usually concerned with a "systems" or "control volume" approach (see Fig. 1.1). The equations for a control volume are presented early in Chapter 1 and are used to derive the more detailed equations of the "field" approach that generally dominates in a second course. Thus, the student who has completed a first course and has some feel for fluid flow and how to solve elementary problems can connect the two methods. In addition there is an attempt to remove the confusion that is often generated by a first course. (Is Bernoulli's equation an expression of conservation of energy or conservation of momentum? What does "lost" energy really mean?)

Computational fluid mechanics is largely ignored in the entire book—mentioned only in connection with a very few specific problems. It is a subject to itself and including it in a book on intermediate fluid mechanics makes the material far too voluminous. At Cornell computational fluid mechanics is taught in a separate course after the students have more than an elementary understanding of fluid mechanics. An attempt, included in a previous version of the notes, to combine the two subjects was a failure.

One goal of the book is to provide a background for studying computational fluid mechanics.

The level of mathematics is that of a second course in calculus. The reader should know the meaning of vector operators (there is a brief explanation in the appendix). I have included the appendix in the initial part of the course to put a diverse group of students on a nearly equal footing. A primary objective is to bring the students to the level that they can read papers in fluid mechanics, and developing a good feel for the notation is important to achieve that goal. The remainder of the book makes use of tensors only where clarity is enhanced and does not force the equations into tensor notation.

Chapters 1, 2, and 3 should be studied in sequence. The remainder of the chapters are largely, though not totally, independent. Chapter 1 treats the basic equations for the motion of a fluid. A thorough understanding of that material is essential if the student is to develop a feel for fluid mechanics and the approximations that are made to solve real problems. I find that the typical student has little idea of the difference between energy and momentum and how these quantities interact in fluid flow. Vorticity is one of the mysteries of fluid mechanics and the beginning graduate student seems to regard the Navier-Stokes equations as something beyond ordinary human understanding. This chapter also demonstrates the utility of the Einstein notation and how notation can influence understanding.

Chapter 2 on similitude and modeling is a continuation of Chapter 1, an extension of the study of the equations of fluid mechanics. It demonstrates that we can discover a great deal about the behavior of fluids without solving the equations and without invoking complicated mathematics as well as how the mathematics complements and extends observation. The table listing common dimensionless numbers should serve as a reference in addition to showing the student the utility of dimensional reasoning through the basic equations.

Chapter 3 introduces ideal fluid flow to the extent that is needed for most calculations. Although complex notation is covered in the last section, it is not central to the chapter and can be omitted if the instructor does not want to take the time to teach complex variables to those students who may not know it already. In my opinion, complex solutions are becoming less valuable because of their inherent limitations and because more problems can now be solved numerically.

Chapter 4 is a continuation of ideal flow solutions for problems with free surfaces. Free surface flow remains an interesting and difficult area of fluid mechanics. Exactly what constitutes free surface flow seems to be an area of confusion in the literature. In this book the definition of free surface flow is that which has a deformable or unknown solution region—the "mathematical definition" as opposed to the "physical definition," a flow in which a boundary adjusts itself according to the flow conditions—thus excluding problems of shallow water hydraulics, all linear wave problems, and some nonlinear wave problems.

Chapter 5 treats flows with zero Reynolds number and those without acceleration. Chapter 6 covers laminar flow. The treatment of boundary layer theory is somewhat superficial since that can be a course in itself. Chapter 7 considers turbulent flow, but again is little more than an introduction. Despite the titles of these chapters, some turbulent flow intrudes into the chapter on laminar flow and vice-versa in order to treat situations that may have either type of flow and to allow the development of comparative analyses.

The remainder of the chapters constitute the applications part of the book. Shallow water flow (Chapter 8) shows solutions with the assumption of hydrostatic pressure so that the equations can be reduced in dimension. This part of the book obviously caters to the interests of civil and agricultural engineers, but parallels with compressible flow are drawn. The equations of open channel flow are derived from the fundamentals of Chapter 1 and the mathematics of hyperbolic equations are presented, as well as the averaging that is necessary to arrive at the working equations. Although sonic velocity may be clear to the student of gas dynamics, critical velocity and critical depth in shallow water hydraulics seem to remain areas of confusion even at this late date.

Chapter 9 on circulation is an introduction to large scale (geophysical) flows. Chapter 10 on waves is a further continuation on ideal flow. Finally, Chapter 11 forms a brief introduction to transport, which is the object of many fluid mechanics calculations.

My original goal in listing books on fluid mechanics in a bibliography was simply to point out to my students the rich literature on the subject. The decision as to where fluid mechanics stopped and some other subject began was especially difficult.

The addition of the occasional historical notes was a change from the course notes that formed a basis for this book. I find the history of fluid mechanics not only interesting but enlightening in that it gives insight to the subject and especially to the tortuous path of development. The students can take heart in the knowledge that some of the early Greats who have their names attached to equations or dimensionless numbers were also confused about the distinction of momentum and energy.

ACKNOWLEDGMENTS

The primary contributors to this book have been my students (too numerous to mention individually). They are the ones who have pointed out errors in the notes and have suggested better methods of presentation. In fact, I have learned much of the fluid mechanics that I know from my students.

Professor Ian Wood (University of Canterbury, New Zealand) sat through the course in 1990 and encouraged me to turn the notes into a book. Professor Wood's intimate knowledge of fluid mechanics has been an inspiration.

Professor Enzo Levi (Universidad Nacional Autónomo de México and Instituto Mexicano de Tecnología de Agua, now deceased) was a catalyst in the historical notes. His book *El Agua Según la Ciencia* (in Spanish, being translated into English by the American Society of Civil Engineers at this writing) is a most interesting history of fluid mechanics and one that shows the interplay between the pioneers.

The following reviewers provided many helpful comments and suggestions: Nick Katopodes, University of Michigan; Louis H. Motz, University of Florida; and Ralph Wurbs, Texas A & M University.

The problems were solved and checked by Dr. T. K. Tsay of National Taiwan University.

Finally, my immediate colleagues at Cornell (Wilfried Brutsaert, Philip L-F. Liu and Gerhard Jirka) have provided an atmosphere conducive to learning and research. A part of them all is contained herein.

James A. Liggett

CORRIGENDA

As a long-time user of text books, I know how annoying errors can be. I have tried to be accurate, but some errors may appear in this book. I suggest that the reader keep a record of errors on these pages. *Please* make a copy of it and send it to me so that errors can be reported to other readers and corrected in future printings. If you send the errors that you find, I will return to you the complete list.

James A. Liggett

Page	Eq. no. Fig. no. Line no.	Incorrect text	Correct text

Page	Eq. no. Fig. no. Line no.	Incorrect text	Correct text

NOTATION

Roman Capital

A	area
B	a general constant
C	a general constant
C_L	lift coefficient
C_D	drag coefficient
C_p	pressure coefficient
C_{ref}	refraction coefficient
C	concentration (mass per unit volume)
CV	control volume
D	when used as a derivative indicates the substantial derivative
	drag force
	Ekman depth of influence
$Đ$	diffusion or dispersion coefficient
$Đ^t$	turbulent diffusion coefficient
$Đ^h$	diffusion coefficient for heat
$Đ^d$	hydrodynamic dispersion coefficient
$Đ^m$	molecular diffusion coefficient
E	energy
E	Ekman number
F	a general function
	indicates dimensions of force
$\vec{F}$	force vector
F	Froude number
H	elevation above a datum
J	as a subscript, refers to a hydraulic jump
K	apparent mass factor
	hydraulic conductivity
K_p	coefficient of compressibility
L	indicates dimensions of length
	a general length
	wave length
M	indicates dimensions of mass
	strength of a doublet
M	Mach number

P	a unit conversion factor
	the wetted perimeter
Pe	Peclet number
Q	flow rate
Q	cavitation number
R	the distance from the origin in a cylindrical coordinate system
R_h	hydraulic radius
R	Reynolds number
Ro	Rossby number
$\mathbf{R}$	reaction rate
S	area
	source term
S_0	bottom slope
S_f	friction slope
St	Strouhal number
T	indicates dimensions of time
	temperature
	wave period
T_s	surface tension
U	a reference velocity
V	volume
	velocity of an object
Ve	Vedernikov number
W	work
	a characteristic vertical velocity
	in wakes, the velocity defect
W_p	pressure work
W	Weber number

Lowercase Roman

a	a general variable
a_{ij}	angle cosine $[= \cos(x_i, x_j)$
b	a general variable
c	wave speed
	as a subscript, critical conditions
c_c	wave speed of capillary waves in gravity
$\tilde{c}$	wave speed of capillary waves without gravity
c_p	specific heat capacity at constant pressure
c_v	specific heat capacity at constant volume

d	distance between plates for Poiseuille flow
	as a subscript in waves, refers to deep water
d_{ij}	the rate of strain tensor
e	energy
	2.71828183
$\vec{e}_i$	the unit vector in the i-direction
f	a general function
	the Coriolis parameter ($f = 2\Omega \sin\phi$)
g	the acceleration of gravity (32.16 ft/sec^2; 9.806 m/s^2)
h	elevation in the Bernoulli equation
	scale factor in curvilinear coordinate systems
	as a subscript, usually means horizontal
h_L	head lost
i	a general coordinate direction as a subscript
	the unit imaginary number $\sqrt{-1}$
j	a general coordinate direction as a subscript
k	a general coordinate direction as a subscript
	wave number
k_{ij}	a constant in the rate-of-strain vs. shear relationship
k_h	heat transfer coefficient
k_m	mass transfer coefficient
l	length
m	mass
	exponent of the hydraulic radius
n	Manning's friction factor
	in porous media, the porosity
$\mathbf{n}$	as a vector or in a derivative, the normal direction to a line or surface
n_c	effective porosity
o	as a subscript, a reference quantity
p	pressure
$\hat{p}$	combination of pressure and gravity, $\hat{p} = p + \rho g h$
q	heat flow
	transport of a substance (mass/area/time)
	strength per unit of length of a line source or per unit of area of a surface source
r	the distance from the origin in a spherical coordinate system
s	arc length along a curve

t	time
	as a vector, the tangential direction to a curve
	as a superscript on dynamic or kinematic viscosity, denotes turbulent eddy viscosity
u	component velocities (u_x = velocity in the x-direction, etc.
u_{gx}, u_{gy}	component velocities in geostrophic flow
u_*	friction velocity ($= \sqrt{\tau/\rho}$)
v_s	specific volume ($= 1/\rho$)
$\tilde{u}$	when dealing with viscous flow, potential flow velocity
u_c	critical velocity
v	as a subscript, usually means vertical
$\tilde{w}$	width of a channel at the water surface
$\hat{w}$	work other than pressure work
x	coordinate direction, as a subscript components (e.g., velocity, u_x)
y	coordinate direction, as a subscript components (e.g., velocity, u_y)
z	coordinate direction, as a subscript components (e.g., velocity, u_z)

Greek Capital

Γ	vortex strength
Δ	a small increment
Π	a dimensionless number
Φ	the potential function
Ψ	the stream function
Ω	speed of rotation of the earth
Ω_{ij}	the rotation tensor

Lowercase Greek

α	a general angle
	energy correction factor
α_L	longitudinal dispersivity
α_T	transverse dispersivity
β	momentum correction factor
	first coefficient in the expansion of the Coriolis parameter ("β plane")
γ	specific weight
δ	a general small quantity
	boundary layer thickness
δ_d	displacement thickness

δ_m	momentum thickness
δ_e	energy thickness
δ_{ij}	the Kronecker delta
e	a general small quantity
e_{ijk}	the alternating unit tensor
ζ	a coordinate direction
	a function describing a free surface
η	the distance to a water surface from the bed of a channel or the equilibrium surface
	a general argument of a function or a coordinate
$\tilde{\eta}$	the maximum depth in a channel
θ	a general angle
	a coordinate angle (longitude) in a cylindrical coordinate system
κ	the von Kármán constant
	the bulk viscosity
λ	an eigenvalue
	size of wall roughness
	with boundary layer calculations, a shape factor
μ	viscosity
μ^t	eddy viscosity
ν	kinematic viscosity
ν^t	turbulent kinematic viscosity
ξ	a coordinate
	mixing length
π	3.14159265
ρ	fluid density
σ	surface tension
	frequency, wave frequency
τ	shear stress
	a time-like coordinate
τ_{ij}	the shear tensor
τ_0 or τ_w	wall shear
$\tau_{sx}, \tau_{sy}, \tau_{Bx}, \tau_{By}$	free surface and bottom shears
v	a general variable
ϕ	a coordinate angle (latitude) in a spherical coordinate system
	a small departure from the free stream potential
ψ	a small departure from the free stream function

ω	angular speed of rotation
ω_{ij}	the rotation tensor
$\vec{\omega}$	the rotation vector

Symbols

$\vec{\nabla}$	the gradient operator
∇^2	the Laplace operator
∇^4	the biharmonic operator
$\cdot$	indicates scalar product of two vectors
$\times$	indicates vector product of two vectors
$:$	indicates scalar product of two tensors
$\vec{}$	arrow overline indicates a vector
$\vec{\underline{}}$	arrow overline and underline indicates a dyad
$\overline{}$	overline indicates an averaged variable
$\overline{\overline{}}$	double overline indicates a doubly averaged variable
$'$	(prime) indicates a derivative with respect to the argument of the function
	indicates a fluctuating quantity or a departure from average
$\{\cdot\}$	a column vector
$\lfloor\cdot\rfloor$	a row vector
$\cos(x_i, x_j)$	the angle cosine between the i-direction and the j-direction ($= a_{ij}$)
$,$	(comma) indicates differentiation (e.g., $u_{,i}$ is the i-component of $\vec{\nabla}u$)
∂_i	indicates differentiation with respect to the i-coordinate
∂V	the boundary of the volume V
∂A	the boundary of area A
$*$	as a subscript or a superscript usually indicates a dimensionless quantity
$\sim$	(tilde) indicates a free surface quantity
$\underline{}$	(underline) indicates the value on the bed (of a channel or body of water)
$\hat{}$	(hat) indicates a different variable
	when applied to pressure, is a combination of pressure and gravity, $\hat{p} = p + \rho g h$
$\breve{}$	(breve) indicates a Laplace transform

FLUID MECHANICS

THE BASIC EQUATIONS OF FLUID MECHANICS

1.1 METHODS OF DESCRIPTION

The solution to a problem may consist of the barest essentials or may be given in great detail. The method of description to be chosen to describe fluid flow depends on the demands of the problem and on the ability to provide a detailed solution. The more detailed descriptions, in general, require more sophisticated analysis and are worked out with greater difficulty than the less detailed descriptions. The necessary approximations as well as the economics of providing a solution may dictate the level of description.

A problem can be analyzed in a system of constant mass or in a system of constant volume. In solid mechanics a system is usually studied that always consists of the same particles and hence is of constant mass. Many fluid mechanics problems can be solved by applying the laws of mechanics and thermodynamics to systems of constant mass. An example of such a problem is that of a piston compressing gas in a cylinder. For most fluid flow problems, however, a system composed of a constant volume rather than a constant mass is more convenient. Using a constant volume, a set of coordinate axes can be defined that are fixed within the volume in order to study the details of how the fluid moves with respect to the coordinates. Since the laws of mechanics and thermodynamics are usually written for constant mass systems, at least as studied in elementary mechanics, a revision of these laws is necessary to apply them to constant volume systems. For the most elementary descriptions, the results are simple. In the case of steady flow, much useful information can be obtained without knowing the details of the flow processes inside the volume.

The two methods of description, constant mass and constant volume, form what is called the systems approach. The systems approach is directed at finding overall results without looking at the details of what goes on inside the system. For some problems, however, more detail is needed, or the systems approach does not provide a solution. The field approach, which attempts to describe the fluid flow at each point in the coordinate system, may provide the solution in such situations. Within the field approach there are two degrees of detail. The less detailed, and the one that is studied most, is the Eulerian method. The Eulerian method leads to equations that give basic flow quantities such as velocity, pressure, and temperature at each point in the flow field as a function of the coordinates of the point and time. The more detailed method, the Lagrangian method, gives the basic flow quantities for each fluid particle in the flow field and the location of all the particles. Thus, the Lagrangian method traces each fluid particle and describes the conditions of that particle, whereas the Eulerian method simply describes conditions at a point without reference to which particle occupies the point.

Hydraulics and elementary fluid mechanics texts treat, primarily, the systems approach. The resulting equations are in integral form or, after carrying out the integration, in algebraic form. The equations that result from the field approach, on the other hand, are differential or partial differential equations. (This rather sweeping statement has many exceptions; integral equations that give very detailed descriptions can be formed.) The levels of description are generalized in Fig. 1.1. In that figure the detail increases from left to right and so do the complexities of computation.

In either system there are only a very few problems in fluid mechanics that can be solved without introducing some degree of approximation. In fact, the subject of fluid mechanics consists of a study of the mathematical methods used to solve the resulting equations and of the types of approximations that must be employed in order to obtain useful results. The initial approximation for most problems concerns the dimensionality of the

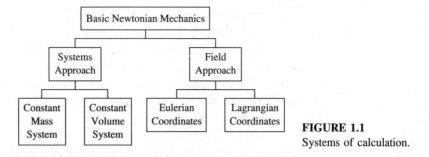

FIGURE 1.1
Systems of calculation.

problem. We live in a three-dimensional, time-dependent world. Through-out most of this century, however, engineering solutions consisted of find-ing one-dimensional approximations. The digital computer has enabled the solution of many multidimensional problems; however, three-dimensional, time-dependent solutions are still rare. One of the primary approximations remains the dimensionality of the flow.

1.2   THE CONTROL VOLUME EQUATION

The equations of mechanics and thermodynamics as related to constant mass systems are familiar. In the present section those relationships are applied to constant volume systems. Such relationships can be expressed generally for any advected fluid property. Figure 1.2a shows a control volume that, by definition, is fixed in space with respect to the coordinate system. At some arbitrary time t, a given mass of fluid occupies this volume. A small time later, at $t + \Delta t$, a portion of this fluid has moved out of the control volume and some of the fluid originally outside the control volume has entered

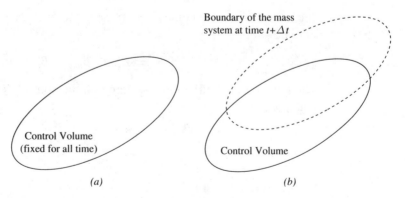

FIGURE 1.2
The control volume and mass system at times t and $t + \Delta t$.

to replace that which has left. The dashed line of Fig. 1.2*b* indicates the boundary of the fluid mass that was inside the control volume at time *t* but has moved downstream at time $t + \Delta t$; that is, the dashed line passes through the particles that occupied the solid line time Δt time units earlier. Thus, the portion of fluid inside the dashed line forms a system of constant mass and can be treated by the familiar laws of mechanics and thermodynamics. In the remainder of this section those laws are related to their counterparts for the constant volume system.

Let *B* represent the total of any extensive property (such as mass, momentum, or energy) inside the system under consideration. Then let *b* represent the amount of *B* per unit mass so that

$$B = \int_V \rho b \, dV \tag{1.1}$$

To differentiate between the quantities of *B* in the control volume and the mass system at different times, the following notation is used:

$B_t = B$ of the mass system at time *t*

$B_{t+} = B$ of the mass system at time $t + \Delta t$

$B'_t = B$ of the control volume at time *t*

$B'_{t+} = B$ of the control volume at time $t + \Delta t$

$B_{out} = B$ that has left the control volume in the time interval Δt

$B_{in} = B$ that has entered the control volume in the time interval Δt

Obviously from the above definitions

$$B_t = B'_t \tag{1.2}$$

since the same mass is involved at the same time. Equating the volumes in Fig. 1.2 yields

$$B_{t+} = B'_{t+} + B_{out} - B_{in} \tag{1.3}$$

The total change in the mass system and the control volume in the time Δt is defined as

$$\Delta B = B_{t+} - B_t \tag{1.4}$$

$$\Delta B' = B'_{t+} - B'_t \tag{1.5}$$

Substituting (1.2) and (1.3) into (1.4) gives

$$\Delta B = B'_{t+} - B'_t + B_{out} - B_{in} \tag{1.6}$$

Dividing (1.6) by Δt yields

$$\frac{\Delta B}{\Delta t} = \frac{\Delta B'}{\Delta t} + \frac{B_{out} - B_{in}}{\Delta t} \tag{1.7}$$

Letting Δt approach zero, (1.7) is written in differential form as

$$\frac{dB}{dt} = \frac{dB'}{dt} + \lim_{\Delta t \to 0} \frac{B_{out} - B_{in}}{\Delta t} \tag{1.8}$$

Equation (1.8) relates the property B' of the control volume to the property B of the mass system. The last term is the net rate at which this property flows through the surface of the control volume. In words this equation reads

$$\begin{bmatrix} \text{The rate of change} \\ \text{of any property of} \\ \text{the mass system} \end{bmatrix} = \begin{bmatrix} \text{The rate of change} \\ \text{of the property in} \\ \text{the control volume} \end{bmatrix} + \begin{bmatrix} \text{The net rate the} \\ \text{property leaves} \\ \text{the control volume} \end{bmatrix}$$

Using (1.1), (1.8) can be written as

$$\frac{d}{dt} \int_V \rho b \, dV = \frac{d}{dt} \int_{CV} \rho b \, dV + \lim_{\Delta t \to 0} \frac{B_{out} - B_{in}}{\Delta t} \tag{1.9}$$

The two integrals in (1.9) are different since the limits on the first (the boundary of the mass system) changes with time whereas the limits on the second are fixed for all time as the boundary of the stationary control volume. Since the limits of the second integral are not a function of time, the time derivative can be moved inside the integral sign,

$$\frac{d}{dt} \int_V \rho b \, dV = \int_{CV} \frac{\partial}{\partial t} (\rho b) \, dV + \lim_{\Delta t \to 0} \frac{B_{out} - B_{in}}{\Delta t} \tag{1.10}$$

Because B is a property that is advected with the fluid, the outflow of B from the control volume is a function only of the velocity on the control surface. From Fig. 1.3 the mass outflow in time Δt from a small area on the control surface is $\rho (\vec{u} \cdot \vec{n}) \Delta A \, \Delta t$ where $\vec{n}$ is the outward unit vector normal to the area ΔA. The quantity of B that flows out of the element of surface

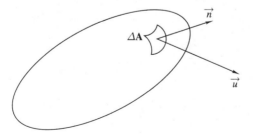

FIGURE 1.3
An incremental area on the control volume.

is then $\rho b(\vec{u} \cdot \vec{n}) \Delta A \, \Delta t$. Summing such areas over the entire surface gives

$$\lim_{\Delta t \to 0} \frac{B_{out} - B_{in}}{\Delta t} = \int_{CS} \rho b(\vec{u} \cdot \vec{n}) dA \qquad (1.11)$$

in which the integral is taken over the surface surrounding the control volume (the control surface, CS). Using (1.11) in (1.10) yields

$$\frac{dB}{dt} = \int_{CV} \frac{\partial}{\partial t}(\rho b) dV + \int_{CS} \rho b(\vec{u} \cdot \vec{n}) dA \qquad (1.12)$$

If the flow is steady, the first integral disappears and the right side references only variables on the boundary of the control volume.

Equation (1.12) is the general control volume integral that relates the constant mass system to the control volume. This equation is used to derive the conservation equations (conservation of mass, momentum, energy, and angular momentum) of fluid mechanics for a constant volume system from those equations that apply to a constant mass system.

The derivation of the more detailed equations of the Eulerian description usually begins with an infinitesimal control volume, thus obtaining differential equations. The derivation is easier and more instructive, however, to proceed from the general control volume equation. The divergence theorem relates volume and surface integrals through the equation

$$\int_V \vec{\nabla} \cdot \vec{u} \, dV = \int_{\partial V} \vec{u} \cdot \vec{n} \, dA \qquad (1.13)$$

where ∂V is the bounding surface of the volume V, $\vec{u}$ is any vector quantity, and $\vec{n}$ is the outward unit normal vector from the surface. The divergence theorem is derived by dividing the volume into infinitesimal cells by planes parallel to the coordinate axes, as shown in Fig. 1.4. The vector $\vec{u}$ is identified with the velocity of an incompressible fluid (although the divergence

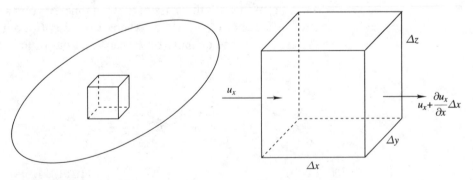

FIGURE 1.4
A small element inside the control volume.

theorem is quite general and $\vec{u}$ can be any vector). The volume rate of flow out of a cell minus the volume rate of flow into the cell is

$$\left(u_x + \frac{\partial u_x}{\partial x}\Delta x\right)\Delta y\,\Delta z + \left(u_y + \frac{\partial u_y}{\partial y}\Delta y\right)\Delta x\,\Delta z + \left(u_z + \frac{\partial u_z}{\partial z}\Delta z\right)\Delta x\,\Delta y$$

$$- (u_x\,\Delta y\,\Delta z + u_y\,\Delta x\,\Delta z + u_z\,\Delta x\,\Delta y) \qquad (1.14)$$

$$= \left(\frac{\partial u_x}{\partial x} + \frac{\partial u_y}{\partial y} + \frac{\partial u_z}{\partial z}\right)\Delta x\,\Delta y\,\Delta z = (\vec{\nabla}\cdot\vec{u})\,\Delta V$$

The sum of the net rate of flow from all the individual cells gives the net rate of flow out of the volume

$$\text{Rate of flow out of the volume} = \int_V \vec{\nabla}\cdot\vec{u}\,dV \qquad (1.15)$$

The rate of flow out of the volume can also be expressed as the sum of the contribution from each element of surface area ΔA, giving

$$\text{Rate of flow out of the volume} = \int_{\partial V} \vec{u}\cdot\vec{n}\,dA \qquad (1.16)$$

Equating (1.15) and (1.16) gives the divergence theorem (1.13). In deriving the divergence theorem, we have assumed that the space derivatives of $\vec{u}$ exist throughout the volume; that is, there is no discontinuity in $\vec{u}$ such as would exist in the case of shock waves. This assumption restricts the application of the resulting differential equations to problems that do not contain such discontinuities.

The divergence theorem is used to convert the surface integral of (1.12) to a volume integral

$$\frac{dB}{dt} = \int_{CV}\left[\frac{\partial}{\partial t}(\rho b) + \vec{\nabla}\cdot(\rho b\vec{u})\right]dV \qquad (1.17)$$

1.3 CONSERVATION OF MASS—THE EQUATION OF CONTINUITY

When dealing with constant mass systems, the law of conservation of mass is so trivial that it is usually not mentioned. In constant volume systems, however, mass must be explicitly conserved. Using the previously derived control volume equations, B is associated with the total mass of the system. From (1.1) b must take on the value of unity. Obviously, the rate of change of mass of a constant mass system is zero, so that $dB/dt = 0$ and (1.12) becomes

$$\int_{CV}\frac{\partial\rho}{\partial t}\,dV + \int_{CS}\rho(\vec{u}\cdot\vec{n})dA = 0 \qquad (1.18)$$

The differential equation is a direct result of (1.17)

$$\int_{CV} \left[\frac{\partial \rho}{\partial t} + \vec{\nabla} \cdot (\rho \vec{u}) \right] dV = 0 \tag{1.19}$$

Because the control volume can have arbitrary limits—the limits of the integral can take on arbitrary values—the control volume could be chosen to surround any nonzero region of the integrand, leading to a violation of (1.19). The conclusion is that, if (1.19) is valid, the integrand must be zero everywhere, so that

$$\frac{\partial \rho}{\partial t} + \vec{\nabla} \cdot (\rho \vec{u}) = 0 \quad \text{or} \quad \frac{\partial \rho}{\partial t} + \frac{\partial}{\partial x_i}(\rho u_i) = 0 \tag{1.20}$$

Again, we have assumed that all first derivatives are continuous in the control volume. In the case of conservation of mass, the implication is that there is no source or sink of fluid in the control volume.

1.4 CONSERVATION OF MOMENTUM—THE EQUATION OF MOTION

The equation of motion is obtained by associating B with the momentum in the system. Momentum is a vector quantity, the product of mass and velocity. Thus, b becomes the velocity vector $\vec{u}$ from (1.1). From Newton's law of motion the rate of change of momentum of a constant mass system is equal to the applied force

$$\frac{d}{dt}\vec{B} = \vec{F} \tag{1.21}$$

Equation (1.12) becomes

$$\vec{F} = \int_{CV} \frac{\partial}{\partial t}(\rho \vec{u}) \, dV + \int_{CS} \rho \vec{u}(\vec{u} \cdot \vec{n}) dA \tag{1.22}$$

in which $\vec{F}$ is the sum of all forces acting on the fluid in the control volume.

To obtain the differential equation, we use (1.17) and shrink the control volume to a point. Dividing by the volume, the left side becomes the force per unit volume $\vec{f}$. The equation states that the rate of change of momentum per unit volume is equal to the force per unit volume $\vec{f}$ so that the differential equation is

$$\vec{f} = \frac{\partial}{\partial t}(\rho \vec{u}) + \vec{\nabla} \cdot (\rho \vec{u}\vec{u}) \tag{1.23}$$

The quantity $\vec{u}\vec{u}$ is a dyadic product $(u_i u_j)$. Equation (1.23) can be simplified by the use of the continuity equation. Rewriting (1.23) gives

$$f_i = \frac{\partial}{\partial t}(\rho u_i) + \frac{\partial}{\partial x_j}(\rho u_i u_j) = u_i \frac{\partial \rho}{\partial t} + \rho \frac{\partial u_i}{\partial t} + \rho u_j \frac{\partial u_i}{\partial x_j} + u_i \frac{\partial}{\partial x_j}(\rho u_j)$$

$$= \rho \frac{\partial u_i}{\partial t} + \rho u_j \frac{\partial u_i}{\partial x_j} \tag{1.24}$$

The extreme right side results from setting $u_i \frac{\partial \rho}{\partial t} + u_i \frac{\partial}{\partial x_j}(\rho u_j)$ to zero as follows from (1.20). For many calculations using numerical methods, it is better to use (1.23) or the second part of (1.24) (the *conservation form* of the momentum equation) than the reduced form [the last part of (1.24)].

The force term needs to be related to the fluid variables such as pressure and velocity. That term represents all the real forces on the system. Herein, we consider only pressure forces, shear forces, and gravity force on the fluid element. The pressure forces per unit volume are easily derived by considering the infinitesimal cube of Fig. 1.5. Only the pressures acting on the faces perpendicular to the x-axis are shown. The net force per unit volume in the x-direction is

$$\frac{1}{\Delta V}\left[p\,\Delta y\,\Delta z - \left(p + \frac{\partial p}{\partial x} \right) \Delta y\,\Delta z \right] = -\frac{\partial p}{\partial x} \tag{1.25}$$

The shear stresses acting in the x-direction on a small cube are shown in Fig. 1.6. The first subscript on τ indicates the axis perpendicular to the face of the cube. The second subscript indicates the direction of the action of the stress. (The normal component τ_{xx} is due only to viscous action and not pressure.) The sign convention is chosen so that the force acts in the positive direction on the face of the cube that is nearest to the origin, or so that the force couples tend to rotate the near face of the cube to the right when looking in the positive direction along the axis normal to the face. The net force per unit volume on the cube in the ith direction due to

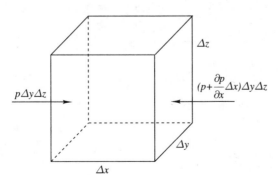

FIGURE 1.5
Pressure forces in the x-direction.

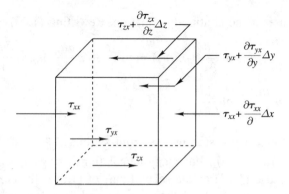

FIGURE 1.6
Shear stresses in the x-direction on the elementary volume.

viscous shear is

$$(f_i)_\tau = -\left(\frac{\partial \tau_{xi}}{\partial x} + \frac{\partial \tau_{yi}}{\partial y} + \frac{\partial \tau_{zi}}{\partial z}\right) = -\frac{\partial \tau_{ji}}{\partial x_j} \qquad (1.26)$$

An alternate point of view is to think of the shear stresses not as stresses but as transporting momentum. A retarding stress would transport momentum out of the volume whereas a stress in the direction of motion transports momentum into the volume. This point of view is useful in determining the signs of the various stresses. It also corresponds closely to the actual physical process since the "shear" on a volume is created by fluid molecules entering the volume at a higher or lower velocity than the average velocity of the particles already in the volume. Figure 1.7 shows the same shear stress components as in Fig. 1.6 but pictures them as momentum transport vectors. In this case the first subscript indicates the direction of transport and the second subscript indicates the component of momentum; for example, τ_{yx} is the component of x-momentum transported in the y-direction. The

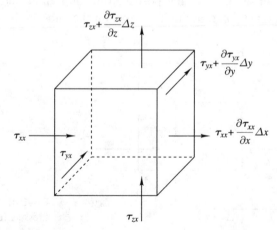

FIGURE 1.7
Momentum transfer in the x-direction to and from the elementary volume.

force in the x-direction on the volume due to viscous shear is the rate at which x-momentum enters the volume, or

$$(f_x)_\tau \Delta V = \tau_{xx} \Delta y\, \Delta z + \tau_{yx} \Delta x\, \Delta z + \tau_{zx} \Delta x\, \Delta y$$

$$- \left[\left(\tau_{xx} + \frac{\partial \tau_{xx}}{\partial x} \Delta x \right) \Delta y\, \Delta z + \left(\tau_{yx} + \frac{\partial \tau_{yx}}{\partial y} \Delta y \right) \Delta x\, \Delta z \right.$$

$$\left. + \left(\tau_{zx} + \frac{\partial \tau_{zx}}{\partial z} \Delta z \right) \Delta x\, \Delta y \right]$$

$$= - \left(\frac{\partial \tau_{xx}}{\partial x} + \frac{\partial \tau_{yx}}{\partial y} + \frac{\partial \tau_{zx}}{\partial z} \right) \Delta V$$

$$(1.27)$$

which is the same as (1.26) for the x-direction.

The gravity force in the i-direction is the weight of the element times the cosine of the angle between the vertical and the i-direction.

$$(f_i)_g = -\rho g \frac{\partial h}{\partial x_i} \tag{1.28}$$

where h is positive upward.

Using (1.25), (1.26), and (1.28) with (1.24), the equation of motion in differential form and written for the i-direction becomes

$$\rho \frac{\partial u_i}{\partial t} + \rho u_j \frac{\partial u_i}{\partial x_j} = -\frac{\partial p}{\partial x_i} - \rho g \frac{\partial h}{\partial x_i} - \frac{\partial \tau_{ji}}{\partial x_j} \tag{1.29}$$

Equation (1.29) is written in the reduced form; the conservation form is obtained by using (1.23) instead of (1.24). A physical interpretation of the terms in (1.29) is given in sections 1.6 and 1.9.

1.5 CONSERVATION OF ENERGY—THE ENERGY EQUATION

The first law of thermodynamics states

$$\begin{bmatrix} \text{The rate of change} \\ \text{of energy of the} \\ \text{mass system} \end{bmatrix} = \begin{bmatrix} \text{The net rate of} \\ \text{heat flow into} \\ \text{the system} \end{bmatrix} - \begin{bmatrix} \text{The rate that work} \\ \text{is being done by} \\ \text{the system} \end{bmatrix}$$

or in equation form

$$\frac{dE}{dt} = \frac{d\tilde{q}}{dt} - \frac{d\hat{w}}{dt} \tag{1.30}$$

The energy in a mass of fluid consists of (1) potential energy (due to elevation above a datum), (2) the kinetic energy of the fluid particles, and (3) the internal thermal energy. The energy for a volume is given by

$$
E = \int_V \rho g h \, dV + \int_V \frac{\rho}{2} u^2 \, dV + \int_V \rho e \, dV
$$

$$
= \left[\begin{array}{c} \text{Potential} \\ \text{energy} \end{array} \right] + \left[\begin{array}{c} \text{Kinetic} \\ \text{energy} \end{array} \right] + \left[\begin{array}{c} \text{Internal} \\ \text{energy} \end{array} \right]
$$

(1.31)

in which h is the elevation above a datum and e is the internal energy per unit mass. In reference to (1.1), B is associated with the total energy and b with the quantity $gh + u^2/2 + e$. The general control volume equation becomes

$$
\frac{d\tilde{q}}{dt} - \frac{d\hat{w}}{dt} = \int_{CV} \frac{\partial}{\partial t} \left[\rho \left(gh + \frac{u^2}{2} + e \right) \right] dV
$$

$$
+ \int_{CS} \rho \left(gh + \frac{u^2}{2} + e \right) (\vec{u} \cdot \vec{n}) \, dA
$$

(1.32)

Notice that pressure does not appear explicitly in (1.32); it is part of the work term $\hat{w}$ and not part of the energy. It is usually convenient to separate the pressure work from $\hat{w}$; therefore, we define

$$
W = \hat{w} - \text{ pressure work } = \hat{w} - W_p
$$

(1.33)

Consider the pressure acting on a small element of area of the mass system as shown in Fig. 1.8. The area is moving with the vector velocity $\vec{u}$. The rate of doing work is the force times the velocity. Integrated over the surface, the total pressure work on the mass system becomes

$$
\frac{dW_p}{dt} = \int_A (\vec{u} \cdot \vec{n}) p \, dA
$$

(1.34)

At time $t = 0$ the area of the mass system and the control volume coincide, so that the integral of (1.34) can be integrated over the surface of the control

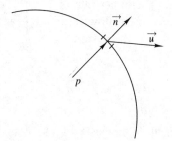

FIGURE 1.8
Pressure acting on a section of the control volume.

volume. Combining (1.33), (1.34), and (1.32) yields

$$\frac{d\tilde{q}}{dt} - \frac{dW}{dt} = \int_{CV} \frac{\partial}{\partial t}\left[\rho\left(gh + \frac{u^2}{2} + e\right)\right]dV$$

$$+ \int_{CS} \rho\left(gh + \frac{u^2}{2} + e + \frac{p}{\rho}\right)(\vec{u}\cdot\vec{n})dA$$

(1.35)

The differential form of the energy equation is limited since, in the work term on the left of (1.36), only the work done by viscous force is considered. Referring to Fig. 1.6, the viscous work in the x-direction is calculated by

$$\frac{dW_{vx}}{dt} = -(\tau_{xx}\Delta y\,\Delta z + \tau_{yx}\Delta x\,\Delta z + \tau_{zx}\Delta x\,\Delta y)u_x$$

$$+ \left[\tau_{xx}u_x + \frac{\partial}{\partial x}(\tau_{xx}u_x)\Delta x\right]\Delta y\,\Delta z$$

$$+ \left[\tau_{yx}u_x + \frac{\partial}{\partial y}(\tau_{yx}u_x)\Delta y\right]\Delta x\,\Delta z + \left[\tau_{zx}u_x + \frac{\partial}{\partial z}(\tau_{zx}u_x)\Delta z\right]\Delta x\,\Delta y$$

$$= \frac{\partial}{\partial x_i}(\tau_{ix}u_x)\Delta V$$

(1.36)

in which W_{vx} is the work done by the element due to viscous forces in the x-direction. The heat flux vector q is defined so that

$$\left[\begin{array}{c}\text{Net rate of energy}\\ \text{into the element}\\ \text{due to conduction}\end{array}\right] = -(\vec{\nabla}\cdot\vec{q})\Delta V$$

(1.37)

The differential equation is obtained by applying the divergence theorem to (1.35) and shrinking the volume to a point

$$-\vec{\nabla}\cdot\vec{q} - \vec{\nabla}\cdot(\tau\cdot\vec{u}) = \frac{\partial}{\partial t}\left(\rho gh + \rho\frac{u^2}{2} + \rho e\right) + \vec{\nabla}\cdot\left(\rho gh + \rho\frac{u^2}{2} + \rho e + p\right)\vec{u}$$

(1.38)

To obtain a more familiar form of the energy equation, it is rewritten as

$$-\frac{\partial q_i}{\partial x_i} - \frac{\partial}{\partial x_i}(\tau_{ij}u_j) = \frac{\partial\rho}{\partial t}\left(gh + \frac{u^2}{2} + e\right)$$

$$+ \rho\frac{\partial}{\partial t}\left(\frac{u^2}{2} + e\right) + g\rho\frac{\partial h}{\partial t} + \left(gh + \frac{u^2}{2} + e\right)\frac{\partial}{\partial x_i}(\rho u_i)$$

(1.39)

$$+ \rho u_i\frac{\partial}{\partial x_i}\left(\frac{u^2}{2} + e\right) + g\rho u_i\frac{\partial h}{\partial x_i} + \frac{\partial}{\partial x_i}(p u_i)$$

From the continuity equation the sum of the first and fourth terms on the right side is zero. Also, the third term is zero since $\partial h / \partial t = 0$. Rearranging terms gives the final equation

$$\rho \frac{\partial}{\partial t} \left(\frac{u^2}{2} + e \right) + \rho u_i \frac{\partial}{\partial x_i} \left(\frac{u^2}{2} + e \right) = - \frac{\partial q_i}{\partial x_i} - \frac{\partial}{\partial x_i} (\tau_{ij} u_j + p u_i) - \rho g u_i \frac{\partial h}{\partial x_i}$$

$$(1.40)$$

A physical interpretation of the terms of (1.40) is given in sections 1.6 and 1.9.

1.6   TIME DERIVATIVES

Several types of time derivatives are used in fluid mechanics. To illustrate their use, consider a simple example. Suppose we are assigned to monitor the air pollution in a large city. There are several ways we could perform this duty. The most obvious is to mount a gauge on a fixed tower. The gauge would then record the rate of change of pollutants while holding the space coordinates fixed or, in other words, observing the derivative $\partial s / \partial t$ (where s stands for smoke).

A second method would be to mount the gauge on an airplane. The rate of change of pollution as seen by the gauge in that case is not only a function of the change at a single position but also of the motion of the airplane. Thus, the gauge observes the change at a single point plus the rate of change in space times the velocity of the airplane, or

$$\frac{ds}{dt} = \frac{\partial s}{\partial t} + u_i^a \frac{\partial s}{\partial x_i}$$

$$(1.41)$$

in which $\vec{u}^a$ is the velocity of the airplane

$$u_i^a = \frac{dx_i^a}{dt}$$

$$(1.42)$$

and x_i^a is the ith coordinate of the airplane.

A third method of measurement would be to mount the gauge in a balloon taking measurements as the balloon floats along with the air currents. In this case the velocity is the same as the velocity of the fluid, so (1.41) becomes

$$\frac{Ds}{Dt} = \frac{\partial s}{\partial t} + u_i \frac{\partial s}{\partial x_i}$$

$$(1.43)$$

The capital D is used to show that the velocity appearing in the equation is the velocity of the fluid—in fact, this type of derivative is defined by (1.43). The derivative is called the *substantial derivative* or the *material derivative* and shows the rate of change referred to an individual fluid particle. This type of differentiation is not confined to scalars and may be applied to vectors and tensors. The substantial derivative of a vector is defined as

$$\frac{D\vec{u}}{Dt} = \frac{\partial \vec{u}}{\partial t} + \frac{1}{2}\vec{\nabla}(\vec{u} \cdot \vec{u}) - \vec{u} \times (\vec{\nabla} \times \vec{u}) \tag{1.44}$$

or in index notation

$$\frac{Du_i}{Dt} = \frac{\partial u_i}{\partial t} + u_j\frac{\partial u_i}{\partial x_j} \tag{1.45}$$

In Cartesian coordinates (1.44) and (1.45) are exactly the same, but the transformation properties are different for curvilinear coordinate systems.

These are not the only kinds of time derivatives used in fluid mechanics. The time rate of change of a property as viewed from a coordinate system that is both moving and rotating with the fluid is given by the *codeformational time derivative* or *Jaumann derivative*. For example, the Jaumann derivative of a tensor is

$$\frac{D^J \tau_{ij}}{D^J t} = \frac{D\tau_{ij}}{Dt} + \frac{1}{2}(\omega_{ik}\tau_{kj} - \omega_{kj}\tau_{ik}) \tag{1.46}$$

where ω_{ij} is the rotation of the fluid and is defined by Eq. (1.54) of the next section. The rate of change with time of a property as viewed from a system that is moving, rotating, and deforming with the fluid is given by the *codeformational time derivative* or *convected derivative* (see Bird et al. 1977).

An example of the substantial derivative appears in the equation for conservation of mass. The second of (1.20) is

$$\frac{D\rho}{Dt} = -\rho\frac{\partial u_i}{\partial x_i}$$

$$\begin{bmatrix} \text{Change of} \\ \text{density of a} \\ \text{fluid particle} \end{bmatrix} = \begin{bmatrix} \text{Expansion} \\ \text{of the fluid} \end{bmatrix} \tag{1.47}$$

For a fluid of constant density, the change in the density of a fluid particle is zero, so that

$$\frac{\partial u_i}{\partial x_i} = 0 \tag{1.48}$$

Such a fluid is often called *incompressible*; however, constant density implies more than incompressibility since the density can change with dissolved matter (such as salt) or with temperature.

Notice that the substantial derivative also appears in Eqs. (1.29) and (1.40). The use of the substantial derivative in these equations affords a good physical interpretation of the terms, as illustrated below. The equation of motion in Cartesian coordinates is

$$\rho \frac{Du_i}{Dt} = -\frac{\partial p}{\partial x_i} - \rho g \frac{\partial h}{\partial x_i} - \frac{\partial \tau_{ji}}{\partial x_j}$$

$$\begin{bmatrix} \text{Mass times} \\ \text{acceleration} \\ \text{of a particle} \end{bmatrix} = \begin{bmatrix} \text{pressure} \\ \text{force} \end{bmatrix} + \begin{bmatrix} \text{gravity} \\ \text{force} \end{bmatrix} + \begin{bmatrix} \text{viscous or} \\ \text{friction} \\ \text{forces} \end{bmatrix} \qquad (1.49)$$

The equation of energy is

$$\rho \frac{D}{Dt} \left(\frac{u^2}{2} + e \right) = -\frac{\partial q_i}{\partial x_i} - \frac{\partial}{\partial x_i}(\tau_{ij} u_j) - \frac{\partial}{\partial x_i}(p u_i) - \rho g u_i \frac{\partial h}{\partial x_i} \quad (1.50)$$

$$\begin{bmatrix} \text{Rate of change} \\ \text{of energy of} \\ \text{a particle} \end{bmatrix} = \begin{bmatrix} \text{Rate of} \\ \text{heat} \\ \text{transfer} \end{bmatrix} + \begin{bmatrix} \text{Work by} \\ \text{viscous} \\ \text{forces} \end{bmatrix} + \begin{bmatrix} \text{Work by} \\ \text{pressure} \\ \text{forces} \end{bmatrix} + \begin{bmatrix} \text{Work by} \\ \text{gravity} \\ \text{forces} \end{bmatrix}$$

1.7 TRANSLATION, ROTATION, AND RATE OF STRAIN

Both the equations of motion and the energy equation contain the shear stress tensor τ. To write these equations in working form, this tensor must be expressed in terms of more fundamental quantities such as the velocity and its derivatives. As a preparation for the derivation of such expressions, we must study the way in which a fluid particle can move.

Consider a point x_i^0 in a fluid where the velocity is $\vec{u}^0$ (see Fig. 1.9). At a neighboring point, the coordinates of which are $x_i^0 + \Delta x$, the velocity is $\vec{u}^0 + \Delta \vec{u}$. Assuming that $\vec{u}$ is a continuous function of the space variables, it can be expanded in a Taylor series about the point x_i^0

$$\vec{u}^0 + \Delta \vec{u} = \vec{u}^0 + \frac{\partial \vec{u}}{\partial x_i} \Delta x_i + \frac{\partial^2 \vec{u}}{\partial x_i^2} \frac{(\Delta x_i)^2}{2!} + \cdots \qquad (1.51)$$

Considering terms of only the first order in Δx_i

$$\Delta u_i = \frac{\partial u_i}{\partial x_j} \Delta x_j \qquad (1.52)$$

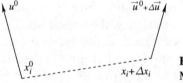

FIGURE 1.9
Movement of neighboring points.

Adding and subtracting terms, (1.52) is rewritten

$$\Delta u_i = \frac{1}{2}\left(\frac{\partial u_i}{\partial x_j} + \frac{\partial u_j}{\partial x_i}\right)\Delta x_j + \frac{1}{2}\left(\frac{\partial u_i}{\partial x_j} - \frac{\partial u_j}{\partial x_i}\right)\Delta x_j \tag{1.53}$$

From the terms of the previous equation the following are defined:

$$\omega_{ij} = \frac{1}{2}\left(\frac{\partial u_i}{\partial x_j} - \frac{\partial u_j}{\partial x_i}\right) \tag{1.54}$$

$$d_{ij} = \frac{1}{2}\left(\frac{\partial u_i}{\partial x_j} + \frac{\partial u_j}{\partial x_i}\right) \tag{1.55}$$

Thus, the tensor $\partial u_i/\partial x_j$ has been split into an antisymmetrical tensor, ω_{ij}, and a symmetrical tensor, d_{ij}.

Consider a rectangular particle of fluid with one corner located at the origin, as shown in Fig. 1.10. The fluid is moving with some velocity that varies in space and with velocity $\vec{u}^0$ at the origin, velocity $\vec{u}^a = \vec{u}^0 + (\partial\vec{u}/\partial y)\Delta y$ in point a and velocity $\vec{u}^c = \vec{u}^0 + (\partial\vec{u}/\partial x)\Delta x$ in point c. Now consider the same particle after some small time Δt, as seen in Fig. 1.11. The point o has moved a distance $\vec{u}^0\Delta t$, the point a has moved a distance $\vec{u}^a\Delta t$, etc. Since these velocities are, in general, slightly

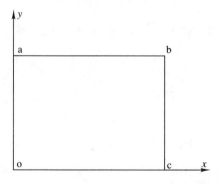

FIGURE 1.10
An elementary volume in the origin.

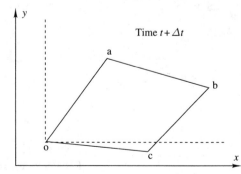

FIGURE 1.11
A general movement of the elementary volume.

different, the particle has become distorted from its original rectangular shape. To see what form the distortion takes, first consider the case where

$$\frac{\partial u_x}{\partial y} = -\frac{\partial u_y}{\partial x} \qquad \text{and} \qquad \frac{\partial u_x}{\partial x} = \frac{\partial u_y}{\partial y} = 0 \qquad (1.56)$$

Because there is no change in the x-velocity in the x-direction, the sides a-b and o-c have neither lengthened nor shortened; similarly, the sides o-a and b-c remain the same length. After the time Δt the particle appears as shown in Fig. 1.12. The point a has traveled further in the x-direction than the point o by the amount $\partial u_x/\partial y \, \Delta y \, \Delta t$, and the point c has traveled further in the y-direction than the point o by the amount $\partial u_y/\partial x \, \Delta x \, \Delta t$. The angle that side o-a makes with the vertical is $\partial u_x/\partial y \, \Delta t$; the angle that side o-c makes with the horizontal is $\partial u_y/\partial x \, \Delta t$.

Thus, under the previously stated assumption, the particle has undergone only a transformation plus a rotation. Extending the analysis to three dimensions, we see that the condition for this type of motion is $\omega_{ij} \neq 0$ and $d_{ij} = 0$. Further examination of the tensor ω_{ij} shows that it describes the rotation of fluid particles.

The vector product of the antisymmetric tensor ω (or any antisymmetric tensor) with a vector $\vec{a}$ can be written

$$\omega \cdot \vec{a} = \vec{\omega} \times \vec{a} \qquad (1.57)$$

where $\vec{\omega}$ is the vector of the tensor ω and is defined as

$$\vec{\omega} = -(\vec{e}_1 \omega_{23} + \vec{e}_2 \omega_{31} + \vec{e}_3 \omega_{12}) \qquad (1.58)$$

Notice that

$$\vec{\omega} = \frac{1}{2} \vec{\nabla} \times \vec{u} \qquad (1.59)$$

The term $\vec{\nabla} \times \vec{u}$ is called the *vorticity* vector.

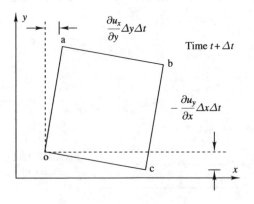

FIGURE 1.12
Rotation of the elementary volume.

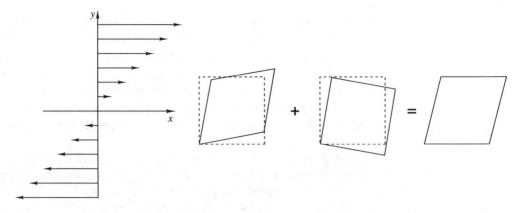

FIGURE 1.13
The shear flow produces pure stress plus pure rotation.

Since ω_{ij} has been identified with the rotation of a fluid element, d_{ij} can be identified as the distortion or *rate of strain* of an element. That is, ω_{ij} reflects the rigid body rotation of a particle whereas d_{ij} reflects the rate of movement of the different points in the particle relative to each other. Thus, the motion of a fluid can be described as

1. a rigid body translation plus
2. a rigid body rotation (antisymmetric tensor) plus
3. a deformation (symmetric tensor).

These effects are illustrated by the simple shear flow shown in Fig. 1.13. A particle near the origin is distorted and rotated as shown to produce this type of flow.

1.8 STRESS AND RATE OF STRAIN

In the equations of motion the shear stress tensor must still be related to the physical aspects of the flow. The basis for the relation is Newton's law of viscosity. In our notation

$$\tau_{yx} = -\mu \frac{du_x}{dy} \tag{1.60}$$

in which μ, the *viscosity* of the fluid, is a constant of proportionality between the viscous stress and the velocity gradient. The viscosity is a property of the fluid and a fundamental constant from the point of view of fluid mechanics. The viscosity could be derived by beginning the derivation of the equations of fluid mechanics starting with quantum theory. As a result a formula would appear that represents viscosity in terms of the "more fundamental" parameters of Schrödinger's equation, which expresses the electrostatic forces between electrons and nuclei. From our point of view, however, Newton's law is an underived postulate. Not all fluids obey Newton's law—perhaps none does under extreme enough conditions. Those that do not are called *non-Newtonian* fluids. Fortunately, air and water, the most common fluids, are Newtonian in the usual range of applications.

Historical Note:[1] The resistance of fluids to motion has long been a puzzling subject. It was known since Aristotle (384–322 BC) that objects heavier than water in the shape of thin plates could float on the surface. Didn't that indicate that water has a static friction—similar to the starting force of a solid on a surface—that must be overcome to set an object into motion? The argument raged into the days of Galileo Galilei (1564–1642). Daniel Bernoulli, while in St. Petersburg (1727–1733) performed experiments in a tube of only 5 mm diameter. He noticed that fluid emerged vertically upward from the open end under a given head at a velocity much less than he predicted. He attributed the phenomenon to the adhesion of fluid particles to the wall of the tube—what we now call the no-slip condition (see section 1.13). Sir Isaac Newton (1642–1727) at Cambridge University was the first to quantify the friction and he published what is essentially (1.60) in his second book, *Mathematical Principles of Natural Philosophy*, in 1686. Although Newton's scientific accomplishments were among the greatest ever known, he professed modesty, exemplified by his famous remark (in a letter to Robert Hooke of February 5, 1676): "If I have seen farther, it is by standing on the shoulders of giants." [Actually, Robert Burton, an English clergyman who lived a generation before Newton wrote a similar statement and attributed it to the poet Lucan (first century) who said "Pigmies placed on the shoulder of giants see more than the giants themselves."]

The minus sign of (1.60) arises due to the sign convention that was adopted for the shear stress tensor. Newton's law was postulated for the very special case when all of the other components of the stress tensor are zero, except τ_{xy}. For the purpose of establishing a general equation, (1.60) needs to be generalized and related to the other components of the stress tensor. In that way the viscous forces will become a function of the deformation of the fluid (the velocity gradients) and the viscosity (or coefficient of viscosity), μ.

---

[1]Although the historical notes come from many sources, Levi (1989) has been a primary reference.

First, it is shown that the stress tensor is symmetrical by taking moments about the lower left corner of the small particle shown in Fig. 1.14. The sum of the moments is equal to the moment of inertia times the angular acceleration, or

$$\tau_{zx} \Delta x \, \Delta y \, \Delta z - \tau_{xz} \Delta y \, \Delta z \, \Delta x = \alpha \int_V \rho r^2 dV \qquad (1.61)$$

where α is the angular acceleration and r is the distance from the lower left corner. The length r is of the order of the length of the sides or smaller, and the integral over the volume is of the order of the length of the sides cubed. Taking V as the order of $(\Delta x_i)^3$, the right hand side is of the order $(\Delta x_i)^5$ whereas the left side is of the order $(\Delta x_i)^3$. Thus, as the lengths of the sides approach zero, the equation can maintain balance only if $\tau_{zx} = \tau_{xz}$. Similar analysis on the other stress components shows that

$$\tau_{ij} = \tau_{ji} \qquad (1.62)$$

Newton's law of viscosity is linear and we will generalize the law while holding to its linear nature. The six distinct components of the stress tensor will be related to the six distinct components of the rate of strain tensor in the manner

$$
\begin{aligned}
\tau_{11} &= k_{11}d_{11} + k_{12}d_{22} + k_{13}d_{33} + k_{14}d_{23} + k_{15}d_{31} + k_{16}d_{12} \\
\tau_{22} &= k_{21}d_{11} + \qquad\qquad\qquad\qquad\qquad\qquad + k_{26}d_{12} \\
\tau_{33} &= k_{31}d_{11} + \qquad\qquad\qquad\qquad\qquad\qquad + k_{36}d_{12} \\
\tau_{23} &= k_{41}d_{11} + \qquad\qquad\qquad\qquad\qquad\qquad + k_{46}d_{12} \\
\tau_{31} &= k_{51}d_{11} + \qquad\qquad\qquad\qquad\qquad\qquad + k_{56}d_{12} \\
\tau_{12} &= k_{61}d_{11} + k_{62}d_{22} + k_{63}d_{33} + k_{64}d_{23} + k_{65}d_{31} + k_{66}d_{12}
\end{aligned}
\qquad (1.63)
$$

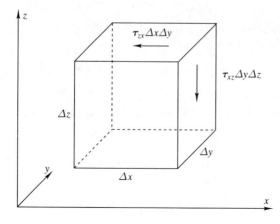

FIGURE 1.14
Shear stresses causing rotation about the y-axis.

The k are not tensors but simply different scalar constants. They represent 36 constants of proportionality that must be related to physical properties of the fluid. We assume that the fluid is isotropic (that is, its properties are independent of direction) so that the k remain the same with any orientation of the coordinate axes.

Consider a coordinate system that coincides with the principal axes of the rate of strain tensor so that

$$d_{ij} = 0 \qquad \text{for } i \neq j \qquad (1.64)$$

For this coordinate system equations (1.63) are reduced to three terms apiece on the right side

$$
\begin{aligned}
\tau_{11} &= k_{11}d_{11} + k_{12}d_{22} + k_{13}d_{33} \\
\tau_{22} &= k_{21}d_{11} + \phantom{k_{12}d_{22}} + k_{23}d_{33} \\
\tau_{33} &= k_{31}d_{11} + \phantom{k_{12}d_{22}} + k_{33}d_{33} \\
\tau_{23} &= k_{41}d_{11} + \phantom{k_{12}d_{22}} + k_{43}d_{33} \\
\tau_{31} &= k_{51}d_{11} + \phantom{k_{12}d_{22}} + k_{53}d_{33} \\
\tau_{12} &= k_{61}d_{11} + k_{62}d_{22} + k_{63}d_{33}
\end{aligned}
\qquad (1.65)
$$

We now choose a different set of principal axes $x_{i'}$, which are rotated 180 degrees about the z-axis as given by the transformation

$$
\begin{Bmatrix} x_{1'} \\ x_{2'} \\ x_{3'} \end{Bmatrix} =
\begin{bmatrix} -1 & 0 & 0 \\ 0 & -1 & 0 \\ 0 & 0 & 1 \end{bmatrix}
\begin{Bmatrix} x_1 \\ x_2 \\ x_3 \end{Bmatrix}
\qquad (1.66)
$$

Transforming the off-diagonal stress tensors into the primed coordinate system gives (because (1.65) should be independent of the coordinate system), for example

$$\tau_{2'3'} = k_{41}d_{1'1'} + k_{42}d_{2'2'} + k_{43}d_{3'3'} \qquad (1.67)$$

This same quantity can be related to the stress in the original coordinate system by the transformation

$$\tau_{2'3'} = a_{2'i}a_{3'j}\tau_{ij} = a_{2'2}a_{3'3}\tau_{23} = -\tau_{23} \qquad (1.68)$$

in which the a_{ij} from the transformation matrix of (1.66) have been used. The transformation of the rate of strain tensor is

$$d_{i'j'} = a_{i'i}a_{j'j}d_{ij} \qquad (1.69)$$

Again using the transformation matrix of (1.66) and substituting for the nine a of (1.69) yield

$$d_{i'i'} = d_{ii} \qquad \text{no sum on } i \text{ or } i' \qquad (1.70)$$

Using this result in (1.67) gives

$$\tau_{2'3'} = k_{41}d_{11} + k_{42}d_{22} + k_{43}d_{33} \qquad (1.71)$$

Comparing (1.71) with the fourth equation of (1.65) indicates that $\tau_{2'3'} = \tau_{23}$, which appears to contradict (1.68). The only way both equations can be satisfied is for $\tau_{2'3'} = \tau_{23} = 0$. Similar rotations about the x-axis and y-axis will give

$$\tau_{ij} = 0 \qquad \text{for} \quad i \neq j \qquad (1.72)$$

showing that the principal axes of the rate of strain tensor are also the principal axes of the stress tensor. For the coordinate system of the principal axes, the original 36 k have been reduced to 9 as follows:

$$\tau_{11} = k_{11}d_{11} + k_{12}d_{22} + k_{13}d_{33}$$
$$\tau_{22} = k_{21}d_{11} + k_{22}d_{22} + k_{23}d_{33} \qquad (1.73)$$
$$\tau_{33} = k_{31}d_{11} + k_{32}d_{22} + k_{33}d_{33}$$

The number of constants can be reduced further by a series of 90-degree rotations, which means that the tensors remain in a coordinate system that coincides with their principal axes. First consider the 90-degree rotation about the x-axis with the transformation matrix

$$a_{i'i} = \begin{bmatrix} 1 & 0 & 0 \\ 0 & 0 & 1 \\ 0 & -1 & 0 \end{bmatrix} \qquad (1.74)$$

The rate of strain tensor transforms as

$$d_{1'1'} = a_{1'i}a_{1'j}d_{ij} = d_{11}$$
$$d_{2'2'} = a_{2'i}a_{2'j}d_{ij} = d_{33} \qquad (1.75)$$
$$d_{3'3'} = a_{3'i}a_{3'j}d_{ij} = d_{22}$$

Equations (1.75) could as well apply to the τ; therefore,

$$\tau_{1'1'} = k_{11}d_{11} + k_{12}d_{22} + k_{13}d_{33} = \tau_{11}$$
$$\tau_{2'2'} = k_{21}d_{11} + k_{22}d_{33} + k_{23}d_{22} = \tau_{33} \qquad (1.76)$$
$$\tau_{3'3'} = k_{31}d_{11} + k_{32}d_{33} + k_{33}d_{22} = \tau_{22}$$

Comparing (1.73) with (1.76) gives the following relationships for some of the k:

$$k_{12} = k_{13} \qquad k_{21} = k_{31} \qquad k_{23} = k_{32} \qquad k_{22} = k_{33} \qquad (1.77)$$

A similar set of equations can be developed by taking a 90-degree rotation about the y-axis

$$a_{i'i} = \begin{bmatrix} 0 & 0 & -1 \\ 0 & 1 & 0 \\ 1 & 0 & 0 \end{bmatrix} \tag{1.78}$$

Applying this rotation the following equations hold in parallel with the above development:

$$d_{1'1'} = a_{1'i}a_{1'j}d_{ij} = d_{33} \quad d_{2'2'} = a_{2'i}a_{2'j}d_{ij} = d_{22} \quad d_{3'3'} = a_{3'i}a_{3'j}d_{ij} = d_{11} \tag{1.79}$$

$$\tau_{1'1'} = k_{11}d_{33} + k_{12}d_{22} + k_{13}d_{11} = \tau_{33}$$
$$\tau_{2'2'} = k_{21}d_{33} + k_{22}d_{22} + k_{23}d_{11} = \tau_{22} \tag{1.80}$$
$$\tau_{3'3'} = k_{31}d_{33} + k_{32}d_{22} + k_{33}d_{11} = \tau_{11}$$

$$k_{11} = k_{33} \qquad k_{12} = k_{32} \qquad k_{21} = k_{23} \qquad k_{13} = k_{31} \tag{1.81}$$

Equations (1.77) and (1.81) form sufficient relationships between the k so that only two independent quantities remain; they are redefined as λ and k

$$\lambda = k_{12} = k_{13} = k_{31} = k_{21} = k_{23} = k_{32} \qquad k = k_{11} = k_{22} = k_{33} \tag{1.82}$$

Using λ and k to rewrite equations (1.73) yields

$$\tau_{11} = kd_{11} + \lambda(d_{22} + d_{33}) = \lambda \frac{\partial u_i}{\partial x_i} + (k - \lambda)d_{11}$$

$$\tau_{22} = kd_{22} + \lambda(d_{11} + d_{33}) = \lambda \frac{\partial u_i}{\partial x_i} + (k - \lambda)d_{22} \tag{1.83}$$

$$\tau_{33} = kd_{33} + \lambda(d_{11} + d_{22}) = \lambda \frac{\partial u_i}{\partial x_i} + (k - \lambda)d_{33}$$

For a general set of coordinates x', y', z' (not in the principal axes of the tensors) $\tau_{i'j'} = a_{i'i}a_{j'j}\tau_{ij} = 0$ but $\tau_{ij} = 0$ for $i \neq j$, leading to

$$\tau_{i'j'} = a_{i'1}a_{j'1}\left[\lambda \frac{\partial u_i}{\partial x_i} + (k - \lambda)d_{11}\right] + a_{i'2}a_{j'2}\left[\lambda \frac{\partial u_i}{\partial x_i} + (k - \lambda)d_{22}\right]$$

$$+ a_{i'3}a_{j'3}\left[\lambda \frac{\partial u_i}{\partial x_i} + (k - \lambda)d_{33}\right] = \lambda \frac{\partial u_{k'}}{\partial x_{k'}}\delta_{i'j'} + (k - \lambda)d_{i'j'} \tag{1.84}$$

Dropping the primes for a general coordinate system

$$\tau_{ij} = \lambda \frac{\partial u_k}{\partial x_k}\delta_{ij} + (k - \lambda)d_{ij} \tag{1.85}$$

Equation (1.85) expresses the information in (1.63) where the 36 constants have been reduced to two. To see how the remaining constants are related to the physical properties of a fluid, consider the simple two-dimensional shear flow between flat plates as shown in Fig. 1.13. Equation (1.60) applies directly and τ_{yx} (together with its symmetric component τ_{xy}) is the only nonzero component of the shear stress tensor. From (1.85)

$$\tau_{yx} = \frac{1}{2}(k - \lambda)\frac{\partial u_x}{\partial y} \tag{1.86}$$

For (1.86) to agree with (1.60)

$$k - \lambda = -2\mu \tag{1.87}$$

For incompressible, unstratified fluids, where $\rho = $ constant and thus $\vec{\nabla} \cdot \vec{u} = 0$, the shear stress tensor becomes

$$\tau_{ij} = -2\mu d_{ij} \tag{1.88}$$

where now the original 36 constants have been reduced to one.

Although we have accomplished our objective of expressing the stress tensor in terms of the rate of strain tensor and the properties of the fluid for an incompressible fluid, there still remains the matter of the other constant, λ. To find the significance of that constant, i is substituted for j everywhere that j appears in (1.85) and the summation convention is applied (an operation called a *contraction*) with the result

$$\tau_{ii} = 3\lambda\frac{\partial u_i}{\partial x_i} - 2\mu\frac{\partial u_i}{\partial x_i} = (3\lambda - 2\mu)\frac{\partial u_i}{\partial x_i} \tag{1.89}$$

In a resting fluid the only normal stress acting on an area is the pressure or, to be more explicit, the *thermodynamic pressure* (such as appears in the equation of a perfect gas, $p = \rho R T$). In a moving fluid, however, there is the additional normal stress, $\tau_{ii}/3$, where $\tau_{ii} = \tau_{xx} + \tau_{yy} + \tau_{zz}$ is given by (1.89). A *mean pressure* is defined as

$$\bar{p} = p + \frac{\tau_{ii}}{3} = p + \frac{3\lambda - 2\mu}{3}\frac{\partial u_i}{\partial x_i} \tag{1.90}$$

For an incompressible fluid where the divergence of the velocity is zero, $p = \bar{p}$. For a general fluid we substitute the substantial derivative into (1.90) and define another property of the fluid, the *second coefficient of viscosity* or the *bulk viscosity* κ to give

$$\bar{p} = p - \frac{3\lambda - 2\mu}{3\rho}\frac{D\rho}{Dt} = p + \frac{\kappa}{\rho}\frac{D\rho}{Dt} \tag{1.91}$$

in which

$$\kappa = \frac{2}{3}\mu - \lambda \tag{1.92}$$

The kinetic theory of gases confirms that this quantity should exist. The bulk viscosity provides damping for volumetric vibrations of a gas. Its magnitude is much smaller than the magnitude of μ for gases and is zero for low density, monatomic gases. It can be large, however, for liquids and is several times the magnitude of μ for water. From the point of view of fluid mechanics, κ is not considered important, except for problems dealing with volumetric vibrations such as the transmission of sound waves, since it does not appear in the equations for incompressible fluids and is small for gases. The values of κ are difficult to obtain either from molecular theory or from experiment.

The final expression for the stress tensor is

$$\tau_{ij} = \left(\frac{2}{3}\mu - \kappa\right)\frac{\partial u_k}{\partial x_k}\delta_{ij} - 2\mu d_{ij} \tag{1.93}$$

or, neglecting the bulk viscosity κ,

$$\tau_{ij} = \frac{2}{3}\mu\frac{\partial u_k}{\partial x_k}\delta_{ij} - 2\mu d_{ij} \tag{1.94}$$

Using (1.94) in the equation of motion (1.29) yields

$$\begin{aligned}
\rho\frac{Du_j}{Dt} &= -\rho g\frac{\partial h}{\partial x_j} - \frac{\partial p}{\partial x_j} - \frac{\partial}{\partial x_i}\left(\frac{2}{3}\mu\frac{\partial u_k}{\partial x_k}\delta_{ij} - 2\mu d_{ij}\right) \\
&= -\rho g\frac{\partial h}{\partial x_j} - \frac{\partial p}{\partial x_j} - \frac{2}{3}\mu\frac{\partial}{\partial x_j}\frac{\partial u_i}{\partial x_i} + \mu\frac{\partial}{\partial x_j}\frac{\partial u_i}{\partial x_i} + \mu\frac{\partial^2 u_j}{\partial x_i \partial x_i}
\end{aligned} \tag{1.95}$$

The final result is the Navier-Stokes equation

$$\rho\frac{Du_i}{Dt} = -\rho g\frac{\partial h}{\partial x_i} - \frac{\partial p}{\partial x_i} + \frac{1}{3}\mu\frac{\partial}{\partial x_i}\frac{\partial u_j}{\partial x_j} + \mu\frac{\partial^2 u_i}{\partial x_j \partial x_j} \tag{1.96}$$

Note that (1.96) consists of three equations, one for each value of the unrepeated index, i. Combined with the continuity equation (1.20) there are four equations in the three u_i and p. If ρ is also an unknown (the case of a compressible, unstratified fluid), an equation of state—a relationship between density and pressure or density, pressure, and temperature—must be added, in which case the energy equation is needed to provide an additional relationship, say a total of 6 equations in u_i, p, ρ, and temperature.

Historical Note: Louis Navier (1785–1836), using the work of Newton, published his equations in 1821. He attributed forces in fluids to a repulsion or attraction between molecules that depends on the difference in relative velocity. His calculation that the velocity in a square tube varies directly with the length of the side and inversely with the length of the tube were verified by the experiment of a few years earlier by Pierre Girard (1765–1836). George Stokes (1819–1903) was apparently the first to define the rotation vector and give physical meaning to the terms of Navier's equations. He published his work "On the theories of the internal friction of fluids in motion, and of the equilibrium and motion of elastic solids" in 1845. Though Navier has a constant of proportionality in his equation that he related to the adhesion of molecules, Stokes called it the "coefficient of viscosity." Stokes was also aware of the bulk viscosity and aware that in most cases it could be neglected without appreciable error.

1.9   THERMAL AND MECHANICAL ENERGIES

An equation for the mechanical energy of a moving fluid can be obtained by multiplying (1.49) by u_i

$$\rho \frac{D}{Dt} \frac{u^2}{2} = -\rho g u_i \frac{\partial h}{\partial x_i} - u_i \frac{\partial p}{\partial x_i} - u_i \frac{\partial \tau_{ij}}{\partial x_j} \tag{1.97}$$

Equation (1.97) is rewritten in the form

$$\frac{\rho}{2} \frac{Du^2}{Dt} = -\frac{\partial}{\partial x_i}(p u_i) - \frac{\partial}{\partial x_i}(\tau_{ij} u_i) - \rho g u_i \frac{\partial h}{\partial x_i} + p \frac{\partial u_i}{\partial x_i} + \tau_{ij} \frac{\partial u_j}{\partial x_i} \tag{1.98}$$

Subtracting (1.98) from (1.50) provides an equation for thermal energy

$$\rho \frac{De}{Dt} = -\frac{\partial q_i}{\partial x_i} - p \frac{\partial u_i}{\partial x_i} - \tau_{ij} \frac{\partial u_j}{\partial x_i} \tag{1.99}$$

Evidently the terms common to (1.98) for mechanical energy and (1.99) for thermal energy are those terms that represent the processes whereby the conversion from one type of energy to another can take place. The common terms can be identified as follows:

$$-p \frac{\partial u_i}{\partial x_i} = \begin{bmatrix} \text{The rate of reversible conversion} \\ \text{to internal energy by compression} \end{bmatrix}$$

$$-\tau_{ij} \frac{\partial u_j}{\partial x_i} = \begin{bmatrix} \text{The rate of irreversible conversion} \\ \text{to internal energy by viscous action} \end{bmatrix}$$

The reversibility of the terms can be determined by examining their signs. The term $p \, \partial u_i / \partial x_i$ can be either positive or negative according to the sign of the divergence, which depends on whether the fluid is expanding or contracting; thus, the flow of energy can be in either direction—from thermal to mechanical or from mechanical to thermal.

To determine the sign of the second quantity, (1.94) is expanded and multiplied by the appropriate derivative to give

$$-\tau_{ij}\frac{\partial u_j}{\partial x_i} = \mu\left(-\frac{2}{3}\frac{\partial u_k}{\partial x_k}\frac{\partial u_j}{\partial x_i}\delta_{ij} + \frac{\partial u_i}{\partial x_j}\frac{\partial u_j}{\partial x_i} + \frac{\partial u_j}{\partial x_i}\frac{\partial u_j}{\partial x_i}\right) \tag{1.100}$$

The right-hand side of (1.100) can be written as a perfect square

$$-\tau_{ij}\frac{\partial u_j}{\partial x_i} = \frac{\mu}{2}\sum_i\sum_j\left(\frac{2}{3}\frac{\partial u_k}{\partial x_k}\delta_{ij} - \frac{\partial u_i}{\partial x_j} - \frac{\partial u_j}{\partial x_i}\right)^2 \tag{1.101}$$

in which the summation convention on the right side has been replaced by specific summation signs. Equation (1.101) indicates that the term $\tau_{ij}\partial u_j/\partial x_i$ is always negative, showing that this term does represent an irreversible conversion of mechanical energy to thermal energy.

If the fluid is incompressible, the divergence of the velocity is zero and thus the reversible term is always zero. For an incompressible fluid there can be a conversion of mechanical to thermal energy but not vice versa. Mechanical energy that is converted to thermal energy is lost to the flow in the sense that it cannot be reconverted to mechanical energy. Thus, "friction" does represent "lost" energy to the flow system.

1.10 PERFECT FLUIDS

In this section the general equations are specialized for a perfect fluid; that is, a fluid with zero viscosity. Setting $\mu = 0$, the continuity, momentum, and energy equations reduce to

$$\frac{\partial \rho}{\partial t} + \frac{\partial}{\partial x_i}(\rho u_i) = 0 \tag{1.102}$$

$$\rho\frac{Du_i}{Dt} = -\rho g\frac{\partial h}{\partial x_i} - \frac{\partial p}{\partial x_i} \tag{1.103}$$

$$\rho\frac{D}{Dt}\left(\frac{u^2}{2} + e\right) = -\frac{\partial}{\partial x_i}(pu_i) - \rho g u_i\frac{\partial h}{\partial x_i} \tag{1.104}$$

The energy equation is written for no heat transfer as well as frictionless flow. Equations (1.102) and (1.103) represent four equations in the five unknowns u_x, u_y, u_z, ρ, and p. The internal energy e appears as an additional unknown in (1.104). An equation of state must be added to have a determinate system. For liquids the equation of state takes the form $\rho = $ constant; for gases the equation of state depends upon the process. An example of an equation of state for gas flow is the isentropic relation

$$\frac{p}{p_0} = \left(\frac{\rho}{\rho_0}\right)^{\gamma} \tag{1.105}$$

in which p_0 and ρ_0 are known constants and $\gamma = c_p/c_v$, the ratio of the specific heat at constant pressure to the specific heat at constant volume (for air $\gamma = 1.405$).

The flow of perfect fluids is usually considered to be irrotational although it is really of constant rotation with the constant often equal to zero. If we consider a spherical fluid particle, we see that all forces are due to pressure and gravity in the absence of shear, and that these forces must act through the center of the particle. Hence, there are no forces that can cause (or stop) rotation. The condition of no rotation is expressed

$$\frac{\partial u_i}{\partial x_j} - \frac{\partial u_j}{\partial x_i} = 0 \tag{1.106}$$

When the flow is irrotational, the equations can be written in terms of a *velocity potential*, Φ, which is defined according to

$$u_i = -\frac{\partial \Phi}{\partial x_i} \tag{1.107}$$

Substitution of (1.107) into (1.106) shows that the irrotationality condition is automatically satisfied or, conversely, that the velocity potential exists only if the flow is irrotational.

To proceed, we assume that the fluid is incompressible in addition to being inviscid. The continuity equation then expresses the zero divergence condition,

$$\frac{\partial u_i}{\partial x_i} = 0 \tag{1.108}$$

Substituting (1.107) into (1.108) gives

$$\frac{\partial^2 \Phi}{\partial x_i \partial x_i} = \nabla^2 \Phi = 0 \tag{1.109}$$

which is known as *Laplace's equation*. The problem of inviscid flow has been reduced to a single equation in one unknown, Φ. Moreover, that equation is linear whereas the original system is nonlinear. The assumptions (or approximations) of irrotationality and incompressibility have resulted in tremendous simplifications. A similar equation for compressible flow is developed later in this section.

Once the velocity potential is known from solving (1.109), the equation of motion provides a means for finding the pressure distribution. Substitution of (1.107), the definition of the potential, into (1.103) gives

$$-\frac{\partial}{\partial t}\frac{\partial \Phi}{\partial x_i} + \frac{\partial \Phi}{\partial x_j}\frac{\partial^2 \Phi}{\partial x_i \partial x_j} = -\frac{1}{\rho}\frac{\partial p}{\partial x_i} - g\frac{\partial h}{\partial x_i} \tag{1.110}$$

The second term of (1.110) can be expressed as

$$\frac{\partial \Phi}{\partial x_j}\frac{\partial^2 \Phi}{\partial x_i \partial x_j} = \frac{\partial}{\partial x_i}\left(\frac{1}{2}\frac{\partial \Phi}{\partial x_j}\frac{\partial \Phi}{\partial x_j}\right) = \frac{\partial}{\partial x_i}\left(\frac{u_j u_j}{2}\right) \tag{1.111}$$

Noting that $u_i u_i = u_1^2 + u_2^2 + u_3^2 = u^2$, (1.110) can be written, after changing the order of differentiation, in the form

$$\frac{\partial}{\partial x_i}\left(-\frac{\partial \Phi}{\partial t} + \frac{1}{2}u^2 + \frac{p}{\rho} + gh\right) = 0 \tag{1.112}$$

Equation (1.112) indicates that the quantity in brackets is invariant in each of the coordinate directions; thus, it can be at most a function of time

$$-\frac{\partial \Phi}{\partial t} + \frac{1}{2}u^2 + \frac{p}{\rho} + gh = f(t) \tag{1.113}$$

If the motion is steady, the right-hand side becomes a constant

$$\frac{1}{2}u^2 + \frac{p}{\rho} + gh = \text{constant} \tag{1.114}$$

which is *Bernoulli's equation* derived under the restrictions of (1) incompressible flow, (2) irrotational flow, and (3) steady flow. (For most flows the irrotationality condition implies the absence of shear stress, so that frictionless flow does not appear in the list of restrictions.) Under these restrictions, (1.114) is a point equation—in contrast to the integral equations of sections 1.3, 1.4, and 1.5, which apply over a volume—since it was derived from a differential equation and it applies to all points in the flow field. This latter statement is important because we will derive another equation in the next section from an energy point of view that looks very much like Bernoulli's equation but is used differently.

An equation similar to (1.114) can be obtained by beginning with the equation of mechanical energy for an incompressible fluid. For frictionless and steady flow (1.97) becomes

$$u_i \frac{\partial}{\partial x_i}\left(\frac{u^2}{2} + \frac{p}{\rho} + gh\right) = 0 \tag{1.115}$$

which states that the scalar product of the velocity and the gradient of the quantity in parentheses is zero. The operator $\vec{u} \cdot \vec{\nabla}(\cdot)$ selects the component of the quantity in parentheses in the direction of the flow, that is, along the streamline. Unless the $\vec{u}_i$ are all zero, the quantity in parentheses is invariant, but we must be careful to recognize that it is constant only along a streamline and can take on different values on different streamlines. Thus, another form of Bernoulli's equation is

$$\frac{u^2}{2} + \frac{p}{\rho} + gh = \text{constant along a streamline} \qquad (1.116)$$

Equation (1.116) is derived under the restrictions of (1) incompressible flow, (2) steady flow, and (3) frictionless flow. The irrotationality condition is not in the derivation. If, however, the flow is irrotational, then the constant becomes universal; it is the same for all streamlines and (1.116) becomes identical to (1.114).

A similar equation can be derived for compressible fluids from (1.104). In gases the gravity force is usually small compared with other forces so that it can be neglected. With that approximation and for steady flow, the energy equation becomes

$$\rho u_i \frac{\partial}{\partial x_i} \left(\frac{u^2}{2} + e \right) = -p \frac{\partial u_i}{\partial x_i} - u_i \frac{\partial p}{\partial x_i} \qquad (1.117)$$

From the continuity equation (1.20)

$$\frac{\partial u_i}{\partial x_i} = -\frac{u_i}{\rho} \frac{\partial \rho}{\partial x_i} \qquad (1.118)$$

Using (1.118) in (1.117) yields

$$u_i \frac{\partial}{\partial x_i} \left(\frac{u^2}{2} + e \right) = -u_i \frac{\partial}{\partial x_i} \left(\frac{p}{\rho} \right) \qquad (1.119)$$

leading to an equation that applies along a streamline

$$\frac{u^2}{2} + e + \frac{p}{\rho} = \text{constant along a streamline} \qquad (1.120)$$

The quantity $e + \frac{p}{\rho}$ is called the *enthalpy* and is often designated by the letter h; we will not use that notation since it could be confused with elevation. The constant can apply to all points in the fluid if the flow is irrotational and if the internal energy is the same for all parts of the resting fluid.

Historical Note: There was not just one Bernoulli who had significant accomplishments in physics; there were five. The equation is named for Daniel Bernoulli (1700–1782). The others were: Johann (1667–1748), Daniel's father; Nikolaus, his brother; Jacob (1654–1705), his uncle; and Johann, a cousin. Although Swiss, the father was a professor in Belgium for a time. Daniel and Nikolaus worked as academics in St. Petersburg, where Daniel began to write a treatise on fluid mechanics in 1729. Daniel, at the insistence of his father, studied medicine and received his degree in that field.

To the great disappointment of Johann, whose field was mathematics, Daniel began to study fluid mechanics. Daniel and Johann, far from cooperating in their many discoveries, were fierce competitors, Johann being extremely jealous of the accomplishments of his son although his own work was notable. The father-son dispute became especially acute when Johann wrote his book *Hydraulics*. It was sent to Leonard Euler (the German mathematician, 1707–1783) in two parts in 1739 and 1740, but was dated 1732. The reason for the earlier date was apparently to preempt Daniel's *Hydrodynamics*, published in 1738. Daniel complained that his father had claimed most of the ideas for his own and those that he had not claimed, he minimized as trivial. In total, the contributions of the Bernoulli family in mathematics were outstanding.

Bernoulli's equation appears in Chapter 13 of *Hydrodynamics* in connection with the flow out of orifices. Although Bernoulli writes of "specific kinetic energy" (kinetic energy per unit of mass), he clearly uses momentum principles to write (Levi, 1989, p. 93–94, equation 3 with a change of sign in the velocity term)

$$\frac{p}{\rho} = gH - \frac{u^2}{2}$$

where p is the pressure in an outflow tube from a reservoir, H is the height of the water surface from the outflow, and u is the velocity in the tube.

1.11 TUBE FLOW

In elementary fluid mechanics the integral form of the conservation equations is often specialized for flow in a tube or pipe. We consider a tube with a control volume that consists of two cross-sections, one where the flow enters the control volume and one where the flow leaves, plus the pipe wall between the cross-sections. The equation of continuity (1.19) then takes on a particularly simple form

$$\rho_1 A_1 U_1 = \rho_2 A_2 U_2 \tag{1.121}$$

in which U is the average velocity defined as

$$U = \frac{1}{A} \int_A u \, dA \tag{1.122}$$

Notice that, for flow into the control volume, designated as flow through section one, $\vec{u} \cdot \vec{n}$ is negative and its integral defines the average velocity.

The equation for the conservation of momentum (1.22) for steady flow becomes

$$\vec{F} = \int_{A_1} \rho \vec{u}(\vec{u} \cdot \vec{n}) dA + \int_{A_2} \rho \vec{u}(\vec{u} \cdot \vec{n}) dA \tag{1.123}$$

in which $\vec{F}$ represents the sum of all real forces on the fluid in the control volume. These forces may consist of the pressure force from the fluid outside the control volume on the fluid inside the control volume, the pressure forces from the pipe walls, the shear on the pipe, gravity, and anything additional such as electrostatic forces or a machine in the control volume. The integral over the pipe boundary A_3 has not been written since it is zero due to the fact that $\vec{u} \cdot \vec{n}$ is zero on that boundary. Defining the *momentum correction factor* β as

$$\beta = \frac{1}{U^2 A} \int_A u^2 \, dA \qquad (1.124)$$

Equation (1.123) can be integrated so that

$$\vec{F} = \beta_2 \rho_2 Q \vec{U}_2 - \beta_1 \rho_1 Q \vec{U}_1 \qquad Q = U_1 A_1 = U_2 A_2 \qquad (1.125)$$

The minus sign on the quantity for section one is due to $\vec{u} \cdot \vec{n}$ being negative for inflow to the section.

An equation that looks very much like the Bernoulli equation, and is often confused with it, can be derived from the integral form of the energy equation (1.35). We assume a constant density, steady flow through a tube (pipe flow). The relevant control volume is the same as that for the momentum equation. Specializing (1.35) for this case produces

$$\frac{d\tilde{q}}{dt} - \frac{dW}{dt} = \int_{A_1} \rho \left(gh + \frac{u^2}{2} + e + \frac{p}{\rho} \right) (\vec{u} \cdot \vec{n}) dA$$

$$+ \int_{A_2} \rho \left(gh + \frac{u^2}{2} + e + \frac{p}{\rho} \right) (\vec{u} \cdot \vec{n}) dA \quad (1.126)$$

To integrate the velocity terms of (1.126), an additional quantity, the *energy correction factor*, is defined as

$$\alpha = \frac{1}{U^3 A} \int_A u^3 \, dA \qquad (1.127)$$

Also, we assume that the pressure is hydrostatic in both cross-sections—that there is no significant acceleration at right angles to the primary flow—resulting in $gh + \frac{p}{\rho} = $ constant in the cross-section. With these assumptions

$$\frac{1}{\rho U A} \left(\frac{d\tilde{q}}{dt} - \frac{dW}{dt} \right) = -\alpha_1 \frac{U_1^2}{2} - \frac{p_1}{\rho} - gh_1 - e_1 + \alpha_2 \frac{U_2^2}{2} + \frac{p_2}{\rho} + gh_2 + e_2$$

$$(1.128)$$

If we assume that there is no work inside the control volume, the second term on the left of (1.128) is zero. Then

$$\alpha_1 \frac{U_1^2}{2} + \frac{p_1}{\rho} + gh_1 = \alpha_2 \frac{U_2^2}{2} + \frac{p_2}{\rho} + gh_2 + gh_L \qquad (1.129)$$

in which the last term represents the rate of loss of mechanical energy between the two sections divided by the mass flow rate. The *head loss* is

$$h_L = \frac{e_2 - e_1}{g} - \frac{1}{g\rho U A} \frac{d\tilde{q}}{dt} \qquad (1.130)$$

Consider the case of no heat transfer; then $e_1 \leq e_2$ since the mechanical energy can only be converted to thermal energy and not vice versa (see section 1.9). Thus, $h_L \geq 0$ and the production of thermal energy is lost to the flow.

Although (1.129) looks very much like Bernoulli's equation, it is entirely different. First, it comes from energy principles whereas Bernoulli's equation comes from the conservation of momentum. Second, it is to be applied over two sections of a finite control volume whereas Bernoulli's equation is to be applied at one or more points in the fluid. To put the energy correction factor in Bernoulli's equation would be nonsense. Third, there has been no assumption regarding viscosity or irrotationality; in fact, the "head loss" term absorbs the conversion of mechanical energy into thermal energy.

1.12   THE VORTICITY EQUATIONS

From the previous section the rotational aspects, or vorticity, of the flow are obviously important. The equation of motion can be rewritten to express the diffusion of vorticity. This formulation has advantages in the solution of problems as well as a contribution to the understanding of vorticity.

Using the definition of the substantial derivative (1.44) and specializing the equation of motion for the case of a constant density fluid yields

$$\frac{\partial \vec{u}}{\partial t} - \vec{u} \times (\vec{\nabla} \times \vec{u}) = -\vec{\nabla}\left(\frac{u^2}{2} + \frac{p}{\rho} + gh\right) + \frac{\mu}{\rho}\nabla^2\vec{u} \qquad (1.131)$$

Equation (1.131) leads immediately to Bernoulli's equation if the flow is inviscid ($\mu = 0$), if there is no rotation ($\vec{\nabla} \times \vec{u} = 0$), and if the flow is steady. For the present purposes we take the curl of (1.131)

$$\vec{\nabla} \times \frac{\partial \vec{u}}{\partial t} - \vec{\nabla} \times \left[\vec{u} \times \left(\vec{\nabla} \times \vec{u}\right)\right] = \frac{\mu}{\rho}\vec{\nabla} \times \nabla^2\vec{u} \qquad (1.132)$$

Using identities 12c and 12e in the Appendix and the equation of continuity for a constant density fluid, $\vec{\nabla} \cdot \vec{u} = 0$,

$$\vec{\nabla} \times \left[\vec{u} \times \left(\vec{\nabla} \times \vec{u}\right)\right] = \left[\left(\vec{\nabla} \times \vec{u}\right) \cdot \vec{\nabla}\right]\vec{u} - (\vec{u} \cdot \vec{\nabla})(\vec{\nabla} \times \vec{u}) \qquad (1.133)$$

Changing to subscript notation, changing the order of differentiation, and using the definition of vorticity (1.59), (1.132) can be written

$$\frac{D\omega_i}{Dt} = \omega_j \frac{\partial u_i}{\partial x_j} + \frac{\mu}{\rho} \frac{\partial^2 \omega_i}{\partial x_j \partial x_j} \tag{1.134}$$

When multiplied by 2, (1.134) becomes the *vorticity transport equation*. If $\mu = 0$ and $\omega = 0$ initially, then

$$\frac{D\omega_i}{Dt} = 0 \tag{1.135}$$

which indicates that if the rotation of a particle is initially zero and the flow is frictionless, the rotation remains zero.

If the flow does not vary with x_3 (two-dimensional flow), the only possible component of rotation is ω_3. In that case the first term on the right of the equals sign of (1.134) is zero and that equation reduces to

$$\frac{D\omega_3}{Dt} = \frac{\mu}{\rho} \frac{\partial^2 \omega_3}{\partial x_j \partial x_j} \tag{1.136}$$

If $\mu = 0$, then for a particle $\omega_3 = $ constant, showing that the rotation for a particle in two-dimensional, inviscid flow is constant.

A general viscous flow starting from rest is initially irrotational, but the rotation cannot remain zero. Velocity gradients near the boundaries (resulting from the no-slip boundary condition; see section 1.13) cause rotation, and (1.134) indicates that the vorticity is diffused inward from the boundaries.

Although the foregoing discussion began by assuming that the density is constant, many of the same considerations hold for a fluid of nonconstant density. For a variable density, inviscid fluid, the equation of motion (1.96) is divided through by ρ. Then we take the curl of both sides and rearrange terms to give

$$\frac{D\vec{\omega}}{Dt} = \left(\vec{\omega} \cdot \vec{\nabla}\right)\vec{u} - \vec{\omega}\left(\vec{\nabla} \cdot \vec{u}\right) + \vec{\nabla}p \times \vec{\nabla}\frac{1}{\rho} \tag{1.137}$$

Although the last term disappears for $\rho = $ constant, in general $\vec{\omega} = 0$ is not a solution to this equation even though the viscosity, μ, has been taken identically equal to zero.

In many problems of the flow of a gas the equation of state can be written in the form

$$\rho = f(p) \tag{1.138}$$

In such a case

$$f \vec{\nabla} \frac{1}{\rho} + \frac{f'}{\rho} \vec{\nabla} p = 0 \tag{1.139}$$

in which $f' = df/dt$, the derivative of (1.138). Equation (1.139) indicates that $\vec{\nabla} \frac{1}{\rho}$ and $\vec{\nabla} p$ are parallel vectors; thus, the quantity $\vec{\nabla} p \times \vec{\nabla} \frac{1}{\rho} = 0$ and $\vec{\omega} = 0$ is a solution of (1.137). The conclusion is that the flow of an inviscid fluid is irrotational if it is irrotational at any time (such as when starting from rest) and if either the density is constant or the density gradient and the pressure gradient are in the same direction.

A *stratified* or *baroclinic* fluid is defined as a fluid of nonconstant density where the pressure and density gradients are not necessarily parallel; otherwise, the fluid is *barotropic*. Examples of stratified fluids appear commonly in the atmosphere and in bodies of water. The atmosphere is often stratified due to variations in moisture content and temperature. The common temperature inversion is an excellent example. The ocean is heavily stratified due to temperature differences and variations in dissolved salt content. Many lakes are effectively stratified from a variation of suspended solids or, more importantly, from temperature differences. In all these cases the flow is rotational (unless the pressure and density gradients happen to coincide as in the rather trivial case of a vertically stratified fluid with no horizontal pressure gradients) even though viscosity is neglected. In such cases, we are forced to give up the simplifications that are introduced by assuming a velocity potential.

1.13 BOUNDARY CONDITIONS

The exact type of boundary conditions that are applied to the Navier-Stokes equations, or to one of the approximations, depends on the particular problem. In the present section only the most general types of boundary conditions are discussed; those that apply to a specific type of problem are deferred for the discussion of that problem.

At every solid-fluid boundary the condition is imposed that the fluid does not penetrate the boundary, except in the case of a porous boundary. In other words, this condition states that the fluid velocity normal to the boundary is equal to the velocity of the boundary. The equation of a bounding surface can be expressed in a functional relationship such as

$$\zeta(x, y, z, t) = 0 \tag{1.140}$$

Taking the substantial derivative of the function produces

$$\frac{D\zeta}{Dt} = \frac{\partial \zeta}{\partial t} + u_i \frac{\partial \zeta}{\partial x_i} = 0 \tag{1.141}$$

A unit vector normal to the surface is $\vec{\nabla}\zeta/|\vec{\nabla}\zeta|$ and the velocity of the surface in the normal direction is

$$u_n = \frac{\vec{\nabla}\zeta \cdot \vec{u}}{|\vec{\nabla}\zeta|} = -\frac{\dfrac{\partial \zeta}{\partial t}}{\sqrt{\left(\dfrac{\partial \zeta}{\partial x}\right)^2 + \left(\dfrac{\partial \zeta}{\partial y}\right)^2 + \left(\dfrac{\partial \zeta}{\partial z}\right)^2}} \qquad (1.142)$$

In the very common case where the position of the boundary is independent of time, the normal fluid velocity, u_n, is zero on $\zeta = 0$.

For a viscous fluid (when the viscous term is not neglected) the condition that the velocity on a solid surface is exactly equal to the velocity of the surface is imposed. Such a statement is really two boundary conditions: $u_n = 0$ and $u_t = 0$, where u_t is the velocity tangential to the surface. The latter condition is called the *no-slip* condition. The no-slip condition results from experimental observation and is confirmed by molecular theory. An exception to the no-slip condition occurs when dealing with highly rarefied gases or problems with extremely small dimensions where the basic continuum theory is not valid.

In the case of inviscid flow, setting $\mu = 0$ in (1.96) has eliminated the highest order derivatives in the equation of motion. That change means that not as many boundary conditions are necessary (or can be applied) for a solution. The consequence is that the no-slip condition cannot be enforced. Thus, the solution of (1.96) is often quite different, even for small values of μ, than the corresponding potential flow solution. The analyst must often apply judgment to decide if the neglect of viscosity has an important effect on the solution of a specific problem.

For free surfaces there are two boundary conditions. The dynamic condition states that the pressure on a fluid-fluid interface (such as air-water) is continuous, or the pressure has a jump specified by conditions at the interface. The kinematic condition is a result of topology and says that particles on the surface remain on the surface. These are derived in Chaps. 4 and 10.

In many cases—numerical solutions especially—boundary conditions must be "invented" in order to truncate the solution domain artificially. These are cases where the solution is desired in a definite region, but the region is not naturally closed. An example might be the flow in a harbor, where the harbor opens onto the sea. The flow in the entire ocean cannot be solved, so the solution domain is truncated at some arbitrary line. Boundary conditions must be imposed along that line. The nature of the boundary conditions depends on the particular problem. Those that have been used in some combination are pressure, tangential velocity, normal velocity, normal pressure gradient, tangential vorticity, and normal vorticity (Gresho, 1991). In free surface flows there is a need to pass disturbances from the solution region without reflection, in which case "radiation boundary conditions" are formulated. Another method is to damp disturbances at the boundary. These alternatives are discussed in connection with the particular problem.

1.14   THE LAGRANGIAN EQUATIONS

Finally, we arrive at the most detailed description of fluid flow, the Lagrangian description. The resulting equations are not often used, at least for a full description, since the detail is rarely necessary and because the nonlinearities appear in an awkward manner, creating difficulties with the solution. For certain types of unsteady flow problems, however, and for other special cases, the Lagrangian equations are useful. The equations are presented herein for a frictionless fluid since they are hopelessly complex otherwise.

In the Lagrangian system the coordinates a, b, c represent the initial (at time t_0) coordinates of a particle and the coordinates x, y, z represent the particle coordinates at any time t. The equations express x, y, z in terms of the independent variables a, b, c, t. The velocities of a particle parallel to the x-, y-, and z-axes are $\partial x/\partial t$, $\partial y/\partial t$, and $\partial z/\partial t$, respectively, in which the partial derivative with respect to time implies that the independent coordinates a, b, c and time are to be held constant. Let F_x, F_y, F_z represent the forces per unit mass, other than pressure force, on a particle in the coordinate directions. Equating the acceleration to the force per unit mass gives

$$\frac{\partial^2 x}{\partial t^2} = -\frac{1}{\rho}\frac{\partial p}{\partial x} + F_x \qquad \frac{\partial^2 y}{\partial t^2} = -\frac{1}{\rho}\frac{\partial p}{\partial y} + F_y \qquad \frac{\partial^2 z}{\partial t^2} = -\frac{1}{\rho}\frac{\partial p}{\partial z} + F_z$$

$$(1.143)$$

The derivatives on the right sides are a bit peculiar since p is differentiated with respect to a dependent variable. This situation is resolved by use of the relations

$$\frac{\partial p}{\partial a} = \frac{\partial p}{\partial x}\frac{\partial x}{\partial a} + \frac{\partial p}{\partial y}\frac{\partial y}{\partial a} + \frac{\partial p}{\partial z}\frac{\partial z}{\partial a}$$

$$\frac{\partial p}{\partial b} = \frac{\partial p}{\partial x}\frac{\partial x}{\partial b} + \frac{\partial p}{\partial y}\frac{\partial y}{\partial b} + \frac{\partial p}{\partial z}\frac{\partial z}{\partial b} \qquad (1.144)$$

$$\frac{\partial p}{\partial c} = \frac{\partial p}{\partial x}\frac{\partial x}{\partial c} + \frac{\partial p}{\partial y}\frac{\partial y}{\partial c} + \frac{\partial p}{\partial z}\frac{\partial z}{\partial c}$$

The first of (1.143) is multiplied by $\partial x/\partial a$, the second by $\partial y/\partial a$, the third by $\partial z/\partial a$, and the three equations are added. The result is that the terms containing the pressure derivatives add to form the terms of the right side of the first of (1.144). Finally, there result the equations of motion

$$\left(\frac{\partial^2 x}{\partial t^2} - F_x\right)\frac{\partial x}{\partial a} + \left(\frac{\partial^2 y}{\partial t^2} - F_y\right)\frac{\partial y}{\partial a} + \left(\frac{\partial^2 z}{\partial t^2} - F_z\right)\frac{\partial z}{\partial a} + \frac{1}{\rho}\frac{\partial p}{\partial a} = 0$$

$$\left(\frac{\partial^2 x}{\partial t^2} - F_x\right)\frac{\partial x}{\partial b} + \left(\frac{\partial^2 y}{\partial t^2} - F_y\right)\frac{\partial y}{\partial b} + \left(\frac{\partial^2 z}{\partial t^2} - F_z\right)\frac{\partial z}{\partial b} + \frac{1}{\rho}\frac{\partial p}{\partial b} = 0$$

$$\left(\frac{\partial^2 x}{\partial t^2} - F_x\right)\frac{\partial x}{\partial c} + \left(\frac{\partial^2 y}{\partial t^2} - F_y\right)\frac{\partial y}{\partial c} + \left(\frac{\partial^2 z}{\partial t^2} - F_z\right)\frac{\partial z}{\partial c} + \frac{1}{\rho}\frac{\partial p}{\partial c} = 0$$

$$(1.145)$$

The Lagrangian form of the equation of continuity is derived by considering a small particle such as that shown in Fig. 1.15. The sides of the small volume at time $t = 0$ are Δa, Δb, and Δc. At a later time the element is deformed due to differences in velocities throughout the volume. After some time the vectors forming the sides of the element are

$$\frac{\partial \vec{r}}{\partial a}\Delta a \qquad \frac{\partial \vec{r}}{\partial b}\Delta b \qquad \frac{\partial \vec{r}}{\partial c}\Delta c$$

If $\vec{a}$ and $\vec{b}$ are the sides of a parallelogram, the area of the parallelogram is $A = |\vec{a} \times \vec{b}|$. If $\vec{c}$ is a vector representing the third side of a parallelepiped of which the parallelogram is the front face, the volume $V = (\vec{a} \times \vec{b}) \cdot \vec{c}$. In the case of the element the volume is

$$\Delta V = \left(\frac{\partial \vec{r}}{\partial a} \times \frac{\partial \vec{r}}{\partial b}\right) \cdot \frac{\partial \vec{r}}{\partial c}\Delta a\,\Delta b\,\Delta c \qquad (1.146)$$

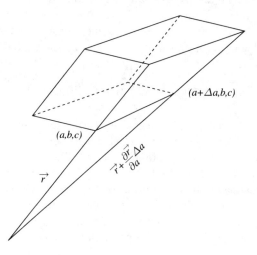

FIGURE 1.15
A particle in the Lagrangian coordinate system.

Equating the mass of the element at time $t = 0$ to a later time

$$\rho \left(\frac{\partial \vec{r}}{\partial a} \times \frac{\partial \vec{r}}{\partial b} \right) \cdot \frac{\partial \vec{r}}{\partial c} = \rho \begin{vmatrix} \dfrac{\partial x}{\partial a} & \dfrac{\partial y}{\partial a} & \dfrac{\partial z}{\partial a} \\ \dfrac{\partial x}{\partial b} & \dfrac{\partial y}{\partial b} & \dfrac{\partial z}{\partial b} \\ \dfrac{\partial x}{\partial c} & \dfrac{\partial y}{\partial c} & \dfrac{\partial z}{\partial c} \end{vmatrix} = \rho_0 \tag{1.147}$$

Equation (1.147) is often written symbolically as

$$\frac{\partial(x, y, z)}{\partial(a, b, c)} = \frac{\rho_0}{\rho} \tag{1.148}$$

The left side of (1.148) is called the *Jacobian* of x, y, z with respect to a, b, c.

Equations (1.145) and (1.148) form four equations in the unknowns x, y, z, p, ρ. Just as in the Eulerian case an equation of state must be added to close the system.

Historical Note: Joseph Louis Lagrange (1736–1813) was a noted French mathematician whose book, *Analytical Mechanics*, was almost a bible of mathematics and mechanics. Lagrange and Leonard Euler (1707–1783) were probably the two greatest mathematicians of the century [together with Pierre Laplace (1749–1827), a French astronomer]. They certainly left their mark on fluid mechanics with the two methods of description of flow. But which method should be attached to which name? In Euler's first paper on hydraulics, "On the movement of water in pipes," we find the equations

$$m\frac{d^2x}{dt^2} = F_x \qquad m\frac{d^2y}{dt^2} = F_y \qquad m\frac{d^2z}{dt^2} = F_z$$

(Newton did not write these equations previously; he expressed the equations of motion in terms of momentum.) These are then expressed in the form of (1.143). Euler's derivation of the equation of continuity begins with a parallelepiped of dimensions $\Delta x, \Delta y, \Delta z$ and expresses the dimensions after a time Δt as

$$\left(1 + \frac{\partial u_x}{\partial x} \Delta t \right), \left(1 + \frac{\partial u_y}{\partial y} \Delta t \right), \left(1 + \frac{\partial u_z}{\partial z} \Delta t \right)$$

Lagrange, on the other hand, introduced the velocity potential in his 1781 paper, "Memorandum on the theory of movement of fluids," essentially an Eulerian concept. According to Levi (1989), the Eulerian and Lagrangian terminology were introduced by the German mathematician Dirichlet in a publication of 1860 after his death. Apparently Dirichlet falsely believed that Lagrange had utilized the Lagrangian method since 1762. Euler, however, had proposed the method in a letter to Lagrange in 1759 as a technique to solve the problem of the propagation of sound. A few months later, in 1760,

Euler developed the total method in another letter to Lagrange. The 1760 letter is published in *Memorie dell'Accademia delli Scienze di Torino*. Thus, it appears clear that the Lagrangian method was developed by Euler, but the attribution in the names will not change.

PROBLEMS

1.1. Derive the Navier-Stokes equations for two-dimensional flow in the same way that the three-dimensional equations were derived. Begin with an infinitesimal volume of sides Δx by Δy.

Note: The two-dimensional Navier-Stokes equations are universally used as in (1.96) with the indices taken as 1 and 2; that is, nothing changes in the third direction. That usage is correct since we really live in a three-dimensional world and are actually making an approximation in representing a flow as two-dimensional. But what would these equations be in an actual two-dimensional world?

1.2. Show that the pressure in a free jet is everywhere zero.

1.3. Show that the Bernoulli sum

$$\frac{u^2}{2} + \frac{p}{\rho} + gh$$

is a constant along a vortex line for steady, inviscid flow. That is, show that under these conditions

$$\omega_i \frac{\partial}{\partial x_i} \left(\frac{u^2}{2} + \frac{p}{\rho} + gh \right) = 0$$

1.4. Compute the vorticity for laminar flow between parallel plates.

1.5. Derive the differential equation of continuity directly by using an infinitesimal control volume and without reference to the control volume integrals.

1.6. Derive the differential equation of motion directly by using an infinitesimal control volume and without reference to the control volume integrals.

1.7. Derive the differential equation of energy directly by using an infinitesimal control volume and without reference to the control volume integrals.

1.8. Derive the equation of continuity in cylindrical coordinates by balancing the inflow, outflow, and the storage in the small volume shown in Fig. 1.16. Show that the result for incompressible flow is simply $\vec{\nabla} \cdot \vec{u} = 0$.

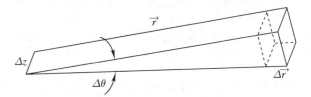

FIGURE 1.16
An element in a cylindrical coordinate system.

1.9. How does a small particle of fluid change in the following types of flows?
- (*a*) Uniform flow; $u_x = $ constant, $u_y = 0$, $u_z = 0$
- (*b*) Source flow; $u_r = C/r, u_z = 0, u_\theta = 0$
- (*c*) Expansion flow; $u_x = f(x), u_y = 0, u_z = 0$
- (*d*) Potential vortex; $u_r = 0, u_\theta = C/r, u_z = 0$
- (*e*) Forced vortex; $u_r = 0, u_\theta = Cr, u_z = 0$

1.10. The rate of strain tensor for its principal axes is

$$\mathbf{d} = \begin{bmatrix} d_{11} & 0 & 0 \\ 0 & d_{22} & 0 \\ 0 & 0 & d_{33} \end{bmatrix}$$

Show that for a small volume with sides Δx, Δy, Δz oriented so that the x-,y-, and z-axes coincide with the principal axes of the tensor $\mathbf{d}$, the rate of change of volume with respect to time is

$$\frac{d}{dt} \Delta V = d_{ii} \Delta V = (d_{11} + d_{22} + d_{33}) \Delta V$$

But d_{ii} is invariant so d_{ii} is the percentage increase in volume per unit time for any axes. The law of conservation of mass states that $d(\rho \Delta V)/dt = 0$. Use these last two equations and the definition of d_{ij} to derive the equation of continuity.

1.11. The decay of velocity in a vortex is given by

$$u_\theta = \frac{\Gamma_0}{4\pi r}\left[1 - \exp\left(-\frac{r^2}{4vt}\right)\right]$$

where Γ_0 and v are constants. Show that this flow satisfies the equation of continuity and give an expression for the vorticity of the flow.

1.12. Obtain the expression for the substantial derivative of a scalar and of the velocity components in cylindrical coordinates.

1.13. Obtain the equation of motion in cylindrical coordinates.

1.14. Obtain the equation of energy in cylindrical coordinates.

1.15. Obtain the equation of continuity in spherical coordinates.

1.16. Obtain the equation of motion in spherical coordinates.

1.17. Obtain the equation of energy in spherical coordinates.

1.18. Obtain (1.97) from (1.96).

1.19. Use the integral form of the continuity equation to show that $\rho A U = $ constant for steady flow in a duct where U is the average velocity and A is the cross-sectional area of the duct.

1.20. For incompressible, two-dimensional flow the stream function Ψ is defined such that

$$u_x = -\frac{\partial \Psi}{\partial y} \qquad u_y = \frac{\partial \Psi}{\partial x}$$

(a) Show that with this definition the equation of continuity is automatically satisfied.

(b) Show that the vorticity transport equation becomes

$$\frac{\partial}{\partial t}(\nabla^2 \Psi) - \frac{\partial \Psi}{\partial y}\frac{\partial}{\partial x}(\nabla^2 \Psi)$$

$$+ \frac{\partial \Psi}{\partial x}\frac{\partial}{\partial y}(\nabla^2 \Psi) = v\nabla^4 \Psi$$

In this form the equation of motion contains only one unknown.

1.21. Show that (1.146) does indeed represent the volume of a parallelepiped with sides Δa, Δb, Δc.

1.22. Show that

$$u_x = -\frac{2xyz}{(x^2 + y^2)^2} \qquad u_y = \frac{(x^2 - y^2)z}{(x^2 + y^2)^2}$$

$$u_z = \frac{y}{x^2 + y^2}$$

are possible components of an incompressible fluid motion. What is the vorticity of this motion?

1.23. Derive the Navier-Stokes equations in four dimensions. To save doing the rotations, assume that (1.29) and (1.85) are correct and derive the four-dimensional equivalent to (1.96). Do you get (1.96) from the four-dimensional equations by assuming that nothing changes in the x_4-direction? Why?

APPENDIX TO CHAPTER 1. SUMMARY OF THE EQUATIONS OF CONTINUITY AND MOTION

Rectangular coordinates
Continuity:

$$\frac{\partial \rho}{\partial t} + \frac{\partial}{\partial x_i}(\rho u_i) = 0$$

Motion:

$$\rho \frac{Du_i}{Dt} = -\rho g \frac{\partial h}{\partial x_i} - \frac{\partial p}{\partial x_i} + \frac{1}{3} \frac{\partial}{\partial x_i} \left(\frac{\partial u_j}{\partial x_j} \right) + \mu \frac{\partial^2 u_i}{\partial x_j \partial x_j}$$

Cylindrical coordinates $R = \sqrt{y^2 + z^2}$ $\quad \theta = \arctan \dfrac{y}{z}$

Continuity:

$$\frac{\partial \rho}{\partial t} + \frac{1}{R} \frac{\partial}{\partial R} (\rho R u_R) + \frac{1}{R} \frac{\partial}{\partial \theta} (\rho u_\theta) + \frac{\partial}{\partial x} (\rho u_x) = 0$$

Motion in R-direction:

$$\rho \left(\frac{\partial u_R}{\partial t} + u_R \frac{\partial u_R}{\partial R} + \frac{u_\theta}{R} \frac{\partial u_R}{\partial \theta} - \frac{u_\theta^2}{R} + u_x \frac{\partial u_R}{\partial x} \right) = -\rho g \frac{\partial h}{\partial R} - \frac{\partial p}{\partial R}$$

$$+ \mu \left\{ \frac{\partial}{\partial R} \left[\frac{1}{R} \frac{\partial}{\partial R} (R u_R) \right] + \frac{1}{R^2} \frac{\partial^2 u_R}{\partial \theta^2} - \frac{2}{R^2} \frac{\partial u_\theta}{\partial \theta} + \frac{\partial^2 u_R}{\partial x^2} \right\}$$

Motion in θ-direction:

$$\rho \left(\frac{\partial u_\theta}{\partial t} + u_R \frac{\partial u_\theta}{\partial R} + \frac{u_\theta}{R} \frac{\partial u_\theta}{\partial \theta} + \frac{u_R u_\theta}{R} + u_x \frac{\partial u_\theta}{\theta x} \right) = -\frac{\rho g}{R} \frac{\partial h}{\partial \theta} - \frac{1}{R} \frac{\partial p}{\partial \theta}$$

$$+ \mu \left\{ \frac{\partial}{\partial R} \left[\frac{1}{R} \frac{\partial}{\partial R} (R u_\theta) \right] + \frac{1}{R^2} \frac{\partial^2 u_\theta}{\partial \theta^2} + \frac{2}{R^2} \frac{\partial u_R}{\partial \theta} + \frac{\partial^2 u_\theta}{\partial x^2} \right\}$$

Motion in x-direction:

$$\rho \left(\frac{\partial u_x}{\partial t} + u_R \frac{\partial u_x}{\partial R} + \frac{u_\theta}{R} \frac{\partial u_x}{\partial \theta} + u_x \frac{\partial u_x}{\partial x} \right) = -\rho g \frac{\partial h}{\partial x} - \frac{\partial p}{\partial x}$$

$$+ \mu \left[\frac{1}{R} \frac{\partial}{\partial R} \left(R \frac{\partial u_x}{\partial R} \right) + \frac{1}{R^2} \frac{\partial^2 u_x}{\partial \theta^2} + \frac{\partial^2 u_x}{\partial x^2} \right]$$

Spherical coordinates $r = \sqrt{x^2 + y^2 + z^2}$ $\quad \theta = \arctan \dfrac{\sqrt{x^2 + y^2}}{z}$

$$\phi = \arctan \frac{y}{x}$$

Continuity:

$$\frac{\partial \rho}{\partial t} + \frac{1}{r^2} \frac{\partial}{\partial r} (\rho r^2 u_r) + \frac{1}{r \sin \theta} \frac{\partial}{\partial \theta} (\rho u_\theta \sin \theta) + \frac{1}{r \sin \theta} \frac{\partial}{\partial \phi} (\rho u_\phi) = 0$$

Motion in r-direction:

$$\rho\left(\frac{\partial u_r}{\partial t} + u_r\frac{\partial u_r}{\partial r} + \frac{u_\theta}{r}\frac{\partial u_r}{\partial \theta} + \frac{u_\phi}{r\sin\theta}\frac{\partial u_r}{\partial \phi} - \frac{u_\phi^2 + u_r^2}{r}\right) = -\rho g\frac{\partial h}{\partial r} - \frac{\partial p}{\partial r}$$

$$+ \mu\left(\nabla^2 u_r - \frac{2}{r^2}u_r - \frac{2}{r^2}\frac{\partial u_\theta}{\partial \theta} - \frac{2}{r^2}u_\theta\cot\theta - \frac{2}{r^2\sin\theta}\frac{\partial u_\phi}{\partial \phi}\right)$$

Motion in θ-direction:

$$\rho\left(\frac{\partial u_\theta}{\partial t} + u_r\frac{\partial u_\theta}{\partial r} + \frac{u_\theta}{r}\frac{\partial u_\theta}{\partial \theta} + \frac{u_\phi}{r\sin\theta}\frac{\partial u_\theta}{\partial \phi} + \frac{u_r u_\theta}{r} - \frac{u_\phi^2\cot\theta}{r}\right)$$

$$= -\frac{\rho g}{r}\frac{\partial h}{\partial \theta} - \frac{1}{r}\frac{\partial p}{\partial \theta} + \mu\left(\nabla^2 u_\theta + \frac{2}{r^2}\frac{\partial u_r}{\partial \theta} - \frac{u_\theta}{r^2\sin^2\theta} - \frac{2\cos\theta}{r^2\sin^2\theta}\frac{\partial u_\phi}{\partial \phi}\right)$$

Motion in ϕ-direction:

$$\rho\left(\frac{\partial u_\phi}{\partial t} + u_r\frac{\partial u_\phi}{\partial r} + \frac{u_\theta}{r}\frac{\partial u_\phi}{\partial \theta} + \frac{u_\phi}{r\sin\theta}\frac{\partial u_\phi}{\partial \phi} + \frac{u_\phi u_r}{r} + \frac{u_\theta u_\phi}{r}\cot\theta\right)$$

$$= -\frac{\rho g}{r\sin\theta}\frac{\partial h}{\partial \phi} - \frac{1}{r\sin\theta}\frac{\partial p}{\partial \theta}$$

$$+ \mu\left(\nabla^2 u_\phi - \frac{u_\phi}{r^2\sin^2\theta} + \frac{2}{r^2\sin\theta}\frac{\partial u_r}{\partial \phi} + \frac{2\cos\theta}{r^2\sin^2\theta}\frac{\partial u_\theta}{\partial \phi}\right)$$

where

$$\nabla^2 = \frac{1}{r^2}\frac{\partial}{\partial r}\left(r^2\frac{\partial}{\partial r}\right) + \frac{1}{r^2\sin\theta}\frac{\partial}{\partial \theta}\left(\sin\theta\frac{\partial}{\partial \theta}\right) + \frac{1}{r^2\sin^2\theta}\left(\frac{\partial^2}{\partial \phi^2}\right)$$

CHAPTER
2

SIMILITUDE
AND
MODELING

2.1 MODELING, DIMENSIONS, AND UNITS

Although the equations of motion of a Newtonian fluid have been known since 1821, no general analytical solution is yet available. Great progress has been made in recent years on numerical solutions, but many unsolved problems remain. The difficulties in solving these equations preclude their use to predict directly the general behavior of many practical fluid systems. In this chapter we show how the equations can be used indirectly to indicate the solution, or if not the solution, the properties of many situations. One method is to find suitable approximations (such as that of a perfect fluid) that can be solved more easily. Another is to use the equations to relate one situation (a model) to another (a prototype) so that the behavior of one can be used to predict the behavior of the other. The present chapter is a continuation of the study of the equations of fluid mechanics and their properties.

Physical modeling is one of the oldest tools of the practical hydraulician. It is generally considered an empirical technique and as much an art as a science. Certainly, those who are skilled at modeling need experience and judgment to design the model and to interpret the results. The equations can go far, however, in dictating proper design of models and in interpreting results. In the best sense the model serves as an analytical machine, a type of computer, that solves the equations of fluid mechanics for certain parameters. Used in that way the mathematics of fluid mechanics dictates the method to use. Art and experience become valuable when approximations are necessary.

The principal tool to be used in the interpretation of model results has been dimensional analysis. Dimensional analysis alone, however, is inadequate, but the equations can serve as a supplement. The basis of dimensional analysis is that each quantity is made up of the "fundamental dimensions" of mass, length, and time and that any relationship (equation) between variables must be homogeneous in these fundamental dimensions. Since mass, length, and time are considered fundamental dimensions, the units attached to those quantities are called fundamental units. All other quantities such as force, velocity, and density are measured in derived units. Newtonian mechanics seems to furnish a good reason for considering mass m as fundamental since it appears in the relationship

$$\vec{F} = m\frac{d\vec{u}}{dt} \tag{2.1}$$

and is seemingly independent of all physical constants such as gravity. We take the point of view, however, that mass is no more fundamental than force. Newton's law of gravitation is

$$F = \frac{kmm'}{r^2} \tag{2.2}$$

in which F is the mutual attraction between two objects of masses m and m' that are a distance r apart. k is a constant of proportionality. Even though k is measured in derived units, (2.2) appears to indicate that k is more fundamental than mass. This point can be carried further by considering non-Newtonian mechanics (relativity) in which the speed of light appears as a fundamental quantity. Thus, we conclude that the choice of "fundamental" quantities or units is simply one of convenience and not dictated by the laws of physics.

It is often convenient to consider force as fundamental instead of mass. Or, in Newtonian mechanics, force, mass, length, and time may be considered as equally fundamental but related according to (2.1). This latter point of view reveals four systems of measurements, those of mass-length-time, force-length-time, mass-force-time, and mass-force-length. The remaining dimensions in each of these systems become, respectively, mass-

TABLE 2.1
Dimensions and units

Term	MLT	FLT	English Gravitational		English Absolute		SI
			MLT	FLT	MLT	FLT	
Force	ML/T^2	F	slug-ft/sec^2	lb$_f$	lb$_m$-ft/sec^2	pdl	N
Mass	M	FT^2/L	slug	lb$_f$-sec^2/ft	lb$_m$	pdl-sec^2-ft	kg
Length	L	L	ft	ft	ft	ft	m
Time	T	T	sec	sec	sec	sec	s

length/(time)^2, force-(time)^2/length, force-(time)^2/mass, and (mass-length/force)$^{1/2}$. In engineering practice the first two systems are in common use.

If the unit of force is taken as fundamental, the resulting system of measure is termed *gravitational* whereas an *absolute* system is one where the unit of mass is considered fundamental. Table 2.1 shows those systems of dimensions and units that are commonly used. The pound-mass (lb$_m$) and pound-force (lb$_f$) are taken as numerically equal, thus furnishing a conversion equation between the absolute and gravitational systems. Also in common use is a system that mixes the absolute and gravitational systems by measuring force in pounds-force and mass in pounds-mass.

The SI system combines the gravitational and absolute system by defining a separate fundamental unit for mass (the kilogram) and force (the Newton). The relative simplicity of the SI system, even aside from the round number conversion factors, is obvious; however, the primary reason for adoption is that it eliminates the multiplicity of units that are in the English system and thus sets a world wide standard.

2.2 THE PI-THEOREM

Dimensional reasoning, the balancing of dimensions in each term of an equation, has long been the basis for defining the important parameters in a functional relationship. The foundation of dimensional analysis rests upon the Buckingham pi-theorem, although this theorem is not often used directly in the process. The pi-theorem rests on the assumption that in any physical process there exists a functional relationship between variables of the form

$$f(x_1, x_2, x_3, \ldots, x_n) = 0 \qquad (2.3)$$

where x_i is a physical variable. Presumably (2.3) could be rearranged to solve for any of the x_i in terms of the other variables. It is such a functional relationship that is sought.

Equation (2.3) must be dimensionally homogeneous; all of the terms must have the same dimensions. In other words there must be a transformation of units so that

$$f(x'_1, x'_2, x'_3, \ldots, x'_n) = 0 \tag{2.4}$$

in which the x'_i correspond to the x_i except that the units of measurement are different, but the function is precisely the same as the function of (2.3). Such a transformation is performed by defining conversion factors P_i for each of the fundamental units. Then

$$x'_i = x_i P_1^{a^{i1}} P_2^{a^{i2}} \cdots P_m^{a^{im}} \qquad \text{no sum on } i \text{ or } m \tag{2.5}$$

where m is the number of fundamental units contained in x_i and the a^{ij} are the exponents on the corresponding units. (Note that the a^{ij} are exponents, not superscripts, on the P whereas the ij are superscripts attached to the a.)

Example 2.1.   The pressure of 0.1 ton/ft^2 is converted to pounds/in^2 as follows:

$$x'_i = 0.1(2000)(12)^{-2} = 1.39 \text{ lb/in}^2 \tag{2.6}$$

In (2.6) $P_1 = 2000$ (the number of pounds in one ton), $P_2 = 12$ (the number of inches in one foot), $a^{11} = 1$, and $a^{12} = -2$. In other words, P_i is the number of primed units in one unprimed unit. In this example the pound-force was considered as a fundamental unit.

Obviously, a product is transformed by transforming each member as follows:

$$x'_1 x'_2 \cdots x'_n = x_1 P_1^{a^{11}} P_2^{a^{21}} \cdots P_m^{a^{m1}} x_2 P_1^{a^{12}} P_2^{a^{22}} \cdots P_m^{a^{m2}} \cdots x_n P_1^{a^{1n}} P_2^{a^{2n}} \cdots P_m^{a^{mn}} \tag{2.7}$$

in which m is the total number of fundamental units (three in a mass-length-time system) and some of the a^{ij} may be zero. The product is *dimensionless* if and only if

$$
\begin{aligned}
a^{11} + a^{12} + a^{13} + \cdots + a^{1n} &= 0 \\
a^{21} + a^{22} + a^{23} + \cdots + a^{2n} &= 0 \\
\vdots \qquad \vdots \qquad \vdots \qquad \ddots \qquad \vdots \qquad \vdots \\
a^{m1} + a^{m2} + a^{m3} + \cdots + a^{mn} &= 0
\end{aligned}
\tag{2.8}
$$

More generally, the transformation of a variable raised to an exponent is

$$(x_i')^k = x_i^k P_1^{ka^{11}} P_2^{ka^{21}} \cdots P_m^{ka^{m1}} \tag{2.9}$$

so that the a are multiplied by the exponent. The equations corresponding to (2.8) that must be satisfied in order that $(x_1')^{k_1}(x_2')^{k_2} \cdots (x_n')^{k_n}$ is dimensionless are

$$a^{11}k_1 + a^{12}k_2 + a^{13}k_3 + \cdots + a^{1n}k_n = 0$$

$$a^{21}k_1 + a^{22}k_2 + a^{23}k_3 + \cdots + a^{2n}k_n = 0 \qquad \text{no sum on } n$$

$$\vdots \qquad \vdots \qquad \vdots \qquad \ddots \qquad \vdots \qquad \vdots \tag{2.10}$$

$$a^{m1}k_1 + a^{m2}k_2 + a^{m3}k_3 + \cdots + a^{mn}k_n = 0$$

or simply

$$a^{ij}k_j = 0 \tag{2.11}$$

A product of variables can be made dimensionless by choosing the k that satisfy (2.11). The matrix is called a dimensional matrix.

Example 2.2. We wish to create a dimensionless product out of the variables representing velocity u acceleration of gravity g and length L in an MLT system. The desired product is

$$u^{k_1} g^{k_2} L^{k_3} = u^{k_1} P_M^0 P_L^{k_1} P_T^{-k_1} g^{k_2} P_M^0 P_L^{k_2} P_T^{-2k_2} L^{k_3} P_M^0 P_L^{k_3} P_T^0 \tag{2.12}$$

in which P_M, P_L, and P_T represent arbitrary conversion factors for mass, length, and time, respectively. The resulting equations corresponding to (2.10) are

$$0 = 0 \qquad k_1 + k_2 + k_3 = 0 \qquad -k_1 - 2k_2 = 0 \tag{2.13}$$

giving two equations in three unknowns. The solution of these equations is obviously not unique. We discard the trivial solution ($k_1 = k_2 = k_3 = 0$) and choose $k_1 = 1$, $k_2 = k_3 = -0.5$. The resulting dimensionless product that corresponds to (2.12) is

$$\Pi = \frac{u}{\sqrt{gL}} \tag{2.14}$$

Even though this result is not unique, the other solutions are not independent in that they are Π raised to some power.

The foregoing development indicates a general method of creating a dimensionless equation when given the implicit functional relationship (2.3). The problem is to find the maximum number of independent dimensionless products Π_i formed from the variables x_i. Then (2.3) is equivalent to

$$F(\Pi_1, \Pi_2, \Pi_3, \ldots, \Pi_p) = 0 \qquad (2.15)$$

where p is a number less than or equal to n. The solution to this problem is simply a solution of (2.11). Notice that, in general, there are m equations (the number of fundamental dimensions) in n exponents (the number of variables). There may be none, one, or several independent (meaning not linearly dependent) solutions to these equations that lead to none, one, or several independent dimensionless groups.

The number of independent Π-groups depends on the rank of the dimensional matrix. The rank of an m by n matrix is defined as the largest nonzero determinant that can be formed from the matrix. Obviously, the rank is not larger than the smaller of m and n. A theorem of linear algebra (see, for example, Hadley, 1961) states that an m by n set of linear, homogeneous equations of matrix rank k leads to $p = n - k$ independent solutions. Thus, the dimensional matrix determines the number of Π-groups in (2.15).

Example 2.3. Consider a flow problem containing a functional relationship

$$f(u, \rho, \mu, D, g) = 0 \qquad (2.16)$$

where u is velocity (L/T), ρ is density (M/L^3), μ is viscosity (M/LT), D is length (L), and g is acceleration (L/T^2). The dimensional matrix is

	u	ρ	μ	D	g
M	0	1	1	0	0
L	1	-3	-1	1	1
T	-1	0	-1	0	-2

The rank of this matrix is 3 as shown by taking the determinant from the last 3 columns

$$\begin{vmatrix} 1 & 0 & 0 \\ -1 & 1 & 1 \\ -1 & 0 & -2 \end{vmatrix} = \begin{vmatrix} 1 & 1 \\ 0 & -2 \end{vmatrix} = -2 \qquad (2.17)$$

Thus, there are $5 - 3 = 2$ independent Π-groups resulting from the equations

$$k_2 + k_3 = 0$$
$$k_1 - 3k_2 - k_3 + k_4 + k_5 = 0 \qquad (2.18)$$
$$-k_1 - k_3 - 2k_5 = 0$$

We are free to choose two of the five k. Choosing $k_1 = 1$ and $k_2 = 1$ results in $k_3 = -1$, $k_4 = 1$, and $k_5 = 0$, giving the dimensionless group

$$\Pi_1 = u^{k_1} \rho^{k_2} \mu^{k_3} D^{k_4} g^{k_5} = \frac{u \rho D}{\mu} \qquad (2.19)$$

Choosing $k_1 = 1$ and $k_2 = 0$ leads to $k_3 = 0$, $k_4 = -0.5$, and $k_5 = -0.5$, giving

$$\Pi_2 = u^{k_1} \rho^{k_2} \mu^{k_3} D^{k_4} g^{k_5} = \frac{u}{\sqrt{gD}} \qquad (2.20)$$

Obviously, Π_1 and Π_2 are independent since they contain different variables. Any other dimensionless group is a combination of Π_1 and Π_2. The five original variables have been reduced to two, which is of great help in doing parametric studies and forming experimental relationships.

The pi-theorem is a mathematical statement in algebra; it allows us to reduce the number of variables in a problem provided we know enough about the problem to choose the relevant variables. The use of the pi-theorem has several advantages:

1. It permits the systematic reduction in the number of variables in a functional relationship by k, the rank of the dimensional matrix.
2. It *may* reveal a variable that has been included in the functional relationship but does not belong, or it *may* reveal the fact that an important variable has been omitted from the functional relationship. Such revelations occur when the dimensional equations cannot be satisfied, for example, if only one of the variables contains mass.

The pi-theorem does not, however, give all of the information that one would like to have. It has three gigantic shortcomings:

1. In contrast to advantage (2) above, dimensional analysis more often will not discover erroneous or omitted variables. The critical decision of which variables are to be included must be made before invoking the pi-theorem and there is little basis, except practical experience, that indicates how this decision should be made.
2. In modeling, the model-prototype relationship dictated by dimensional analysis is usually impossible to create, yet dimensional analysis gives no indication of what approximations can or should be made.
3. Distorted models (different scales in the coordinate directions) are often useful and necessary; however, dimensional analysis does not indicate how these should be constructed or operated.

2.3 SIMILITUDE

The very essence of modeling is to create a flow that is "dynamically similar" to that in the prototype. By dynamically similar we mean that the same dimensionless equations are to be solved under the same boundary conditions in both systems. Practical considerations, of course, usually dictate that the model must be much smaller than the prototype. By performing certain measurements on the model (that is, by using the model as an analog computer to solve the equations), the prototype behavior is inferred after applying the appropriate "scale factors." Thus, from the engineering point of view, the model must behave like the prototype, at least in those areas that are of interest. From the point of view of fluid mechanics the model will behave like the prototype if the equations of motion and the boundary conditions that apply to the prototype transform under suitable changes of scale of length, time, and mass to the equations and boundary conditions that apply to the model. These transformations are precisely those used in the last section

$$x_i' = x_i \frac{L_M}{L_P} \qquad t' = t \frac{T_M}{T_P} \qquad m' = m \frac{M_M}{M_P} \qquad (2.21)$$

in which the primed quantities refer to the model and L_M/L_P, T_M/T_P, and M_M/M_P are the scale factors of length, time, and mass, respectively.

Rather than formally transform the equations in each case, the equations are written in dimensionless form so that they are valid under all scale changes of length, time, and mass. The dimensionless variables are defined as

$$x_i^* = \frac{x_i}{L_0} \quad u_i^* = \frac{u_i}{U_0} \quad t^* = \frac{U_0}{L_0}t \quad p^* = \frac{p}{p_0} \quad h^* = \frac{h}{L_0} \quad \rho^* = \frac{\rho}{\rho_0} \quad (2.22)$$

where L_0, U_0, p_0, and ρ_0 are dimensional constants of reference and are characteristic of the flow system. Under the above transformation, the equations of continuity (1.20) and motion (1.96) become

$$\frac{U_0\rho_0}{L_0} \frac{\partial \rho^*}{\partial t^*} + \frac{U_0\rho_0}{L_0} \frac{\partial}{\partial x_i^*} \left(\rho^* u_i^*\right) = 0 \qquad (2.23)$$

$$\frac{U_0^2\rho_0}{L_0} \rho^* \frac{Du_i^*}{Dt^*} = -\rho_0 g \rho^* \frac{\partial h^*}{\partial x_i^*} - \frac{p_0}{L_0} \frac{\partial p^*}{\partial x_i^*} + \frac{1}{3} \frac{\mu U_0}{L_0^2} \frac{\partial}{\partial x_i^*} \frac{\partial u_j^*}{\partial x_j^*} + \frac{\mu U_0}{L_0^2} \frac{\partial^2 u_i^*}{\partial x_j^* \partial x_j^*}$$
$$(2.24)$$

Dividing (2.23) by $U_0\rho_0/L_0$ yields

$$\frac{\partial \rho^*}{\partial t^*} + \frac{\partial}{\partial x_i^*} \left(\rho^* u_i^*\right) = 0 \qquad (2.25)$$

Dividing (2.24) by $U_0^2\rho_0/L_0$ gives

$$\rho^* \frac{Du_i^*}{Dt^*} = -\frac{gL_0}{U_0^2} \rho^* \frac{\partial h^*}{\partial x_i^*} - \frac{p_0}{\rho_0 U_0^2} \frac{\partial p^*}{\partial x_i^*} + \frac{\mu}{\rho_0 U_0 L_0} \left(\frac{1}{3} \frac{\partial}{\partial x_i^*} \frac{\partial u_j^*}{\partial x_j^*} + \frac{\partial^2 u_i^*}{\partial x_j^* \partial x_j^*}\right)$$
$$(2.26)$$

The no-slip boundary condition is

$$U_0 u_i^* = 0 \qquad \text{or} \qquad u_i^* = 0 \qquad (2.27)$$

which is preserved under scale transformation. Any condition at infinity will also be preserved under all transformations of length, mass, and time.

The fact that there is a single length transformation indicates that all linear dimensions of the prototype must be in the same scale in the model (that is, a "scale model"). Also, if the flow is unsteady, it must be assumed that the time history of the model and prototype is the same under the transformation. No reference quantities appear in the continuity (2.25), indicating that the same equation applies to model and prototype under any transformation. The dimensionless numbers that appear in (2.26) are

$$\boldsymbol{F} = \frac{U_0}{\sqrt{gL_0}} \qquad C_p = \frac{p_0}{\frac{1}{2}\rho_0 U_0^2} \qquad \boldsymbol{R} = \frac{\rho_0 U_0 L_0}{\mu} \qquad (2.28)$$

in which $\boldsymbol{F}$ is the *Froude number*, C_p is the *pressure coefficient*, and $\boldsymbol{R}$ is the *Reynolds number*. The Froude number is a relative measure of the ratio of inertial terms to gravity terms; a fountain that shoots water high into the air has a large Froude number. The pressure coefficient measures the ratio of the pressure term to the inertial terms. The Reynolds number is a guide to the ratio of the inertial terms to the viscous terms; a fluid with a Reynolds number much less than unity flows like the proverbial molasses in January. If these numbers are identical in both model and prototype, the model and prototype are dynamically similar. That is, the two flows obey exactly the same equations, the same boundary conditions, and the same initial conditions (for unsteady flows that have identical histories) and thus should have the same solutions. Such similarity criteria form the real justification for all models and analogs. Only by obtaining identical equations and boundary conditions in both prototype and analog can complete dynamic similarity be assured.

The above considerations dictate the scales of length, time, and mass. For complete similitude

$$(\boldsymbol{F})_P = (\boldsymbol{F})_M \qquad (C_p)_P = (C_p)_M \qquad (\boldsymbol{R})_P = (\boldsymbol{R})_M \qquad (2.29)$$

where the subscripts P and M indicate prototype and model. Assuming that both model and prototype are in the same gravitational field, the first of (2.29) gives

$$\frac{\left(U_0^2\right)_M}{\left(U_0^2\right)_P} = \frac{(L_0)_M}{(L_0)_P} \qquad (2.30)$$

If a length scale is chosen, (2.30) dictates a velocity (time) scale. From the last of (2.29)

$$\frac{(\mu)_M}{(\mu)_P} = \frac{(\rho_0)_M}{(\rho_0)_P} \frac{(U_0)_M}{(U_0)_P} \frac{(L_0)_M}{(L_0)_P} \qquad (2.31)$$

Using (2.30) to eliminate the velocity ratio from (2.31), the ratio of viscosities is

$$\frac{(\mu)_M}{(\mu)_P} = \frac{(\rho_0)_M}{(\rho_0)_P} \left(\frac{(L_0)_M}{(L_0)_P} \right)^{3/2} \qquad (2.32)$$

Equation (2.32) indicates that, except for the trivial case of equal length scales, the viscosity and density ratios cannot both be unity. This fact creates a problem for the engineer. A supply of fluids with properties such that viscosity and density can be varied arbitrarily does not exist. Moreover, water and air are often the only practical modeling fluids for economic reasons. Thus, complete dynamic similarity is often unattainable and we must seek either incomplete similitude or a suitable approximation. Fortunately, the equations can serve as a guide in finding approximations so that the essential features of a flow are modeled properly and, as explained in the next section, providing complete similitude is not always necessary. The same analysis also serves in finding suitable approximations for the analytical or numerical solution of the equations.

> **Historical Note:** William Froude (1810–1879) was an English engineer who contributed much to hydraulics and fluid mechanics. He was a keen observer of fluid mechanics phenomena, especially waves and free surface flow. The Froude number—and Reynolds number—were named in 1919 by a German professor, Moritz Weber (1871–1951); however, Froude never utilized this dimensionless group. Nevertheless, he deserves the honor.

2.4 INCOMPLETE SIMILITUDE

Equations (2.29) give necessary and sufficient conditions (except as noted in Secs. 2.5 and 2.9 below) for complete similitude. Complete similitude is often not possible but, fortunately, exact reproduction of the desired phenomena is not always necessary. A lack of similitude is not an approximation but instead represents a situation in which the desired phenomena are dynamically similar but other features of the flow are not. To illustrate incomplete similitude, only constant density fluids are considered in this section; compressible fluids are treated in Sec. 2.8.

Consider the flow of a liquid in a pipe. The general pressure level obviously has nothing to do with the flow pattern or resistance to the flow (assuming sufficiently high pressures; we do not account for cavitation in this section). This fact is evident if we write the equations of motion as

$$\rho \frac{Du_i}{Dt} = -\rho g \frac{\partial h}{\partial x_i} - \frac{\partial}{\partial x_i}(p - p_0) + \mu \frac{\partial^2 u_i}{\partial x_j \partial x_j} \qquad (2.33)$$

If a dimensionless pressure is defined as

$$p^* = \frac{p - p_0}{\rho_0 U_0^2} \qquad (2.34)$$

and the other dimensionless quantities are defined as in (2.22), the dimensionless equation of motion becomes

$$\rho^* \frac{Du_i^*}{Dt^*} = -\frac{\rho^*}{F^2} \frac{\partial h^*}{\partial x_i^*} - \frac{\partial p^*}{\partial x_i^*} + \frac{1}{R} \frac{\partial^2 u_i^*}{\partial x_j^* \partial x_j^*} \qquad (2.35)$$

The pressure coefficient no longer appears and similitude does not depend on it. In deriving (2.35) no approximations have been made, but the requirement of similitude has been relaxed. The new criteria require that velocity, resistance, and other quantities are reproduced in the model but pressures are not. Since we are usually concerned with the flow pattern and not the absolute pressures, this definition of similitude can be regarded as completely satisfactory for most purposes. Although the pressure coefficient does not appear, the pressure is still present in the equations and the pressure gradient, as opposed to the absolute pressure, is modeled correctly.

Also, the flow pattern in an enclosed system is not affected by the Froude number or by gravity forces in general. This fact is illustrated by subtracting the hydrostatic contribution from the pressure and defining the dimensionless pressure as

$$p^* = \frac{(p + \rho_0 g h) - (p_0 + \rho_0 g L_0)}{\rho_0 U_0^2} \qquad (2.36)$$

The resulting dimensionless equations of motion are

$$p^* \frac{Du_i^*}{Dt^*} = \frac{1 - \rho^*}{F^2} \frac{\partial h^*}{\partial x_i^*} - \frac{\partial p^*}{\partial x_i^*} + \frac{1}{R} \frac{\partial^2 u_i^*}{\partial x_j^* \partial x_j^*} \qquad (2.37)$$

For a constant density fluid, $\rho^* = \rho/\rho_0 = 1$, (2.37) becomes

$$\frac{Du_i}{Dt^*} = -\frac{\partial p^*}{\partial x_i^*} + \frac{1}{R} \frac{\partial^2 u_i^*}{\partial x_j^* \partial x_j^*} \qquad (2.38)$$

which indicates that the Reynolds number is the only dimensionless quantity that must be the same in model and prototype. Again, no approximations have been made but the requirement for similitude has been relaxed. In using (2.38) as a modeling criterion, the pressures in the prototype and the pressure distribution created by a gravity field are not reproduced in the model. Yet the general flow patterns in the model and the prototype will be the same if all the other criteria, including equivalent Reynolds numbers, are met.

The transformations represented by (2.34) and (2.36) are valid for any incompressible flow, but care must be exercised when dealing with certain flows. If the pressure should fall below the vapor pressure of the liquid, cavitation results and ρ is no longer a constant as has been assumed. Also, these transformations do not apply to flows with a free surface because in that case some of the similarity criteria enter through the boundary conditions; see Secs. 2.5, 2.9, and 2.10.

The difference between incomplete similitude and approximate similitude is important to understand. As we have defined incomplete similitude, no approximations have been made. Instead, the phenomena of interest are modeled whereas those aspects not of interest are not similar and cannot be scaled from model to prototype. In the examples, if the density is not constant or if some of the similarity criteria enter through the boundary conditions, incomplete similitude may not represent even a remote approximation to that which is desired. Approximations of similitude, as opposed to incomplete similitude, are treated in Sec. 2.6.

2.5   FROUDE AND WEBER MODELING

For flows with a free surface some of the similarity criteria enter through the boundary conditions and the transformation (2.36) is not valid because gravity and pressure level are critical to the problem. Thus, the Froude number criterion will remain important to the satisfaction of similitude. The free surface boundary condition is usually taken as

$$p = p_a \tag{2.39}$$

in which p_a is atmospheric pressure (or other imposed pressure on the free surface and may be a variable). Equation (2.39) is not preserved under the transformation (2.36). Instead, the dimensionless pressure is defined as

$$p^* = \frac{p - p_a}{\rho_0 U_0^2} \tag{2.40}$$

which gives for the special case of $p_a = $ constant

$$\rho^* \frac{Du_i^*}{Dt^*} = -\frac{1}{F^2} \frac{\partial h^*}{\partial x_i^*} - \frac{\partial p^*}{\partial x_i^*} + \frac{1}{R} \frac{\partial^2 u_i^*}{\partial x_j^* \partial x_j^*} \tag{2.41}$$

in which the Froude number still appears.

If surface tension is considered, the pressure is not continuous across the surface and the boundary condition is given, approximately, by (Lamb, 1945)

$$p_a - p = -\sigma \frac{d^2\eta}{ds^2} \tag{2.42}$$

where σ is the surface tension parameter (with dimensions of F/L), s is distance along the free surface, and η is the elevation of the free surface as a function of x and y and is normalized in the same way as h. Using (2.40), the free surface boundary condition becomes

$$p^* = \frac{\sigma}{L_0 \rho_0 U_0^2} \frac{d^2\eta^*}{ds^{*2}} \tag{2.43}$$

Equation (2.43) is preserved under the mass, length, and time transformations only if the *Weber number*,

$$W = \frac{L_0 \rho_0 U_0^2}{\sigma} \tag{2.44}$$

is the same for both model and prototype. The Weber number measures the ratio of inertial forces to surface tension forces; a liquid with a low Weber number is likely to dribble down the side of a bottle when poured.

Both (2.39) and (2.42) show that consideration of the boundary conditions is important in the transformations. If the Weber number is very large, then (2.43) is approximated by $p^* = 0$, which is (2.39) with $p_a = 0$ (using "gauge pressure"), and thus surface tension is not important; however, that is an approximation and the analyst must decide what constitutes a large Weber number and, hence, suitable accuracy. Exact or incomplete similitude requires that the Weber number remains in the transformation if surface tension is present at all.

Historical Note: The Weber number is named for Moritz Weber (1871–1951), who was a professor of naval mechanics at the Polytechnic Institute of Berlin. Weber named both the Froude and Reynolds number. There were two other notable Webers in fluid mechanics, Ernst Heinrich Weber (1795–1878) and his brother, Wilhelm Eduard Weber (1804–1891). Both were active in wave mechanics.

2.6 APPROXIMATE SIMILITUDE

Generally three quantities, the Froude number, the pressure coefficient, and the Reynolds number, are required to have equal values in model and prototype for complete similitude to exist. In addition, the boundary conditions (or additional terms in the equation of motion) may require other similarity criteria. Even the type of incomplete similitude in which the desired flow pattern is precisely reproduced is often difficult or impossible to obtain. For example, if water is to be used in both the model and the prototype, the density and the viscosity are (approximately) the same in both model and prototype. (Although these quantities vary somewhat with temperature in water, such variation is far too small to be useful in most cases.) If the Reynolds number is to be made equivalent, the velocity ratio is

$$\frac{(U_0)_M}{(U_0)_P} = \frac{(\mu)_M (\rho_0)_P (L_0)_P}{(\mu)_P (\rho_0)_M (L_0)_M} = \left[\frac{(\mu)_M (\rho_0)_P}{(\mu)_P (\rho_0)_M}\right] \frac{(L_0)_P}{(L_0)_M} \tag{2.45}$$

The quantity in brackets is fixed by the model and prototype fluids (usually very close to one). If the Froude number is to be made equivalent, then

$$\frac{(U_0)_M}{(U_0)_P} = \sqrt{\frac{(L_0)_M}{(L_0)_P}} \tag{2.46}$$

assuming the model and prototype are in the same gravitational system. Obviously, (2.45) and (2.46) conflict except in the trivial case where $(L_0)_M = (L_0)_P$. The problem is further complicated if the Weber number enters or if the flow is compressible so that the pressure coefficient remains in the problem.

The situation outlined above is the usual one that faces the engineer. If the engineer is attempting to solve a problem that is complex enough to require modeling, it is often complex enough so that even the type of incomplete similitude described in Sec. 2.4 cannot be obtained. In that case the engineer must settle for *approximate similitude*. That is, the objective of exactly modeling the desired phenomena must be abandoned and the model has to be constructed so as to obtain the best practical approximation. In doing so, the engineer must decide:

1. Can the desired phenomena be reproduced with sufficient accuracy even though a portion of the equation of motion is neglected? If the answer is no, then

2. Can the neglected factor be calculated or modeled separately?

The equations of motion and the boundary conditions provide a clue to question (1) whereas the solution to these equations may provide the answer to question (2).

An example of approximate similitude is given in the last section. From (2.43) it is obvious that if the surface tension parameter σ is small enough or if the length scale is large enough, then the Weber number is large and this equation is a good approximation to (2.39). In modeling practice the Weber number is made large by making $(L_0)_M$ large or by using a surface reactant that makes σ small.

In most models with a free surface (weirs, spillways, jets, waves, etc.) the Froude number must be made equivalent in model and prototype since the boundary condition (2.39) is essential. This consideration usually precludes modeling the viscous action. From (2.41) an exception might be made if the Froude number is large enough, a condition not often encountered in practice, unfortunately.

Viscous effects can sometimes be calculated separately. The drag on ship models is made up of wave drag (modeled accurately by the Froude law) and viscous drag. Both drag forces are important so one cannot be neglected in favor of the other. The viscous drag can be calculated using boundary layer theory. Thus, the wave drag on the model is found from the total drag (measured) minus the viscous drag (calculated). The wave drag is scaled by means of the Froude number to the prototype. Then the prototype drag can be found from the wave drag (scaled from the model) plus the viscous drag (calculated). The process is expressed as follows:

$$(\text{Wave drag})_M = (\text{Measured drag})_M - (\text{Calculated viscous drag})_M$$

$$(\text{Drag})_P = (\text{Scaled wave drag})_P + (\text{Calculated viscous drag})_P$$

In many other cases where the Froude number is used as the modeling criterion, viscous effects can be ignored or calculated. Some such cases are discussed in Secs. 2.7, 2.8, and 2.10.

2.7 THE REYNOLDS NUMBER

All fluids possess a finite viscosity even though the viscosity of water, air, and many other common fluids is small. A small viscosity leads to a large Reynolds number (see Eq. 2.28). Since it is the reciprocal of the Reynolds number that appears in the equation of motion, a large Reynolds number indicates that the viscous terms are small compared with some of the other terms in the equation. Although it would seem at first glance that the viscous terms can be neglected if the Reynolds number is large, such is not the case. The Reynolds number multiplies the terms in the equation that contain the highest order space derivatives. Neglecting these derivatives would change the order of the equation, and hence the character of the equation, leading to solutions that may not approximate reality or the true solution, even in the case of a large Reynolds number.

A discussion of such a change in the character of a differential equation would necessitate an investigation at considerable depth of the theory of such equations. The following example is given to illustrate this phenomenon. Consider the equation

$$m\frac{d^2x}{dt^2} + kx = 0 \tag{2.47}$$

under the boundary conditions that x and dx/dt are finite and not equal for $t = 0$. Equation (2.47) represents a linear, single degree of freedom vibration of a mass m on a spring with stiffness k. Suppose that a solution is desired for the case that m (the coefficient of the highest order derivative) is near zero. If m is greater than zero, the solution is

$$x = A \sin\sqrt{\frac{k}{m}}t + B \cos\sqrt{\frac{k}{m}}t \tag{2.48}$$

As m approaches zero, the solution shows oscillations of greater and greater frequency. If, however, m becomes less than zero, the solution changes character entirely to

$$x = A \exp\left(\sqrt{\frac{k}{|m|}}t\right) + B \exp\left(-\sqrt{\frac{k}{|m|}}t\right) \tag{2.49}$$

Now as m approaches zero (from the negative side), x grows very large for all finite t and the solution shows no oscillations. Obviously, the solution to (2.47) is radically different depending on the sign of m, but in either case the absolute value of m is small.

If damping is included, the equation becomes

$$m\frac{d^2x}{dt^2} + p\frac{dx}{dt} + kx = 0 \tag{2.50}$$

in which p is the damping coefficient. This simple equation provides an example closer to the case of the Navier-Stokes equations. For $m \equiv 0$ the solution is

$$x = A \exp\left(-\frac{k}{p}t\right) \tag{2.51}$$

and no oscillation occurs but instead x approaches zero exponentially with time. Moreover, with $m > 0$ the solution to (2.50) depends on two boundary conditions to determine the two constants that result from the integration. With $m \equiv 0$, (2.51) contains only one constant and only one boundary condition is required or may be given. Thus, the character of the equation and its solution are radically different for m small but finite and for m identically equal to zero.

In the case of the Navier-Stokes equations, there are also two boundary conditions that must be applied,

$$\vec{u} \cdot \vec{n} = 0 \quad\text{and}\quad \vec{u} \cdot \vec{t} = 0 \tag{2.52}$$

These set both the normal and tangential velocities to zero on (immobile) boundaries. Both of these conditions are often important in the solution of a problem. If the highest order derivatives of the equation are discarded so that the equation becomes of lower order, both conditions cannot be satisfied. The one that is discarded is the tangential condition since it would obviously be nonsense for the fluid to pass through a solid wall. As in the case of the vibration equation, the discard of a term, and the subsequent elimination of a boundary condition, may have a profound effect on the solution.

Many cases exist, however, where the viscous terms must be discarded to obtain a solution, at least an economical solution. The subject of potential flow (except flow in porous media) is concerned with these cases. Fortunately, some such solutions give valid, if approximate, results as verified by experiment. There is, however, little solid, theoretical basis for discarding the viscous terms and when a solution is found in this manner the ultimate criterion of its validity must be experimental verification.

Often the results of experimentation with a model can be scaled (approximately) to a prototype even though the Reynolds numbers are not the same. The following four criteria must be satisfied:

1. The Reynolds number must be large enough so that the viscous terms are small compared with other terms in the equation.
2. The Reynolds number must be such that the flow is either laminar in both model and prototype or turbulent in both model and prototype.
3. Separation (stall) must not occur (except at sharp corners so that the flow remains similar) in either the model or the prototype. At the time of the model design, the experimenter may not know if separation occurs, but if it does, the occurrence of separation cannot be determined from the model experiments unless the Reynolds criterion is observed. (Even the Reynolds criterion may not be enough because free stream turbulence and boundary disturbances must also be modeled.) A separate calculation can often indicate roughly whether or not separation takes place.
4. The modeler must discard all quantitative viscous effects such as viscous drag, heating in the boundary layer, etc.

Even if all precautions are taken, certain effects are not modeled properly. Typical of these is the entrainment of air on spillway models. Air entrainment appears to be largely a viscous effect and is not well reproduced in Froude models.

The previous section described a number of situations in which it was necessary to model according to the Froude law and, consequently, often impossible to satisfy the Reynolds criterion. Fortunately, good results can be obtained in many such cases. In most wave problems, for example, solid boundaries are remote, viscous damping is *very* small, and the flow pattern can be modeled well without satisfying the Reynolds criterion. In general the Reynolds criterion is abandoned in models of flow over dams, flow in open channels, spillways, some turbine tests, and some compressible flows. Often, special provisions can be made to compensate for viscous effects even in the absence of Reynolds modeling. Some such considerations are discussed in Sec. 2.10.

When the Reynolds number is used as a modeling criterion, one must ensure that the surface roughness of the model and the prototype is similar. Actually, this similarity is implied by the requirement for geometric similarity, but it is not easily achieved. Resistance depends on *form* as well as on the size of the surface roughness. Often the problem of form is avoided by assuming a "random" roughness. In any case, the surfaces of a (smaller) model should be smoother than those of the prototype (with the exception of distorted models, discussed below).

In the case of flow around objects, free stream turbulence should be scaled from model to prototype when modeling with the Reynolds number. This fact is a result of the equation of motion when the time-dependent terms are included. Turbulent flow is characterized by an unsteady, pseudorandom fluctuation of the velocity. Thus, (2.26) can be used to represent turbulent flow if the velocity is taken to be the true, unsteady velocity rather than the usual average over a short time period. Since the flow is unsteady when viewed this way, the initial conditions must transform properly under the transformation of mass, length, and time. The modeler cannot obtain similar initial conditions, but in a quasi-steady flow (a flow where a short time average of the actual velocity is steady) these conditions are exhibited in the turbulent structure, leading to the requirement that the free stream turbulence be similar.

Historical Note: Osborne Reynolds (1842–1912) held a chair in civil and mechanical engineering at Owens College (later Victoria University) in Manchester, England. He was basically an engineer but worked in a variety of fields including mechanics, fluid mechanics, thermodynamics, and electricity. In 1883 Reynolds presented his most famous experiment in an article "An experimental investigation of the circumstances which determine whether the motion of water shall be direct or sinuous, and of the law of resistance in parallel channels." In that amazing paper Reynolds characterized much of the transition from laminar to turbulent flow that we understand today. He fully realized the influence of viscosity, velocity, and size, all quantities that are present in the Reynolds number. Making use of the Navier-Stokes equations, he wrote what is currently called the Reynolds number as $\rho U^2/(\mu U/D)$ (our notation), which he interpreted as the ratio of inertial forces to viscous forces. He made an analogy with the movement of an army of troops. The larger the army and the more rapid the movement, the greater the probability of disorder. The discipline of the troops and the viscosity in the liquid tend to create ordered motion.

2.8 COMPRESSIBLE FLOWS

The transformation (2.36), leading to (2.37), is used to develop the modeling criteria for compressible flow. In (2.37) both the Froude and the Reynolds numbers appear. In addition to the equations for the motion of a fluid, an equation of state must be considered. It is a relationship between pressure and density. We will consider it of the form

$$p = f(\rho) \tag{2.53}$$

Using the dimensionless pressure as defined in (2.36)

$$\rho_0 U_0^2 p^* - \rho_0 g h + p_0 + \rho_0 g L_0 = f(\rho_0 \rho^*) \tag{2.54}$$

To proceed further, the equation of state is taken as the isentropic relationship

$$p = \kappa \rho^\gamma \tag{2.55}$$

where κ is a constant and γ is the ratio of specific heats of the gas (γ is dimensionless). After dividing (2.54) by p_0 and normalizing h

$$\frac{\rho_0 U_0^2}{p_0} p^* - \frac{\rho_0 g L_0}{p_0}(h^* - 1) + 1 = \rho^{*\gamma} \tag{2.56}$$

The constant κ has been removed by taking $\kappa = p_0/\rho_0^\gamma$.

The speed of sound in a gas is given by

$$c = \sqrt{\frac{dp}{d\rho}} \tag{2.57}$$

Using (2.55) in (2.57) gives

$$c = \sqrt{\gamma \frac{p}{\rho}} \tag{2.58}$$

Using (2.58) written for the reference quantities to eliminate p_0 in (2.56)

produces

$$\gamma \frac{U_0^2}{c_0^2} p^* - \frac{\gamma U_0^2}{c_0^2} \frac{g L_0}{U_0^2} (h^* - 1) + 1 = \rho^{*\gamma} \qquad (2.59)$$

The Mach number is $M = u/c$ and the reference Mach number is $M_0 = U_0/c_0$ where c_0 is calculated at reference pressure and density. If the Mach number is a measure of the ratio of inertial terms to elastic (compressible) terms, then M/F is a measure of gravity terms to elastic terms. Using the Mach and Froude numbers in (2.59) gives

$$\gamma M_0^2 p^* - \frac{\gamma M_0^2}{F^2} (h^* - 1) + 1 = \rho^{*\gamma} \qquad (2.60)$$

Thus, complete similarity requires that the Froude number, the Mach number, the Reynolds number, and γ all be equal in the model and prototype.

For many cases of approximate similitude the Mach number is small,

$$\gamma M_0^2 \ll 1 \qquad \text{and} \qquad \frac{\gamma M_0^2}{F^2} \ll 1 \qquad (2.61)$$

leading to the conclusion from (2.60) that $\rho^* \approx 1$. Also, the term $(1-\rho^*)/F^2$ in the equation of motion is small so that all similarity criteria disappear except the Reynolds number. Experiment indicates that if the Mach number is low enough (less than, say, 0.1 to 0.3) the flow behaves approximately as an incompressible flow.

The Froude criterion is usually neglected in any compressible flow since gravity forces are assumed to be small. As is indicated above, the Froude number disappears along with the Mach number if the Mach number is low enough and the Froude number is not too small. For high Mach number and reasonable temperature (the temperature determines the pressure-density ratio from the perfect gas law, $p/\rho = RT$) the velocity must be large; thus, the Froude number is high and the ratio $\gamma M_0^2/F^2$ remains small. The same consideration can be noted from (2.56) where, for a gas, ρ_0 is very small, making the number $\rho_0 g L_0/p_0$ small. Also, if the Froude number is large due to a large velocity, the term $(1 - \rho^*)/F^2$ in the equation of motion is small, and the Froude number can be neglected.

The Mach number may be small in certain situations and yet the Froude number is important. In dealing with atmospheric flows, for example, the atmosphere may be stratified, making gravity forces important. At the same time the normalizing length L_0 is large, perhaps equal to a substantial part of the height of the atmosphere or the length of atmospheric waves. Under some such conditions the compressibility of the atmosphere can become important even for small Mach numbers.

The Reynolds number is often neglected in the equation of motion for compressible fluids for the same reasons as those cited in Sec. 2.7. The same sorts of inaccuracies occur in the compressible case plus such difficult-to-model phenomena as shock wave thickness and separation due to shock wave–boundary layer interaction.

Often, especially in aerodynamics, the equations for compressible flow neglect gravity and viscosity and can be linearized as a result. In such cases the results can be transformed from model to prototype for subsonic flow with geometric similarity as the only modeling criterion; the transformation is affine and depends on the Mach number. The equations demonstrating that fact are presented in the chapter on potential flow.

2.9 CAVITATION MODELING

The similarity criteria previously derived for an incompressible liquid are valid provided that the pressure at any point does not fall below the threshold of the vapor pressure. An equation of state for a liquid can be written as

$$\rho = \rho_0 \quad \text{for} \quad p > p_v \qquad \rho = \rho_v \quad \text{for} \quad p = p_v \qquad (2.62)$$

where p_v is the vapor pressure of the liquid (a function of temperature) and ρ_v is the density of the vapor at the vapor pressure. For most practical cases the pressure cannot become less than the vapor pressure over any substantial region unless all of the liquid in the system is vaporized. (Water can sustain tension or negative pressure under certain circumstances. This fact is important in flow through capillaries.)

If the transformation (2.34) is used, the dimensionless equation of state is

$$\rho^* = 1 \quad \text{for} \quad p^* > -\frac{p_0 - p_v}{\rho_0 U_0^2} \qquad \rho^* = \frac{\rho_v}{\rho_0} \quad \text{for} \quad p^* = -\frac{p_0 - p_v}{\rho_0 U_0^2}$$
$$(2.63)$$

The dimensionless number that has appeared is the *cavitation number*,

$$Q = \frac{p_0 - p_v}{\frac{1}{2}\rho_0 U_0^2} \qquad (2.64)$$

which must be equal in the model and the prototype. For complete similarity the ratio ρ_v/ρ_0 should also be equivalent in model and prototype. From (2.37) the Froude criterion must be satisfied. Fortunately, all similarity criteria can be met, except the Reynolds criterion, by scaling the velocities according to

$$\frac{(U_0)_M}{(U_0)_P} = \sqrt{\frac{(L_0)_M}{(L_0)_P}} \tag{2.65}$$

The pressures are scaled according to

$$\frac{(p_0 - p_v)_M}{(p_0 - p_v)_P} = \frac{(L_0)_M (\rho_0)_M}{(L_0)_P (\rho_0)_P} \tag{2.66}$$

If the model uses the same fluid as the prototype and is tested at the same temperature, the density ratio is unity and the vapor pressures are equivalent. In that case the pressure in the model should be adjusted according to

$$(p_0)_M = (p_v)_M + (p_0 - p_v)\frac{(L_0)_M}{(L_0)_P} \tag{2.67}$$

Often the models are tested at a different temperature, which has a marked effect of the vapor pressure and allows the test to be carried out at a lower pressure. The fluid in the model should not have an unusually low content of dissolved gas since some liquids can sustain large tensions (high negative pressures) if their gas content is low enough.

2.10   DISTORTED MODELS

Rivers, harbors, lakes, and oceans cover large horizontal lengths or areas (horizontal scales) while the depths (vertical scales) are comparatively shallow. A uniform modeling scale that reduces the horizontal dimensions to a reasonable size makes the vertical dimensions unreasonably small. These models are necessarily constructed according to the Froude law and, if the depth becomes too small, the flow may be laminar whereas it is turbulent in the prototype. In such a case there is no hope for even approximate similitude. In addition the surface tension can have an effect on small scale models that is not present to a similar extent in the larger prototype.

Engineers have solved the problem by using different horizontal and vertical scales in river and harbor models. The use of different scales violates the initial assumption of a scale model. The procedure is justifiable if: (1) the Reynolds number criterion can be neglected to a good approximation, and (2) the vertical accelerations in the flow are small so that the pressure is nearly hydrostatic throughout. The latter condition leads to a simplification of the vertical equation of motion (the essential approximation of shallow water theory; see Chap. 8) written as

$$p - p_a = \rho g(\eta - z) \tag{2.68}$$

in which η is the liquid depth measured from a datum located below the free surface and the coordinate z is positive in the upward direction and opposite to the pull of gravity. Equation (2.68) is used to replace the pressure with η as a dependent variable. It also assumes that the flow is primarily two-dimensional, that the variables are functions of the horizontal coordinates only.

Consider first the continuity equation. The flow is one of constant density so that the relevant equation is simply

$$\frac{\partial u_i}{\partial x_i} = \frac{\partial u_x}{\partial x} + \frac{\partial u_y}{\partial y} + \frac{\partial u_z}{\partial z} = 0 \qquad (2.69)$$

To eliminate the vertical dependence, (2.69) is integrated from the solid lower boundary, $z = b(x, y)$, to the free surface $z = \eta$ (see Sec. 8.1.2)

$$\int_b^\eta \left(\frac{\partial u_x}{\partial x} + \frac{\partial u_y}{\partial y} + \frac{\partial u_z}{\partial z} \right) dz = \frac{\partial}{\partial x}(\bar{u}_x d) + \frac{\partial}{\partial y}(\bar{u}_y d) - u_x(\eta)\frac{\partial \eta}{\partial x} + u_x(b)\frac{\partial b}{\partial x}$$

$$- u_y(\eta)\frac{\partial \eta}{\partial y} + u_y(b)\frac{\partial b}{\partial y} + u_z(\eta) - u_z(b) \qquad (2.70)$$

in which $d = \eta - b$ is the depth. The derivative $\partial b/\partial x$ represents the slope of the bottom in the x-direction. Similarly, $u_x(b)$ is the x-velocity at the bottom and $u_x(\eta)$ is the x-velocity at the surface. The vertically averaged velocities in the x- and y-directions are $\bar{u}_x$ and $\bar{u}_y$. The free surface and bottom boundary conditions are (without derivation at this point; they are derived in Chap. 4)

$$u_z(\eta) = \frac{\partial \eta}{\partial t} + u_x(\eta)\frac{\partial \eta}{\partial x} + u_y(\eta)\frac{\partial \eta}{\partial y} \qquad u_z(b) = u_x(b)\frac{\partial b}{\partial x} + u_y(b)\frac{\partial b}{\partial y}$$

$$(2.71)$$

These conditions guarantee that the steady velocity is tangential to the surfaces and that the unsteady velocity is of the right magnitude to account for the vertical rise or fall of the free surface. The no-slip condition on the solid boundary does not apply since the viscous terms have been neglected. Applying (2.71) to (2.70)

$$\frac{\partial d}{\partial t} + \frac{\partial}{\partial x_i}(d\bar{u}_i) = 0 \qquad \text{sum on } i \text{ from 1 to 2} \qquad (2.72)$$

Neglecting the viscous terms and using (2.68), the equation of motion is also integrated between the bottom and the free surface. With the approximation that the velocity is constant in the vertical, the result is

$$\frac{D\bar{u}_i}{Dt} = -g\frac{\partial d}{\partial x_i} \qquad i = 1,\ 2 \qquad (2.73)$$

where i now takes on only two values, 1 and 2. Equations (2.72) and (2.73) constitute three equations in the two $\bar{u}_i$ and the depth, d.

Separate horizontal and vertical length scales are defined as L_H and L_V. The dimensionless variables are redefined as

$$x_i^* = \frac{x_i}{L_H} \qquad \bar{u}_i^* = \frac{\bar{u}_i}{U_0} \qquad t^* = \frac{U_0}{L_H}t \qquad d^* = \frac{d}{L_V} \qquad (2.74)$$

and the dimensionless equations become

$$\frac{\partial d^*}{\partial t^*} + \frac{\partial}{\partial x_i^*}(d^*\bar{u}_i^*) = 0 \qquad \text{sum on } i \text{ from 1 to 2} \qquad (2.75)$$

$$\frac{D\bar{u}_i^*}{Dt^*} = -\frac{1}{F^2}\frac{\partial d^*}{\partial x_i^*} \qquad i = 1,\ 2 \qquad (2.76)$$

where the Froude number is now defined as

$$F = \frac{U_0}{\sqrt{gL_V}} \qquad (2.77)$$

Equations (2.75) and (2.76) indicate that similitude depends only on the Froude criterion, which is based on the characteristic vertical dimension. The characteristic horizontal length does not appear in the equations. The free surface boundary condition, $p = p_a$, is automatically satisfied from the hydrostatic condition (2.68) since $\eta = z$ on the free surface.

In distorted river models the amplification of scale in the vertical direction creates a problem in making the fluid resistance of the model similar to that of the prototype. The larger depth means that there is less friction than would otherwise be the case, leading to velocities that are unrealistically high in the model. The resistance problem is solved by creating an artificial roughness or turbulence-generating mechanism in the model. There are empirical methods for calculating the approximate roughness necessary in the model, but the correct roughness is actually set by trial and error in the verification process. The roughness adjustment does have some theoretical justification for its application to similitude if the equation of motion is written with the velocities divided into their time-averaged component and their fluctuating component. The major contribution to resistance then comes from the *Reynolds stresses* (see Sec. 7.2.2)

$$\tau_{ij} = \rho\overline{u_i'u_j'} \qquad (2.78)$$

where u_i' represents the unsteady (fluctuating) velocity component and the overbar indicates a short time average. The artificial roughness, which may take the form of screens, grids, or projections from the bottom, tends to create a greater level of turbulence and, thus, scales this part of the equation of motion. Since there is no precise way to determine just what sort of roughness is necessary, the models must be verified; that is, they are tested against known responses in the prototype and adjusted until they predict these responses to sufficient accuracy. After verification the model assumes that similitude has been achieved and the model can be used to predict unknown occurrences in the prototype.

2.11 A FEW DIMENSIONLESS NUMBERS

Many dimensionless parameters other than those cited in the above sections are used in fluid mechanics. Additional numbers enter through other terms in the equations of motion, for example, in terms that appear when the equations are written in a rotating coordinate system. Many enter through other equations that are combined with the equations of fluid motion for the solutions of certain problems, such as the dispersion of a pollutant. The following list is not exhaustive but does express most of the numbers used in engineering problems with common fluids in the absence of electrical or magnetic fields. Many of the numbers have alternate definitions that may depend on the particular problem. They are summarized in Table 2.2.

Some numbers relate to restricted problems; the

$$\text{Dean number} = \boldsymbol{De} = \frac{UR}{\nu}\sqrt{\frac{R}{r}} \tag{2.79}$$

applies to viscous flow in a pipe of radius R around a bend of radius r in a fluid of kinematic viscosity ν. In addition to the Froude number, a common quantity used in open channel flow or overland flow is

$$\text{kinematic flow number} = \boldsymbol{k} = \frac{SLg}{U^2} \tag{2.80}$$

in which S is the (dimensionless) slope of the bed of an open channel and L is a characteristic length along the channel. It is used as a measure of the relative importance of the friction (or gravity) terms and the dynamic terms in the equation for open channel flow. A high kinematic flow number means that the dynamic (acceleration) terms are relatively unimportant.

The ratio of inertial terms to elastic terms can be measured by the

$$\text{Cauchy number} = \boldsymbol{Ca} = \frac{\rho U^2}{K} = \boldsymbol{M}^2 \tag{2.81}$$

in which K is the elasticity of the fluid. The Cauchy number is the square of the Mach number.

TABLE 2.2
Dimensionless numbers

Name		Definition	Meaning	Usage
Bond number	Bo	$\dfrac{\rho g L^2}{\sigma}$	weight/surface tension	drops
Brinkman number	Br	$\dfrac{\mu U^2}{k_h \Delta T}$	viscous transport/heat flow	viscous heating
capacity coefficient		$\dfrac{Q}{\omega L^3}$	flow/rotating volume	rotating machinery
capillary number	Ca	$\dfrac{\mu U}{\sigma}$	viscous force/surface tension	capillary effects
cavitation number	Q	$\dfrac{p_0 - p_v}{\frac{1}{2}\rho_0 U_0^2}$	excess over vapor pressure/dynamic pressure	cavitation
Cauchy number	Ca	$\dfrac{\rho U^2}{K}$	inertial terms/elastic terms	compressible flow
Courant number	Cr	$(u \pm c)\dfrac{\Delta t}{\Delta x}$	disturbance speed/relative grid spacing	numerical calculations
Darcy number	Da	$\dfrac{\beta_T g k_e L \Delta T}{\text{Đ}\nu}$	thermal expansion, diffusivity, and viscosity	heat transfer in porous media
Dean number	De	$\dfrac{U R}{\nu}\sqrt{\dfrac{R}{r}}$	centrifugal force/viscous force	flow in curved conduits
Deborah number	De	$\dfrac{\mu_0 U}{\tau L}$	viscous shear/total shear	non-Newtonian flow
drag coefficient	C_D	$\dfrac{F_D}{\frac{1}{2}A\rho U^2}$	drag force/dynamic pressure force	flow around object
Eckert number	Ec	$\dfrac{U^2}{c_p \Delta T}$	kinetic energy/thermal energy	heat transfer in a flowing fluid
Ekman number	Ek	$\dfrac{\nu}{\omega L^2}$	viscous (turbulent) terms/rotational terms	rotating flow
Euler number	E	$\dfrac{\rho U^2}{\Delta p}$	dynamic pressure/pressure on object	general flow
Fourier number	Fo	$\dfrac{\nu}{UL}$	kinematic viscosity or diffusivity/advection	transport and flow

TABLE 2.2
Dimensionless numbers

Name		Definition	Meaning	Usage
Froude number	F	$\dfrac{U}{\sqrt{g\eta_0}}$	flow speed/wave speed	open channel flow
Görtler number	$Gö$	$\dfrac{\rho U \delta_m}{\mu}\sqrt{\dfrac{\delta_m}{R_0}}$	momentum transport/viscous stress	flow over curved surfaces
Graetz number	Gz	$\dfrac{\rho U A c_p}{k_h L}$	heat gained in flow through a pipe/heat transfer	heat transfer in flow in pipes
Grashof number	Gr	$\dfrac{\rho^2 \beta_T g L^3 \Delta T}{\mu^2}$	heat transfer/heat by friction	heat transfer
head coefficient		$\dfrac{P}{\rho Q \omega^2 L^2}$	power/rotational energy	rotating machinery
Jacob number	Ja	$\dfrac{c_t \rho_1 \Delta T}{h \rho_v}$	heat transfer/heat of vaporization	evaporation of bubbles
Keulegan–Carpenter number	K	$\dfrac{UT}{D}$	scale of flow/width of object	drag in a wave field
kinematic flow number	k	$\dfrac{SLg}{U^2}$	down-slope gravity terms/dynamic terms	open channel flow
Knudsen number	Kn	$\dfrac{\lambda}{L}$	molecular distance/characteristic length	checks continuum assumption
lift coefficient	C_L	$\dfrac{F_L}{\frac{1}{2}A\rho U^2}$	lift force/dynamic pressure force	flow around airfoils
Mach number	M	$\dfrac{U}{c}$	flow speed/speed of sound	compressible flow
Nusselt number	Nu	$\dfrac{k_h L}{Đ}$	heat transfer/diffusivity	heat transfer
Péclet number	Pe	$\dfrac{UL}{Đ}$	advection/diffusion	mass or heat transport
grid Péclet number	Pe	$\dfrac{U\Delta x}{Đ}$	advection/diffusion	numerical calculations
power coefficient		$\dfrac{P}{\rho\omega^3 L^5}$	power output/rotational energy	rotating machinery

TABLE 2.2
Dimensionless numbers

Name		Definition	Meaning	Usage
Prandtl number	Pr	$\dfrac{\nu}{Ð}$	kinematic viscosity/thermal diffusivity	heat transport
Rayleigh number	Ra	$\dfrac{g\beta_T\,\Delta T\,L^3}{\nu Ð}$	thermal expansion, diffusivity and viscosity	mixed convection
Reynolds number	R	$\dfrac{UL}{\nu}$	inertia terms/viscous terms	viscous flow
Richardson number	Ri	$\dfrac{\frac{g}{\rho}\frac{\partial\rho}{\partial z}}{\left(\frac{\partial u}{\partial z}\right)^2}$	bouyancy gradient/velocity gradient	stratified flow
Rossby number	Ro	$\dfrac{U}{\omega L}$	velocity/rotational velocity	geophysical flows
Ruark number	Ru	$\dfrac{\Delta p}{\rho U^2}$	pressure on object/dynamic pressure	reciprocal of Euler number
Schmidt number	Sc	$\dfrac{\nu}{Ð}$	kinematic viscosity/diffusivity	mass transport
Sherwood number	Sh	$\dfrac{k_m L}{Ð}$	mass transfer/diffusivity	mass transfer
specific speed		$\dfrac{\rho^{3/4}\omega Q^{5/4}}{P^{3/4}}$		rotating machinery
Stanton number	St	$\dfrac{k_h}{\rho U c_p}$	energy flux/flow energy	heat transfer
Stokes number	St	$\dfrac{U\mu}{L^2\Delta\gamma}$	viscous forces/submerged weight	falling particles
Strouhal number	St	$\dfrac{\omega D}{U}$	time scale of flow/oscillating period	flow around cylinders, vortices
Taylor number	Ta	$\dfrac{\omega L^2}{\nu}$	rotational terms/viscous (turbulent) terms	geophysical flows
Ursell parameter	Ur	$\dfrac{HL^2}{4\pi^2\eta_0^3}$	wave amplitude, wave length, water depth	wave mechanics
Weber number	W	$\dfrac{L_0\rho_0 U_0^2}{\sigma}$	dynamic pressure force/surface tension force	surface tension

72

There are several numbers applicable to the transport of heat or of substances. The

$$\text{Prandtl number} = \textbf{Pr} = \frac{\nu}{D} \tag{2.82}$$

is the ratio of the kinematic viscosity of a fluid to the thermal diffusivity (both of which are molecular quantities) and is a property of the fluid. It is a measure of the efficiency that a fluid transports momentum relative to heat in the absence of turbulence. Tables of Prandtl numbers are found in Bird, et al. (1960). For monatomic gases the Prandtl number is 2/3, for diatomic gases it is 14/19, for water it is approximately 7, and for air it is approximately 0.71. A turbulent Prandtl number is also used but is a property of the particular flow and not of the fluid. The Graetz number, used for problems of heat flow through tubes, is a combination of the Prandtl number and the Reynolds number

$$\text{Graetz number} = \textbf{Gz} = \frac{\rho U A c_p}{k_h L} = \frac{\pi D}{4L} \textbf{R} \, \textbf{Pr} \tag{2.83}$$

in which A is the area of the tube, L is its length, D is its diameter, k_h is the thermal conductivity, and c_p is the heat capacity of the fluid at constant pressure. The Graetz number does not consider heating by fluid friction. If that is included, the Brinkman number becomes important; it is

$$\text{Brinkman number} = \textbf{Br} = \frac{\mu U^2}{k_h \Delta T} \tag{2.84}$$

Also, the Eckert number is a combination of the Prandtl number and the Brinkman number,

$$\text{Eckert number} = \textbf{Ec} = \frac{U^2}{c_p \Delta T} = \frac{\textbf{Br}}{\textbf{Pr}} \tag{2.85}$$

The capillary number is another combination,

$$\text{capillary number} = \textbf{Ca} = \frac{\mu U}{\sigma} = \frac{\textbf{W}}{\textbf{R}} \tag{2.86}$$

The Bond number is the ratio between the weight of a drop of liquid ($\rho g L^3$) and the force due to surface tension (σL)

$$\text{Bond number} = \textbf{Bo} = \frac{\rho g L^2}{\sigma} \tag{2.87}$$

The Jacob number treats a bubble in a liquid with vapor inside,

$$\text{Jacob number} = \textbf{Ja} = \frac{c_t \rho_1 \Delta T}{h \rho_v} \tag{2.88}$$

where c_t is the heat capacity of the liquid at constant temperature, ρ_v is the density of the fluid vapor, ρ_1 is the density of the liquid, ΔT is the difference in temperature in the bubble and liquid, and h is the enthalpy of vaporization.

Bringing in the temperature, the

$$\text{Grashof number} = \boldsymbol{Gr} = \frac{\rho^2 \beta_T g L^3 \Delta T}{\mu^2} \tag{2.89}$$

where ΔT is the temperature difference between a solid and a flowing fluid and β_T is a coefficient of thermal expansion. It is used in heat transfer problems. The

$$\text{Nusselt number} = \boldsymbol{Nu} = \frac{k_h L}{\mathcal{D}} \tag{2.90}$$

in which $\mathcal{D}$ is thermal diffusivity. The

$$\text{Rayleigh number} = \boldsymbol{Ra} = \frac{g \beta_T \Delta T L^3}{\nu \mathcal{D}} \tag{2.91}$$

The Darcy number (or Darcy-Rayleigh number) applies to heat transfer in flow through porous media; it is

$$\text{Darcy number} = \boldsymbol{Da} = \frac{\beta_T g k_e L \Delta T}{\mathcal{D} \nu} \tag{2.92}$$

in which k_e is the permeability of the medium and $\mathcal{D}$ is the thermal diffusivity of the medium. The

$$\text{Stanton number} = \boldsymbol{St} = \frac{k_h}{\rho U c_p} = \frac{\boldsymbol{Nu}}{\boldsymbol{R}\,\boldsymbol{Pr}} \tag{2.93}$$

where β_T is a coefficient of volumetric expansion with temperature, T. The Péclet number measures the relative advection and diffusion,

$$\text{Péclet number} = \boldsymbol{Pe} = \frac{U L}{\mathcal{D}} = \boldsymbol{Pr}\,\boldsymbol{R} \tag{2.94}$$

where $\mathcal{D}$ is a coefficient of diffusion. A high Péclet number indicates that a substance is transported by velocity rather than diffused through molecular (or sometimes turbulent) processes. In the advection-diffusion equation the diffusion terms are multiplied by the reciprocal of the Péclet number (see Sec. 11.1), but these terms contain the highest order derivatives. The result is that the advection-diffusion equation becomes difficult to solve for high Péclet numbers, especially using numerical methods. The Navier-Stokes equations are analogous, but the transported quantity is momentum. In numerical calculation with the advection-diffusion equation, a grid Péclet number is defined in which the characteristic length is the length of a finite grid spacing, Δx, so that

$$\text{grid Péclet number} = \frac{U \Delta x}{D} \tag{2.95}$$

The grid Péclet number is useful for numerical calculation with parabolic equations; the Courant number,

$$\text{Courant number} = \boldsymbol{Cr} = (u \pm c)\frac{\Delta t}{\Delta x} \tag{2.96}$$

is used for hyperbolic equations (and sometimes for parabolic equations without c) where u is a local velocity and c is the speed of a small disturbance.

The

$$\text{Schmidt number} = \boldsymbol{Sc} = \frac{\nu}{D} = \frac{\boldsymbol{Pe}}{\boldsymbol{R}} \tag{2.97}$$

and represents the same phenomena as the Prandtl number except that the molecular diffusivity of heat is replaced by the molecular diffusivity of a general substance. Tables of the Schmidt number are found in Davies (1972). The

$$\text{Sherwood number} = \boldsymbol{Sh} = \frac{k_m L}{D} \tag{2.98}$$

in which k_m is a mass transfer coefficient. The Fourier number is

$$\text{Fourier number} = \boldsymbol{Fo} = \frac{T D}{L^2} \quad \text{or} \quad \frac{T \nu}{L^2} \quad \text{or} \quad \frac{D}{UL} \quad \text{or} \quad \frac{\nu}{UL} \tag{2.99}$$

Relative transport is also measured by the

$$\text{Knudsen number} = \boldsymbol{Kn} = \frac{\lambda}{L} \tag{2.100}$$

in which λ is the molecular mean free path and L is a characteristic length. If the Knudsen number is large, molecular collisions with boundaries dominate intermolecular collisions. In turbulent flow the

$$\text{turbulent Knudsen number} = \frac{\text{turbulent eddy size}}{\text{scale length}} \tag{2.101}$$

Several numbers express the ratio of forces on a body with the dynamic terms of the equation. The

$$\text{Euler number} = \boldsymbol{E} = \frac{\rho U^2}{\Delta p} \tag{2.102}$$

where Δp is the difference in pressure in the free stream (far from an object) and the pressure on the body. It is twice the reciprocal of the pressure coefficient. Sometimes the Euler number is defined as $\boldsymbol{E} = \sqrt{1/C_p}$. The Ruark number, $\boldsymbol{Ru}$, is the reciprocal of the Euler number. Coefficients similar to the pressure coefficient are

$$\text{drag coefficient} = C_D = \frac{F_D}{\frac{1}{2}A\rho U^2} \qquad (2.103)$$

$$\text{lift coefficient} = C_L = \frac{F_L}{\frac{1}{2}A\rho U^2} \qquad (2.104)$$

in which F_D is the drag force on an object, F_L is the lift force (perpendicular to the free stream), and A is an area. In the case of drag, the area is usually that which is projected from the object perpendicular to the free stream; in the case of lift, the area is that of the chord plane of a wing.

In oscillating flows the

$$\text{Strouhal number} = \boldsymbol{St} = \frac{\omega D}{U} \qquad (2.105)$$

where ω is the frequency of the oscillation and D is a characteristic dimension. The drag on vertical cylinders in a wave field consists of an inertial part due to the oscillating flow and a part due to fluid velocity. The ratio of these quantities is expressed in the

$$\text{Keulegan-Carpenter number} = \boldsymbol{K} = \frac{UT}{D} \qquad (2.106)$$

in which T is the wave period and D is the diameter of the cylinder. In wave mechanics the Ursell parameter

$$\text{Ursell parameter} = \boldsymbol{Ur} = \frac{HL^2}{4\pi^2\eta_0^3} \qquad (2.107)$$

governs the choice of an approximate theory. The crest to trough height is H, L is the wave length, and η_0 is the depth of the undisturbed water. The Ursell number determines the relative importance of frequency dispersion and nonlinearity.

For stratified flows the

$$\text{Richardson number} = \boldsymbol{Ri} = \frac{\dfrac{g}{\rho}\dfrac{\partial\rho}{\partial z}}{\left(\dfrac{\partial u}{\partial z}\right)^2} \qquad (2.108)$$

in which u is a local velocity and z is measured vertically or normal to the direction of stratification. The Richardson number is often found in alternate forms. Non-Newtonian fluids may have several dimensionless numbers.

The

$$\text{Deborah number} = \boldsymbol{De} = \frac{\mu_0 U}{\tau L} \tag{2.109}$$

in which μ_0 is the highest value of viscosity (which decreases with increasing shear rate) of the fluid and τ is the shear stress.

Several numbers involve rotating flow, or the equations of motion written in coordinates that rotate (a "noninertial" system). This type of flow is especially important since a coordinate system fixed with respect to the earth is rotating (see Sec. 9.1). The

$$\text{Rossby number} = \boldsymbol{Ro} = \frac{U}{\omega L} \tag{2.110}$$

in which ω is the speed of rotation. It measures the relative importance of the rotational terms and the inertial terms. In the case of the rotating earth ω is one revolution per day (approximately) and the characteristic length L must be large for the rotational terms to have much influence. The relative importance of the viscous terms and the rotational term is measured by the

$$\text{Ekman number} = \boldsymbol{Ek} = \frac{\nu}{\omega L^2} \tag{2.111}$$

Although many dimensionless numbers are combinations of other dimensionless numbers, few are duplicated. The reciprocal of the Ekman number is the

$$\text{Taylor number} = \boldsymbol{Ta} = \frac{\omega L^2}{\nu} \tag{2.112}$$

Vorticity or rotation can occur spontaneously on curved surfaces. Flow over a concave plate often breaks up into Görtler vortices if the Görtler number is greater than about 0.3. It is defined as

$$\text{Görtler number} = \boldsymbol{G\ddot{o}} = \frac{\rho U \delta_m}{\mu} \sqrt{\frac{\delta_m}{R_0}} \tag{2.113}$$

where δ_m is the momentum thickness of the boundary layer and R_0 is the radius of curvature of the plate.

The fall or rise of a particle under gravity is allied to the

$$\text{Stokes number} = \frac{U \mu}{L^2 \Delta \gamma} \tag{2.114}$$

in which L is taken as the radius of a sphere with volume equivalent to that of the particle and $\Delta \gamma$ is the difference in specific weight of the object and the fluid.

Turbomachinery has a large set of dimensionless numbers. A few are the

$$\text{power coefficient} = \frac{P}{\rho \omega^3 L^5} \tag{2.115}$$

in which P is the power output or consumption and ω is the rotating speed, the

$$\text{capacity coefficient} = \frac{Q}{\omega L^3} \tag{2.116}$$

where Q is the volume rate of flow through the machine, the

$$\text{head coefficient} = \frac{P}{\rho Q \omega^2 L^2} \tag{2.117}$$

and the

$$\text{specific speed} = \frac{\rho^{3/4} \omega Q^{5/4}}{P^{3/4}} \tag{2.118}$$

2.12   CONCLUSION

The dimensionless equations of fluid mechanics are particularly important since their solution is independent of the units of measurement and independent of the scale of the problem. From these equations stem two important considerations:

1. The dimensionless equations highlight the important effects in any flow problem and indicate which terms are small enough to be neglected if the complete equations cannot be solved.
2. The similarity and dissimilarity between two related flow systems, model and prototype, are indicated by the dimensionless equations.

In reality our knowledge concerning both of these points would only be a fraction of what it is if we did not rely to a considerable extent on the empirical facts resulting from experiment. Experimental result, and not the equations alone, shows the situations in which viscosity can be neglected. Even in those cases where the equations indicate the correct approximation, the magnitude of the error, and sometimes the verification of the approximation, must come from experiment.

Remarkably, such definite conclusions can be obtained from the general equations of fluid mechanics with little reference to specific problems. Although experimentation is essential in fluid mechanics, the equations are an important guide to experiment, even if they cannot be solved directly in the case of a specific problem.

PROBLEMS

2.1. One approximation to the equation of motion involves neglecting the nonlinear inertial terms of the substantial derivative, $\rho Du_i/Dt$.
   (a) Under what conditions would such an approximation be valid?
   (b) Assuming that this approximation is valid, what are the conditions on similitude? Explain.

2.2. Assume that the bulk modulus of a certain fluid is a function of the pressure and density, $\beta = \beta(p, \rho)$.
   (a) How many dimensionless parameters do you expect from this equation in an MLT system?
   (b) How many dimensionless parameters do you expect from this equation in an FLT system?
   (c) Form the dimensionless parameters for (a) and (b).

2.3. A combustion chamber of a rocket engine is designed for a pressure of 250 psia and a temperature of 5000°R.
   (a) How would a 1:10 scale model be operated?
   (b) If the prototype is expected to develop 20,000 lb thrust, what thrust should the model develop? Assume the same atmospheric conditions.

2.4. In rotating machinery the volumetric flow rate is given as $Q = Q(H, P, F, \rho, \beta, T, \eta, D, N)$ where H is the head, P is the power, F is the thrust, ρ is the density, β is the bulk modulus of the fluid, T is the torque, η is the efficiency, D is the rotor diameter, and N is the rotational speed.
   (a) How many independent dimensionless groups are expected from this functional relationship?
   (b) Form all possible dimensionless groups. A type of Mach number should be included. Other dimensionless groups are a torque coefficient (torque divided by something) and a thrust coefficient.

2.5. At certain wind speeds a cylindrical object such as a flagpole or smokestack is subject to severe vibration from the air separating first from one side then from the other. What criteria should be used to model this phenomenon? Include the structural considerations.

2.6. Derive a dimensionless energy equation analogous to (2.26). Explain the significance of the dimensionless groups.

2.7. (a) A spillway 30 m high is to be modeled with a model 3 m high. What is the ratio of model to prototype flow per unit width, q_M/q_P?
   (b) A valve is to be modeled with a scale of 1:10. What is the ratio of model flow to prototype flow, Q_M/Q_P?

2.8. A river system is to be modeled with scale ratios of 1:1000 horizontal and 1:10 vertical. Manning's equation (in SI units) for steady flow in a wide channel is

$$U = \frac{1}{n}d^{2/3}\sqrt{S}$$

in which U is the average velocity, n is the roughness coefficient, d is the depth, and S is the slope of the river. If $n_M = 0.5\,n_P$, what should be the slope ratio, S_M/S_P?

CHAPTER
3

POTENTIAL
FLOW
(FUNDAMENTALS)

Potential flow is one of the most useful and most used approximations to the equations of fluid mechanics. The mathematical theory is well developed. It can easily be, and has been, the subject of an entire book. In this writing it is divided into two parts. The first part is intended to be an introduction with enough explanation and solution techniques to allow the reader to solve most of the elementary problems. The second part expands the subject into some of the more mathematical solutions and includes moving boundary problems.

3.1 APPLICATIONS

This chapter begins by listing some of the applications of potential flow. Due to the approximation of zero viscosity, the problems that are adequately approximated are severely limited. They include, however, important applications in aerodynamics and hydrodynamics.

3.1.1 Incompressible, Inviscid Flow

The definition of the velocity potential as given in (1.107) is repeated here

$$u_i = -\frac{\partial \Phi}{\partial x_i} \tag{3.1}$$

As indicated in Chap. 2, this definition makes sense only if the fluid is irrotational

$$\vec{\nabla} \times \vec{u} = 0 \tag{3.2}$$

and, conversely, if the potential exists, the fluid must be irrotational. Using (3.1) in the equation of continuity for a fluid of constant density (where the velocity is divergence free) results in Laplace's equation

$$\frac{\partial^2 \Phi}{\partial x_i \partial x_i} = 0 \tag{3.3}$$

Thus, the study of potential flow is a study of the solution of Laplace's equation (sometimes called the potential equation).

The equation of motion under the restrictions of irrotationality and steady flow gives a single equation, Bernoulli's equation, for the pressure (Sec. 1.10)

$$\frac{u^2}{2} + \frac{p}{\rho} + gh = \text{constant} \tag{3.4}$$

in which u is the magnitude of the velocity, $u^2 = u_i u_i$, p is pressure, ρ is density, g is the acceleration of gravity, and h is elevation.

3.1.2 Compressible, Linearized Flow

Compressible flow falls under the category of potential flow if certain additional approximations are made. The equation of mechanical energy for steady flow and no friction is (from Sec. 1.9)

$$\rho u_i \frac{\partial}{\partial x_i} \frac{u^2}{2} + \rho g u_i \frac{\partial h}{\partial x_i} + u_i \frac{\partial p}{\partial x_i} = 0 \tag{3.5}$$

The last derivative is written

$$\frac{\partial p}{\partial x_i} = \frac{dp}{d\rho} \frac{\partial \rho}{\partial x_i} = c^2 \frac{\partial \rho}{\partial x_i} \tag{3.6}$$

where p is assumed to be a function of ρ, and c is the speed of sound given by

$$c = \sqrt{\left(\frac{dp}{d\rho}\right)_s} \tag{3.7}$$

where the subscript s signifies that entropy is to be held constant when taking the indicated derivative. Using (3.7) in (3.5) gives

$$u_i u_j \frac{\partial u_j}{\partial x_i} + g u_i \frac{\partial h}{\partial x_i} + u_i \frac{c^2}{\rho} \frac{\partial \rho}{\partial x_i} = 0 \tag{3.8}$$

Using the equation of continuity for the steady flow of a compressible fluid in (3.8) yields

$$\frac{1}{c^2}\left(\frac{\partial \Phi}{\partial x_i}\frac{\partial \Phi}{\partial x_j}\frac{\partial^2 \Phi}{\partial x_i \partial x_j}+g\frac{\partial \Phi}{\partial x_i}\frac{\partial h}{\partial x_i}\right)=\frac{\partial^2 \Phi}{\partial x_i \partial x_i} \tag{3.9}$$

To simplify (3.9) further approximation is made that the primary velocity is in the x-direction and is much greater than the velocity in the other coordinate directions, or

$$u_x^2 \gg u_y^2 + u_z^2 \tag{3.10}$$

Equation (3.10) applies to "thin airfoil theory." A perturbation potential Φ'' is introduced such that

$$\Phi = \Phi' + \Phi'' \qquad u_x = -\frac{\partial \Phi'}{\partial x}-\frac{\partial \Phi''}{\partial x} \qquad u_y = -\frac{\partial \Phi''}{\partial y} \qquad u_z = -\frac{\partial \Phi''}{\partial z} \tag{3.11}$$

in which Φ' represents the undisturbed potential (meaning that, in flow around an object, Φ' is the potential that would occur if the object were not present) and is a function of x only. The assumption (3.10) indicates that the derivatives of Φ'' are much smaller than the derivatives of Φ'. Using (3.11), (3.9) becomes, after neglecting the elevation term,

$$\frac{1}{c^2}u_i u_j\left(\frac{\partial^2 \Phi'}{\partial x_i \partial x_j}+\frac{\partial^2 \Phi''}{\partial x_i \partial x_j}\right)=\frac{\partial^2 \Phi'}{\partial x_i \partial x_i}+\frac{\partial^2 \Phi''}{\partial x_i \partial x_i} \tag{3.12}$$

Equation (3.12) must be satisfied by the undisturbed potential, Φ', so that

$$\frac{1}{c^2}u_i u_j\frac{\partial^2 \Phi'}{\partial x_i \partial x_j}=\frac{\partial^2 \Phi'}{\partial x_i \partial x_i} \tag{3.13}$$

Subtracting (3.13) from (3.12) and replacing the velocity terms by the derivatives of the potential gives an equation for the disturbance potential

$$\frac{1}{c^2}\left(\frac{\partial \Phi'}{\partial x_i}+\frac{\partial \Phi''}{\partial x_i}\right)\left(\frac{\partial \Phi'}{\partial x_j}+\frac{\partial \Phi''}{\partial x_j}\right)\frac{\partial \Phi''}{\partial x_i \partial x_j}=\frac{\partial^2 \Phi''}{\partial x_i \partial x_i} \tag{3.14}$$

The products of the derivatives of Φ'' are neglected, leading to

$$\frac{1}{c^2}\frac{\partial \Phi'}{\partial x_i}\frac{\partial \Phi'}{\partial x_j}\frac{\partial^2 \Phi''}{\partial x_i \partial x_j}=\frac{\partial^2 \Phi''}{\partial x_i \partial x_i} \tag{3.15}$$

Now Φ' is a function of x only so that the left side potential terms are

$$\frac{\partial \Phi'}{\partial x_i}\frac{\partial \Phi'}{\partial x_j}\frac{\partial^2 \Phi''}{\partial x_i \partial x_j}=u_x^2\frac{\partial^2 \Phi''}{\partial x^2} \tag{3.16}$$

and (3.15) becomes

$$(1 - M^2)\frac{\partial^2 \Phi''}{\partial x^2} + \frac{\partial^2 \Phi''}{\partial y^2} + \frac{\partial^2 \Phi''}{\partial z^2} = 0 \tag{3.17}$$

in which $M = u/c$ is the Mach number and is approximated by $M \approx u_x/c$ from (3.10). We assume that M is a constant and is equal to the free stream (undisturbed) Mach number.

With the approximation that M is constant throughout, (3.17) is linear, but it is still not Laplace's equation. If M is greater than unity (supersonic flow), the character of (3.17) is entirely different than Laplace's equation and cannot serve as an approximation to Laplace's equation under any circumstances. If M is less than one, (3.17) can become Laplace's equation under the affine transformation (a linear transformation in which there are no singularities and all finite points transform to finite points) defined by

$$\xi = x \qquad \eta = y\sqrt{1 - M^2} \qquad \zeta = z\sqrt{1 - M^2} \qquad \Phi'' = \lambda\phi \tag{3.18}$$

where λ is an arbitrary constant. The result of the transformation is

$$\frac{\partial^2 \phi}{\partial \xi^2} + \frac{\partial^2 \phi}{\partial \eta^2} + \frac{\partial^2 \phi}{\partial \zeta^2} = 0 \tag{3.19}$$

Equation (3.19) indicates that a subsonic, compressible flow can be approximated by an incompressible flow, if the disturbance is small, by stretching the dimensions at right angles to the main stream by the factor $\sqrt{1 - M^2}$. Thus, this sort of compressible flow becomes a potential flow. Obviously, the approximations are restrictive.

3.1.3 Flow through Porous Media

Most flow through natural soils or artificial media used for filtering or heat exchange is governed by Darcy's law

$$\vec{u} = -K\vec{\nabla}\Phi \tag{3.20}$$

in which $\vec{u}$ is the specific discharge vector (or macroscopic velocity; the average fluid velocity in the pores is the specific discharge divided by the porosity), $\Phi = h + p/(\rho g)$ (the head) and K is a constant of proportionality called the *hydraulic conductivity*. In anisotropic media K becomes a symmetric tensor. Darcy's law is generally nonderivable, but one can make statistical arguments based on laminar flow for its validity. The macroscopic velocity is not the real fluid velocity between grains of a porous medium but is the velocity that results from dividing the volumetric flow rate through a small area (including both pores and solid) at right angle to the velocity vector by the area. In order for Darcy's law to be valid, one must be able to define three length scales: a typical dimension of any problem must be large compared to the diameter of the grains in a granular medium, a diameter, that, in turn, must be large compared to the mean free path of the fluid molecules. That is, the (molecular scale) $\ll$ ("Darcy" scale) $\ll$ (problem scale).

Assuming a divergence-free flow and a constant hydraulic conductivity, Darcy's law gives

$$\vec{\nabla} \cdot \vec{u} = -K \nabla^2 \Phi = 0 \tag{3.21}$$

Thus, this type of flow satisfies Laplace's equation and is a potential flow. The flow between the grains—on a scale smaller than the Darcy scale—is not potential flow but is viscous. Neither is "non-Darcy flow" where Darcy's law does not apply, as in high Reynolds number flows in macropores, for example. Nevertheless, most porous media flow is assumed to obey Darcy's law and is potential flow. For porous media the potential has a physical meaning, the head, as opposed to the mathematical abstraction used in hydrodynamic flows. In (3.1) the minus sign in the definition of the potential is arbitrary, but in (3.20) the minus sign is necessary in order that the velocity be positive in the downhill direction of the head.

3.1.4 Waves and Free Surface Flow

The flow of an incompressible fluid with a free surface is a special case of the application cited in Sec. 3.1.1. The free surface, however, causes a great complication. The free surface boundary condition is nonlinear and thus creates all of the complications of any nonlinear problem. Of greater difficulty is the fact that the position of the surface is part of the solution to the problem and at the same time it is a boundary of the solution domain. That means that the problem is to be solved in an unknown and constantly changing domain, a fact that in itself represents a type of nonlinearity as is shown in the next chapter. Free surface problems also occur in flow in porous media.

> **Historical Note:** Pierre Simon Laplace (1749–1827) was a child prodigy who became a professor of mathematics at the Ecole Militaire in Paris at the age of 18. His most famous work was the five volumes of *Celestial Mechanics*. He was a contemporary of Lagrange, with whom he collaborated on the theory of waves. Laplace was also known for work on the theory of sound. Today the Laplace equation applies to a broad spectrum of physical problems.

3.2 THE STREAM FUNCTION

For two-dimensional and axially symmetric flows the stream function Ψ is a useful companion to the potential function. It is defined

$$u_x = -\frac{\partial \Psi}{\partial y} \qquad u_y = \frac{\partial \Psi}{\partial x} \qquad (3.22)$$

The definition of the stream function means that the two-dimensional equation of continuity is automatically satisfied since

$$\frac{\partial u_x}{\partial x} + \frac{\partial u_y}{\partial y} = -\frac{\partial^2 \Psi}{\partial x \partial y} + \frac{\partial^2 \Psi}{\partial y \partial x} \qquad (3.23)$$

For the stream function to exist, the flow must be divergence free just as it must be rotation free for the potential function to exist. For a rotation-free flow the stream function satisfies Laplace's equation

$$\frac{\partial u_y}{\partial x} - \frac{\partial u_x}{\partial y} = \frac{\partial^2 \Psi}{\partial x^2} + \frac{\partial^2 \Psi}{\partial y^2} = 0 \qquad (3.24)$$

In the case of axially symmetric flow, the stream function is defined as

$$u_R = \frac{1}{R}\frac{\partial \Psi}{\partial x} \qquad u_x = -\frac{1}{R}\frac{\partial \Psi}{\partial R} \qquad (3.25)$$

The continuity equation is automatically satisfied although the irrotationality condition does not lead to Laplace's equation in cylindrical coordinates (see Sec. 3.9).

The lines $\Psi = $ constant are *streamlines*, defined such that the velocity vector is always tangent to the line. If we take the total derivative of Ψ

$$d\Psi = \frac{\partial \Psi}{\partial x}dx + \frac{\partial \Psi}{\partial y}dy \qquad (3.26)$$

then divide by dx while holding Ψ constant (so that $d\Psi = 0$)

$$\left(\frac{\partial y}{\partial x}\right)_\Psi = \frac{u_y}{u_x} \qquad (3.27)$$

where the subscript on the derivative indicates the quantity that is to be held constant. Equation (3.27) shows that the ratio of a change of y and x along the line $\Psi = $ constant is the same as the ratio of the velocities in the coordinate directions; thus, the resultant velocity vector is tangent to the line $\Psi = $ constant.

In the case of the potential function the total derivative is

$$d\Phi = \frac{\partial \Phi}{\partial x}dx + \frac{\partial \Phi}{\partial y}dy \qquad (3.28)$$

and

$$\left(\frac{\partial y}{\partial x}\right)_{\Phi} = -\frac{u_x}{u_y}$$
(3.29)

Since $(\partial y/\partial x)_\Phi$ is the negative reciprocal of $(\partial y/\partial x)_\Psi$, the lines $\Phi =$ constant, the *equipotential lines*, are at right angles to the streamlines.

The difference between two stream functions is the quantity (in units of L^2/T for the two-dimensional case) of flow between the streamlines. Defining q as the line integral (Fig. 3.1)

$$q = \int_{P_1}^{P_2} (u_x dy - u_y dx)$$
(3.30)

makes q the quantity of flow crossing any line that connects P_1 and P_2. Using the definition of Ψ from (3.22) in (3.30) gives

$$q = -\int_{P_1}^{P_2} \left(\frac{\partial \Psi}{\partial y}dy + \frac{\partial \Psi}{\partial x}dx\right) = \Psi_1 - \Psi_2$$
(3.31)

which indicates that the integral is independent of the path and is the difference of the two stream functions. The axisymmetric case is similar; the flow crossing the area between two circles at radii of R_1 and R_2 is the difference of the stream function in those two radii multiplied by 2π.

The stream function can always be obtained from the velocity potential, and vice versa, if both exist. Combining (3.1) and (3.22) gives

$$\frac{\partial \Phi}{\partial x} = \frac{\partial \Psi}{\partial y} \qquad \frac{\partial \Phi}{\partial y} = -\frac{\partial \Psi}{\partial x}$$
(3.32)

which can be solved for either Φ or Ψ, whichever is unknown. In such a solution there is always an arbitrary constant. Obviously, a flow does not

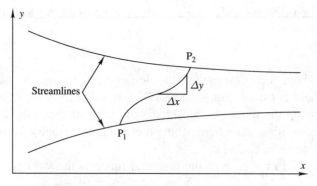

FIGURE 3.1
Integration along a curve between two streamlines.

change by adding a constant to either the potential or the stream function since the velocity depends on the derivatives of those quantities. If the flow is not irrotational and if Ψ is given, then the solution of the two equations (3.32) cannot be obtained so that Φ is determined within an arbitrary constant. That is, the Φ obtained from the first of (3.32) will not be the same as the Φ obtained from the second and the two cannot be made compatible by adjustment of the arbitrary function of x from the integration of the first equation and the arbitrary function of y from the integration of the second equation.

Example 3.1. Given $\Psi = x - x^3y/2$, find Φ. From (3.22) and (3.32) $u_x = -\partial\Phi/\partial x = x^3/2$ and $u_y = -\partial\Phi/\partial y = 1 - 3x^2y/2$; thus, mass is conserved since $\vec{\nabla} \cdot \vec{u} = 0$. Integrating the first equation of (3.32) yields $\Phi = -x^4/8 + f_1(y)$; integrating the second equation of (3.32) yields $\Phi = -y + 3x^2y^2/4 + f_2(x)$. Even though we are free to choose the two functions $f_1(x)$ and $f_2(y)$, the two expressions for Φ cannot be made equivalent. The flow is not irrotational and Φ does not exist.

Example 3.2. Given $\Psi = xy$, find Φ. The velocities are $u_x = -x$ and $u_y = y$. The velocity potential from u_x is $\Phi = x^2/2 + f_1(y)$; from u_y it is $\Phi = -y^2/2 + f_2(x)$. The proper choices of f_1 and f_2 produce $\Phi = (x^2 - y^2)/2 + \text{constant}$.

Historical Note: The concept of streamlines was introduced by Jean Charles Borda (1733–1799) in his studies of flow through orifices. Borda, a French mathematician and military engineer, was a contemporary of Laplace and Lagrange. His work on orifices produced the "Borda mouthpiece," a tube inserted into the side of a tank so that the entrance of water in the tube is away from the wall of the tank. Using momentum principles, Borda was able to compute the size of the "vena-contracta." Both the stream function and potential were introduced by Lagrange.

3.3 FLOW NETS

Both the potential and the stream function satisfy Laplace's equation and the lines of constant potential are always at right angles to the streamlines (except at singular points). Using this property, a solution can be constructed by sketching a "flow net" consisting of equipotential lines and streamlines. The prerequisites for the sketch are a large piece of paper with an accurate drawing of the physical problem showing the boundary conditions, a soft pencil, and a good eraser. The equipotential lines are then sketched while ensuring that the conditions along Dirichlet boundaries (where the potential is given) are satisfied. (Note, however, that the velocities depend on the derivatives of the potential; the pattern is independent of the absolute value of the potential). The equipotential lines must meet the no-flow Neumann boundaries at right angles. Streamlines are then constructed so that they cross the equipotential lines at right angles everywhere. Using the same interval for the streamlines and equipotential lines, a series of curvilinear "squares" should appear.

After the initial try some of the streamlines and equipotential lines will be crowded or too far apart, and the "squares" will not have equal sides. The sketch is then adjusted to improve the situation. (Apparently the important tool is the eraser.) After the sketch looks reasonably good, it can be further improved by drawing the diagonals to the squares. These diagonals form another curvilinear set of lines that should cross at right angles. They will indicate where further improvement can be made in the sketch. After a period of trial and error, a reasonable flow net appears. At the least, flow net sketching can yield a rough solution and familiarity with the method gives the analyst a feel for the final result, which will lead to the discovery of error in other techniques.

Velocities along the streamlines can be found by numerical differentiation. Pressures can be found from Bernoulli's equation in the case of hydrodynamic flows or from the head and elevation in the case of flow in porous media. The numerical differentiation necessary to find the velocities magnifies any errors in the solution for the potential and stream function and thus, if velocities are necessary, the original solution should be done as accurately as possible. The method is best illustrated by an example.

Example 3.3. Consider the flow under a dam as illustrated in Fig. 3.2. The boundary conditions are:

Line ① is a constant head (Dirichlet) boundary where $\Phi = H_1$ and $\partial \Psi / \partial n = \partial \Psi / \partial y = 0$.

Line ② is a constant head boundary where $\Phi = H_2$ and $\partial \Psi / \partial y = 0$.

Line ③ is a streamline (Neumann) boundary where $\partial \Phi / \partial n = \partial \Phi / \partial y = 0$ and $\Psi = $ constant.

Line ④ is a streamline boundary where $\partial \Phi / \partial y = 0$ and $\Psi = $ constant.

First, the streamlines are sketched. They must be vertical at the constant head boundaries and horizontal at the line of symmetry under the center of the dam. Their spacing on the constant head lines is initially a guess.

Second, the equipotential lines are drawn. The one on the line of symmetry is vertical. The equipotential lines must begin and end on streamline boundaries and intersect those boundaries at right angles. At infinity the potential spacing tends to infinity and the lines are nonexistent.

The upstream and downstream corners of the dam represent singular points where the equipotential lines and streamlines meet at other than a right

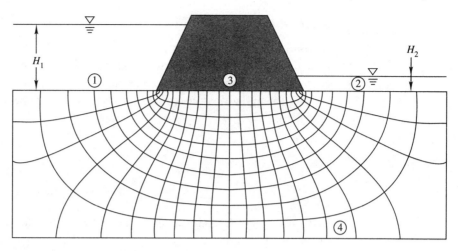

FIGURE 3.2
Flow under a dam on a permeable foundation.

angle. The largest errors in the final solution are likely to be in the vicinity of these corners.

After the initial sketch, the flow net is improved by the methods cited above. The figure gives the final result.

3.4 THE FUNDAMENTAL SINGULARITIES

Analytical solutions to potential flow problems can, in principle, be obtained as combinations of three fundamental solutions. In two cases these solutions are singular, meaning that Laplace's equation is not satisfied at a point called a singular point.

3.4.1 Parallel Flow

For the rather trivial case where the streamlines are straight and parallel and the velocity is everywhere constant, the potential is

$$\Phi = -U_i x_i + \text{constant} \tag{3.33}$$

in which U_i is the free stream velocity in the ith direction. The corresponding stream function for two-dimensional parallel flow is

$$\Psi = -U_x y + U_y x + \text{constant} \tag{3.34}$$

In both (3.33) and (3.34) the constant is arbitrary since the velocities depend only on the derivatives of the potential and the stream function.

3.4.2 Source (Sink) Flow

In source flow all the streamlines meet in a point from which the fluid is flowing (Fig. 3.3). The equipotential lines are circles with the point as the center. In two dimensions the potential is the solution to Laplace's equation written for cylindrical coordinates without θ or z dependence

$$\frac{\partial}{\partial R}\left(R\frac{\partial \Phi}{\partial R}\right) = 0 \qquad (3.35)$$

or

$$\Phi = -\frac{Q}{2\pi}\ln R + \text{constant} \qquad (3.36)$$

in which R is the distance from the point. If the source is located at the point x_0, y_0 then

$$R = \sqrt{(x - x_0)^2 + (y - y_0)^2} \qquad (3.37)$$

The quantity of flow coming from the source can be obtained by integrating the radial velocity around any curve that contains the source. If the curve is a circle centered on the source

$$u_R = -\frac{\partial \Phi}{\partial R} = \frac{Q}{2\pi R} \qquad Q = \int_0^{2\pi} u_R R \, d\theta \qquad (3.38)$$

Thus, Q is the rate of fluid coming from the source (in dimensions of L^2/T). The corresponding stream function is

$$\Psi = -\frac{Q}{2\pi}\arctan\frac{y - y_0}{x - x_0} + \text{constant} \qquad (3.39)$$

If Q is negative, the point is called a *sink*.

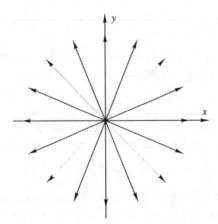

FIGURE 3.3
A two-dimensional point source.

In three dimensions the potential for the source is the solution of the Laplace's equation in spherical coordinates without θ or ϕ dependence

$$\frac{\partial}{\partial r}\left(r^2\frac{\partial \Phi}{\partial r}\right) = 0 \tag{3.40}$$

or

$$\Phi = \frac{q}{4\pi r} + \text{constant} \tag{3.41}$$

in which $r = \sqrt{(x - x_0)^2 + (y - y_0)^2 + (z - z_0)^2}$. Again, Q is the rate of flow (now in dimensions L^3/T) from the source as can be shown by integrating the radial velocity on the surface of a sphere surrounding the source.

Sources and sinks may be distributed singularities. A *line source* on the x-axis is shown in Fig. 3.4. In general a line source is defined in three dimensions as

$$\Phi = \frac{1}{4\pi}\int_C \frac{q}{\sqrt{(x - \xi)^2 + (y - \eta)^2 + (z - \zeta)^2}}\,ds \tag{3.42}$$

where q is the strength per unit length of the line C, which is described by a function

$$f(\xi, \eta, \zeta) = 0 \tag{3.43}$$

A similar definition is available for a source distributed over an area. The *surface source* is

$$\Phi = \frac{1}{4\pi}\int_A \frac{w}{\sqrt{(x - \xi)^2 + (y - \eta)^2 + (z - \zeta)^2}}\,dA \tag{3.44}$$

in which w is the strength per unit area.

3.4.3 Vortex Flow

For a vortex the streamlines and equipotential lines are reversed from the source flow. The two-dimensional potential and stream function

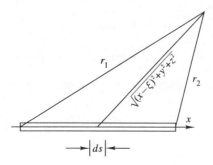

FIGURE 3.4
A three-dimensional line source.

are

$$\Phi = -\frac{\Gamma}{2\pi} \arctan \frac{y - y_0}{x - x_0} + \text{constant} \qquad (3.45)$$

$$\Psi = \frac{\Gamma}{2\pi} \ln R + \text{constant} \qquad (3.46)$$

in which Γ represents the strength of the vortex and is more fully defined in Sec. 3.10. Vortices, as well as sources, can be distributed along lines or over areas.

3.5   SUPERPOSITION

The great advantage of a linear equation is that solutions can be superposed (added) to obtain new solutions. In this way elementary solutions can be put together to form other solutions that satisfy the boundary conditions to a particular problem. Assume that Φ_1 and Φ_2 both satisfy Laplace's equation. Then a third solution is

$$\Phi_3 = a\Phi_1 + b\Phi_2 \qquad (3.47)$$

where a and b are constants. That Φ_3 is a solution is obvious from

$$\nabla^2 \Phi_3 = a\nabla^2 \Phi_1 + b\nabla^2 \Phi_2 \qquad (3.48)$$

Nearly all potential flow solutions use this important property.

> **Example 3.4   The Rankine half-body.**   Flow around a projectile-like object can be calculated by the superposition of parallel and source flow. In two dimensions with the source placed in the origin
>
> $$\phi = -Ux - \frac{Q}{2\pi} \ln R \qquad (3.49)$$
>
> The flow picture is shown in Fig. 3.5. The width of the body can be changed by altering the constants U and Q; larger U or smaller Q makes the body thinner.
> Using (3.49) the velocities are
>
> $$u_x = U + \frac{Q}{2\pi} \frac{x}{R^2} \qquad u_y = \frac{Q}{2\pi} \frac{y}{R^2} \qquad (3.50)$$
>
> The point where the nose of the body crosses the x-axis is a *stagnation point* in which the velocity is zero. Obviously, the y-velocity is zero everywhere on the x-axis from symmetry. From the first of (3.50) the x-velocity is zero at
>
> $$x = -\frac{Q}{2\pi U} \qquad (3.51)$$

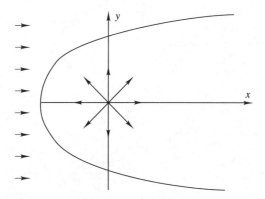

FIGURE 3.5
A Rankine half-body.

The body is most easily plotted by defining the stream function

$$\Psi = -Uy - \frac{Q}{2\pi}\theta \qquad \theta = \arctan\frac{y}{x} \tag{3.52}$$

Substituting (3.51) for x in (3.52) gives the value of the streamline (stream function) through the stagnation point (where $\theta = \pi$ and $y = 0$, so that $\Psi = -Q/2$). The body is then plotted by using this value in (3.52) and solving for y as a function of x or, alternatively, assuming a value for θ and finding the corresponding value of y.

3.6 CIRCULATION, ROTATION, AND VORTICITY

The ideas of rotation and vorticity were introduced in the derivation of the Navier-Stokes equations and later seen in the vortex potential. In this section these quantities are brought together and defined with circulation.

3.6.1 Circulation and Vorticity

Consider a plane area bounded by curve ∂A as shown in Fig. 3.6. The *circulation* about this area is defined as the (counterclockwise) line integral

$$\Gamma = \oint_{\partial A} \vec{u} \cdot \vec{t} \, ds \tag{3.53}$$

in which $\vec{u}$ is the velocity vector and $\vec{t}$ is a unit vector tangent to the curve. If the area is divided into a fine mesh with sides of length Δx and Δy, the circulation about one rectangle of the mesh is

$$\Delta\Gamma = u_x\Delta x + \left(u_y + \frac{\partial u_y}{\partial x}\Delta x\right)\Delta y - \left(u_x + \frac{\partial u_x}{\partial y}\Delta y\right)\Delta x - u_y\Delta y$$

$$= \left(\frac{\partial u_y}{\partial x} - \frac{\partial u_x}{\partial y}\right)\Delta x\Delta y$$

$$\tag{3.54}$$

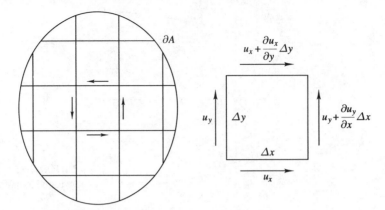

FIGURE 3.6
Circulation around a curve and an area.

The circulation about the total area is the sum of the circulations about the mesh elements that comprise the area

$$\Gamma = \sum \Delta\Gamma = \sum \left(\frac{\partial u_y}{\partial x} - \frac{\partial u_x}{\partial y} \right) \Delta x \, \Delta y \qquad (3.55)$$

Notice that the contributions from adjacent meshes cancel along the common boundaries, leaving the external boundaries as the only real contributors. If we take the length of the sides of the mesh to zero while increasing the number of elements, the summation can be replaced by an integral

$$\oint_{\partial A} \vec{u} \cdot \vec{t} \, ds = \int_A \left(\frac{\partial u_y}{\partial x} - \frac{\partial u_x}{\partial y} \right) dA \qquad (3.56)$$

in which A is the area contained in the boundary ∂A. Notice that the quantity in parentheses is the magnitude of the vector $\vec{\nabla} \times \vec{u}$, which is twice the vorticity of the fluid element, defined by (1.59). Equation (3.56) can be generalized to

$$\oint_{\partial A} \vec{u} \cdot \vec{t} \, ds = \int_A \left(\vec{\nabla} \times \vec{u} \right) \cdot \vec{n} \, dA \qquad (3.57)$$

which holds for any plane area if $\vec{u}$ is differentiable in A and on ∂A. Although (3.57) was derived for a plane area, a similar argument leads to the same equation for a three-dimensional area. In the general case (3.57) is called *Stokes' theorem*.

Vorticity can be related to rotation by considering a small circular element as shown in Fig. 3.7. Suppose that the fluid in this element is rotating with constant angular velocity ω about the center. Taking the line

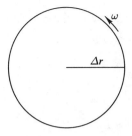

FIGURE 3.7
Circulation about a small circular area.

integral about the element yields

$$\Gamma = (\omega \Delta r) 2\pi \, \Delta r = 2\omega A \tag{3.58}$$

Thus, the vorticity of the element is twice the angular velocity of rotation as expressed by (1.59).

3.6.2 Vorticity

In source flow the equation of continuity, and hence Laplace's equation, is not satisfied at a point. In a similar way the irrotationality condition is not satisfied at a point in vortex flow; the space derivatives do not exist at the point of a vortex and Stokes' theorem (3.57) is not valid for an area containing a vortex.

In two-dimensional flow circulation about a point can exist, but the circulation around areas not containing the singularity is zero. Consider a vortex placed in the origin as shown in Fig. 3.8. The velocity potential for this case is

$$\Phi = -\frac{\Gamma}{2\pi}\theta \tag{3.59}$$

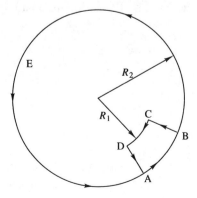

FIGURE 3.8
Circulation about a vortex.

in which $\theta = \arctan(y/x)$. The velocity in the θ-direction is

$$u_\theta = -\frac{1}{R}\frac{\partial \Phi}{\partial \theta} = \frac{\Gamma}{2\pi R} \tag{3.60}$$

Applying (3.53) over the path ABEA

$$\Gamma = \int_0^{2\pi} \frac{\Gamma}{2\pi R} R \, d\theta = \Gamma \tag{3.61}$$

showing that the Γ in (3.59) is the same as the Γ that appears in the potential and stream functions for vortex flow, Eqs. (3.45) and (3.46).

The circulation about an area not containing the singular point remains zero as shown by applying (3.53) over the path ABCDA of Fig. 3.8

$$\Gamma = \int_{\theta_A}^{\theta_B} \frac{\Gamma}{2\pi R} R \, d\theta + \int_{R_2}^{R_1} u_R \, dR + \int_{\theta_C}^{\theta_D} \frac{\Gamma}{2\pi R} R \, d\theta + \int_{R_1}^{R_2} u_R \, dR$$
$$= 0 \tag{3.62}$$

The first and third terms cancel while the second and fourth terms are zero since $u_R = 0$.

In cases where the flow field contains physical boundaries, the circulation is zero in an irrotational flow field about any closed curve—from (3.57)—if the area contained in that curve is not interrupted by a physical boundary. Geometrical considerations make this statement quite different for two- and three-dimensional flow. In two-dimensional flow about a stationary body, such an uninterrupted area cannot be defined for a curve that encircles the object as indicated by Fig. 3.9. Thus, any two-dimensional object can have circulation around it.

A three-dimensional object cannot have circulation around it since an area can be defined that is not interrupted by the solid (see Fig. 3.10). An exception is an object that has a hole through it. In that case circulation can exist around a curve that passes through the hole since the area defined by the curve is always interrupted by the solid. An example is the torus shown in Fig. 3.11.

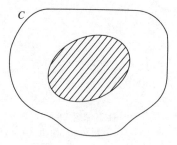

FIGURE 3.9
Circulation about a two-dimensional object.

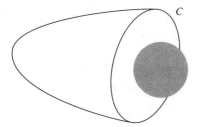

FIGURE 3.10
A three-dimensional object inside a sock.

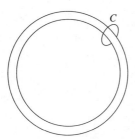

FIGURE 3.11
A multiply connected three-dimensional space.

If an uninterrupted area can be defined for any arbitrary closed curve, the space is said to be *simply connected* (Fig. 3.10). If there exists a curve where an uninterrupted area cannot be defined, the space is *multiply connected* (Figs. 3.9 and 3.11).

3.6.3   Vortex Lines and Vortex Motion

The vortex has been defined only in two-dimensional flow where it is represented by a point and where the vorticity vector points at right angles to the plane of the flow. A three-dimensional view of the two-dimensional vortex would show that it is an infinitely long line in the z-direction. In general three-dimensional flow, such a line can exist anywhere in the flow and can be of any shape. The vortex line consists of the series of locations where the rotation vector, $\vec{\omega} = \vec{\nabla} \times \vec{u}/2$, is not zero. Such a line is called a *vortex filament*. The general properties of a vortex filament are developed in this section.

Consider a vortex filament surrounded by a sheath as shown in Fig. 3.12. The flow is everywhere irrotational except on the filament. The line integral around the boundary of the sheath is zero from the application of Stokes' theorem

$$\oint_A^B \vec{u} \cdot \vec{t}\, ds + \oint_B^C \vec{u} \cdot \vec{t}\, ds + \oint_C^D \vec{u} \cdot \vec{t}\, ds + \oint_D^A \vec{u} \cdot \vec{t}\, ds = 0 \qquad (3.63)$$

As point A approaches D and point B approaches C, the first and third

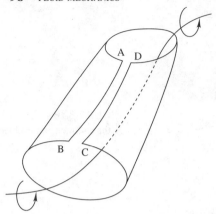

FIGURE 3.12

A vortex filament surrounded by a sheath

integrals of (3.63) cancel so that

$$\oint_D^A \vec{u} \cdot \vec{t}\, ds = \oint_C^B \vec{u} \cdot \vec{t}\, ds = \Gamma \qquad (3.64)$$

which shows that the circulation about a vortex filament is constant, an expression of *Helmholtz' first theorem*. Stokes' theorem also implies that a vortex filament is everywhere continuous and nonending, which expresses *Helmholtz' second theorem*. This fact is apparent since the vorticity vector $\vec{\omega}$ satisfies an equation of continuity for vorticity

$$\vec{\nabla} \cdot \vec{\omega} = \frac{1}{2} \vec{\nabla} \cdot (\vec{\nabla} \times \vec{u}) = 0 \qquad (3.65)$$

showing that, like a constant density fluid, vorticity is divergence free. A vortex filament may either end at a boundary (but not in the fluid) or it may be in the form of a closed path. Actually, the vorticity need not be confined to a single line; it may be distributed over a finite area, but the *vortex tube* formed by that area must be continuous.

An equation for the velocity field induced by a vortex filament is derived next. That equation is consistent with the two-dimensional definition. In the two-dimensional case (see Fig. 3.13) the velocity at point P is $u = \Gamma/(2\pi R)$. The contribution of the segment ds is such that

$$\frac{\Gamma}{2\pi R} = \oint_{-\infty}^{\infty} f(r, \beta)\, ds \qquad (3.66)$$

Recognizing that $s = R \tan \beta$ and that $r = R \sec \beta$ for the infinite straight

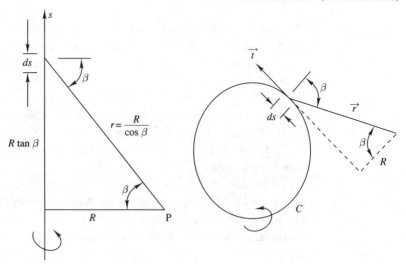

FIGURE 3.13
Velocity induced by a vortex filament.

line filament, (3.66) becomes

$$\frac{\Gamma}{2\pi R} = \oint_{-\pi/2}^{\pi/2} f(r,\beta) R \sec^2 \beta \, d\beta \qquad (3.67)$$

A function that satisfies (3.67) is

$$f(r,\beta) = \frac{\Gamma}{4\pi} \frac{\cos \beta}{r^2} \qquad (3.68)$$

For positive Γ the velocity at point P is at right angles to both the plane of the vortex line and to r; therefore, (3.66) is generalized to

$$\vec{u} = \frac{\Gamma}{4\pi} \oint_C \frac{\vec{t} \times \vec{r}}{r^3} ds \qquad (3.69)$$

where curve C represents the vortex filament. Equation (3.69) is the *Biot-Savart* law. It is important not only in fluid mechanics but also in electrical engineering where it is used to calculate the magnetic field in the neighborhood of a current carrying wire. In electrical engineering it is called *Ampere's Law*.

3.6.4 Vortex Sheets

The concepts of vorticity and vortex filaments are used in two ways in fluid mechanics. The first is to use vorticity to replace some physical object, just as the source is used to create, for example, a half-body. In aerodynamics vorticity is used to replace airfoils and to calculate the lift on wings.

Vorticity used in this way is called *bound vorticity* since the filaments are fixed in space. In a general fluid motion there often exists *free vorticity*, which moves freely with the fluid and is itself governed by the laws of fluid motion. That is, once vorticity is generated in a fluid, it persists and is carried along with the fluid. The study of fluid turbulence makes wide use of free vorticity. Whereas bound vorticity is an artificial concept used to satisfy the boundary condition of a problem, free vorticity occurs naturally in fluid motion. Examples abound in the atmosphere (tornadoes, hurricanes, water spouts, dust devils, willi-willies).

Consider the flow near the downstream edge of a thin body as shown in Fig. 3.14. The streams on the top and bottom of the object have unequal velocities, which results in a discontinuity of velocity at the trailing edge. Obviously, the rotation of a fluid particle is not zero in the plane of the discontinuity. The area of rotation may be confined to an infinitesimally thin plane or it may be spread over a zone of thickness a. The phenomenon can be represented by a *vortex sheet* as shown in Fig. 3.15. The vortex sheet is conceptually an infinite number of vortex filaments, each having infinitesimal strength, placed side by side to form a surface. The strength per unit length (normal to the filaments) of the vortex sheet is γ. The velocity induced in the fluid is

$$\vec{u} = \frac{1}{4\pi} \int_A \frac{\gamma}{r^3} \left(\vec{t} \times \vec{r} \right) dA \qquad (3.70)$$

in which $\vec{t}$ is still defined as the unit tangent vector to the filaments (as shown in Fig. 3.15) and the integration is over the entire vortex sheet. If the filaments are doubly infinite in the z-direction (two-dimensional

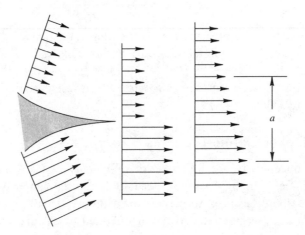

FIGURE 3.14
Flow at different velocities joining at the trailing edge of a thin body.

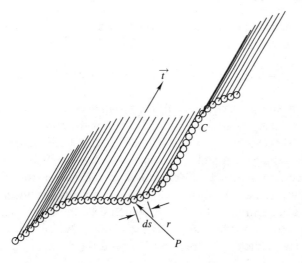

FIGURE 3.15
The edge of a vortex sheet

flow),

$$\vec{u} = \frac{1}{2\pi} \oint_C \frac{\gamma}{r^2} (\vec{e}_z \times \vec{r}) \, ds \tag{3.71}$$

where C is the curve formed by the edge of the vortex sheet as shown in Fig. 3.15.

In aerodynamics the vortex sheet can be used to represent two-dimensional wings. For thin airfoils the mean camber line is replaced by a vortex sheet. The strength of the distribution of vorticity is determined such that the boundary conditions are satisfied. The free-vortex sheet behind the wing moves with the fluid and becomes distorted.

3.7 THE DOUBLET

Consider a two-dimensional superposition of parallel flow, source flow, and sink flow as shown in Fig. 3.16. The source and sink are made equal so

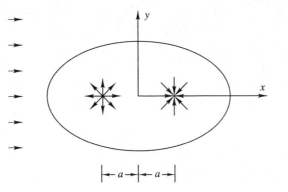

FIGURE 3.16
The Rankine body.

that the fluid emitted by the source is absorbed by the sink. The result is a *Rankine body* that has the potential

$$\Phi = -Ux - \frac{Q}{4\pi}\{\ln[(x+a)^2 + y^2] - \ln[(x-a)^2 + y^2]\} \qquad (3.72)$$

The resulting curve is much like the half-body except that it is closed. The curve can be plotted by methods similar to those used with the half-body.

The length and width of the Rankine body depend on the magnitude of the source and sink, their spacing, and the free stream velocity. If the source and sink are moved closer to the origin, the body becomes more nearly circular, but does not become a round cylinder until the distance between them goes to zero, in which case the source and sink cancel each other. To avoid the mutual cancellation, the magnitude of the source and sink are made to grow infinitely large as the distance between them becomes infinitesimal. A quantity M is defined such that

$$M = 2Qa = \text{constant} \qquad (3.73)$$

The potential for the Rankine body is written

$$\Phi = -Ux - \frac{Q}{4\pi}\ln\frac{(x+a)^2 + y^2}{(x-a)^2 + y^2} \qquad (3.74)$$

As a becomes small, we neglect the terms of order a^2. Factoring out $x^2 + y^2$ from the numerator and denominator yields

$$\Phi = -Ux - \frac{Q}{4\pi}\ln\frac{1 + \dfrac{2ax}{x^2 + y^2}}{1 - \dfrac{2ax}{x^2 + y^2}} \qquad (3.75)$$

The argument of the logarithm can be expanded in the form

$$\frac{1+\varepsilon}{1-\varepsilon} = 1 + 2\varepsilon + \cdots \qquad (3.76)$$

and

$$\ln(1 + 2\varepsilon) = 2\varepsilon - \frac{(2\varepsilon)^2}{2} + \cdots \qquad (3.77)$$

where higher order terms in ε have been neglected. Applying (3.76) and (3.77) to (3.75) with $\varepsilon = 2ax^2/(x^2 + y^2)$ gives

$$\Phi = -Ux - \frac{Q}{4\pi}\left(\frac{4ax}{x^2 + y^2} + \cdots\right) \approx -Ux - \frac{M}{2\pi}\frac{x}{R^2} \qquad (3.78)$$

The potential of (3.78) represents the superposition of parallel flow and a double source or *doublet*. The stream function for the flow is

$$\Psi = -Uy\left(1 - \frac{M}{2\pi U}\frac{1}{R^2}\right) \tag{3.79}$$

For the streamline with $\Psi = 0$, we can rearrange (3.79) to yield

$$R^2 = \frac{M}{2\pi U} = R_0^2 \tag{3.80}$$

where R_0 is the radius of the cylinder. The flow pattern is shown in Fig. 3.17. The general potential and stream function for flow around a circular cylinder of radius R_0 with its center in the origin are

$$\Phi = -Ux\left(1 + \frac{R_0^2}{R^2}\right) \qquad \Psi = -Uy\left(1 - \frac{R_0^2}{R^2}\right) \tag{3.81}$$

The potential flow around the cylinder is not usually realistic since separation occurs for high—or even moderate—Reynolds numbers. This case is a prime example of the damage done by neglecting the viscous terms of the Navier-Stokes equations. An examination of the pressure distribution on the cylinder would show that the pressure is a maximum at the upstream stagnation point. It then decreases in the downstream direction on the upstream part of the cylinder. On the downstream portion of the cylinder the pressure should again increase to the stagnation pressure, which would occur at the rear stagnation point. However, the friction of the solid on the fluid decreases the momentum in the *boundary layer* (Chap. 6), the layer of fluid very close to the solid. The fluid does not have sufficient momentum to

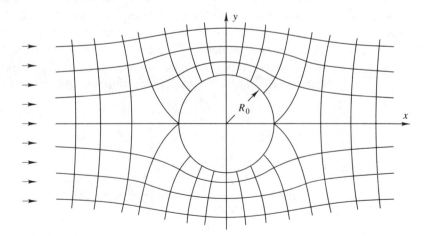

FIGURE 3.17
Flow around a circular cylinder.

overcome the adverse pressure gradient on the rear part of the cylinder and so it "separates" from the solid. The result is a *wake* behind the cylinder. The ideal flow pattern shown in the figure is doubly symmetric, whereas in separated flow it is not. The flow in the wake is unsteady even if the approaching flow is steady. In ideal flow there is no drag on the cylinder whereas in the actual flow there can be considerable drag.

3.8   THE METHOD OF IMAGES

One of the computational advantages—but a disadvantage in the sense of realism—of potential flow is that the no-slip condition is not applied to solid surfaces; the only boundary condition is that the normal derivative of the potential disappears on the surface. A wall can be represented by creating a symmetrical flow pattern about the position of the wall. For example, the flow around a cylinder close to a wall can be solved by placing an image cylinder an equal distance on the other side of the wall as shown in Fig. 3.18. By symmetry the streamline midway between the cylinders is straight and can represent the straight wall. In the case where the cylinders are represented by doublets, this solution is not perfect since the image cylinder distorts the closed streamline that forms the real cylinder, making it not perfectly round. It is, however, a good approximation if the distance from the cylinder to the wall is somewhat greater than the radius of the cylinder. If the cylinder is represented as a solid instead of by a doublet, the method of images does not have this problem.

A cylinder in a channel (between two walls) can be treated in a similar fashion with two image cylinders. Actually, the image cylinder on one side distorts the wall on the other side. That distortion can be fixed by creating another image cylinder. For a wall to remain perfectly straight, the wall must be a line of symmetry. The idea in the present case is to obtain symmetry about both walls. That can be done (Fig. 3.19) by putting in an infinite number of images so that, using doublets for the cylinders, the potential

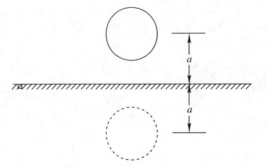

FIGURE 3.18
Flow around a cylinder near a wall.

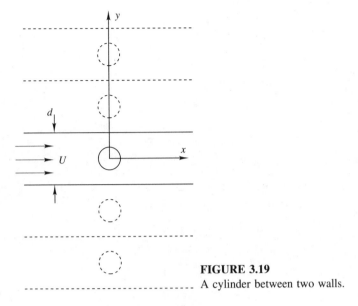

FIGURE 3.19
A cylinder between two walls.

becomes

$$\Phi = -Ux\left[1 + \sum_{i=-\infty}^{\infty} \frac{R_0^2}{x^2 + (y - id)^2}\right] \tag{3.82}$$

In using equations such as (3.82) for calculation, the limits on the summation are taken sufficiently large so that the wall distortion does not unduly influence the result.

The same methods can be used for the solution of many symmetrical and antisymmetrical flow patterns. The applications are numerous. The method of images permits the calculation of the *wall effect* in wind tunnel and water tunnel experiments. Since nearly all experimental apparatus have walls that may not be present in the prototype, the calculation of this effect is important.

3.9 AXISYMMETRIC FLOWS

In the case of axisymmetric flow we define (Fig. 3.20)

$$R = \sqrt{y^2 + z^2} \tag{3.83}$$

The potential is defined as it is in three-dimensional flow and satisfies Laplace's equation written in cylindrical coordinates (without the θ term)

$$\frac{\partial}{\partial R}\left(R\frac{\partial \Phi}{\partial R}\right) + R\frac{\partial^2 \Phi}{\partial x^2} = 0 \tag{3.84}$$

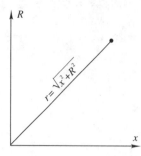

FIGURE 3.20
Cylindrical coordinates.

The stream function must satisfy the irrotationality condition written in cylindrical coordinates

$$\frac{\partial^2 \Psi}{\partial x^2} + \frac{\partial^2 \Psi}{\partial R^2} - \frac{1}{R}\frac{\partial \Psi}{\partial R} = 0 \tag{3.85}$$

A definition that satisfies (3.85) is

$$u_R = \frac{1}{R}\frac{\partial \Psi}{\partial R} \qquad \text{and} \qquad u_x = -\frac{1}{R}\frac{\partial \Psi}{\partial R} \tag{3.86}$$

The stream function definition does not lead to Laplace's equation through the irrotationality condition as is the case with two-dimensional flow. The equations remain linear and the principle of superposition is still valid.

3.9.1   The Fundamental Singularities

The potential and stream function for parallel flow are

$$\Phi = -Ux + \text{constant} \qquad \Psi = -U\frac{R^2}{2} + \text{constant} \tag{3.87}$$

Source flow with the singularity in the origin is described by

$$\Phi = -\frac{Q}{4\pi}\frac{1}{\sqrt{x^2 + R^2}} + \text{constant} \qquad \Psi = -\frac{Q}{4\pi}\frac{x}{\sqrt{x^2 + R^2}} + \text{constant} \tag{3.88}$$

No axisymmetric flow comparable to the two-dimensional vortex exists. The doublet can be derived in the same manner as for the two-dimensional flow; that is, take the source at $x = -\varepsilon$ and the sink at $x = \varepsilon$, and let ε approach zero while the quantity $Q\varepsilon$ is kept constant. The result is

$$\Phi = -\frac{M}{4\pi}\frac{x}{(x^2 + R^2)^{3/2}} + \text{constant} \qquad \Psi = \frac{M}{4\pi}\frac{R^2}{(x^2 + R^2)^{3/2}} + \text{constant} \tag{3.89}$$

The constant, M, is related to the radius of a sphere in a parallel flow field as

$$r_s = \left(\frac{M}{2\pi U}\right)^{1/3} \tag{3.90}$$

3.9.2 The Line Source

The singularities can be superposed to form objects such as the axisymmetric Rankine body. For a more general method of defining shapes consider the line source of intensity q per unit of length as shown in Fig. 3.21. From (3.88) the stream function is

$$\Psi = \frac{1}{4\pi} \int_{x_1}^{x_2} \frac{x - \xi}{\sqrt{(x - \xi)^2 + R^2}} q \, d\xi + \text{constant} \tag{3.91}$$

The constant is conveniently chosen in order to make the stream function zero on the negative x-axis so that

$$\Psi = \frac{1}{4\pi} \int_{x_1}^{x_2} \left[1 + \frac{x - \xi}{\sqrt{(x - \xi)^2 + R^2}} \right] q \, d\xi \tag{3.92}$$

The surface of the body and the negative x-axis form the zero streamline.

Adding the parallel flow, the equation with the stream function set equal to zero serves as an integral equation for the source distribution that is used to represent an arbitrarily shaped body; that is, q is found such that

$$\frac{1}{4\pi} \int_{x_1}^{x_2} \left[1 + \frac{x - \xi}{\sqrt{(x - \xi)^2 + R^2}} \right] q \, d\xi = U \frac{R^2}{2} \tag{3.93}$$

when R is specified as a function of x for the surface of the body. The points x_1 and x_2 are, in general, the leading and trailing edges of the body.

Usually, it is not possible to solve (3.93) analytically and numerical methods must be used. A common method is to set q equal to a constant

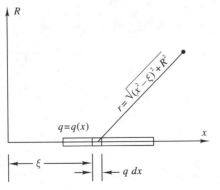

FIGURE 3.21
A line source.

in the range $x_i \le x < x_{i+1}$. In this range the left side of (3.93) becomes

$$\frac{q}{4\pi} \int_{x_1}^{x_2} \left[1 + \frac{x - \xi}{\sqrt{(x - \xi)^2 + R^2}} \right] d\xi = \frac{q}{4\pi} \left[(x_2 - x_1) - (r_2 - r_1) \right] \quad (3.94)$$

in which $r_i = \sqrt{(x - x_i)^2 + R^2}$. If N short lengths of line source are put together to form the body, the equation for the stream function becomes

$$\Psi = -U \frac{R^2}{2} + \sum_{i=1}^{N} \frac{q_i}{4\pi} \left[(x_{i+1} - x_i) - (r_{i+1} - r_i) \right] \quad (3.95)$$

To satisfy (3.95) we can choose at least $M \ge 2N$ points on the body at which the stream function is to be set equal to zero (see Fig. 3.22) and solve for both the q_i and the x_i. The difficulty is that equations (3.95) are nonlinear using this method. Alternately, we can choose, somewhat arbitrarily, the x_i and solve a linear set of equations for the q_i. Then (3.95) is to be satisfied at the M points (in the least squares sense if $M > N$) giving M equations in the q_i. If the x_i are chosen, x_1 and x_{N+1} are set a short distance from the leading and trailing edges respectively.

Actually, (3.95) simplifies somewhat for a closed body. In that case

$$\sum_{i=1}^{N} q_i (x_{i+1} - x_i) = 0 \quad (3.96)$$

because the sum of the positive and negative sources is zero since no fluid

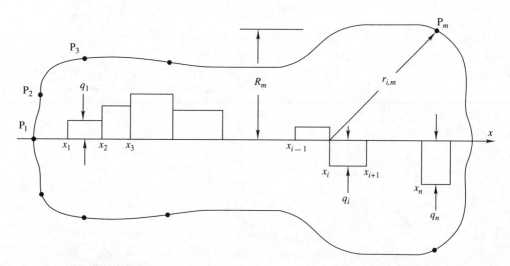

FIGURE 3.22
An axisymmetric body formed by line sources.

passes through the boundary of the body. Thus, referring to Fig. 3.22

$$U\frac{R_m^2}{2} + \frac{1}{4\pi} \sum_{i=1}^{N} q_i(r_{i+1,m} - r_{i,m}) = 0 \qquad (3.97)$$

where $r_{i,m}$ represents the distance from the point x_i to the point P_m. If $M = N$ the system of equations is determinate. In choosing the points P_m, either the leading or trailing edge, but not both, can be used, if both are used, one of the equations may be redundant.

The fact that (3.97) is satisfied does not guarantee an acceptable solution to a problem. Some shapes cannot be adequately represented by sources on the axis of symmetry, regardless of how many sources are used. The final solution should be checked to ensure that the body shape does not vary widely from what is intended between the points P_i. Also, the function $q_i(x)$ should behave in a reasonable manner. Some adjustments can be made by choosing the values of x_i or solving for the x_i. As in any set of nonlinear equations, however, the solution may be unstable. An unstable solution is likely in the case that the object cannot be represented by sources. A sign of instability may be that the boundary streamline passes through the designated points but oscillates unreasonably between the points.

3.9.3 Line Doublets

Singularities other than sources can be used to represent body shapes. If doublets are used, (3.93) becomes

$$\frac{1}{4\pi} \int_{x_1}^{x_2} \frac{R^2}{[(x-\xi)^2 + R^2]^{3/2}} q \, d\xi = U\frac{R^2}{2} \qquad (3.98)$$

where q is now the intensity per unit length of the doublet. In a similar way sources or doublets may be distributed along a line perpendicular to the x-axis to obtain a solution for flow around objects that have their largest dimension normal to the flow.

3.10 SURFACE SOURCES

The use of singularities on the axis of symmetry is a relatively simple way to calculate some flows, but the method frequently fails to produce the desired result. More robust numerical techniques that can be used more generally are available. In particular, sources can be distributed anywhere in the flow field. The most effective positions to place sources are on the surface of the body. In two dimensions a line source is simply placed on the surface of the body so that the potential in any point p is

$$\Phi_{\mathrm{p}} = -\frac{1}{2\pi} \int_{\partial A} q \ln R_{\mathrm{pQ}} \, ds \qquad (3.99)$$

in which Q is an integration point on the surface of the object and R_{pQ} is the distance between points p and Q. The source strength must be calculated so that the boundary conditions are satisfied.

Equation (3.99) can be differentiated to yield

$$\frac{\partial \Phi}{\partial n_{\mathrm{P}}} = -\frac{1}{2\pi} \int_{\partial A} \frac{q}{R_{\mathrm{PQ}}} \frac{\partial R_{\mathrm{PQ}}}{\partial n_{\mathrm{P}}} ds + \alpha q(\mathrm{P}) \qquad (3.100)$$

in which α is the boundary angle at point P ($\alpha = \pi$ on a smooth boundary) and point p has been renamed P when it is on the boundary. The subscript on n is to emphasize that the derivative is taken with respect to the outer normal at P. Writing (3.99) for a boundary point when the boundary condition is Dirichlet (Φ is known) and writing (3.100) at Neumann boundaries ($\partial \Phi / \partial n$ known) provides integral equations for the unknown source strength q. The integral of (3.100) is not easy to evaluate as it contains a nonintegrable singularity at P, where the integration point and the evaluation point come together. It can, however, be evaluated in the sense of the Cauchy principal value.

Equations similar to (3.99) and (3.100) can be derived using vorticity instead of sources. These techniques have been widely used in both two and three dimensions to find complex flow fields, but are best known for computing flow around airplanes and other objects and for groundwater flow. The current manifestation is the so-called boundary integral equation method or the boundary element method (Liggett and Liu, 1983; Brebbia and Dominguez, 1992). Such methods have been known since the 1930s, but widespread implementation had to await the invention of the modern digital computer.

3.11 UNSTEADY FLOW

In this section the case of a moving object in a resting fluid is considered. The potential is determined in the usual way except that its derivatives must go to zero at infinity in order that there be no fluid motion at infinity. The boundary condition on the object is

$$\vec{\nabla}\Phi \cdot \vec{n} = -\vec{V} \cdot \vec{n} \qquad (3.101)$$

in which $\vec{V}$ is the velocity of the object or, more exactly, the velocity of the point on the object where the boundary condition is to be applied. A potential function that satisfies these conditions can usually be deduced from the steady flow potential about the same object.

3.11.1 The Moving Sphere

Consider a sphere moving in a stationary fluid with velocity $\vec{V}$ in the positive x-direction as shown in Fig. 3.23. The potential for steady flow around a sphere is

$$\Phi = -Ux - \frac{Ur_s^3}{2} \frac{x}{(x^2 + R^2)^{3/2}} + \text{constant} \tag{3.102}$$

in which r_s is the radius of the sphere. The unsteady potential should not contain a term for parallel flow since the fluid far from the sphere is stationary. The doublet must be oriented in the direction opposite to the oncoming fluid from the point of view of the sphere so its sign must be reversed from its previous use. If the center of the sphere is at the point $x = 0$ at time $t = 0$, then it will be at the point

$$\xi = \int_0^t V \, d\tau \tag{3.103}$$

in subsequent times. Using these clues, the unsteady potential is found to be

$$\Phi = \frac{Vr_s^3}{2} \frac{x - \xi}{[(x - \xi)^2 + R^2]^{3/2}} + \text{constant} \tag{3.104}$$

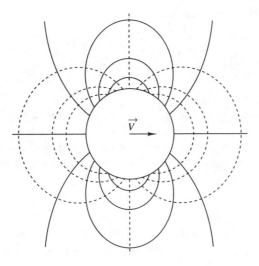

FIGURE 3.23
Streamlines (solid) and equipotential lines (dashed) for a moving sphere. Compare to Fig. 3.17.

3.11.2   Virtual Mass

The force required to accelerate an object through a fluid is greater than that required to accelerate the same object in a vacuum since some of the surrounding fluid must be accelerated at the same time. From Newton's law of motion the force required to accelerate an object in the ξ-direction at the rate of $d^2\xi/dt^2$ is

$$F_i = m_0 \frac{d^2\xi_i}{dt^2} + \rho \int_V \frac{Du_i}{Dt} dV \qquad (3.105)$$

where m_0 is the mass of the object. The last term of (3.105) represents the mass times the acceleration of all the fluid particles summed over the fluid-filled space.

Equation (3.105) can be simplified by changing the integral to a surface integral through the use of the divergence theorem (1.13). Using the definition of the substantial derivative and the equation of continuity for an incompressible fluid

$$\frac{Du_i}{Dt} = \frac{\partial u_i}{\partial t} + u_j \frac{\partial u_i}{\partial x_j} = \frac{\partial u_i}{\partial t} + \frac{\partial}{\partial x_j}(u_i u_j) \qquad (3.106)$$

The time derivative can be written as

$$\frac{\partial u_i}{\partial t} = -\frac{\partial}{\partial t}\frac{\partial \Phi}{\partial x_i} = -\frac{\partial}{\partial x_i}\frac{\partial \Phi}{\partial t} = -\vec{\nabla} \cdot \left(\vec{e}_i \frac{\partial \Phi}{\partial t} \right) \qquad (3.107)$$

The last term of (3.106) is

$$\frac{\partial}{\partial x_j}(u_i u_j) = \vec{\nabla} \cdot (u_i \vec{u}) \qquad (3.108)$$

Using the divergence theorem

$$\int_V \frac{Du_i}{Dt} dV = \int_{\partial V} \left[-\frac{\partial \Phi}{\partial t}(\vec{e}_i \cdot \vec{n}) + u_i(\vec{u} \cdot \vec{n}) \right] dA \qquad (3.109)$$

Using (3.109) in (3.105)

$$F_i = m_0 \frac{d^2\xi_i}{dt^2} + \rho \int_{\partial V} \left[-\frac{\partial \Phi}{\partial t}(\vec{e}_i \cdot \vec{n}) + u_i(\vec{u} \cdot \vec{n}) \right] dA \qquad (3.110)$$

The integration is carried out over the entire boundary of the fluid-filled space, which includes the boundary of the object as well as the boundaries that may confine the fluid. The unit normal vector $\vec{n}$ is positive outward from the region occupied by the fluid.

Consider again the case of a sphere moving along the x-axis in an infinite fluid. The boundary integration is carried out over the surface of the sphere and the infinite boundary, but it will be zero (due to zero velocity) over the infinite boundary. Note, in particular, that the potential goes to zero as r^{-3}. The time derivative of Φ in (3.104) is

$$
\frac{\partial \Phi}{\partial t} = -\frac{V^2 r_s^3}{2} \left\{ \frac{1}{[(x - \xi)^2 + R^2]^{3/2}} - \frac{3(x - \xi)^2}{[(x - \xi)^2 + R^2]^{5/2}} \right\}
$$
$$
+ \frac{\dot{V} r_s^3}{2} \frac{x - \xi}{[(x - \xi)^2 + R^2]^{3/2}}
\tag{3.111}
$$

in which $\dot{V} = dV/dt$. The space derivative of Φ is

$$
\frac{\partial \Phi}{\partial x} = \frac{V r_s^3}{2} \left\{ \frac{1}{[(x - \xi)^2 + R^2]^{3/2}} - \frac{3(x - \xi)^2}{[(x - \xi)^2 + R^2]^{5/2}} \right\}
\tag{3.112}
$$

On the sphere $x - \xi = r_s \cos \theta$, $\vec{e}_x \cdot n = -\cos \theta$, and $\vec{u} \cdot \vec{n} = -V \cos \theta$ [using (3.101), see Fig. 3.24]. Since the unit normal vector is out of the fluid and into the sphere

$$
\frac{\partial \Phi}{\partial t}(\vec{e}_x \cdot \vec{n}) = \frac{1}{2}[V^2(1 - 3\cos^2 \theta) - \dot{V} r_s \cos \theta] \cos \theta
\tag{3.113}
$$

and

$$
u_x(\vec{u} \cdot \vec{n}) = \frac{V^2}{2} \cos \theta (1 - 3\cos^2 \theta)
\tag{3.114}
$$

On the sphere the element of area is $2\pi r_s \sin \theta (r_s \Delta \theta)$. The integral over the area of the sphere is

$$
I = 2\pi r_s^2 \int_0^\pi \frac{\dot{V} r_s}{2} \cos^2 \theta \sin \theta d\theta = \dot{V} \left[\frac{1}{2} \left(\frac{4}{3} \pi r_s^3 \right) \right]
\tag{3.115}
$$

The quantity in brackets is one-half of the volume of the displaced fluid.

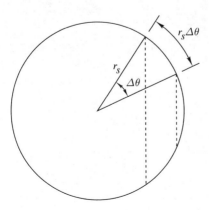

FIGURE 3.24
Boundary integration on a sphere.

Equation (3.110) can be written in the case of the sphere as

$$F_i = \left(m_0 + \frac{1}{2}m_f\right)\frac{d^2\xi_i}{dt^2} \tag{3.116}$$

in which m_f is the mass of the displaced fluid.

The same principle holds for the motion of any object through a fluid. In general, the equation of motion for the object is

$$F_i = (m_0 + Km_f)\frac{d^2\xi_i}{dt^2} \tag{3.117}$$

where K is dependent on the shape of the object. The factor Km_f is called the *virtual mass* and $m_0 + Km_f$ is the *apparent mass*. For ellipsoids with major axis $2a$ and minor axis $2b$, the value of K is given by Table 3.1, where K_1 is the virtual mass factor when the ellipsoid is moving parallel to the major axis and K_2 is the virtual mass factor when the ellipsoid is moving parallel to the minor axis.

3.11.3 Kinetic Energy

The apparent kinetic energy of a moving mass is the sum of the kinetic energy of the mass and the kinetic energy of the moving fluid

$$KE = \frac{1}{2}m_0V^2 + \frac{\rho}{2}\int_V u^2 dV = \frac{1}{2}V^2(m_0 + Km_f) \tag{3.118}$$

In the following we calculate the kinetic energy of a general object made up of doublets that are aligned on the x-axis and moving in the x-direction. The square of the velocity is

$$u^2 = \vec{\nabla}\Phi \cdot \vec{\nabla}\Phi = \vec{\nabla} \cdot (\Phi\vec{\nabla}\Phi) - \Phi\nabla^2\Phi = \vec{\nabla} \cdot (\Phi\vec{\nabla}\Phi) \tag{3.119}$$

Using the divergence theorem

$$\int_V \vec{\nabla} \cdot (\Phi\vec{\nabla}\Phi)dV = \int_{\partial V} \Phi\vec{\nabla}\Phi \cdot \vec{n}\, dA \tag{3.120}$$

TABLE 3.1
Apparent mass factors for ellipsoids

a/b	1	2	3	4	5	8	∞
K_1	.5	.210	.122	.082	.059	.029	0
K_2	.5	.702	.803	.860	.895	.945	1

Again, considering an object immersed in fluid of infinite extent, the area integral has to be evaluated only over the object; the integrand is zero at infinity since both Φ and $\vec{\nabla}\Phi$ go strongly to zero at infinity. We take V as constant and consider the potential

$$\Phi = \frac{1}{4\pi} \sum_{i=1}^{N} \frac{M_i(x - \xi_i - Vt)}{[(x - \xi_i - Vt)^2 + R^2]^{3/2}} \tag{3.121}$$

Since the velocity of the object V is constant, the integral (3.118) has the same value for all time and can most easily be evaluated at $t = 0$.

The unsteady potential is divided into a steady potential Φ_s and a parallel flow potential Φ_p. At $t = 0$

$$\Phi_s = Vx + \frac{1}{4\pi} \sum_{i=1}^{N} \frac{M_i(x - \xi_i)}{[(x - \xi_i)^2 + R^2]^{3/2}} \qquad \Phi_p = Vx \tag{3.122}$$

Equation (3.120) becomes

$$\int_{\partial V} \Phi \vec{\nabla}\Phi \cdot \vec{n}\, dA = \int_{\partial V} (\Phi_s - \Phi_p)\vec{\nabla}(\Phi_s - \Phi_p) \cdot \vec{n}\, dA$$

$$= \int_{\partial V} (\Phi_s - \Phi_p)\vec{\nabla}\Phi_p \cdot \vec{n}\, dA \tag{3.123}$$

since $\vec{\nabla}\Phi_s \cdot \vec{n} = 0$ from the boundary condition on the object. Using $\vec{\nabla}\Phi_p \cdot \vec{n} = V\vec{e}_x \cdot \vec{n}$

$$\int_{\partial V} \Phi_p \vec{\nabla}\Phi_p \cdot \vec{n}\, dA = V^2 \int_{\partial V} x(\vec{e}_x \cdot \vec{n})\, dA = -V^2 \forall \tag{3.124}$$

in which $\forall$ is the volume of the object.

To evaluate the other part of (3.123) consider a very large sphere of radius r_∞ surrounding the object. Then

$$-\int_{A} \Phi_s \vec{\nabla}\Phi_p \cdot \vec{n}\, dA = -\int_{A+S} \Phi_s \vec{\nabla}\Phi_p \cdot \vec{n}\, dA + \int_{S} \Phi_s \vec{\nabla}\Phi_p \cdot \vec{n}\, dA \tag{3.125}$$

where A is the area of the object and S is the area of the sphere surrounding the object. From the divergence theorem

$$\int_{A+S} \Phi_s \vec{\nabla}\Phi_p \cdot \vec{n}\, dA = \int_V \vec{\nabla} \cdot \Phi_s \vec{\nabla}\Phi_p \, dV$$
$$= \int_V (\vec{\nabla}\Phi_s \cdot \vec{\nabla}\Phi_p) dV$$
$$= \int_V \vec{\nabla} \cdot \Phi_p \vec{\nabla}\Phi_s \, dV \qquad (3.126)$$
$$= \int_{A+S} \Phi_p \vec{\nabla}\Phi_s \cdot \vec{n}\, dA$$

Therefore,

$$-\int_A \Phi_s \vec{\nabla}\Phi_p \cdot \vec{n}\, dA = -\int_A \Phi_p \vec{\nabla}\Phi_s \cdot \vec{n}\, dA + \int_S (\Phi_s \vec{\nabla}\Phi - \Phi_p \vec{\nabla}\Phi_s) \cdot \vec{n}\, dA$$
$$(3.127)$$

The first integral on the right of (3.127) is zero from the boundary condition. On the large sphere

$$\vec{\nabla}\Phi_p \cdot \vec{n} = \frac{\partial \Phi_p}{\partial r_\infty} = V\cos\theta \qquad (3.128)$$

and

$$\vec{\nabla}\Phi_s \cdot \vec{n} = \frac{\partial \Phi_s}{\partial r_\infty} = V\cos\theta + \frac{1}{4\pi}\sum_{i=1}^N M_i \left(\frac{\cos\theta}{r_\infty^3} - \frac{3r_\infty \cos\theta}{r_\infty^4} \right)$$
$$(3.129)$$
$$= V\cos\theta - \frac{\cos\theta}{2\pi r_\infty^3}\sum_{i=1}^N M_i$$

In obtaining (3.129) we used the relation for large r_∞

$$\frac{x-\xi}{r_\infty} = \frac{x}{r_\infty} - \cos\theta \qquad (3.130)$$

Combining (3.128) and (3.129) gives

$$(\Phi_s \vec{\nabla}\Phi_p - \Phi_p \vec{\nabla}\Phi_s) \cdot \vec{n} = \frac{3}{4}V\frac{\cos^2\theta}{\pi r_\infty^2}\sum_{i=1}^N M_i \qquad (3.131)$$

On the sphere an element of area is $dA = 2\pi r_\infty^2 \sin\theta d\theta$. Using the latter relationship with (3.129), (3.127) becomes

$$-\int_{\partial V} \Phi_s \vec{\nabla}\Phi_p \cdot \vec{n}\, dA = \frac{3}{2}V\sum_{i=1}^N M_i \int_0^\pi \cos^2\theta \sin\theta\, d\theta = V\sum_{i=1}^N M_i$$
$$(3.132)$$

Combining (3.124) and (3.132) gives the kinetic energy of an object made up of doublets moving through a stationary fluid as

$$KE = \frac{1}{2}V^2 \left[m_0 + \rho \left(\frac{1}{V} \sum_{i=1}^{N} M_i \right) - \rho \forall \right] \tag{3.133}$$

The apparent mass as computed by (3.133) is the same as that computed for a sphere if a single doublet of strength $M = 2\pi V r_s^3$ is used.

3.12   COMPLEX NOTATION

The use of complex variables is an elegant way to solve many potential flow problems in two dimensions. It is powerful because it allows the mapping of a flow region into a different region where the flow is easier to solve.

There are, however, several disadvantages to the complex notation. The most restrictive is that the solutions are limited to two dimensions with no possibility for extension to three dimensions. For analytical solutions curved boundaries are difficult to map (with some exceptions for specific curves), leading again to the necessity of numerical solutions.

3.12.1   Basics

The usual complex notation is that

$$z = x + iy \tag{3.134}$$

in which x and y are the coordinate directions and $i = \sqrt{-1}$. Consider first a function of a complex variable, $f(z)$, and the derivatives of that function. By definition

$$\frac{df}{dz} = \lim_{\Delta z \to 0} \frac{f(z + \Delta z) - f(z)}{\Delta z} \tag{3.135}$$

However, Δz can tend toward zero in an infinite number of ways. First, we could take Δx toward zero and then Δy, or we could first let Δy go to zero then Δx, or we could specify that Δx and Δy go toward zero simultaneously along any chosen line. All of these processes should give the same derivative if the function $f(z)$ is *analytic* (or *holomorphic* or *regular*). Thus,

$$\frac{df}{dz} = \lim_{\Delta x \to 0} \frac{f(z + \Delta x) - f(z)}{\Delta x} = \lim_{\Delta y \to 0} \frac{f(z + i\Delta y) - f(z)}{i\Delta y} \tag{3.136}$$

But these functions are the definitions of the partial derivatives

$$\frac{\partial f}{\partial x} = \lim_{\Delta x \to 0} \frac{f(z + \Delta x) - f(z)}{\Delta x} \qquad \frac{1}{i}\frac{\partial f}{\partial y} = \lim_{\Delta y \to 0} \frac{f(z + i\Delta y) - f(z)}{i\Delta y} \tag{3.137}$$

and from (3.136)

$$\frac{\partial f}{\partial x} = \frac{1}{i}\frac{\partial f}{\partial y} \tag{3.138}$$

Take the function of the complex variable as

$$f(z) = W = \Phi + i\Psi \tag{3.139}$$

Then the partial derivatives are

$$\frac{\partial W}{\partial x} = \frac{\partial \Phi}{\partial x} + i\frac{\partial \Psi}{\partial x} \qquad \frac{\partial W}{\partial y} = \frac{\partial \Phi}{\partial y} + i\frac{\partial \Psi}{\partial y} \tag{3.140}$$

Using (3.138) to relate the two parts of (3.140)

$$\frac{\partial \Phi}{\partial x} + i\frac{\partial \Psi}{\partial x} = -i\frac{\partial \Phi}{\partial y} + \frac{\partial \Psi}{\partial y} \tag{3.141}$$

Equating real and imaginary parts

$$\frac{\partial \Phi}{\partial x} = \frac{\partial \Psi}{\partial y} \qquad \frac{\partial \Phi}{\partial y} = -\frac{\partial \Psi}{\partial x} \tag{3.142}$$

which are the *Cauchy-Riemann equations*. Differentiating the first with respect to x, the second with respect to y, and equating the mixed derivatives leads to Laplace's equation

$$\frac{\partial^2 \Phi}{\partial x^2} + \frac{\partial^2 \Phi}{\partial y^2} = 0 \tag{3.143}$$

Similarly, if the first of the Cauchy-Riemann equations is differentiated with respect to y and the second is differentiated with respect to x

$$\frac{\partial^2 \Psi}{\partial x^2} + \frac{\partial^2 \Psi}{\partial y^2} = 0 \tag{3.144}$$

Obviously, Φ can be identified with the velocity potential and Ψ with the stream function; W becomes the *complex potential*. The velocities in the x- and y-directions are

$$\frac{dW}{dz} = \frac{\partial W}{\partial x} = \frac{\partial \Phi}{\partial x} + i\frac{\partial \Psi}{\partial x} = -u_x + iu_y$$

$$\frac{dW}{dz} = \frac{1}{i}\frac{\partial W}{\partial y} = -i\frac{\partial \Phi}{\partial y} + \frac{\partial \Psi}{\partial y} = iu_y - u_x \tag{3.145}$$

Thus, the derivative of the complex potential gives the velocities in complex form. Any function of a complex variable can be used to represent a potential flow pattern. The properties of continuity and irrotationality are automatically satisfied.

Historical Note: Augustin Louis de Cauchy (1789–1857) was a French engineer. He was advised by Lagrange and Laplace to devote himself to mathematics due to failing health (Rouse and Ince, 1957). His subsequent contributions to mathematics were outstanding. In fluid mechanics Cauchy was largely known for wave mechanics, along with Poisson (1781–1840). In complex variables Cauchy was responsible for the Cauchy-Riemann equations and defined the complex potential with the stream function as the imaginary part and the potential function as the real part. We owe a large part of the theory of complex variables to Georg Friedrich Bernhard Riemann (1826–1866). He received his Ph. D. from Göttingen with the thesis "Foundations for a General Theory of Functions of a Complex Variable." Riemann was responsible for developing the theory of Riemann surfaces. He eventually became a professor in mathematics at Göttingen, contributing to a number of fields but especially to geometry and topology.

3.12.2 The Fundamental Singularities

The fundamental singularities are listed in complex notation. Parallel flow is

$$W = -Uz + \text{constant} \tag{3.146}$$

Source flow is

$$W = -\frac{Q}{2\pi} \ln z + \text{constant} \tag{3.147}$$

Vortex flow is

$$W = i\frac{\Gamma}{2\pi} \ln z + \text{constant} \tag{3.148}$$

The doublet is

$$W = -\frac{M}{2\pi z} \tag{3.149}$$

which leads to flow around a cylinder

$$W = -Uz - U\frac{R_0^2}{z} \tag{3.150}$$

3.12.3 Conformal Mapping

The ability to map a flow into a simpler region is the primary reason for the use of complex variables. Any function of a complex variable can be considered a mapping function. If the complex variable ζ is a function of z, $\zeta = \zeta(z)$ where $\zeta = \xi + i\eta$, then every point in the z-plane has a corresponding point in the ζ-plane and vice versa. A complex variable is often more easily represented in cylindrical coordinates, so let

$$z = x + iy = re^{i\theta} \qquad \zeta = \xi + i\eta = Re^{i\phi} \qquad (3.151)$$

Consider a few simple mapping functions as examples. First, take

$$\zeta = z^2 = r^2 e^{2i\theta} \qquad (3.152)$$

The z- and ζ-planes are shown in Fig. 3.25. A point such as z_1 maps into the point ζ_1 in which the distance from the origin is squared and the angle from the x-axis is doubled. Figure 3.26 shows an area in the z-plane that is mapped into the ζ-plane. The mapping function can be generalized

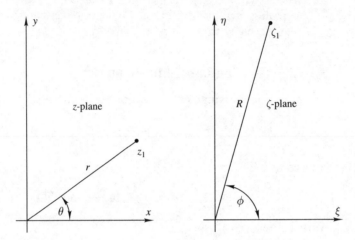

FIGURE 3.25
Mapping of a point in the z-plane into the ζ-plane.

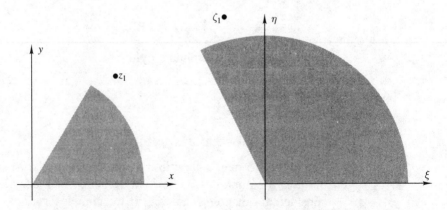

FIGURE 3.26
A sector of a circle mapped into the ζ-plane.

somewhat by taking

$$\zeta = z^n = r^n e^{in\theta} \tag{3.153}$$

in which n is some number. To study the utility of this mapping function, take a parallel flow in the z-plane where the complex potential is given by (3.146). The real axis is a streamline and can be considered a wall. Figure 3.27 shows the real axis in the ζ-plane for the value of n somewhat greater than and somewhat less than unity. This mapping function maps the parallel flow into the flow about a sharp corner. The velocity in the ζ-plane is

$$\frac{dW}{d\zeta} = \frac{\dfrac{dW}{dz}}{\dfrac{d\zeta}{dz}} = -\frac{U}{nz^{n-1}} = -\frac{1}{n}U\zeta^{(1-n)/n} \tag{3.154}$$

which is used to study flow near a sharp corner. For $n > 1$, the velocity at the corner goes to infinity; for $n < 1$, the velocity at the corner goes to zero. As with all transformations the angles are preserved from one plane to the next except at singularities. In the case of the flow around a corner, the origin is such a singularity.

Next consider the mapping function

$$\zeta = z + \frac{R_0^2}{z} \tag{3.155}$$

which looks much like the complex potential for flow around a cylinder. The function is

$$\zeta = re^{i\theta} + \frac{R_0^2}{r}e^{-i\theta} = \left(r + \frac{R_0^2}{r}\right)\cos\theta + i\left(r - \frac{R_0^2}{r}\right)\sin\theta \tag{3.156}$$

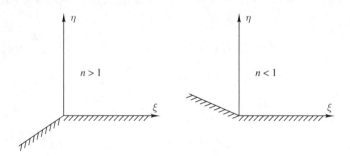

FIGURE 3.27
Mapping of the real axis into the ζ-plane.

When $r = R_0, \xi = 2R_0 \cos \theta$, and $\eta = 0$; therefore, the circle of radius R_0 in the z-plane maps into the real axis between the points $\xi = -2R_0$ and $\xi = 2R_0$. For a somewhat larger circle where $R_1 > R_0$, take ξ and η as implied by the real and imaginary parts of (3.156), square, and add to obtain

$$\frac{\xi^2}{\left(R_1 + \dfrac{R_0^2}{R_1}\right)^2} + \frac{\eta^2}{\left(R_1 - \dfrac{R_0^2}{R_1}\right)^2} = 1 \tag{3.157}$$

which is the equation of an ellipse in the ζ-plane. This mapping function provides a method of finding the flow around an ellipse. We simply use the complex potential for flow around a circle in the z-plane (3.150) and apply the mapping. The velocities can be found in the same manner as in the flow around a corner.

The transformation (3.155) is known as the Joukowski transformation. It is used to find the flow around airfoils because if the circle in the z-plane is taken with its center somewhat off of the origin, the profile in the ζ-plane has an airfoil-like shape. Although it is an elegant method, the difficulty is that we must use the shape that results from the transformation and not necessarily the shape we might want.

3.12.4 Cauchy and Blasius

The use of complex variables also provides a simple method of finding forces on objects. Before deriving the relevant equation, consider a complex function integrated around a closed path in the z-plane

$$I = \oint_{\partial B} f(z)dz = \oint_{\partial B} [f_r(z) + if_i(z)]\,dz \tag{3.158}$$

in which f_r represents the real part of the function and f_i is the imaginary part. Further dividing the complex increment into real and imaginary parts, $dz = dx + i\,dy$, gives

$$I = \oint_{\partial B} (f_r\,dx - f_i\,dy) + i \oint_{\partial B} (f_r\,dy + f_i\,dx) \tag{3.159}$$

Stokes' theorem (3.57) can be written in two dimensions

$$\oint_{\partial B} (u_x\,dx + u_y\,dy) = -\int_B \left(\frac{\partial u_x}{\partial y} - \frac{\partial u_y}{\partial x}\right) dA \tag{3.160}$$

Taking f_r as u_x and f_i as $-u_y$, the real part of (3.159) is

$$\oint_{\partial B} (f_r\,dx - f_i\,dy) = -\int_B \left(\frac{\partial f_r}{\partial y} + \frac{\partial f_i}{\partial x}\right) dA \tag{3.161}$$

Then taking f_r as u_y and f_i as u_x, the imaginary part of (3.159) is

$$\oint_{\partial B} (f_r \, dy + f_i \, dx) = \int_B \left(\frac{\partial f_r}{\partial x} - \frac{\partial f_i}{\partial y} \right) dA \qquad (3.162)$$

The right sides of (3.161) and (3.162) are both zero from the Cauchy-Riemann equations; therefore, a line integral of any complex function is zero around a closed curve provided that the function is regular both on and inside the curve (no singular points), which is *Cauchy's theorem.* Further, if some closed curve C_2 surrounds another closed curve C_1, the integral of a complex function around C_2 is equal to the integral around C_1 provided that there are no singularities between the two curves, although there may be singularities within C_1 making both the integrals not equal to zero.

Consider the integral

$$I = \oint_{\partial B} \frac{f(z)}{z - z_0} dz \qquad (3.163)$$

in which $f(z)$ is everywhere regular inside ∂B and the point z_0 is inside ∂B. Since the entire integrand is not regular in ∂B, the integral is not necessarily zero. Take a circle ∂b of radius r_0 about z_0. Then

$$\oint_{\partial B} \frac{f(z)}{z - z_0} dz - \oint_{\partial b} \frac{f(z)}{z - z_0} dz = 0 \qquad (3.164)$$

On the circle take $z - z_0 = r_0 e^{i\theta}$ and $dz = ir_0 e^{i\theta} d\theta$. Then let the radius r_0 go to zero so that $f(z)$ becomes $f(z_0)$. The integral becomes

$$I = f(z_0) \oint_{\partial b} \frac{dz}{z - z_0} = if(z_0) \oint_0^{2\pi} d\theta = 2\pi if(z_0) \qquad (3.165)$$

The integrals with higher order singularities of the same type can be found by differentiating (3.165) with respect to z_0

$$\oint_{\partial B} \frac{f(z)dz}{(z - z_0)^2} = 2\pi if'(z_0) \qquad (3.166)$$

If the denominator is raised to the nth power, then

$$(n - 1)! \oint_{\partial B} \frac{f(z)dz}{(z - z_0)^n} = 2\pi if^{n-1}(z_0) \qquad (3.167)$$

where the superscript on f indicates the derivative of order $n - 1$.

The complex force on an object (Fig. 3.28) is

$$F = F_r + iF_i = -\oint_{\partial B} p(\sin\theta + i\cos\theta)ds \qquad (3.168)$$

Using trigonometric identities, putting the result in exponential form, and

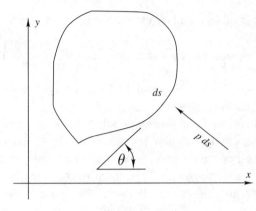

FIGURE 3.28
Force on an object caused by pressure.

then taking $ds = e^{-i\theta} dz$ produces

$$F_r - F_i = -i \oint_{\partial B} p e^{-2i\theta} dz \tag{3.169}$$

Using Bernoulli's equation without the gravity term gives

$$p = \text{constant} - \frac{\rho}{2} \left(u_x^2 + u_y^2 \right) \tag{3.170}$$

The constant can be ignored because it does not contribute to the force on an object. The square of the velocity is

$$u_x^2 + u_y^2 = e^{-2i\theta} \left(\frac{dW}{dz} \right)^2 \tag{3.171}$$

Finally, the force on the object is

$$F_r - iF_i = \frac{i}{2}\rho \oint_{\partial B} \left(\frac{dW}{dz} \right)^2 dz \tag{3.172}$$

which is the *Blasius theorem*. The moments about the origin on the object can be derived as

$$M_r - iM_i = -\frac{i}{2}\rho \oint_{\partial B} \left(\frac{dW}{dz} \right)^2 z\, dz \tag{3.173}$$

Example 3.5   Flow around a cylinder with circulation.   The potential is

$$W = -Uz - U\frac{R_0^2}{z} + \frac{i\Gamma}{2\pi} \ln z \tag{3.174}$$

The derivative is

$$\frac{dW}{dz} = -U + U\frac{R_0^2}{z^2} + \frac{i\Gamma}{2\pi z} \tag{3.175}$$

From the Blasius and Cauchy theorems the force is all in the imaginary (y) direction

$$F = -i\rho U \Gamma \tag{3.176}$$

3.12.5 The Schwarz-Christoffel Theorem

A general method of mapping that would take any region and map it into the real axis would make complex variables extremely useful. Unfortunately such a method is not available. However, we can find a function that will map any polygon into the real axis. Its disadvantage is that it cannot always be expressed in terms of elementary functions.

Consider the equation

$$\frac{dz}{d\zeta} = A \, (a_1 - \zeta)^{-\alpha_1/\pi} \, (a_2 - \zeta)^{-\alpha_2/\pi} \cdots (a_n - \zeta)^{-\alpha_n/\pi} \tag{3.177}$$

in which $\alpha_1, \alpha_2, \ldots, \alpha_n$, shown in Fig. 3.29, sum to 2π. The values of a_i are real and are indicated in the figure as points on the ξ-axis in the ζ-plane. The constant A is, in general, complex. The equation is intended to map the interior of the polygon in the z-plane into the upper half of the ζ-plane so that point a_1' maps into a_1, a_2' maps into a_2, etc. Figure 3.30 shows the mapping in the neighborhood of point a_1' and a_1. In each plane a small arc is taken about the point so that we can consider the mapping without moving exactly to the singularity. In the limit the radii of these arcs will be taken to zero.

We wish to show that as a point in the z-plane moves from $1'$ to $2'$ to $3'$ to $4'$ to $5'$ it moves from 1 to 2 to 3 to 4 to 5 in the ζ-plane. For this

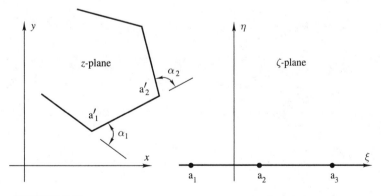

FIGURE 3.29
Mapping the interior of the polygon in the z-plane into the upper half of the ζ-plane.

purpose the relevant part of (3.177) is

$$\frac{dz}{d\zeta} = C\,(a_1 - \zeta)^{-\alpha_1/\pi} \qquad (3.178)$$

where the constant C has replaced A for the present purposes. Obviously,

$$\zeta_1 = \zeta_2 - \Delta\zeta \qquad \Delta\zeta = \Delta\xi_1 \qquad \Delta z = \Delta b e^{-i\theta_1} \qquad (3.179)$$

in which $\Delta\xi_1$ and Δb are real quantities as shown on Fig. 3.30. Using these relationships in (3.178) and approximating the integral of the right side by the trapezoidal rule

$$\Delta b e^{-i\theta_1} = \frac{C}{2}\left[(a_1 - \xi_2)^{-\alpha_1/\pi} + (a_1 - \xi_1)^{-\alpha_1/\pi}\right]\Delta\xi_1 \qquad (3.180)$$

Both of the quantities in parentheses are positive. Writing C as $ce^{i\beta}$, (3.180) becomes

$$\Delta b e^{-i\theta_1} = \frac{ce^{i\beta}}{2}\left[|a_1 - \xi_2|^{-\alpha_1/\pi} + |a_1 - \xi_1|^{-\alpha/\pi}\right]\Delta\xi_1 \qquad (3.181)$$

The latter equation can be used to calculate β, the argument of C, since

$$-i\theta_i = i\beta \qquad (3.182)$$

The magnitude of C (or c) is dependent on the position of the origin in each plane and is irrelevant to the present discussion.

Passing from point 4 to point 5

$$\Delta z = \Delta b e^{-i\theta_2} = \frac{ce^{i\beta}}{2}\left[(a_1 - \xi_5)^{-\alpha_1/\pi} + (a_1 - \xi_4)^{-\alpha_1/\pi}\right]\Delta\xi_4 \qquad (3.183)$$

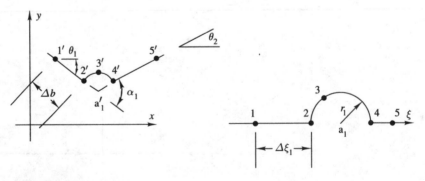

FIGURE 3.30
Detail of the mapping of Fig. 3.29 near one corner.

This time both quantities in brackets are negative. Instead of (3.181) we get

$$\Delta b e^{-i\theta_2} = \frac{ce^{i\beta}}{2}\left[|a_1 - \xi_2|^{-\alpha_1/\pi} + |a_1 - \xi_1|^{-\alpha_1/\pi}\right]e^{-i\alpha}\Delta\xi_1 \qquad (3.184)$$

in which we have used $-1^{-\alpha_1/\pi} = e^{i\pi(-\alpha_1/\pi)} = e^{-i\alpha_1}$. Equating arguments on the left and right sides gives

$$-i\theta_2 = i\beta - i\alpha_1 \qquad (3.185)$$

From the geometry of the z-plane $\theta_2 = \theta_1 + \alpha_1$. Thus, the β of (3.182) and (3.185) are the same. In passing around the arc 2-3-4 the argument in the z-plane has changed by an amount α_1. A similar result is obtained for the other corners of the polygon.

The mapping function is from (3.177)

$$z = A \int \frac{d\zeta}{(a_1 - \zeta)^{\alpha_1/\pi}(a_2 - \zeta)^{\alpha_2/\pi}\cdots(a_i - \zeta)^{\alpha_i/\pi}} + B \qquad (3.186)$$

The quantities A and B are complex constants of the mapping that are arbitrary from the point of view of conformality. A affects both the scale of the mapping and the orientation of the polygon in the z-plane. The constant of integration, B, determines the location of the origin in the z-plane. The a_i are real and represent the locations of the corners in the ζ-plane. Three of the a_i can be chosen arbitrarily in that we have some choice on where to map the corners in the ζ-plane; the others are determined by the mapping. It is often convenient to choose one of the a_i (say a_j) as infinity and also choose $A = Ca_j^{\alpha_j/\pi}$ where C is now the arbitrary constant. Then a term in the denominator becomes

$$\left(\frac{a_j - \xi}{a_j}\right)^{-\alpha_j/\pi} \qquad (3.187)$$

As a goes to infinity, this term become unity and drops out of the equation.

Example 3.6 Flow into a channel through a slit. Figure 3.31 shows an infinitely long channel into which flow enters from a source in the origin. The z- and ζ-planes are shown in Fig. 3.31 where z_1 maps into ζ_1, z_2 maps into ζ_2, etc. However, z_1 is at negative infinity and z_3 is at positive infinity. The angles in the z-plane are

$$\alpha_1 = \pi \qquad \alpha_2 = 0 \qquad \alpha_3 = \pi \qquad \alpha_4 = 0$$

We take

$$\zeta_1 = 0 \qquad \zeta_2 = 1 \qquad \zeta_3 = \infty$$

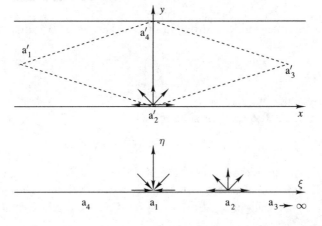

FIGURE 3.31
Flow through a slit
mapped into the upper
half-plane.

The Schwarz-Christoffel mapping becomes

$$z = A \int (\zeta - 0)^{-1}(\zeta - 1)^{0}(\zeta - \zeta_4)^{0} \, d\zeta + B = A \int \frac{d\zeta}{\zeta} + B$$

Choosing $B = 0$ makes $\zeta = 1$ when $z = 0$ and the mapping is simply

$$z = A \ln \zeta$$

To determine A, consider the point $z = ih = A \ln \zeta_4$. Then

$$\zeta_4 = e^{ih/A} = \cos \frac{h}{A} + i \sin \frac{h}{A}$$

Since ζ_4 is real, h/A is either zero or π. Because we do not want a channel of zero height, $h/A = \pi$. Then $A = h/\pi$ and $\zeta_4 = -1$. Finally, the transformation is

$$z = \frac{h}{\pi} \ln \zeta$$

The complex potential must now be written. In the z-plane there is a source in the origin. The flow from the source goes to sinks at $z = \infty$ and $z = -\infty$. The two sinks at $\pm\infty$ must be one-half as strong as the source. From the mapping function

The sink at $z = -\infty$ maps into $\zeta = 0$,

the source at $z = 0$ maps into $\zeta = 1$, and

the sink at $z = \infty$ maps into $\zeta = \infty$.

The complex potential in the ζ-plane becomes

$$W(\zeta) = \frac{Q}{2\pi} \ln(\zeta - 1) - \frac{Q}{4\pi} \ln \zeta + \text{constant}$$

in which Q is the strength of the source. The sink at $z = \infty$ is ignored since the flow becomes parallel a long distance from the origin. The velocity in the ζ-plane is obtained by differentiation following (3.145). The mapping function can then be used to find the velocity in the z-plane. The result is

$$u_x - iu_y = -\frac{Q}{4h}\frac{e^{\pi z/h} + 1}{e^{\pi z/h} - 1}$$

Note that for $z = ih$ the numerator is zero and the point $x = 0$, $y = h$ is a stagnation point.

In the previous example it was easy to find the explicit mapping in terms of elementary functions since two of the angles were zero and a third was eliminated by taking the point to infinity in the ζ-plane. Although one point can always be taken to infinity, the other angles do not usually drop out. When just two terms remain in the denominator of (3.186), the mapping can be expressed in terms of elliptic integrals. When more than two terms remain, numerical integration is usually necessary.

PROBLEMS

3.1. Show that for axisymmetric flow the quantity of flow between two streamlines (actually, concentric stream surfaces) is the difference in the value of the stream function times 2π on the surfaces.

3.2. Given the following values of either Φ or Ψ, find the other, if possible. If it is not possible, why not?
(a) $\Phi = 2xy$
(b) $\Phi = y^2 - x^2$
(c) $\Psi = xy + x^3 - 4y^3/3$

3.3. Show that for the three-dimensional source (see Eq. 3.41) Q is the rate of outflow.

3.4. Find the potential for a line source on the x-axis in two dimensions, then show the two-dimensional equation for a series of line sources of constant intensity [the two-dimensional counterpart to (3.95)].

3.5. Derive the equation for the potential of a doublet in three dimensions [the first of (3.89)].

3.6. Find the radius of a sphere composed of parallel flow and a doublet; that is, derive (3.90).

3.7. Equation (3.94) comes from (3.93) with q set equal to a constant in the range $x_i \leq x < x_{i+1}$.

(a) Verify (3.94).
(b) Derive a similar expression with q varying linearly between x_i and x_{i+1}. The q_i in (3.95) should then be the values of q at the x_i.

3.8. Use (3.98) in place of (3.93) to derive an equation similar to (3.94).

3.9. The velocity potential for flow in a 90-degree corner is $\Phi = A(x^2 - y^2)$ where A is a constant. Give expressions for the velocity components of this flow. Obtain the expression for the stream function. Sketch a few streamlines.

3.10. A tornado is sometimes represented by the superposition of a vortex and a sink. Using that representation derive an expression for the pressure in a tornado vs. distance from the center.

3.11. Show that for the frictionless flow of an unstratified fluid the rotation of each particle is constant. First, take the substantial derivative of (3.53) to obtain

$$\frac{D\Gamma}{Dt} = \frac{D}{Dt}\oint_C \vec{u}\cdot\vec{t}\,ds$$

Then use the equation of motion and Stokes'

theorem to get

$$\frac{D\Gamma}{Dt} = \int_A \left(\vec{\nabla} \frac{1}{\rho} \times \vec{\nabla} p \right) \cdot \vec{n}\, dA$$

The integral is zero if the gradients of the density and pressure are parallel, showing that the rate of change of rotation of a particle is zero.

3.12. Two sources of equal strength are located on the y-axis at $y = h$ and $y = -h$. For a parallel flow of velocity U

(a) Find the potential for the flow.

(b) Under what conditions do the stagnation points occur on the x-axis?

(c) Sketch the possible flow configurations.

3.13. A rotating cylinder has lift because the flow around the cylinder is no longer symmetrical about the x-axis. The rotation of the cylinder drags some fluid along, causing circulation. Consider a cylinder 1 m in diameter that is rotating at 20 revolutions per second in a flow of 20 m/s. Assume that the rotation is 50 percent efficient in producing circulation.

(a) Locate the stagnation points on the cylinder.

(b) Calculate the lift on the cylinder by integrating the pressure over the surface.

3.14. Compute the maximum velocity on the surface of a sphere and on the surface of a cylinder in a parallel flow. If Bernoulli's equation is used to compute the pressure at the point of maximum velocity, would you expect the result to be realistic for a real fluid? Why?

3.15. Consider the flow around a sharp corner. The extreme angle is one in which the flow turns 180°. That is a practical case because it represents a sheetpile cutoff wall in flow through porous media. Show that the velocity is singular at the point of the 180° turn and goes to infinity as the reciprocal of the square root of the distance from the tip of the cutoff wall.

3.16. Plot a few streamlines for the flow around an ellipse with major axis of 10 and minor axis

of 4. What is the maximum velocity in the flow field? Find an equation for the lift on the ellipse for an arbitrary circulation.

3.17. To study the flow in the neighborhood of a bent wall (Fig. 3.32), the area above the wall is mapped into the upper half of the ζ-plane. Find the mapping function and plot a few streamlines for $n < 1$ and for $n > 1$. Find the velocity at the corner in both cases.

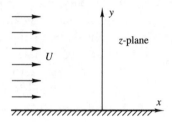

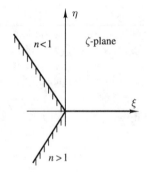

FIGURE 3.32

3.18. Find the mapping function with point A′ mapped into the origin of the z-plane B′ and mapped into $z = 1$ (Fig. 3.33). Then find the velocity in point C for flow above the wall.

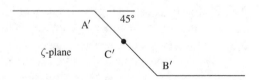

FIGURE 3.33

3.19. Plot a few streamlines in the vicinity of the bend in the channel of Fig. 3.34. Make the widths of both parts of the channel equal. The length of each leg can extend to infinity.

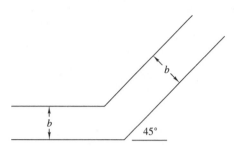

FIGURE 3.34

3.20. Consider the flow around a circular cylinder in the proximity of a wall (Fig. 3.35).

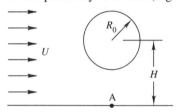

FIGURE 3.35

(a) Develop an expression for the pressure at point A on the wall in terms of the free stream pressure p_∞, the free stream velocity U, the radius R_0 of the cylinder, and the distance H from the wall to the center of the cylinder.

(b) Is this expression exact or does it contain approximations? If it contains approximations, what are they?

(c) Assume that you are to perform this experiment in the laboratory at a Reynolds number R of 50,000, where $R = 2UR/\nu$. The pressure at point A is to be measured by a piezometer. Would you expect the measured pressure to be higher or lower than that which you calculated? Why?

3.21. Repeat Prob. 3.20 for flow around a sphere in the proximity of a wall.

3.22. Consider the bathtub vortex flow through a horizontal orifice (Fig. 3.36). At some distance R_1 from the orifice the depth of the flow is measured as h_1 and the velocity is u_θ in the tangential direction. The outflow Q is also known.

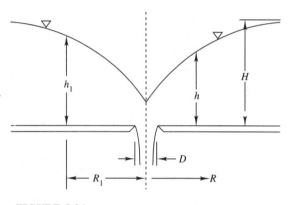

FIGURE 3.36

(a) Find an equation in terms of the above quantities that gives the height H of the water surface at infinity. (Hint: Use a potential vortex but not a potential sink.)

(b) Find the height h of the water surface as a function of distance R from the center of the orifice. State what assumptions have been made, if any.

3.23. Consider the flow under a simple sheet pile dam on an infinite porous medium (Fig. 3.37). Solve the problem by mapping the z-plane and the W-plane into the real axis of the ζ-plane. Find a general expression for the velocity components u_x and u_y in the z-plane. In particular, what are the velocities at points B, C, and D? What is the flow rate under the dam?

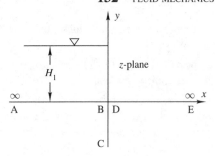

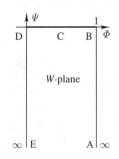

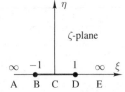

FIGURE 3.37

3.24. Consider the flow over the bent wall shown in Fig. 3.38.

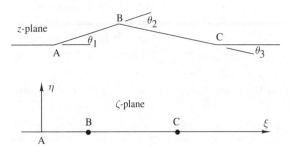

FIGURE 3.38

(a) Give an analytic expression in integral form for the function $z = z(\zeta)$ that trans-

forms the upper part of the z-plane into the upper half of the ζ-plane.

(b) Plot a streamline in the z-plane.

3.25. Compare the shape of two half-bodies. Body A is produced by a source of intensity m located at $x = 0$, $R = 0$ in a parallel flow with constant velocity U. Body B is produced by a line source stretching from $x = -b$ to $x = b$ (See Fig. 3.39). Assume that $2bs = m$.

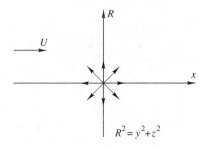

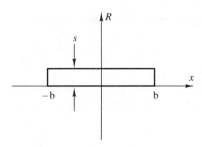

FIGURE 3.39

(a) Compare the radius of the half-body for $x = 0$ and

(b) $x = \infty$ in each case.

(c) Compute the location of the stagnation point in each case.

(d) Compute the final widths of the bodies.

3.26. Superpose a vortex of intensity Γ and a source of intensity s at the same point. Determine the equation of the streamlines and make a sketch of the flow. Find an expression for the pressure that is valid throughout the flow in terms of the pressure at infinity p_∞.

3.27. Derive equations similar to (3.99) and (3.100) using surface vorticity instead of surface sources.

CHAPTER
4

POTENTIAL FLOW (FREE SURFACES)

Free surface flows are among the most challenging of all potential flows. By definition, the position of the free surface, and hence the geometry of the solution region, are part of the solution. Boundary conditions must be applied along the unknown free surface and the problem cannot be solved without those conditions. It's similar to the old joke that "You can't get there from here—you have to go somewhere else first." The dilemma is that we don't know where to apply the boundary condition and cannot solve the problem until we do, but the free surface shape—the domain of solution—awaits the solution. Of course, methods exist, explained herein, to avoid that dilemma.

There are two definitions of free surface flow:

1. Physical definition: Free surface flow occurs in a liquid where one or more of the boundaries is not physically constrained but can adjust itself to conform to the flow conditions.

2. Mathematical definition: Free surface flow occurs in a deformable solution region in which the shape and size of the region is part of the solution.

We adopt the second (mathematical) definition. This definition excludes linear wave problems where the (linearized) boundary conditions are applied on the equilibrium free surface (a fixed, known boundary), not the actual free surface (see Chap. 10). It also excludes shallow water hydraulics where the vertical coordinate is suppressed and the depth becomes a normal dependent variable (see Chap. 8). The vertical coordinate must remain in a free surface problem as an independent variable. Included are nonlinear wave problems, jets, cavities, flow over weirs and spillways, sluice gate flow, free surfaces in porous media (except Dupuit problems), freeze-thaw and solidification problems (the "Stephan problem"), extrusion, and molding. Many of these applications are adequately described by the potential approximation.

4.1   THE EQUATIONS OF FREE SURFACE FLOW

The free surface makes a problem nonlinear in a peculiar way. Usually the free surface boundary conditions themselves are nonlinear even though the governing equation may be linear. Also, the fact that these conditions must be applied to the actual free surface is a type of nonlinearity. The free surface problem can be steady or unsteady. One of the peculiarities is that the steady problem is often more difficult than the unsteady problem.

Applications to free surface problems are numerous. Wave problems are perhaps the most common, but all flows of liquids that do not completely fill containers are free surface problems. Thus, natural flows of liquids contain a free surface.

4.1.1   The Kinematic Boundary Condition

Consider first the steady, two-dimensional problem shown in Fig. 4.1. Since the flow is steady, the free surface is a streamline. The ratio of the vertical and horizontal velocities must equal the slope of the

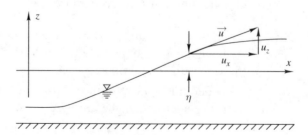

FIGURE 4.1
Steady, two-dimensional free surface.

surface

$$\frac{u_z}{u_x} = \frac{d\eta}{dx} \tag{4.1}$$

in which η is the elevation of the free surface and the z-axis is vertical. Extending (4.1) to three dimensions

$$u_z = u_x \frac{\partial \eta}{\partial x} + u_y \frac{\partial \eta}{\partial y} \tag{4.2}$$

Consider next an unsteady flow with a horizontal free surface that is rising vertically (Fig. 4.2). The rate of rise is

$$\frac{\partial \eta}{\partial t} = u_z \tag{4.3}$$

Combining the steady and the unsteady parts, (4.2) and (4.3), gives

$$u_z = \frac{\partial \eta}{\partial t} + u_x \frac{\partial \eta}{\partial x} + u_y \frac{\partial \eta}{\partial y} \qquad \text{on } z = \eta \tag{4.4}$$

Since (4.4) applies on the free surface, and only on the free surface, it serves as a boundary condition. Because both η and the velocities are dependent variables, (4.4) is a nonlinear equation.

Another way to look at the kinematic condition is to apply the postulate that a particle on the free surface remains on the free surface. Actually, that fact can be proven using topology theory. Since by definition the position of the free surface is at $z = \eta$,

$$\frac{D}{Dt}(z - \eta) = -\frac{\partial \eta}{\partial t} + \vec{u} \cdot \vec{\nabla}(z - \eta) = -\frac{\partial \eta}{\partial t} - u_x \frac{\partial \eta}{\partial x} - u_y \frac{\partial \eta}{\partial y} + u_z$$

$$= -\frac{\partial \eta}{\partial t} + \frac{\partial \Phi}{\partial x}\frac{\partial \eta}{\partial x} + \frac{\partial \Phi}{\partial y}\frac{\partial \eta}{\partial y} - \frac{\partial \Phi}{\partial z} = 0 \tag{4.5}$$

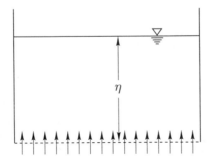

FIGURE 4.2
The simplest unsteady free surface flow.

In performing the gradient in (4.5), remember that $\eta = \eta(x, y, t)$ and so its derivative with respect to z is zero. Thus, (4.5) leads to the same result as (4.4).

In two dimensions it is convenient to work in a coordinate system that is normal and tangential to the free surface (Fig. 4.3). The derivatives of the potential are

$$\frac{\partial \Phi}{\partial x} = -\frac{\partial \Phi}{\partial n} \sin \beta + \frac{\partial \Phi}{\partial s} \cos \beta$$

$$\frac{\partial \Phi}{\partial z} = \frac{\partial \Phi}{\partial n} \cos \beta + \frac{\partial \Phi}{\partial s} \sin \beta$$

(4.6)

in which n is the normal coordinate, s is the tangential coordinate, and β is the angle the surface makes with the horizontal xy-plane. Using (4.6) in the two-dimensional version of (4.5) and using $\partial \eta / \partial x = \tan \beta$

$$\frac{\partial \eta}{\partial t} = -\frac{1}{\cos \beta} \frac{\partial \Phi}{\partial n} \qquad \text{on } z = \eta \text{ in two dimensions} \qquad (4.7)$$

4.1.2   The Dynamic Condition

Another condition comes from the fact that the pressure is specified on the free surface. Often the specified pressure is taken as a constant, but more generally it is a variable and a pressure discontinuity can exist across the surface due to surface tension. The difference in pressure is expressed approximately as

$$p - p_a = -\sigma \vec{\nabla} \cdot \vec{\nabla} \eta \qquad \text{on } z = \eta \qquad (4.8)$$

in which σ is the surface tension parameter (dimensions of F/L) and the vector operation is an approximation for the surface curvature if the slope of the surface is not too great.

Using (4.8) in the Bernoulli equation, the pressure can be eliminated

$$-\frac{\partial \Phi}{\partial t} + \frac{1}{2} \frac{\partial \Phi}{\partial x_i} \frac{\partial \Phi}{\partial x_i} + g\eta = -\frac{p_a}{\rho} + \text{constant} \qquad \text{on } z = \eta \qquad (4.9)$$

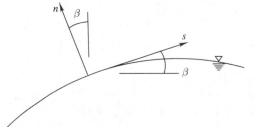

FIGURE 4.3
The coordinate system normal and tangential to the free surface

in which σ has been taken as zero. The constant in the Bernoulli equation can always be set to zero by redefining the potential. For example, if the constant is B, the potential can be defined as

$$\Phi = \Phi' - Bt \tag{4.10}$$

in which case (4.9) would be written in terms of Φ' and the constant would cancel. The cancellation of the constant again illustrates that the flow field results from the spatial gradients of the potential and not from the potential (or pressure) level.

Equation (4.9) serves as the second free surface boundary condition. Since η is an additional unknown (in addition to the velocities), the problem requires two boundary conditions at the free surface. Normally the equation is written to give the change of Φ with respect to time at a point that moves vertically with the free surface whereas the usual partial derivative is written for Φ at a fixed point that does not move. The needed derivative is written

$$\frac{\partial \Phi}{\partial t} = \left(\frac{\partial \Phi}{\partial t} \right)_{x,y} - \frac{\partial \Phi}{\partial z} \frac{\partial \eta}{\partial t} \tag{4.11}$$

in which the subscript x, y indicates that the derivative is to be taken with x and y held constant, but that the z-coordinate is free to vary. We define this derivative to apply to the moving free surface so that it expresses the rate of change of Φ with respect to time following the free surface through time.

Equation (4.11) is used in (4.9). Then with the help of (4.6) and assuming that $p_a = 0$ (or redefining the constant)

$$-\left(\frac{\partial \Phi}{\partial t} \right)_{x,y} + \frac{1}{2} \frac{\partial \Phi}{\partial x_i} \frac{\partial \Phi}{\partial x_i} + g\eta + \frac{\partial \Phi}{\partial z} \frac{\partial \eta}{\partial t} = B \qquad \text{on } z = \eta \tag{4.12}$$

in which B is the Bernoulli constant. Returning to two dimensions we use the velocity relationship $(\partial \Phi / \partial x)^2 + (\partial \Phi / \partial z)^2 = (\partial \Phi / \partial s)^2 + (\partial \Phi / \partial n)^2$ and the kinematic equation (4.7) to obtain

$$\left(\frac{\partial \Phi}{\partial t} \right)_{x,y} = -B + \frac{1}{2} \left[\left(\frac{\partial \Phi}{\partial s} \right)^2 - \left(\frac{\partial \Phi}{\partial n} \right)^2 - 2 \frac{\partial \Phi}{\partial s} \frac{\partial \Phi}{\partial n} \tan \beta \right] + g\eta$$

$$\text{on } z = \eta \text{ in two dimensions}$$

$$\tag{4.13}$$

Equation (4.13) is written in s, n-coordinates because that is a convenient formulation for the calculation of many problems.

4.1.3   Calculations

Consider a calculation of the unsteady problem. An initial condition must give the complete state of the problem, which includes the position of the free surface, the normal derivative to the free surface, and the potential at all points in the domain. Of course, the conditions on the solid boundaries (that $\partial \Phi / \partial n = -U_n$ where U_n is the velocity of the boundary normal to itself; usually, $U_n = 0$) must also be given at the initial time and for all subsequent times. Given the initial conditions, (4.7) specifies how the free surface moves. The time derivative, expanded in a Taylor series, should give an approximation to the elevation of the free surface for a short period after the initial time, say Δt. Equation (4.12) or (4.13) will give an approximation for the potential after a short time. At Δt the geometry of the domain is known, the potential is known on the free surface, and the potential or its normal derivative is known on the boundaries (from the boundary conditions), so Laplace's equation can be solved for the potential everywhere. Then the results for Δt can be used as new initial conditions to advance the solution another increment in time.

The last paragraph gives a brief sketch of a numerical method of solving the free surface problem. In practice a number of details still remain to be resolved. How should (4.7) and (4.12) be formulated to give an accurate solution at time Δt? The solution of the Laplace equation is not trivial, especially since the solution domain will be irregular and constantly changing with each time step. The movement of the free surface is some sort of approximation valid only for short time steps. How do we keep these approximations from accumulating until there is a gross error in the solution? All these questions fall within the category of numerical methods.

4.1.4   The Steady State

In most problems the elimination of time as an independent variable is a large simplification. That may or may not be the case with free surface flows. The Bernoulli condition for steady state becomes

$$B - \frac{1}{2}\left[\left(\frac{\partial \Phi}{\partial s}\right)^2 - \left(\frac{\partial \Phi}{\partial n}\right)^2 - 2\frac{\partial \Phi}{\partial s}\frac{\partial \Phi}{\partial n}\tan \beta \right] - g\eta = 0 \tag{4.14}$$

on $z = \eta$ in two dimensions

which is simply (4.13) without the time derivative. The difficulty is that, unlike the unsteady case, the shape of the solution domain is not given. The analyst must make an initial guess of the position of the free surface. The problem is then solved using the guess to define the solution domain and applying the condition that $\partial \Phi / \partial n = 0$ on the free surface. Equation (4.14) is then checked. If the initial guess is not correct, (i.e., if the analyst is human) (4.14) will not be satisfied. Some strategy must then be developed to refine the guess for the free surface in an iterative manner until (4.14) is satisfied to acceptable accuracy.

In the iterative procedure there is no guarantee that the solution will converge due to the nonlinearity of the boundary condition. In fact, the solution may not be unique; sometimes more than one configuration of the free surface will satisfy the conditions. In such cases the "correct" configuration depends on the evolution of the flow, which is completely neglected in the steady state formulation.

Because of the difficulty sometimes encountered by tackling a steady state problem directly, the solution can be obtained in some problems by extending a time-dependent solution under steady boundary conditions for a long period. If the solution is done numerically, computational time can be excessive. The redeeming feature is that if we don't care about the solutions at intermediate times, the time step can be longer than would be necessary for the accuracy of the unsteady problem. That feature is not always applicable, however, since numerical stability may force shorter time steps. Also, if the evolution of the solution is inaccurate, the ambiguity due to nonuniqueness may not be resolved. More importantly, if the problem is undamped, then waves—artifacts from an inaccurate guess of the initial conditions or inaccurate calculations—can reflect from the boundaries and contaminate the solution forever so that a true steady state is never achieved.

In two-dimensional, steady calculations we can take advantage of the fact that the free surface is a streamline. The free surface conditions then become

$$\Psi = \text{constant} \qquad \text{on } z = \eta \tag{4.15}$$

$$\frac{1}{2}\left(\frac{\partial \Psi}{\partial n}\right)^2 + g\eta = B \qquad \text{on } z = \eta \tag{4.16}$$

An iterative solution becomes somewhat simpler since the free surface is a Dirichlet boundary (Ψ given explicitly). However, the same fundamental difficulties remain.

4.2 POROUS MEDIA

In flow through porous media the boundary conditions are slightly different than in hydrodynamic flows.

4.2.1 The Kinematic Condition.

For flow in a granular medium, (4.5) is still a valid kinematic condition, but care must be taken as to what is used for the velocity. The velocity must be the actual fluid velocity and not the specific discharge or the flux as is commonly used in porous media calculation. Thus, (4.5) is changed to

$$\frac{D}{Dt}(z - \eta) = -\frac{\partial \eta}{\partial t} + \frac{1}{n_e}\vec{u} \cdot \vec{\nabla}(z - \eta) = 0 \tag{4.17}$$

in which n_e is the *effective porosity*. In a porous medium porosity is defined as the volume of voids divided by the total volume of a representative sample. If the sample is filled with water and drained, the volume of water that emerges is less than the volume of voids due to dead end voids and gas (air) that cannot escape because of capillary forces that are greater than the gravity force that is driving the drainage. The volume of water that drains represents the proportion of the total volume that is mobile. Thus, the effective porosity is the mobile volume divided by the total volume of the sample. Similarly, a representative average true velocity of the fluid is the specific discharge or flux divided by the effective porosity.

Using Darcy's law in (4.17) to replace the velocity gives

$$\frac{\partial \eta}{\partial t} = \frac{K}{n_e}\left(\frac{\partial \Phi}{\partial x}\frac{\partial \eta}{\partial x} + \frac{\partial \Phi}{\partial y}\frac{\partial \eta}{\partial y} - \frac{\partial \Phi}{\partial z}\right) \tag{4.18}$$

The hydraulic conductivity and the effective porosity can be made a part of the time variable. We define an "effective time"

$$t_* = \frac{K}{n_e}t \tag{4.19}$$

(Note that this time is not dimensionless; it has dimensions of length and can be made dimensionless by dividing by a characteristic length.) A unit normal vector to the free surface is

$$\vec{n} = \frac{\vec{\nabla}(z - \eta)}{\left|\vec{\nabla}(z - \eta)\right|} \tag{4.20}$$

Consider the scalar product (remembering that η is a function of x, y, and t but not of z)

$$\vec{\nabla}\Phi \cdot \vec{\nabla}(z - \eta) = \vec{\nabla}\Phi \cdot \vec{e}_z - \left[\vec{\nabla}\Phi \cdot \left(\vec{e}_x\frac{\partial \eta}{\partial x} + \vec{e}_y\frac{\partial \eta}{\partial y}\right)\right]$$

$$= \frac{\partial \Phi}{\partial z} - \frac{\partial \Phi}{\partial x}\frac{\partial \eta}{\partial x} - \frac{\partial \Phi}{\partial y}\frac{\partial \eta}{\partial y} \tag{4.21}$$

Also

$$\left|\vec{\nabla}(z - \eta)\right| = \sqrt{1 + \left(\frac{\partial \eta}{\partial x}\right)^2 + \left(\frac{\partial \eta}{\partial y}\right)^2} \tag{4.22}$$

Using (4.18), (4.19), and (4.21)

$$\frac{\partial \eta}{\partial t_*} = -\left|\vec{\nabla}(z - \eta)\right| \frac{\partial \Phi}{\partial n} \qquad \text{on } z = \eta \qquad (4.23)$$

In two dimensions (the xz-coordinate system)

$$\left|\vec{\nabla}(z - \eta)\right| = \sqrt{1 + \left(\frac{\partial \eta}{\partial x}\right)^2} = \frac{1}{\cos \beta} \qquad \text{in two dimensions} \qquad (4.24)$$

and (4.23) becomes

$$\frac{\partial \eta}{\partial t_*} = -\frac{1}{\cos \beta} \frac{\partial \Phi}{\partial n} \qquad \text{on } z = \eta \text{ in two dimensions} \qquad (4.25)$$

which, except for the effective time, is the same as (4.7).

4.2.2 The Pressure Condition

The "dynamic" condition for porous media also states that the pressure is zero on the free surface, which is equivalent to making the head equal to the elevation

$$\Phi = \eta \qquad \text{on } z = \eta \qquad (4.26)$$

As simple as this equation looks, it must be handled with care when taking derivatives because $\Phi = \Phi(x, y, z, t)$ whereas $\eta = \eta(x, y, t)$. The derivative $\partial \Phi / \partial t$ implies the change of Φ with respect to time at a fixed point, whereas $\partial \eta / \partial t$ is the change of the elevation of the free surface while holding the horizontal coordinates x and y constant. The equivalents are

$$\frac{\partial \eta}{\partial t} = \left(\frac{\partial \Phi}{\partial t}\right)_{x,y} \qquad \text{on } z = \eta \qquad (4.27)$$

Equation (4.23) becomes

$$\left(\frac{\partial \Phi}{\partial t_*}\right)_{x,y} = -\left|\vec{\nabla}(z - \eta)\right| \frac{\partial \Phi}{\partial n} \qquad \text{on } z = \eta \qquad (4.28)$$

and (4.25) is similar

$$\left(\frac{\partial \Phi}{\partial t_*}\right)_{x} = -\frac{1}{\cos \beta} \frac{\partial \Phi}{\partial n} \qquad \text{on } z = \eta \text{ in two dimensions} \qquad (4.29)$$

4.2.3 The Seepage Surface

In porous media calculations one additional type of boundary is present that does not appear in hydrodynamic calculations. When the free surface reaches the edge of the medium, as shown in Fig. 4.4, some of the fluid

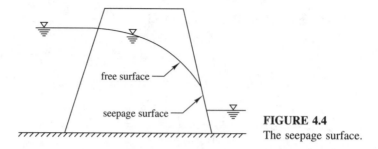

FIGURE 4.4
The seepage surface.

leaves the medium and runs down its edge, the *seepage surface*. From the point of view of flow in the medium, the fluid is crossing a boundary and escapes from the solution domain. The condition on this boundary is that the pressure is zero, or

$$\Phi = z \qquad \text{on the seepage surface} \tag{4.30}$$

The flux across the boundary, $\partial\Phi/\partial n$, becomes the unknown.

The point where the seepage surface intersects the free surface is a part of the free surface. Its vertical location is prescribed by $\Phi = \eta$ and it moves as a free surface point except that it is constrained to move along the edge of the medium. The free surface can be shown to be tangent to the edge of the medium at their intersection (unless the edge of the medium is an overhang, in which case the free surface is vertical). However, the tangential condition is seldom enforced in numerical calculations because experience has shown that it makes little difference.

At the intersection of the free surface and the seepage surface, the normal derivative vanishes, $\partial\Phi/\partial n = 0$, regardless of whether the problem is steady or unsteady. Thus, the movement of the point cannot be calculated by the equations for the other free surface nodes; instead, the pressure condition is used to locate it.

4.3   FAR-FIELD CONDITIONS

In many fluid flow problems the solution region is extended to infinity in one or more directions. That is especially true of problems that occur in nature. Ocean waves often originate far from the region of interest or propagate far beyond the region of interest. Channel flows, atmospheric flows, many aerodynamics problems, groundwater aquifers, and other flow situations extend from the region of interest a distance that is great compared to the dimensions of the problem. On the other hand, numerical—and some analytical—calculations must take place in a finite region. Therefore, each problem of this type must have "numerical boundaries" that have nothing to do with the physical boundaries. We need to choose numerical boundaries in such a way that their placement and the conditions that apply along them do not cause unacceptable errors in the solution.

In many problems the proper choice is simply to place numerical boundaries far enough from the region of interest so that whatever conditions are applied along these boundaries do not affect the solution. The classical method of deciding how far away is to begin the solution with the boundaries close to the region of interest, solve the problem, move the boundaries farther away, solve again and repeat until there is no significant change in the solution. That technique is somewhat expensive due to the multiple solutions and dangerous because it is difficult to decide when the sum of "insignificant" changes adds up to a significant change. It is, however, effective in most cases. In porous media, for example, the entire problem is so highly damped that the solution is often not affected materially by boundary conditions at some distance, if those boundary conditions are stated intelligently.

The truncation of the solution region in hydrodynamic problems is often difficult. Hydrodynamic potential flow problems are completely undamped so that any disturbance created by error in a boundary condition can travel without attenuation to the region of interest. Waves represent a typical example. A moving object in the free surface of the ocean causes a series of waves that propagate away from the object and to infinity. But if the solution region is limited, those waves will soon reach the artificial boundaries. If the boundary conditions do not permit all of the wave energy to pass through, some will be reflected and contaminate the desired solution. Moving the boundaries further away in a two-dimensional problem simply lengthens the required time for the energy to return.

In three-dimensional problems the situation is somewhat better because wave energy that was generated by the moving object diminishes in proportion to the reciprocal of the distance from the object; that is, the energy simply spreads over a larger and larger area the further from the object it travels. Moreover, the reflected energy will also spread and is less likely to be important unless the reflected energy from all the boundaries is focused back to the region of interest. On the other hand, three-dimensional problems are expensive to compute and pocketbook considerations dictate keeping the computational boundaries as close to the object as possible. In the case of close boundaries, neither the propagated energy nor the reflected energy has much space in which it can diminish.

There are two solutions to the problem of artificial boundaries. The first is to provide *radiation boundary conditions*, which allow the energy to flow freely through the boundaries without reflection. Unfortunately, it is not always easy to write and apply such conditions and it is generally impossible in nonlinear problems. The second solution is to provide far-field damping that dissipates the energy so that it cannot reflect from the boundaries.

The difficulties of the finite solution region are not confined to numerical solutions; experiments also suffer from unwanted boundary contamination. A wave tank consists of a wave maker that generates the wave that impinges on the region of interest, and a downstream section through which flows the wave energy after the wave has passed the region of interest. Because the wave tank is finite, special care must be taken at the downstream end. Conventionally, the wave energy is absorbed by a beach or an arrangement of energy absorbing materials since it cannot be passed from the tank. The downstream end of the tank is not the only problem; the wave maker itself may reflect energy that was reflected from the region of interest back toward the wave maker.

Numerical methods are more flexible than experimental arrangements in designing far-field boundary conditions. In some problems those conditions can be written so that the requirements are well satisfied; in others the question of how to treat the far-field boundary remains an unsolved problem.

The damping of energy can be easily accomplished by a simple addition to the dynamic free surface condition

$$\left(\frac{\partial \Phi}{\partial t}\right)_{x,y} = -B + \frac{1}{2}\left[\left(\frac{\partial \Phi}{\partial s}\right)^2 - \left(\frac{\partial \Phi}{\partial n}\right)^2 - 2\frac{\partial \Phi}{\partial s}\frac{\partial \Phi}{\partial n}\tan \beta\right]$$
$$+ g\eta - \lambda(x, y)\Phi \qquad \text{on } z = \eta \tag{4.31}$$

in which λ is a damping factor that is set equal to zero in the region of interest. To see how the term works, consider the ordinary differential equation

$$\frac{d\Phi}{dt} = -\lambda\Phi \tag{4.32}$$

that has the solution

$$\Phi = e^{-\lambda t} \tag{4.33}$$

which indicates that Φ decreases exponentially with time. Even with damping some care must be taken to prevent reflection. If damping is imposed suddenly at some point, there will be energy reflected from the damping region. In the physical wave tank that problem is solved by using "progressive damping" (LeMehaute, 1972), which consists of light damping followed by progressively stronger damping in the downstream direction. The same idea can be applied to numerical solutions by making λ progressively larger in the downstream direction (Betts and Mohamad, 1982).

Even with the use of a damping term, difficulty remains at the section of the incoming wave. The incident waves are generated by a "numerical wave maker," from which there may be reflections in the same way that a physical wave maker would have reflections. As yet, no generally satisfactory method exists for damping the waves that are produced by the wave maker while leaving the waves reflected by the wave maker undamped.

4.4   THE USE OF COMPLEX VARIABLES

For both hydrodynamic flows and flows in porous media, free surfaces can be mapped into simpler regions, using complex variables, by virtue of the boundary conditions. As in all complex variable methods, the technique is limited to two dimensions and it is useful in steady flow.

4.4.1   Hydrodynamics

For steady flows the free surface boundary condition is the Bernoulli relationship

$$\frac{u^2}{2} + \frac{p}{\rho} + gz = \text{constant} \tag{4.34}$$

In many cases we can capture the essential part of the flow if gravity is neglected. Although that is a rather drastic approximation, and one that is totally unacceptable for many problems, it is necessary to obtain a workable method. Neglecting gravity, (4.34) becomes

$$\frac{u^2}{2} + \frac{p}{\rho} = \text{constant} \tag{4.35}$$

Since p is constant on the free surface, u is also constant. Using the ζ-plane as the physical plane, the *hodograph transformation* is

$$z = -\frac{1}{U}\frac{dW}{d\zeta} \tag{4.36}$$

in which U is a characteristic velocity and W is the complex potential in the ζ-plane. Because $dW/d\zeta$ is the complex velocity and because the magnitude, but not direction, of the velocity is constant, $|z|$ in (4.36) is constant and the free surface maps into a circular arc.

Example 4.1   Flow out of an orifice in the end of a channel.   The ζ- and z-planes are indicated in Fig. 4.5. The two-dimensional channel of width b_1 has a velocity of U_∞ far upstream. In the end of the channel there is a slit of width b_2. A jet issues from the slit and has a velocity of U_j far downstream (in the absence of gravity). The problem is to find the shape of the jet and

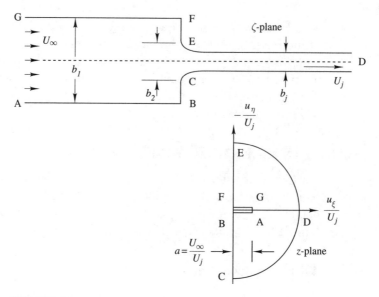

FIGURE 4.5
The ζ-plane and z-plane for flow out of a slit in the end of a channel.

the final width, b_j. Using (4.36) with the characteristic velocity as U_j, the z-plane is as shown in the figure. The separate points are defined:

Points G and A are far upstream where the velocity is U_∞; hence,

$$\frac{u_\xi}{U_j} = \frac{U_\infty}{U_j} = x \qquad \text{and} \ u_\eta = 0 = y$$

Points B and F are in the corners, which are stagnation points, $u_\xi = u_\eta = 0 = x = y$, and the points map into the origin.

At points C and E the pressure is the free surface pressure and thus the velocity is the velocity in the jet. The velocity must be tangent to the walls, so $u_\xi = 0 = x$ and $u_\eta/U_j = \pm 1 = y$.

Point D is far downstream where the velocity is U_j; hence, $u_\xi/U_j = 1 = x$ and $u_\eta = 0$.

The free surface connects points C and E with point D and is a circular arc in order that $|z| = 1$ everywhere on the free surface.

The next task is to write the complex potential in the z-plane such that the lines ABCD and GFED are streamlines. At points A and G there is a source that creates the flow in the channel. The flow from the source is absorbed in the sink at point D. Simply using those singularities, however, does not fulfill the requirements that the imaginary axis and the circle are streamlines. The imaginary axis can be made a streamline by symmetry; for every singularity created on the positive x-axis there is a corresponding singularity on the negative x-axis. The sink at D would take in one-half of its flow from inside the circle and one-half from outside and, thus, must be double the strength of the source at AG. The other one-half of the flow into the sink can come from a source at $x = 1/a$ (Fig. 4.6). The complex potential is then

$$W(z) = \frac{\mu}{2\pi} \ln \frac{(z-a)(z+a)\left(z-\frac{1}{a}\right)\left(z+\frac{1}{a}\right)}{(z-1)^2(z+1)^2} \qquad (4.37)$$

On the circle $z = e^{i\theta}$ that, when used in (4.37), gives

$$W(z) = \frac{\mu}{2\pi} \ln \frac{\cos 2\theta - \left(a^2 + \frac{1}{a^2}\right)}{-2\sin^2\theta} \qquad (4.38)$$

Both the numerator and the denominator of the logarithm term are negative and so $W(z)$ is real. Since $W(z) = \Phi + i\Psi$, and the imaginary part of $W(z)$ is a constant (zero) on the circle, the circle must be a streamline. The real and imaginary axes are also streamlines from symmetry considerations.

Differentiating (4.37) gives

$$\frac{dW}{dz} = \frac{\mu}{2\pi}\left[\frac{1}{z-a} + \frac{1}{z+a} + \frac{1}{z-\frac{1}{a}} + \frac{1}{z+\frac{1}{a}} - 2\left(\frac{1}{z-1} + \frac{1}{z+1}\right)\right] \qquad (4.39)$$

From (4.36)

$$d\zeta = -\frac{1}{U_j}\frac{dW}{dz}\frac{dz}{z} \qquad (4.40)$$

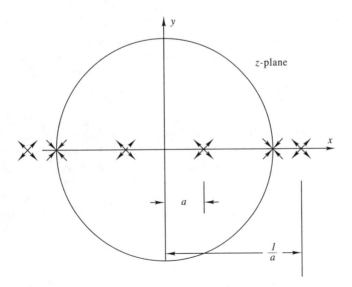

FIGURE 4.6
Location of the singularities for the problem of Fig. 4.5.

Using (4.39) in (4.40) and integrating gives the transformation

$$\zeta = -\frac{1}{U_j}\frac{\mu}{2\pi}\left(\frac{1}{a}\ln\frac{z-a}{z+a} + a\ln\frac{z-\frac{1}{a}}{z+\frac{1}{a}} - 2\ln\frac{z-1}{z+1}\right) + \text{constant} \qquad (4.41)$$

The constant of integration and the distance a are still unknown. They are determined by the conditions

$$\zeta = \frac{ib_1}{2} \quad \text{at } z = 0 \qquad \text{and} \qquad \zeta = -\frac{ib_2}{2} \quad \text{at } z = i \qquad (4.42)$$

The flow out of a slit in an infinite tank is a special case of this problem. In that case $u_\infty \to 0$ and $b_1 \to \infty$ but in such a manner that $U_\infty b_1 = Q$, where Q is a constant and is the flow out of the slit. Consequently, $a = 0$ and the mapping function is

$$\zeta = -\frac{Q}{\pi U_j}\left(\frac{1}{z} + \ln\frac{z-1}{z+1}\right) + \text{constant} \qquad (4.43)$$

Taking $z = i$ and setting $\zeta = -ib_2/2$ gives the constant as

$$\text{constant} = i\left[\frac{b_2}{2} + \frac{Q}{\pi U_j}\left(\frac{\pi}{2} - 1\right)\right] \qquad (4.44)$$

Using (4.44) in (4.43), making $z = 1$, and finding the imaginary part of ζ gives the final width of the jet as

$$\frac{b_j}{b_2} = \frac{\pi}{\pi + 2} = 0.611 \qquad (4.45)$$

which is the contraction coefficient for a jet issuing from a slit.

Historical Note: A more practical problem is the contraction coefficient for a jet issuing from a round orifice. The three-dimensional—or more accurately, axisymmetric—problem cannot be solved analytically. Experiment appeared to confirm that the two-dimensional value also applies to the round orifice. Using a numerical method—the forerunner of what is now known as the boundary integral equation method but necessarily using few elements due to the lack of a computer—Trefftz' (1916) calculations also supported that supposition. Garabedian (1956), a mathematician, computed a value of 0.58 in a remarkable paper. He recognized that the value did not conform to what had been accepted or to the data of Lansford (1934). By extrapolating a part of Lansford's data and dismissing the remainder as inaccurate due to experimental difficulties, he justified his calculation. He also reinterpreted Trefftz' calculation to conform to his value and speculated that a bias toward the accepted value had influenced Trefftz. Garabedian's calculation either went unnoticed among engineers or was not believed. With contrary evidence at hand and with Garabedian's rationalizations, it is not difficult to dismiss the result. Hunt (1968), using an integral technique similar to that of Trefftz, but this time on a computer so that greater accuracy could be achieved, calculated a contraction coefficient of 0.58. (Apparently, Hunt did not know of Trefftz' or Garabedian's works.) Careful experiments, including those by Hunt, confirmed the lower value, which is now accepted.

Example 4.2   Flow around a flat plate.   Consider the flow in the ζ-plane of Fig. 4.7. The flat plate is represented by the line BAC. A potential flow that is symmetric with respect to the plate (no separation) could be formed, but the velocities at the ends of the plate would be infinite. The realistic flow is shown in the figure. There the flow separates from the ends of the plate and leaves a cavity behind the plate. The mapping into the z-plane is also shown in Fig. 4.7. The source at $\zeta = -\infty$ and the sink at $\zeta = \infty$ both are placed into the point $z = 1$ on the real axis.

Following the last example, the source is placed at the point $z = 1 - \epsilon$ and a sink of double strength is placed at $z = 1$. The circle must be a streamline, requiring another source at the point $z = \epsilon/(1 - \epsilon)$. The symmetrical singularities are also placed on the negative z-axis, giving a complex potential as

$$W(z) = \frac{\mu}{2\pi} \ln \frac{\left[z - \left(1 - \frac{\epsilon}{1-\epsilon}\right)\right]\left[z + \left(1 - \frac{\epsilon}{1-\epsilon}\right)\right][z - (1 - \epsilon)][z + (1 - \epsilon)]}{(z - 1)^2(z + 1)^2}$$

$$(4.46)$$

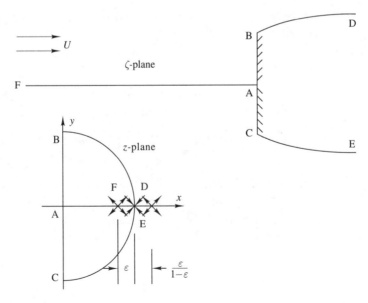

FIGURE 4.7
The ζ-plane and z-plane for free streamline flow around a plate.

We now take ϵ to zero but let μ go to infinity in such a way that $2\mu\epsilon^2 = M$, a constant. Then (4.46) becomes

$$W(z) = -\frac{M}{\pi} \frac{z^2}{(z^2 - 1)^2} \tag{4.47}$$

The latter equation can be used to compute the shape of the free streamline, the velocities, and the pressures in the flow. A drag force can be found by integrating the pressure over the front and rear of the plate. The pressure on the rear of the plate is assumed as constant and equal to the pressure of the free streamlines.

4.4.2   Porous Media

A great variety of mapping solutions are used to solve porous media problems in two dimensions. The book by Polubarinova-Kochina (1952) is a comprehensive source for such applications. These are not included herein. Instead, we mention only a simple technique that can have a variety of uses. The complex potential, when written as any complex function, can be considered as a potential flow

$$W = \Phi + i\Psi = f(z) \tag{4.48}$$

where both Φ and Ψ are solutions to Laplace's equations. Inverting the function

$$z = f^{-1}(W) = x + iy \tag{4.49}$$

indicates that z can be written as a function of W. Using the argument of Sec. 3.12, the real and imaginary parts of z must satisfy Laplace's equation,

$$\frac{\partial^2 x}{\partial \Phi^2} + \frac{\partial^2 x}{\partial \Psi^2} = 0 \qquad \frac{\partial^2 y}{\partial \Phi^2} + \frac{\partial^2 y}{\partial \Psi^2} = 0 \tag{4.50}$$

The solution region in the W-plane is usually much more regular than in the z-plane. More importantly, the solution region in the W-plane is fixed and known whereas the location of a free surface in the z-plane is unknown. The boundary conditions, however, may not be as easy to write in the W-plane.

Example 4.3   Flow through an underdrained dam.   Consider the problem shown in Fig. 4.8. The problem is to find the position of the free surface under

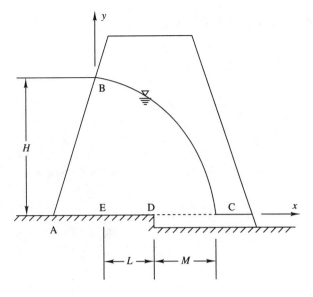

FIGURE 4.8
Flow through an underdrained dam.

steady conditions. The boundary conditions are

$$\Phi = H \quad \text{and} \quad \frac{\partial \Psi}{\partial n} = 0 \quad \text{on } x = (y - H)\cot\alpha$$

$$\frac{\partial \Phi}{\partial n} \quad \text{and} \quad \Psi = 0 \quad \text{on} \quad y = 0, \quad -H\cot\alpha < x < L$$

$$\Phi = y \quad \text{and} \quad \Psi = 1 \quad \text{on the free surface}$$

$$\Phi = 0 \quad \text{and} \quad \frac{\partial \Psi}{\partial y} = 0 \quad \text{on } y = 0, \quad L < x < L + M$$

(4.51)

The W-plane appears as shown in Fig. 4.9. Expressed in terms of y, the

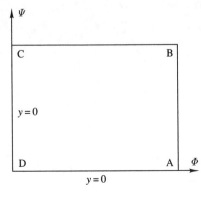

FIGURE 4.9
The potential plane for flow through an underdrained dam.

boundary conditions are

$$y = 0 \quad \text{on} \quad \Phi = 0, \quad 0 < \Psi < 1; \quad y = 0 \quad \text{on} \quad \Psi = 0, \quad 0 < \Phi < H;$$

$$y = \Phi \quad \text{on} \quad \Psi = 1, \quad 0 < \Phi < H;$$

$$-\frac{\partial y}{\partial \Phi} = \frac{\partial y}{\partial \Psi} \cot \alpha \quad \text{on} \quad \Phi = H, \quad 0 < \Psi < 1$$

(4.52)

This problem can be solved for y by any convenient method. Once the solution for y is obtained, the problem for x can be solved. The easiest method is to use the Cauchy-Riemann equations

$$\frac{\partial x}{\partial \Phi} = \frac{\partial y}{\partial \Psi} \qquad \frac{\partial x}{\partial \Psi} = -\frac{\partial y}{\partial \Phi}$$

(4.53)

Using the first of (4.53), the x along the free surface is

$$x(\Phi, \Psi) = x_B + \int_H^\Phi \frac{\partial y}{\partial n} d\Phi$$

(4.54)

Example 4.4 Interface between fresh and salt water in a confined aquifer. The problem is shown in Fig. 4.10. A canal across an island supplies water to an underground aquifer. The water seeping from the canal prevents the aquifer from being contaminated by the salt water. It is a free surface problem because the location of the interface between the fresh and salt water is unknown. The mapping into the W-plane is shown on the right of the figure. The values of Φ at points B and C are assumed known. The free surface is a streamline and the pressure at point A is zero. We assume that the pressure in the salt water is hydrostatic, which gives the head (Φ) in terms of elevation at all points on the free surface. The problem can be solved for y in the manner of the previous example.

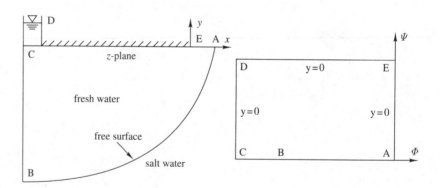

FIGURE 4.10
Flow of fresh water from a canal on an island.

PROBLEMS

4.1. Carry out the details to obtain (4.21).

4.2. Find a Fourier series solution to the linear wave in a box as shown in Fig. 4.11.

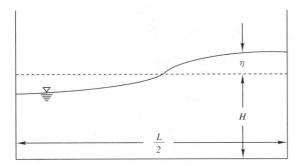

FIGURE 4.11

4.3. The linearized wave equations are

$$\frac{\partial \Phi}{\partial t} = g\eta \qquad \frac{\partial \eta}{\partial t} = -\frac{\partial \Phi}{\partial z}$$

Damping can be added, as in (4.31), if these equations are written

$$\frac{\partial \Phi}{\partial t} = g\eta - \lambda\Phi \qquad \frac{\partial \eta}{\partial t} = -\frac{\partial \Phi}{\partial z}$$

Eliminating η

$$\frac{\partial^2 \Phi}{\partial t^2} + \lambda\frac{\partial \Phi}{\partial t} = -g\frac{\partial \Phi}{\partial z}$$

Show that using the latter equation as a free surface condition leads to damped waves.

4.4. Find the contraction coefficient for flow, Q_0, out of a slit between plates that are at an angle as shown in Fig. 4.12.

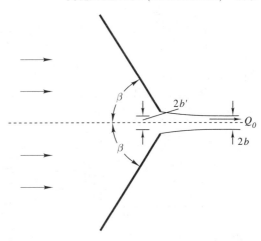

FIGURE 4.12

4.5. Find the drag on a flat plate expressed in terms of the free stream velocity and the length of the plate normal to the flow. Find the drag coefficient

$$C_D = \frac{\text{drag}}{\frac{1}{2}\rho U^2 A} \qquad (4.2)$$

where A is the area of the plate. In two dimensions A becomes the width of the plate.

4.6. Find the transformation and the potential function for the flow around a bent plate as shown in Fig. 4.13. An intermediate mapping will reduce this problem to the one of Example 4.2.

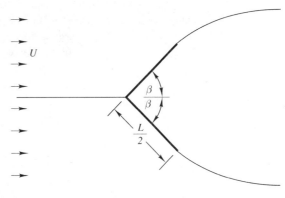

FIGURE 4.13

4.7. Find the drag on the plate of Prob. 4.6 expressed in terms of the free stream velocity and the half-length of the plate.

4.8. Fluid flows out of a convergent nozzle as shown in Fig. 4.14.

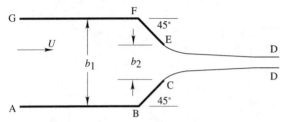

FIGURE 4.14

(*a*) Show the hodograph.

(*b*) Find the potential in the hodograph plane.

(*c*) Find the final width of the jet.

4.9. A cup-type anemometer is constructed as shown in the top of Fig. 4.15. The drag coefficients on the cups are as shown in the other two sketches.

(*a*) Make whatever approximations are necessary and calculate the speed of rotation.

(*b*) Carefully sketch the streamline pattern for the cups in each of the extreme positions (as shown in the figure) and explain why the drag coefficient is greater with the wind vector pointing into the cup. Assume steady flow.

(*c*) Explain how the drag coefficient could be computed in each case using ideal flow.

(*d*) What additional information is needed to compute the speed of rotation more accurately?

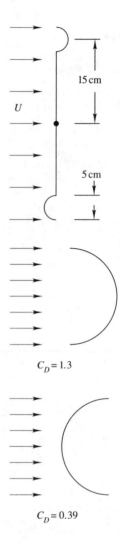

FIGURE 4.15

CHAPTER 5

INERTIALESS
FLOW

Potential theory is concerned with those flows in which the viscous terms can be neglected. In the present chapter we deal with flows where the viscous terms are primary but the acceleration terms are set to zero. Two cases are considered: (1) flows in which acceleration is exactly zero due to the assumptions of the flow regime, and (2) flows in which the acceleration terms can be neglected to a good approximation even though they are not exactly zero. Inertialess flow refers to one of these two cases; we do not mean that the fluid actually has no inertia or that the acceleration term would be zero under different circumstances. Unlike potential flow, the highest order derivatives are not neglected and thus the same sort of problem with the boundary conditions is not present. Since the nonlinear part of the equations is the neglected inertia, we might expect that both analytical and numerical solutions hold no surprises. It turns out, however, that a gremlin called Stokes' paradox is lurking around the corner just to make life interesting.

5.1 TUBE FLOW

First, we consider the flow in a closed conduit. The flow is assumed to be *established*; that is, the entrance and exit regions and other zones where substantial acceleration may take place are sufficiently far away from the region of interest that they have no effect. The definition of the length of the entrance region is not precise, but it is usually from 20 to 60 times the characteristic diameter of the tube. The following analysis excludes regions of acceleration.

5.1.1   Couette and Poiseuille Flow

Consider the flow between parallel flat plates as is shown in Fig. 5.1. The plates are assumed to extend to infinity in both directions and $u_y \equiv 0$. Also, the fluid has constant density. Under these circumstances the equation of continuity (1.20) is automatically satisfied. The equation of motion (1.96) in the x-direction becomes

$$\frac{\partial^2 u_x}{\partial y^2} = \frac{\rho g}{\mu}\frac{\partial h}{\partial x} + \frac{1}{\mu}\frac{\partial p}{\partial x} \tag{5.1}$$

The equation of motion in the y-direction is a statement of hydrostatic pressure. If the pressure term and the gravity term are taken together, the orientation of the plates is obviously immaterial.

Integrating (5.1) twice with respect to y produces

$$u_x = \frac{1}{\mu}\frac{\partial \hat{p}}{\partial x}\frac{y^2}{2} + yf_1(x) + f_2(x) \qquad \hat{p} = p + \rho g h \tag{5.2}$$

in which $f_1(x)$ and $f_2(x)$ are arbitrary functions resulting from the integration of the partial derivative. Because the velocity does not change in

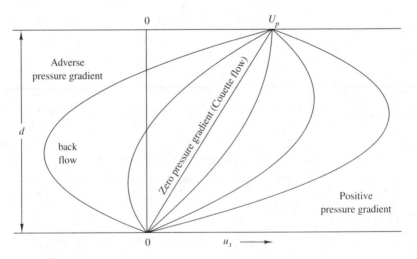

FIGURE 5.1
Flow between parallel flat plates.

the x-direction, f_1 and f_2 must be constants at most. The no-slip boundary conditions are

$$u_x = 0 \quad \text{at } y = 0 \qquad \text{and} \qquad u_x = U_p \quad \text{at } y = d \qquad (5.3)$$

Applying (5.3) to (5.2) gives

$$u_x = -\frac{d^2}{2\mu}\frac{\partial\hat{p}}{\partial x}\left(\frac{y}{d} - \frac{y^2}{d^2}\right) + U_p\frac{y}{d} \qquad (5.4)$$

If $\partial\hat{p}/\partial x = 0$ and $U_p \neq 0$ (*Couette flow*), the velocity varies linearly between the plates. For $U_p = 0$ and $\partial\hat{p}/\partial x \neq 0$ the velocity profile is a parabola centered halfway between the plates and the flow is called *Poiseuille flow*. Several velocity profiles for combinations of the two cases are shown in Fig. 5.1.

For $U_p = 0$ the pressure gradient, $\partial\hat{p}/\partial x$, is a parameter in all of the above equations. The velocity (and the quantity of flow) is linearly dependent on the pressure gradient, which must be impressed on the flow from an outside source. The pressure gradient is balanced by the shear forces on the plates.

Historical Note: Jean Louis Poiseuille (1799–1869) was a French physician who was interested in blood flow. He performed many experiments in small tubes on the order of a millimeter in diameter. The result of the experiments was an equation of the form

$$Q = \left(a_1 + a_2T + a_3T^2\right)\frac{pD^4}{L}$$

in which T is temperature, p is the pressure difference applied to the ends of the tube, D is the diameter, L is the length, and the a_i are empirical coefficients to be determined by the experiment. By including temperature, Poiseuille took account of the viscosity as well as the diameter. His experiments were noted for their accuracy, especially the innovative techniques to measure accurately the diameter of the tubes. The variation in flow with the fourth power of the diameter is known as *Poiseuille's Law*.

5.1.2 Axisymmetric Tube Flow

Round pipes are obviously the most typical carrier of fluids. Because pipes are axisymmetric, the equation of motion is taken as

$$\frac{1}{R}\frac{\partial}{\partial R}\left(R\frac{\partial u_x}{\partial R}\right) = \frac{1}{\mu}\frac{\partial\hat{p}}{\partial x} \qquad (5.5)$$

in which x is the coordinate along the pipe and R is the distance at right angles to the x-axis. Applying the no-slip condition and integrating twice

produces

$$u_x = -\frac{1}{4\mu}\frac{\partial \hat{p}}{\partial x}\left(R_0^2 - R^2\right) \tag{5.6}$$

in which R_0 is the radius of the pipe. The resulting velocity profile is a parabola of revolution with the maximum velocity at the centerline of the pipe. Note that the pressure gradient is always negative for flow in the positive x-direction.

The friction on the pipe walls is a function of the pressure gradient. The friction force per unit length is

$$2\pi R \tau_0 = A\frac{\partial \hat{p}}{\partial x} \tag{5.7}$$

where τ_0 is the shear on the pipe wall and $A = \pi R_0^2$ is the pipe area. Other relationships include the flow rate

$$Q = \int_0^{R_0} 2\pi R u_x\, dR = -\frac{\pi}{8\mu}\frac{\partial \hat{p}}{\partial x}R_0^4 \tag{5.8}$$

and the average velocity

$$U = \frac{Q}{A} = -\frac{1}{8\mu}\frac{\partial \hat{p}}{\partial x}R_0^2 \tag{5.9}$$

From (1.129) the *head loss* per unit length of pipe is

$$\frac{h_L}{L} = -\frac{1}{\rho g}\frac{\partial \hat{p}}{\partial x} = \frac{8\mu U}{\rho g R_0^2} \tag{5.10}$$

The head loss for pipe flow is often calculated by the Darcy-Weisbach equation

$$h_L = f\frac{L}{D}\frac{U^2}{2g} \tag{5.11}$$

where $D = 2R_0$ is the diameter of the pipe and f is the *Darcy friction factor*. Combining (5.10) and (5.11)

$$f = \frac{64\mu}{UD\rho} = \frac{64}{\boldsymbol{R}} \tag{5.12}$$

which indicates that the friction factor depends only on the Reynolds number, $\boldsymbol{R} = UD\rho/\mu$.

5.1.3   Flow in an Arbitrary Cross-section

For tube flow in an arbitrarily shaped cross-section

$$\frac{\partial^2 u_x}{\partial y^2} + \frac{\partial^2 u_x}{\partial z^2} = \frac{1}{\mu}\frac{\partial \hat{p}}{\partial x} \tag{5.13}$$

which is Poisson's equation for the velocity distribution in the tube for flow in the x-direction. For any given flow the right side acts as a constant. In all cases of Poisson's equation where the right side is harmonic (is a solution of Laplace's equation), the equation can be transformed into Laplace's equation. In this case define $\tilde{u}$ such that

$$\frac{\partial^2 \tilde{u}}{\partial y^2} + \frac{\partial^2 \tilde{u}}{\partial z^2} = \frac{\partial^2 u_x}{\partial y^2} + \frac{\partial^2 u_x}{\partial z^2} + 2C_1 + 2C_2 \qquad (5.14)$$

Then $\tilde{u}$ is a solution to Laplace's equation if

$$2C_1 + 2C_2 = -\frac{1}{\mu}\frac{\partial \hat{p}}{\partial x} \qquad (5.15)$$

Thus,

$$\frac{\partial^2 \tilde{u}}{\partial y^2} + \frac{\partial^2 \tilde{u}}{\partial z^2} = 0 \qquad (5.16)$$

which can be solved analytically or numerically in the same way as Laplace's equation for any other problem.

Example 5.1. Consider the problem of flow in a rectangular pipe as shown in Fig. 5.2. The problem is to find the velocity distribution, the wall shear and the head loss. Defining $\tilde{u}$ as

$$\tilde{u} = u_x - \frac{1}{2\mu}\frac{\partial \hat{p}}{\partial x}y^2 = u_x - Ky^2$$

the boundary conditions are

$$\tilde{u} = -Ky^2 \qquad \text{on the pipe walls}$$

The problem can be solved by numerical methods or by application of Fourier series. The latter method (see, for example, Churchill, 1987) yields the

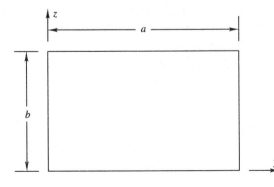

FIGURE 5.2
Cross-section of a rectangular pipe.

solution

$$\tilde{u} = \frac{1}{2\mu} \frac{\partial \hat{p}}{\partial x} b^2 \sum_{n=1}^{\infty} \left\{ \left[\frac{(-1)^n}{n\pi} \left(\frac{2}{n^2\pi^2} - 1 \right) - \frac{2}{n^3\pi^3} \right] \frac{\sin \frac{n\pi y}{b}}{\sinh \frac{n\pi a}{b}} \right.$$

$$\left. \cdot \left(\sinh \frac{n\pi x}{b} + \sinh \frac{n\pi (a-x)}{b} \right) - \frac{1}{n\pi} \left(\cos \frac{n\pi a}{b} + 1 \right) \frac{\sinh \frac{n\pi y}{a}}{\sinh \frac{n\pi b}{a}} \sin \frac{n\pi x}{a} \right\}$$

which gives the velocity distribution. The wall shear comes from the derivative of u_x and the flow rate from the integral of u_x over the cross-section. Given the flow rate, the above equation can be considered an equation in the pressure gradient or the head loss.

5.2 STOKES' FLOW

Obviously, the number of practical flows where the acceleration terms are exactly zero is limited. Most interesting flows have curved streamlines, which means that the flow has a pressure gradient normal to the direction of flow and, thus, accelerations both normal to and along the direction of flow. The approximation of *creeping flow* or *slow flow* neglects the acceleration terms even though they are not identically zero. The basis of the approximation is that the Reynolds number is low, making the viscous terms large compared to the acceleration terms [see (1.96)]. The Reynolds number may be small due to a small density, a small characteristic length, a small velocity, or a very viscous fluid. In the typical example of flow in porous media, the characteristic length, which is the diameter of channels between the grains of a granular medium, is small. Other examples include the movement of small particles through water or air, including aerosols; lubrication, which uses a viscous oil in the small gaps of bearings; motions of viscous fluids such as hot asphalt, sewage sludge, and molasses; processes such as the extrusion of materials; and swimming of micro-organisms (including sperm). These types of fluid motion are sufficiently common to be of engineering importance. In this section we consider only the motion of constant density fluids.

Although the basic approximation is often justified by a low Reynolds number, it really represents the flow of a fluid without mass, an inertialess flow. The same approximation is valid for a fluid with zero density ($\rho = 0$), which means that the density is not a parameter. The drag on a particle moving in air would be the same as the drag on the same particle moving in water if the viscosities of air and water were the same. (The viscosities are actually two orders of magnitude different.)

5.2.1 The Equations of Slow Flow

Neglecting the inertial terms, (1.96) becomes

$$\frac{\partial^2 u_i}{\partial x_j \partial x_j} = \frac{\rho g}{\mu} \frac{\partial h}{\partial x_i} + \frac{1}{\mu} \frac{\partial p}{\partial x_i} \tag{5.17}$$

The pressure gradient has remained in the equation since it could become large. For example, in the simple tube flow of a viscous liquid, a large pressure gradient may be required to drive the liquid through the tube.

In vector notation (5.17) becomes

$$\nabla^2 \vec{u} = \frac{1}{\mu} \vec{\nabla} \hat{p} \tag{5.18}$$

Taking the divergence of (5.18) and changing the order of differentiation produces

$$\nabla^2 \left(\vec{\nabla} \cdot \vec{u} \right) = \frac{1}{\mu} \vec{\nabla} \cdot \vec{\nabla} \hat{p} = 0 \tag{5.19}$$

Thus, the pressure satisfies Laplace's equation. Taking the curl of (5.18) and changing the order of differentiation yields

$$\nabla^2 \left(\vec{\nabla} \times \vec{u} \right) = 2\nabla^2 \vec{\omega} = 0 \tag{5.20}$$

which shows that the vorticity satisfies Laplace's equation also. The vorticity equation (1.134) leads immediately to (5.20) for zero density flow.

If the flow is two-dimensional, the use of the stream function simplifies the equations. Using the stream function in (5.18) gives

$$\frac{\partial \hat{p}}{\partial x} = -\mu \nabla^2 \frac{\partial \Psi}{\partial y} \qquad \frac{\partial \hat{p}}{\partial y} = \mu \nabla^2 \frac{\partial \Psi}{\partial x} \tag{5.21}$$

Differentiating the first of (5.21) with respect to y and the second with respect to x, and subtracting the results gives the *biharmonic equation* (or *bipotential equation*) for the stream function

$$\nabla^4 \Psi \equiv \frac{\partial^4 \Psi}{\partial x^4} + 2 \frac{\partial^4 \Psi}{\partial x^2 \partial y^2} + \frac{\partial^4 \Psi}{\partial y^4} = 0 \tag{5.22}$$

All of the equations of slow flow are linear, a fact which not only greatly facilitates their solution, but also leads to useful mathematical theorems.

5.2.2 Properties of Slow Flow

The most obvious aspect of the above equations is that time does not appear. In the equations of constant density fluid flow, time appears only in the acceleration terms and is discarded with those terms. Unsteady creeping flow is a succession of steady states. It changes instantaneously from one steady state to another. The only case in which the flow can be unsteady is when the boundary conditions are unsteady; there is no unsteady flow under steady boundary conditions.

In making these statements, we have assumed that changes in time are not so rapid as to invalidate the neglect of inertia. Previously, the dimensionless time was defined as $t_* = tU_0/L_0$. We could define the dimensionless time by a characteristic frequency, $t_* = \omega t$. Then the time-dependent inertia term would be

$$\rho\frac{\partial u}{\partial t} = \rho\omega U_0\frac{\partial u_*}{\partial t_*} \tag{5.23}$$

and the Reynolds number would be defined

$$R = \frac{\rho_0\omega_0 L_0^2}{\mu} \tag{5.24}$$

rather than defined in (2.28). For the slow flow approximation to be valid, the Reynolds number of (5.24) must be small, which may not be the case if ω is large.

The biharmonic equation (5.22) has many of the same properties of Laplace's equation. One of these is symmetry. The flow about a symmetrical object will produce symmetrical streamlines. In that sense the flow is reversible. If the pressure gradient were reversed for any given flow, the streamline pattern would be exactly the same, but the velocity along the streamlines would be reversed. There is no wake behind objects or obstructions under the slow flow approximation. Also, a symmetrical object will have an antisymmetrical pressure distribution.

5.2.3 Slow Flow in Infinite Domains

For a large number of engineering problems the Stokes' approximation yields good solutions. For example, Stokes derived the formula that is a classical solution for the drag on a sphere in an infinite domain (Lamb, 1945, p. 597). It begins by seeking a pressure function that satisfies Laplace's equation, reduces to the free stream pressure far from the sphere, and decreases in the direction of the flow. Such a function is

$$\hat{p}_\infty - \hat{p} = C\frac{x}{r^3} \tag{5.25}$$

in which $\hat{p}_\infty$ is the free stream pressure and C is a constant. Using (5.25) in (5.18) gives

$$\nabla^2 u_x = \frac{C}{\mu} \left(\frac{3x^2}{r^5} - \frac{1}{r^3} \right) \tag{5.26}$$

$$\nabla^2 u_y = \frac{C}{\mu} \frac{3xy}{r^5} \tag{5.27}$$

$$\nabla^2 u_z = \frac{C}{\mu} \frac{3xz}{r^5} \tag{5.28}$$

The solutions to the above equations under the boundary condition that $u_x = u_y = u_z = 0$ on $r = r_0$ (the no-slip condition on the surface of the sphere) are

$$u_x = \frac{1}{4} U \left[\frac{3r_0 x^2}{r^3} \left(\frac{r_0^2}{r^2} - 1 \right) - \left(\frac{r_0^3}{r^3} + 3\frac{r_0^2}{r^2} - 4 \right) \right] \tag{5.29}$$

$$u_y = \frac{3}{4} U \frac{r_0 xy}{r^3} \left(\frac{r_0^2}{r^2} - 1 \right) \tag{5.30}$$

$$u_z = \frac{3}{4} U \frac{r_0 xz}{r^3} \left(\frac{r_0^2}{r^2} - 1 \right) \tag{5.31}$$

$$\hat{p} = \hat{p}_\infty - \frac{3}{2} \frac{\mu U r_0 x}{r^3} \tag{5.32}$$

Integrating the pressure and shear over the surface of the sphere gives the drag

$$D = 6\pi \mu r_0 U \tag{5.33}$$

in which U is the velocity of the sphere through the fluid. This equation has been checked experimentally and is valid for Reynolds numbers less than 0.2, approximately. It has been used in a great many applications, which include the movement of aerosols and the settling of dust particles in the atmosphere or in liquids. Equation (5.33) is known as *Stokes' law* and is often expressed in terms of the drag coefficient

$$C_D = \frac{24}{R} \tag{5.34}$$

where R is the Reynolds number, $R = 2U\rho r_0/\mu$.

An easier problem would appear to be the two-dimensional counterpart where the drag on a cylinder in a infinite domain is sought. Surprisingly, there is no solution to this problem using the Stokes' equations. This fact is called *Stokes' paradox*. In the two-dimensional case an equation equivalent to (5.25) is

$$\hat{p}_\infty - \hat{p} = C\frac{x}{r^2} \tag{5.35}$$

and the counterparts of (5.26)–(5.28) are

$$\nabla^2 u_x = \frac{C}{\mu}\left(\frac{2x^2}{r^4} - \frac{1}{r^2}\right) \tag{5.36}$$

$$\nabla^2 u_y = \frac{C}{\mu}\frac{2xy}{r^4} \tag{5.37}$$

As r goes to infinity, u_x must approach the free stream velocity in a manner such as

$$u_x = U + \frac{A}{r} + \cdots \tag{5.38}$$

where the additional terms are of order $1/r^2$ or smaller. Differentiating (5.38)

$$\nabla^2 u_x = \frac{A}{r^3} + \cdots \tag{5.39}$$

But (5.36) contains a larger term, $1/r^2$. Thus, the condition at infinity cannot be met (except for the trivial case of $C = 0$). The conclusion is that no two-dimensional counterpart to the creeping flow solution around a sphere exists. The mathematics appear to indicate that, even for a very long cylinder moving in an infinite fluid, the flow around the ends is so important that it enables the motion. Such a notion is so counter intuitive that we are immediately suspicious. Suspicion pays dividends in that we do need to take a closer look at the mathematics.

In fact the three-dimensional creeping flow solution over a sphere does not conform to experiment at large distances from the sphere, but that does not invalidate the flow near the sphere or the calculated drag force. Whitehead (1889) found that he could not fit the Stokes' solution to the proper boundary conditions at infinity while maintaining the no-slip condition on the sphere (the *Whitehead paradox*).

An explanation of the Stokes' paradox is given by Oseen and appears in Lamb (1945, p. 608). The explanation is that although the inertial forces are everywhere small, the viscous forces become even smaller at large distances from the sphere. The ratio of inertial forces to viscous forces increases in proportion with distance from the sphere and tends to infinity at large distances. The argument is that the neglect of the inertia terms in the equation of motion may be valid in the vicinity of the sphere but not in the far-field. A correction made by Oseen retains the $(\vec{u} \cdot \nabla)\vec{u}$ term in the equation of motion but linearizes it by replacing one of the velocities by the free stream velocity so that it becomes $(\vec{U} \cdot \nabla)\vec{u}$. That change, made largely on an intuitive basis, corrects the far-field flow in three dimensions and permits a two-dimensional solution. Oseen corrects (5.33) to

$$D = 6\pi\mu r_0 U \left(1 + \frac{3}{8}R\right) \tag{5.40}$$

A more rational analysis of the problem is given by Proudman and Pearson (1957) who used matched asymptotic expansions to obtain

$$D = 6\pi\mu r_0 U \left[1 + \frac{3}{8}R + \frac{9}{40}R \ln R + O(R^2)\right] \tag{5.41}$$

The lesson is that, although a small Reynolds number correctly indicates that the inertial terms can be neglected in bounded regions, it does not carry the same implications in infinite regions.

There is an interesting parallel to the Stokes' paradox. Consider a spherical bubble with zero internal (absolute) pressure under water. The pressure of the water outside of the bubble will cause it to implode (causing very high pressures at the moment of collapse). The two-dimensional counterpart would be a round cylinder. It turns out, however, that the collapse of the cylinder requires an infinite amount of energy and hence the implosion cannot take place. Of course, this is a hypothetical phenomenon that cannot be checked by experiment since it would be impossible to create a perfectly round cylinder under water where there is a vacuum in the cylinder. On the other hand, the collapse of the spherical cavity is realistic in that the phenomenon occurs in the cavitation process.

Drag on objects of many different shapes can be calculated either by numerical methods or analytic solutions. An interesting result of inertialess flow is that the drag on an object is larger than the drag on the largest sphere that can be fitted entirely inside the object but smaller than the drag on the smallest sphere that completely surrounds the object.

5.3 HELE-SHAW FLOW

One of the most practical applications of creeping flow is the three-dimensional flow between parallel plates (Fig. 5.3). The apparatus is called a *Hele-Shaw model*. It is used for two distinctly different purposes. The greatest use is probably the simulation of two-dimensional flow in porous media, but it can also be used to simulate hydrodynamic flows. If an object (such as a cylinder) is inserted between the plates, the streamline pattern around the object, assuming slow flow, is similar to the streamline pattern about

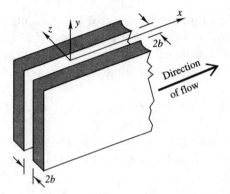

FIGURE 5.3
A Hele-Shaw apparatus.

the same object in potential flow. Remarkably, this example of viscous flow can be used to simulate frictionless flow.

5.3.1 The Hele-Shaw Equations

The analysis begins by defining a coordinate system with the xy-plane midway between the plates and the z-axis normal to the plates. From the analysis of Couette flow, the established velocity distribution between the plates is a parabola

$$u_x = \hat{u}_x \left(1 - \frac{z^2}{b^2} \right) \qquad u_y = \hat{u}_y \left(1 - \frac{z^2}{b^2} \right) \qquad u_z = 0 \qquad (5.42)$$

where the width of the gap between the plates is $2b$ (a constant) and $\hat{u}_x$ and $\hat{u}_y$ are the maximum (centerline) velocities in the x- and y-directions and are themselves functions of x and y. The equation of continuity is

$$\frac{\partial \hat{u}_i}{\partial x_i} = \frac{\partial \hat{u}_x}{\partial x} + \frac{\partial \hat{u}_y}{\partial y} = 0 \qquad (5.43)$$

The equations of motion are

$$\frac{\partial \hat{p}}{\partial x} = \mu \left[\left(1 - \frac{z^2}{b^2} \right) \left(\frac{\partial^2 \hat{u}_x}{\partial x^2} + \frac{\partial^2 \hat{u}_x}{\partial y^2} \right) - \frac{2}{b^2} \hat{u}_x^2 \right] \qquad (5.44)$$

$$\frac{\partial \hat{p}}{\partial y} = \mu \left[\left(1 - \frac{z^2}{b^2} \right) \left(\frac{\partial^2 \hat{u}_y}{\partial x^2} + \frac{\partial^2 \hat{u}_y}{\partial y^2} \right) - \frac{2}{b^2} \hat{u}_y \right] \qquad (5.45)$$

$$\frac{\partial \hat{p}}{\partial z} = 0 \qquad (5.46)$$

Equations (5.44) and (5.45) lead to

$$\nabla^2 \hat{u}_x = \nabla^2 \hat{u}_y = 0 \qquad (5.47)$$

because $\hat{u}_x$, $\hat{u}_y$, and $\hat{p}$ are not functions of z. The only way these equations can balance for an arbitrary value of z is for the terms multiplying z to be zero. The remainder of the equations gives

$$\frac{\partial \hat{p}}{\partial x_i} = -\frac{2\mu}{b^2}\hat{u}_i \qquad i = 1, 2 \tag{5.48}$$

Equation (5.48) expresses the fact that the velocity can be derived from a potential if the potential is expressed as

$$\Phi = \frac{b^2}{2\mu}\hat{p} \tag{5.49}$$

The centerline velocity can then be used to replace the velocity in a two-dimensional potential flow. When used to simulate flow through porous media, the hydraulic conductivity becomes

$$K = \frac{\rho g b^2}{2\mu} \tag{5.50}$$

There are some limitations on the simulation of potential flow in a Hele-Shaw apparatus. First, the flow is assumed to be established everywhere. If the plates are closely spaced, the flow becomes established in a short distance. But if the flow around an object is to be studied, the no-slip condition on the cylinder holds for the model flow whereas it does not for potential flow. The pressure must be positive (*i.e.*, greater than absolute zero) and single valued, which is not the case in some potential flows, especially those with sources and sinks.

The simulation of two-dimensional flow in porous media is usually realistic. Changes in hydraulic conductivity—inhomogeneity—can be simulated by changing the distance between the plates; however, the Hele-Shaw flow in the immediate vicinity of the change in conductivity may differ slightly from reality due to the distance that is required for the flow to become re-established after the disturbance caused by the change in width of the gap.

5.3.2   The Approximation

Because the inertial terms have been neglected, the Reynolds number of the flow must remain low. It is not always obvious how the Reynolds number should be calculated. Based on half of the distance between the plates the Reynolds number is

$$\boldsymbol{R}_p = \frac{Ub\rho}{\mu} \tag{5.51}$$

In those places where the streamlines are nearly straight and parallel, the plate Reynolds number can be large, up to the limit of laminar flow, without violating the approximations. The Reynolds number for flow around an object with a typical dimension of L is

$$\boldsymbol{R}_o = \frac{UL\rho}{\mu} \tag{5.52}$$

The largest viscous term is $\mu \partial^2 u_x / \partial z^2$ and this term should remain large compared to the acceleration terms. We form the ratio

$$\frac{\rho u_x \dfrac{\partial u_x}{\partial x}}{\mu \dfrac{\partial^2 u_x}{\partial z^2}} = O\left(\frac{\dfrac{\rho U^2}{L}}{\dfrac{\mu U}{b^2}}\right) = O\left(\frac{\rho U L}{\mu}\frac{b^2}{L^2}\right) = O\left(\frac{\rho U b}{\mu}\frac{b}{L}\right) \tag{5.53}$$

in which O is to be read "the order of." The criterion for a valid approximation should be

$$\boldsymbol{R}_* = \boldsymbol{R}_o \frac{b^2}{L^2} = \boldsymbol{R}_p \frac{b}{L} \ll 1 \tag{5.54}$$

Actually, experiments indicate that values of $\boldsymbol{R}_*$ up to near unity produce good results.

5.4 LUBRICATION

Hydrodynamic lubrication is a rather specialized subject that can occupy a career. The introduction herein is designed to illustrate that it fits into the category of inertialess flows in most cases. The gap between the solid walls of a bearing can be extremely small, sometimes on the order of 0.01 mm. Thus the Reynolds number, based on the gap width, is small enough to justify the use of inertialess flow. Complications include three-dimensional flow, cavitation in the liquid lubricant, distortion of the "rigid" surface of the bearing, differential heating, and variation of viscosity due to heating. The goal of most calculations is to find the weight that can be supported without breaking down the lubrication, causing metal-on-metal contact, and to find the resistance in the bearing.

A general analysis in simple geometry is that of the slider bearing (Fig. 5.4). For the purpose of analysis we assume that the bearing block is held stationary and the lower surface moves. The no-slip condition applied to the moving surface brings lubricant into the gap. The result is a combination of Couette and Poiseuille flow where the horizontal velocity distribution is given by (5.4). The equations of momentum in the x- and

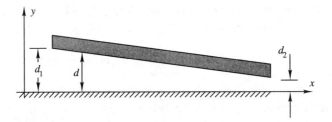

FIGURE 5.4
A slider bearing.

y-directions reduce to

$$\frac{\partial p}{\partial x} = \mu \frac{\partial^2 u_x}{\partial z^2} \qquad \frac{\partial p}{\partial y} = \frac{\partial^2 u_y}{\partial z^2} \tag{5.55}$$

The velocity distribution is

$$u_x = -\frac{d^2}{2\mu}\frac{\partial \hat{p}}{\partial x}\left(\frac{z}{d} - \frac{z^2}{d^2}\right) + U_p\frac{z}{d} \qquad u_y = -\frac{d^2}{2\mu}\frac{\partial \hat{p}}{\partial y}\left(\frac{z}{d} - \frac{z^2}{d^2}\right) \tag{5.56}$$

where $d = d(x)$ and represents the variable height of the gap. Integrating the equation of continuity

$$\int_0^d \frac{\partial u_x}{\partial x}dz + \int_0^d \frac{\partial u_y}{\partial y}dz + \int_0^d \frac{\partial u_z}{\partial z}dz$$

$$= \frac{\partial}{\partial x}\int_0^d u_x dz + U\frac{\partial d}{\partial x} + \frac{\partial}{\partial y}\int_0^d u_y dz + u_z(d) = 0 \tag{5.57}$$

in which the last term represents the vertical velocity of the slider. Substitution of the velocity distribution into (5.57) produces *Reynolds' equation of lubrication*

$$\frac{\partial}{\partial x}\left(d^3 \frac{\partial \hat{p}}{\partial x}\right) + \frac{\partial}{\partial z}\left(d^3 \frac{\partial \hat{p}}{\partial z}\right) = 12\mu\left(U\frac{\partial d}{\partial x} + u_z(d)\right) \tag{5.58}$$

for the pressure distribution. The boundary conditions on the pressure are that it is $\hat{p}_0$ at the edges of the slider.

Discounting the y-direction, the analysis in one dimension is

$$d^3\frac{d\hat{p}}{dx} = 12\mu Ud + C \tag{5.59}$$

where C is a constant of integration. Equation (5.59) is often written so that the constant of integration is $12U d_m$ where $d_0 > d_m > d_1$ (Fig. 5.4) such that the pressure gradient is zero where $d = d_m$,

$$d^3 \frac{d\hat{p}}{dx} = 12\mu U \,(d - d_m) \tag{5.60}$$

At that point the flow is entirely Couette—including the linear velocity distribution—by definition. Writing d as a linear function of x, $d = d_0 + (d_1 - d_0)\,x/L$, integrating once again and applying the boundary conditions yields

$$\hat{p} = \frac{12\mu U L}{d_1 - d_0} \left\{ \frac{d_m/2}{\left[d_0 + \dfrac{(d_1 - d_0)\,x}{L} \right]^2} - \frac{1}{d_0 + \dfrac{(d_1 - d_0)\,x}{L}} \right\} + C_0 \tag{5.61}$$

The constant of integration, C_0, and the value of d_m are chosen so that $\hat{p} = \hat{p}_0$ at each end of the slider. Note that if $d_0 = d_1$, there is no pressure gradient and the flow is Couette flow. A positive pressure results if $d_0 > d_1$ allowing the slider to sustain a load. The pressure varies inversely as the square of the gap distance, so that for very small gaps the pressures can be very high. The load on the slider is determined from the integral of the pressure over the surface of the slider.

PROBLEMS

5.1. Solve the problem of viscous flow in an elliptical conduit of major axis $2a$ and minor axis $2b$. The equation of an ellipse is

$$\frac{y^2}{a^2} + \frac{z^2}{b^2} = 1$$

Using the transformations of (5.14) and (5.15), $\tilde{u}$ on the wall can be made a constant if C_1 and C_2 are chosen such that $K = C_2/C_1$ in which K is used in the same sense as in Example 5.1. Show that $\tilde{u} = $ constant can satisfy both the boundary conditions and the governing equation. Find the flow rate and average velocity in terms of the pressure gradient. Find the head loss. Show that the circular pipe is a special case with $a = b$.

5.2. Write a computer program to sum the series of Example 5.1 for a square conduit ($a = b$). For $a = b = 1$ find the maximum velocity in the cross-section.

5.3. Solve the problem of uniform viscous flow down a plane with a free surface. The free surface is a streamline. The viscous shear on the plane is balanced by the gravity force. For a given flow rate and slope, the depth of the uniform flow is an unknown.

5.4. For Couette flow with no pressure gradient, a differential plate movement of 30 m/s, and distance between the plates of 0.2 cm, find the shear on the plates (fluid temperature 9°C) when the fluid is
(a) air
(b) water
(c) glycerine
(d) "40-weight" engine oil

5.5. For Poiseuille flow (Fig. 5.1 with $U_p = 0$), a distance between plates of 0.2 cm, and a fluid temperature of 9°C, find the pressure gradient necessary to produce a maximum velocity of 30 m/s for

(a) air

(b) water

(c) glycerine

(d) "40-weight" engine oil

5.6. Solve Prob. 5.5 for flow in a 0.2 cm diameter pipe.

5.7. An infinitely long rod of radius R_1 is moving along the centerline of a pipe of radius R_2 in the lengthwise direction. Assume that the pipe contains a viscous liquid and that the pressure gradient is zero. Find an equation for the drag per foot of length.

5.8. Show that the velocity distribution in a triangular duct with equal sides is

$$u_x = -\frac{\rho g \sqrt{3}}{6\mu a}\frac{d}{dx}(p + \rho g h)\left(z + \frac{a}{2\sqrt{3}}\right)$$

$$\cdot \left(z + y\sqrt{3} - \frac{a}{\sqrt{3}}\right)\left(z - y\sqrt{3} - \frac{a}{\sqrt{3}}\right)$$

in which a is the length of the sides, the y-axis is parallel to one on the sides, and the origin is the centroid.

5.9. For slow flow around a sphere calculate the drag by first finding the dissipation of energy in the fluid-filled space.

5.10. Carry out the details of the shear and pressure integration to arrive at (5.33).

5.11. Complete the formulation for the pressure under a slider bearing by finding the constants C_0 and d_m of (5.61). Then find the expression for the upward force on the slider due to the fluid pressure.

CHAPTER
6

LAMINAR FLOW

The flows treated in this chapter are those that contain both viscosity, unlike potential flow, and the inertia terms, unlike inertialess flow. They do not include the most common of natural flows, those with turbulence. The Reynolds number, or the disturbance level, is assumed to remain low enough so that the flow is orderly and laminar. The primary tool is the great idea of Prandtl in the beginning of the twentieth century—boundary layer theory. Boundary layer theory allowed the separation of many flows into the potential and viscous components. The potential component is relatively easy to solve, some approximations can be made to solve the viscous component in many circumstances, and the two components can then be matched for an overall solution. That trick does not, however, allow the solution of all problems and is not a substitute for the complete solution to the Navier-Stokes equations. It does address the primary difficulty in practical solutions, especially numerical solutions: the resolution of phenomena that take place at vastly different scales.

The fact that there is a solution to the equations of motion that fits the boundary conditions of a particular problem does not mean that the solution is realistic. Primarily, one must ask, "Is the solution stable?" The lack of stability means that turbulence will probably result, at least in most practical cases. That question is not asked in this chapter, but it is one of the most important in dealing with any flow situation.

This chapter is meant to give a brief overview of boundary layer theory. The total subject is far too large to include with other features of fluid mechanics. Although only the laminar boundary layer is included, many of the ideas carry over to the turbulent boundary layer, a more common engineering problem.

6.1 STAGNATION FLOW

An "exact" solution exists to the flow of a viscous fluid in the neighborhood of a stagnation point. (The term "exact" is used even though a numerical calculation must be used in the final stages.) This solution is valuable in piecing together the parts of fluid flow around an object and serves as a starting point for many boundary layer calculations. The potential flow approaching a blunt object is replaced by a viscous flow in the neighborhood of the stagnation point. A short distance away from the stagnation point, the exact solution of this section can be used to begin an appropriate boundary layer calculation.

6.1.1 Two-dimensional Stagnation Flow

Consider the flow approaching a flat plate at right angles as shown in Fig. 6.1. Although the plate is assumed to extend to infinity in both di-

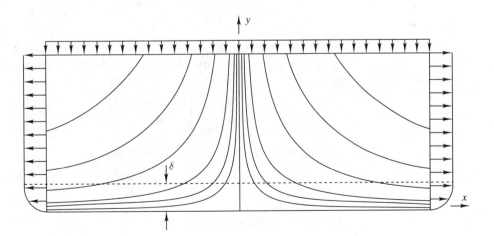

FIGURE 6.1
Stagnation flow on a flat plate.

rections, the center of attention is the stagnation point. The potential flow velocities are

$$\tilde{u}_x = Cx \qquad \tilde{u}_y = -Cy \tag{6.1}$$

in which C is a constant and the potential flow velocities are designated with a tilde in order to differentiate them from velocities that are computed using other equations or approximations. The pressure is calculated by applying Bernoulli's equation with the potential flow velocities

$$\hat{p}_0 - \hat{p} = \frac{\rho}{2}(\tilde{u}_x^2 + \tilde{u}_y^2) = \frac{1}{2}\rho C^2(x^2 + y^2) \qquad \hat{p} = p + \rho gh \tag{6.2}$$

in which $\hat{p}_0$ is the stagnation pressure at $x = y = 0$.

Far from the plate the viscous flow should behave very much like potential flow. As is shown in Fig. 6.1, the velocity in the x-direction is a function of y near the plate and the no-slip condition has been invoked. We assumed that

$$u_x = x\frac{d}{dy}f(y) \tag{6.3}$$

in which $f(y)$ is yet to be determined. The function df/dy presumably is one that affects the flow near the plate but approaches C for values of y far from the plate. From the equation of continuity

$$u_y = -\int \frac{\partial u_x}{\partial x}dy = -\int \frac{df}{dy}dy = -f(y) \tag{6.4}$$

The equation of motion in the x-direction [see (1.96)] is

$$u_x\frac{\partial u_x}{\partial x} + u_y\frac{\partial u_x}{\partial y} = -\frac{1}{\rho}\frac{\partial \hat{p}}{\partial x} + \frac{\mu}{\rho}\left(\frac{\partial^2 u_x}{\partial x^2} + \frac{\partial^2 u_x}{\partial y^2}\right) \tag{6.5}$$

If we can assume that $\partial \hat{p}/\partial x$ is given by differentiating (6.2), then using (6.3), the equation of motion in the x-direction becomes (after dividing by x)

$$\rho\left(\frac{df}{dy}\right)^2 - \rho f\frac{d^2 f}{dy^2} = \rho C^2 + \mu\frac{d^3 f}{dy^3} \tag{6.6}$$

The equation of motion in the y-direction is

$$u_x\frac{\partial u_y}{\partial x} + u_y\frac{\partial u_y}{\partial y} = -\frac{1}{\rho}\frac{\partial \hat{p}}{\partial y} + \frac{\mu}{\rho}\left(\frac{\partial^2 u_y}{\partial x^2} + \frac{\partial^2 u_y}{\partial y^2}\right) \tag{6.7}$$

Again, using (6.2) and (6.3)

$$\rho f\frac{df}{dy} = -\frac{\partial \hat{p}}{\partial y} - \mu\frac{d^2 f}{dy^2} \tag{6.8}$$

The function $f(y)$ is determined by (6.6) whereas (6.8) is used to find the pressure $\hat{p} = F(y)$. The boundary conditions on (6.6) are

$$f = 0 \quad \text{and} \quad \frac{df}{dy} = 0 \quad \text{at} \quad y = 0 \qquad \frac{df}{dy} = C \quad \text{at} \quad y = \infty \qquad (6.9)$$

Unfortunately, there is no apparent analytical solution to (6.6); it can be solved numerically. The equation is nondimensionalized with

$$y_* = \frac{y}{\sqrt{\dfrac{\mu}{C\rho}}} \qquad f_* = \frac{f}{\sqrt{\dfrac{C\mu}{\rho}}} \qquad (6.10)$$

so that

$$\frac{d^3 f_*}{dy_*^3} + f_* \frac{d^2 f_*}{dy_*^2} - \left(\frac{df_*}{dy_*}\right)^2 + 1 = 0 \qquad (6.11)$$

The transformed boundary conditions are

$$f_* = 0 \quad \text{and} \quad \frac{df_*}{dy_*} = 0 \quad \text{at} \quad y_* = 0 \qquad \frac{df_*}{dy_*} = 1 \quad \text{at} \quad y = \infty \qquad (6.12)$$

The boundary layer equations operate in a similar fashion; the y-equation gives a statement about pressure and the velocity is solved by the x-equation. Schlichting (1960) gives the solution to (6.11) and (6.12) in tabular form. It is shown graphically in Fig. 6.2 by the line where $m = 1$.

6.1.2 Axially Symmetric Stagnation Flow

If the flow divides evenly about a point instead of a line—picture a large, round plate—the same sort of solution can be obtained. The same picture applies with the normal to the plate as the x-direction and R as the radial distance from the x-axis. The potential flow solution is

$$\tilde{u}_R = CR \qquad \tilde{u}_x = -2Cx \qquad (6.13)$$

The actual velocities are assumed to be

$$u_R = Rf'(x) \qquad u_x = -2f(x) \qquad (6.14)$$

in which the prime denotes differentiation. The equation of motion in the R-direction is

$$\rho \left(\frac{df}{dx}\right)^2 - 2\rho f \frac{d^2 f}{dx^2} = \rho C^2 + \mu \frac{d^3 f}{dx^3} \qquad (6.15)$$

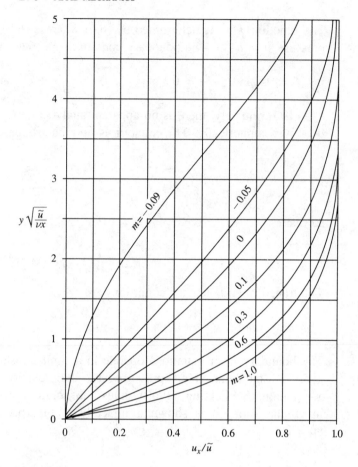

FIGURE 6.2
Solution to Eqs. (6.11) and (6.12).

with the boundary conditions

$$f = 0 \quad \text{and} \quad \frac{df}{dx} = 0 \quad \text{at} \quad x = 0 \qquad \frac{df}{dx} = C \quad \text{at} \quad x = \infty \quad (6.16)$$

After normalization the dimensionless differential equation is

$$\frac{d^3 f_*}{dx_*^3} + 2 f_* \frac{d^2 f_*}{dx_*^2} - \left(\frac{df_*}{dx_*} \right)^2 + 1 = 0 \qquad (6.17)$$

The solution to (6.17) is also tabulated in Schlichting (1960). The primary difference between two-dimensional and axisymmetric stagnation flow is the divergence of the streamlines in the latter. The divergence implies a thinning of the boundary layer and, indeed, the axisymmetric boundary layer is thinner (by about 80 percent) than the two-dimensional counterpart.

6.2 THE BOUNDARY LAYER APPROXIMATION

As the name implies, the boundary layer is a thin region near a solid boundary where the viscous terms are important. The idea was introduced by Prandtl in 1904 and is considered the most important single idea of modern fluid mechanics. Up until that time, practical viscous flow problems were impossible to solve and potential flow solutions often were far from reality for many applications. The idea of the boundary layer showed how potential flow could be put to practical use and at the same time the viscous region is simplified enough to obtain solutions. Using boundary layer theory, the no-slip condition can be satisfied on solid objects. Although the viscous layer may be thin (extremely thin for high Reynolds numbers), it often controls the flow pattern. Even if the student of fluid mechanics does not perform detailed boundary layer calculation, knowledge of the idea and its limitations is important since boundary layer phenomena cause many diferent flow patterns.

Not all fluid flows can be approximated by a potential region connected to a boundary layer. In many flows, including many of engineering importance, the viscosity materially affects large regions or all of the flow domain; no boundary layer approximation is adequate. Such cases include both laminar and turbulent flows. Hence, the boundary layer idea is not a solution to all fluid mechanics problems but instead is a step toward greater command of the subject. If the viscosity materially affects the entire flow, a numerical solution to the Navier-Stokes equations may be easier than in the case of thin boundary layers. The boundary layer phenomenon is an example of the differences in scale—important changes taking place over a length small compared to a typical problem length—that make numerical analysis difficult. Many, many engineering and physics solutions suffer from great scale ranges in the same problem.

Consider the flow about an object. The introduction of boundary-fitted coordinates, such as is shown in Fig. 6.3, makes the solution more conve-

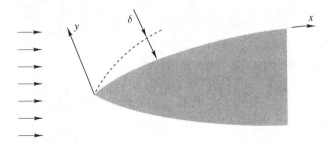

FIGURE 6.3
Coordinates for the boundary layer equations.

nient. The y-axis is normal to the boundary and the x-axis is along the boundary with the origin at the leading edge. We assume immediately that the curvature of the boundary is small compared to the thickness of the boundary layer so that we don't have to worry about using a curvilinear coordinate system; that is,

$$\frac{\delta}{\mathbb{R}} \ll 1 \tag{6.18}$$

in which δ is the boundary layer thickness and $\mathbb{R}$ is the radius of curvature at the point where δ is measured. The definition of boundary layer thickness is that $y = \delta$ at a point where the velocity is some high percentage (say 99 percent) of the velocity that would exist in the potential flow solution around the object. For $y > \delta$ the velocity is, for practical purposes, that calculated by potential flow theory.

The equations of motion are simplified by studying the size of the terms and discarding those terms that are small. The velocities are scaled by using the potential flow velocity $\tilde{u}_x$ or the free stream velocity U. Lengths are scaled according to L, a characteristic dimension of the object, or $\mathbb{R}$, the radius of curvature. Remembering that the x-coordinate is parallel to the body, the y-velocity must be small near the body due to the condition $\vec{u} \cdot \vec{n} = 0$. Terms are normalized with the relationships

$$x^* = \frac{x}{L} \qquad y^* = \frac{y}{\Delta} \qquad u_x^* = \frac{u_x}{U} \qquad u_y^* = \frac{u_y}{U}\frac{L}{\Delta}$$

$$t^* = t\frac{U}{L} \qquad p^* = \frac{p}{\rho U^2} \qquad h^* = \frac{h}{L} \tag{6.19}$$

in which Δ is a characteristic boundary layer thickness. The coordinate normal to the object is scaled on the boundary layer thickness, which is assumed small. Also the velocity normal to the body, u_y, is scaled according to the free stream velocity multiplied by a large quantity. The equation of continuity is

$$\frac{\partial u_x^*}{\partial x^*} + \frac{\partial u_y^*}{\partial y^*} = 0 \tag{6.20}$$

and the equations of motion are

$$\frac{\partial u_x^*}{\partial t^*} + u_x^*\frac{\partial u_x^*}{\partial x^*} + u_y^*\frac{\partial u_x^*}{\partial y^*} = -\frac{\partial p^*}{\partial x^*} - \frac{gL}{U^2}\frac{\partial h^*}{\partial x^*}$$

$$+ \frac{\mu}{\rho U L}\left(\frac{\partial^2 u_x^*}{\partial x^{*2}} + \frac{L^2}{\Delta^2}\frac{\partial^2 u_x^*}{\partial y^{*2}}\right) \tag{6.21}$$

$$\frac{\partial u_y^*}{\partial t^*} + u_x^* \frac{\partial u_y^*}{\partial x^*} + u_y^* \frac{\partial u_y^*}{\partial y^*} = -\frac{L^2}{\Delta^2} \frac{\partial p^*}{\partial y^*} - \frac{gL^3}{U^2 \Delta^2} \frac{\partial h^*}{\partial y^*}$$

$$+ \frac{\mu}{\rho U L} \left(\frac{\partial^2 u_y^*}{\partial x^{*2}} + \frac{L^2}{\Delta^2} \frac{\partial^2 u_y^*}{\partial y^{*2}} \right) \qquad (6.22)$$

If the scaling quantities are chosen correctly, (6.20) shows that $\partial u_x / \partial x$ is of the same size as $\partial u_y / \partial y$. If these terms could not balance each other in the equation of continuity, mass would not be conserved. In other words, a term cannot be eliminated from the equation of continuity.

Next we assume that the Reynolds number $(\rho U L / \mu)$ is large, of order of magnitude L^2 / Δ^2. With this assumption, the next-to-last-term of (6.22) is small (of order Δ^2 / L^2) and the last term is of order one. The large term in the equation is the pressure term (and the elevation term, which always accompanies the pressure term). Because the pressure term is the only large term in the equation, it cannot be balanced by any other term. The conclusion is that

$$\frac{\partial p^*}{\partial y^*} + \frac{gL}{U^2} \frac{\partial h^*}{\partial y^*} = O \left(\frac{\Delta^2}{L^2} \right) \qquad (6.23)$$

which states that the pressure variation across the boundary layer is small. *The pressure is imposed by the potential flow around the object and the boundary layer—as long as it is thin—has negligible influence.* This conclusion in the form of (6.23) is the first of the boundary layer equations, and the equation of motion in the y-direction has been reduced to this simple statement.

In (6.21) the assumption that the Reynolds number is of order L^2 / Δ^2 leads to the conclusions that all terms except the next-to-last are of order one and that the next-to-last term can be discarded as small compared to the others. The final form of the boundary layer equations is

$$\frac{\partial u_x}{\partial x} + \frac{\partial u_y}{\partial y} = 0 \qquad (6.24)$$

$$\frac{\partial u_x}{\partial t} + u_x \frac{\partial u_x}{\partial x} + u_y \frac{\partial u_x}{\partial y} = -\frac{1}{\rho} \frac{\partial p}{\partial x} - g \frac{\partial h}{\partial x} + \frac{\mu}{\rho} \frac{\partial^2 u_x}{\partial y^2} \qquad (6.25)$$

$$\frac{1}{\rho} \frac{\partial p}{\partial y} + g \frac{\partial h}{\partial y} = 0 \qquad (6.26)$$

The equations have been simplified, but not as much as might have been hoped. The equation of motion in the x-direction has lost only one term and is still nonlinear. The unknowns have been reduced to the velocities in each of the directions, u_x and u_y, by considering the pressure, and its variation along the body, as fixed by the potential flow solution. The usually imposed boundary conditions are

$$u_x = 0 \quad \text{and} \quad u_y = 0 \quad \text{at} \quad y = 0 \qquad u_x = \tilde{u}_x \quad \text{at} \quad y = \infty \qquad (6.27)$$

There are three boundary conditions, which is consistent with the two equations where one of the equations is of second order (effectively equivalent to a single equation with a third-order derivative). The first two conditions express the fact that the velocity normal to the body is zero and also that the tangential velocity is zero, the no-slip condition. The last condition states that the velocity becomes the potential flow velocity far from the object (and "far from the object" is at the edge of the boundary layer in the scaled equations—distance as perceived by a mite). The complete statement that is necessary to solve a problem must include the velocity profile at the upstream point. Since the boundary conditions of (6.27) have been applied only to the top and bottom boundaries, the upstream boundary must also have its conditions.

The boundary layer equations are of the parabolic (or diffusion) type. Any change on the body, a very small bump, for example, will affect the flow in the entire boundary layer from the point of the change downstream. An unsteady change on the body is transmitted across the boundary layer instantaneously, but it is advected downstream by the fluid velocity. The unsteady change is not felt upstream.

The assumptions of small boundary layer thickness and of large Reynolds number form limitations to the theory. Either or both of these assumptions may not be the case for any particular problem. In the wake behind an object the area of disturbed (that is, nonpotential) flow is thick and the boundary layer approximation is not valid. At low Reynolds numbers the equations are not valid and the boundary layer may become thick. The boundary layer tends to grow with distance and can become so thick that the approximation is poor. In pipe flow, for example, the boundary layer grows so thick that the entire flow in the pipe is affected by the no-slip condition; there is no potential flow.

Finally, the boundary layer equations can be written without a parameter. Substituting the Reynolds number, (6.25) contains a single parameter,

$$\frac{\partial u_x^*}{\partial t^*} + u_x^* \frac{\partial u_x^*}{\partial x^*} + u_y^* \frac{\partial u_x^*}{\partial y^*} = -\frac{\partial \hat{p}^*}{\partial x^*} + \frac{1}{R} \frac{\partial^2 u_x^*}{\partial y^{*2}} \qquad (6.28)$$

That parameter can be eliminated by a scaling change

$$y^+ = y^* \sqrt{R} \qquad u_y^+ = u_y^* \sqrt{R} \qquad (6.29)$$

Using the potential flow relationship

$$\tilde{u}_x \frac{\partial \tilde{u}_x}{\partial x} = -\frac{1}{\rho} \frac{\partial \hat{p}}{\partial x} \tag{6.30}$$

Using (6.30) to replace the pressure, (6.28) becomes

$$u_x^* \frac{\partial u_x^*}{\partial x^*} + u_y^+ \frac{\partial u_x^*}{\partial y^+} = \tilde{u}_x^* \frac{\partial \tilde{u}_x^*}{\partial x^*} + \frac{\partial^2 u_x^*}{\partial y^{+2}} \tag{6.31}$$

which has the advantage that the equations need to be solved only once for a given geometry regardless of the Reynolds number (provided, of course, that the Reynolds number is in the proper range for laminar flow). In terms of the dimensional variable

$$y = y^+ \frac{\Delta}{\sqrt{R}} \tag{6.32}$$

Because y^+ is independent of the Reynolds number, (6.32) indicates that the boundary layer thickness (where $y = \delta$) becomes smaller as the Reynolds number grows. Obviously, a smaller boundary layer thickness means that the velocity gradient becomes larger in the y-direction, leading to higher shear forces on the object.

Historical Note: Ludwig Prandtl (1875–1953) spent most of his professional career as a professor at the University of Göttingen. Before Göttingen he was on the faculty of the Polytechnic Institute of Hanover where he built a recirculating flume in which he studied the flow around a cylinder, making the flow visible by means of particles. He presented his finding to the Third International Congress of Mathematics where it was not well received because his paper was said to have neither important mathematical developments nor practical results. It did, however, introduce the concept of a boundary layer. Perhaps the idea would have died but for Felix Klein (1849–1925), a respected mathematician at Göttingen. By 1913 Prandtl was operating the first closed-circuit wind tunnel at Göttingen in which he studied aerodynamics. This time was the beginning of the golden age of discovery in aerodynamics. The Englishman, Frederick William Lanchester (1868–1946) had just published his book *Aerial Flight* (1908). Prandtl's wind tunnel exposed many phenomena that Lanchester had predicted in theory. Those were the times of discovery, and Prandtl was the right person at the right place.

Prandtl went on to make many contributions to fluid mechanics. His study of turbulence showed great insight. But perhaps his greatest contribution was the output of students from Göttingen. These include von Kármán, Blasius, Betz, Tollmien, Schiller, Tietjens, Schlichting, Nikuradse, von Mises, Lötz, and many others.

6.3 SEPARATION (STALL)

Most flows of engineering importance that involve air or water are at sufficiently high Reynolds numbers that the boundary layer phenomenon is important. In many cases where the flow boundaries are curved, and even sometimes when they are not, the flow separates from the boundary and creates a *wake*. This phenomenon causes the drastic difference between many real flows and the potential solution.

6.3.1 Definition of Separation

Consider the flow around the top half of the elliptical object as shown in Fig. 6.4. Several velocity profiles are shown but the thickness of the boundary layer is exaggerated in order to make it visible. Recalling that the pressure on the object is obtained from the potential flow solution, the point where the object is widest is close to the point of minimum pressure. Over the forward part of the object the pressure decreases in the downstream direction. This decrease in pressure tends to increase the velocity in the boundary layer. On the downstream side of the midpoint the potential flow pressure increases in the downstream direction with the consequence that the velocity in the boundary layer is retarded. The no-slip condition has dictated that the velocity near the body is already small. There are two opposing forces in this part of the boundary layer. The shear from the flow in the outer part of the boundary layer tends to move the boundary layer fluid in the downstream direction whereas the pressure gradient has the opposite effect. In this battle for direction the pressure gradient has the advantage since it acts throughout the boundary layer whereas the free stream shear is applied only to the outside part. The pressure gradient is often strong enough to cause reverse (upstream) flow near the object. The zone of reverse flow is called the separated area. The word *stall* is also used to describe the phenomenon.

The *point of separation* is that point on the body where the shear and the y-gradient of velocity are zero, that is,

$$\tau_{yx} = \mu \left(\frac{\partial u_x}{\partial y} \right)_{y=0} = 0 \qquad (6.33)$$

The *zone of separation* is that portion of the body where the shear is in the upstream direction or

$$\left(\frac{\partial u_x}{\partial y} \right)_{y=0} < 0 \qquad (6.34)$$

As is evident from Fig. 6.4, the separated velocity profiles contain a point of inflection where the second derivative of the velocity with respect to the distance from the body (the curvature of the velocity profile) is zero

$$\frac{\partial^2 u_x}{\partial y^2} = 0 \tag{6.35}$$

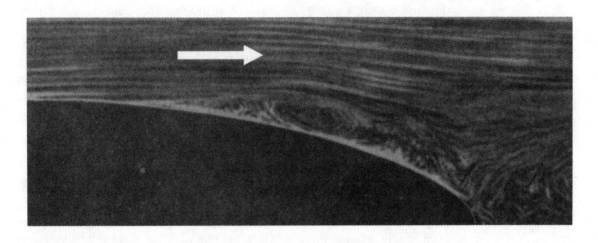

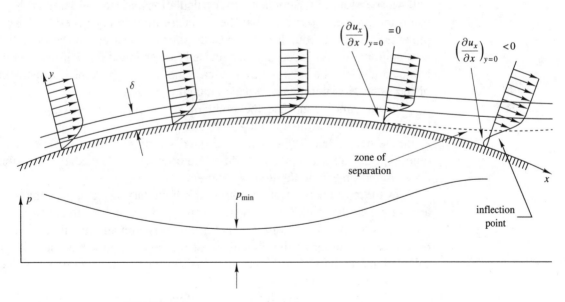

FIGURE 6.4

The boundary layer on an elliptical object (photo from Ludwig Prandtl, *Essentials of Fluid Dynamics*, 1952, p. 140, by permission of Blackie Academic & Professional, an imprint of Chapman & Hall).

From the boundary layer, equation (6.25) is written for a point on the object (where $u_x = u_y = 0$) as

$$\left(\frac{\partial^2 u_x}{\partial y^2} \right)_{y=0} = \frac{1}{\mu} \frac{\partial \hat{p}}{\partial x} \tag{6.36}$$

If the pressure gradient is negative (pressure decreasing in the downstream direction), then $\partial^2 u_x / \partial y^2$ is negative at the boundary and throughout the boundary layer; there is no inflection point and the velocity profile is what one would expect. If the pressure gradient is positive, then $\partial^2 u_x / \partial y^2$ is positive at the boundary and up to the inflection point of the velocity profile. The fact that the velocity profile contains an inflection point is important in the calculations of the stability of the flow. Velocity profiles with inflection points tend to be a source of instability leading to turbulence.

6.3.2 Some Consequences of Separation

The primary cause of separation is the adverse pressure gradient. Wherever the pressure increases in the downstream direction, separation is a likely possibility. Not every adverse pressure gradient will cause separation, but even mildly adverse pressure gradients that act over a considerable length will cause separation. Boundary layer calculations, at least in laminar flow, can often predict separation, but the exact location is not necessarily well predicted. Because boundary layer calculations are based on the potential flow solution, that solution must be revised for separated flows if the overall computation is to be accurate. In any case the flow in the zone downstream of the separation point cannot be computed accurately.

In one important case the point of separation can be located with certainty. In flows around objects with sharp corners, the separation occurs at the corners. That includes flows around buildings, many vehicles, and other types of obstructions. Such flows are often easier to calculate than the flow around objects without sharp corners.

In the portion of separated flow, the boundary layer approximations are not generally valid. Primarily, the region of flow that is affected by viscosity and the no-slip condition is not necessarily thin and its thickness can become of the order of the dimension of the object. Thus, boundary layer calculations, as based on the equations of the last section, must terminate at the point of separation.

After separation the effective shape of the object from the point of view of the potential flow calculation is greatly altered. Because the boundary layer is no longer thin, the pressure on the boundary is not the pressure that would be calculated by potential flow. The pressure in the separated region is often less than the potential flow calculation would indicate. On blunt objects it may be near the free stream pressure instead of approaching the stagnation pressure. Since this is the portion of the flow on the rearward side of the object, the lower pressure, when integrated over the surface, causes a lower than expected force in the direction of the oncoming flow. This force is a part, perhaps the major part, of the drag on the object. Efforts to reduce drag on a wide variety of objects from vehicles to golf balls involve the elimination or reduction of the size of the zone of separation.

The cause of separation can be viewed as a lack of momentum in the boundary layer that otherwise would overcome the adverse pressure gradient. One of the methods to delay or prevent separation is to infuse momentum into the boundary layer. The most common approach is to encourage the boundary layer to be turbulent. In turbulent flow the shear (transverse momentum transport) is much greater than in laminar flow. The infusion of momentum into the boundary layer increases the velocity gradient near the boundary and consequently increases the viscous shear on the boundary. If the goal is to prevent drag, then that increase in shear, causing additional drag, is usually overwhelmed by the decrease in the size of the separated region and the consequent reduction in the pressure drag.

Turbulence generation is accomplished by using a rough boundary to "trip" the laminar boundary layer. The choice of the most effective mechanism depends on the Reynolds number. Even when the boundary layer is already turbulent, the artificial increase of boundary layer momentum can decrease separation, as is accomplished by the rough surface on a golf ball, for example. Another example is the "vortex generator" used on many commercial jet airplanes. The vortex generator consists of one or more rows of small, flat pieces of metal projecting upward from the top of the wing. These pieces of metal increase the turbulence in the boundary layer (the Reynolds number is so high that the boundary layer is already turbulent), infuse additional momentum into the boundary layer, and delay separation. The additional drag on the projections and skin friction on the wing is considered a worthwhile price for the prevention of separation.

When applied to airfoils, separation is usually called stall. Lift on an airfoil comes about because of an increased pressure on the underside and a decreased pressure on the top. On the top of the airfoil the pressure decreases from the free stream pressure to a minimum then increases in the downstream direction from that minimum to the trailing edge. An adverse pressure gradient is formed on a major part of the top of most airfoils. If separation occurs, the pressure in the separated region is much higher than it would otherwise be, resulting in a decrease in lift and an increase in drag. Fig. 6.5 shows the lift and drag coefficients, C_L and C_D, as defined by the

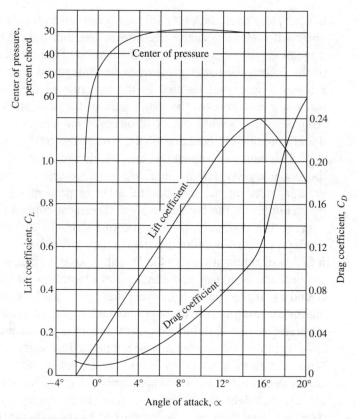

FIGURE 6.5
Lift and drag coefficients for an airfoil. Stall occurs at an angle of attack of approximately 15.5°.

equations

$$\text{Lift} = C_L\left(\frac{1}{2}\rho U^2 A\right) \qquad \text{Drag} = C_D\left(\frac{1}{2}\rho U^2 A\right) \tag{6.37}$$

where A is the area of a horizontal projection of the wing. At high angles of attack, α, the flow separates from a portion of the top of the wing, causing a large increase in the drag coefficient and a decrease in the lift coefficient. Not surprisingly, the airplane cannot fly in this regime and the phenomenon has caused a great many crashes. Some artificial devices such as vortex generators or leading edge cuffs can delay stall, but they cannot prevent it altogether. It is the responsibility of the pilot (and of the warning systems) to avoid the high angles of attack that induce stall. (Frequently, newspaper articles have reported the cause of a crash as engine failure simply because the reporter misunderstood the meaning of the word "stall.")

In many other situations of engineering interest separation can have important consequences. In rivers sediment tends to collect in the *backwater* or separated regions of bends or on the downstream side of other obstructions. Flows over buildings contain separated regions that cause forces with structural importance. Although separation most often has negative consequences, it can sometimes be put to beneficial use. In the example of rivers, separation can be used to exercise some control over the deposition of sediment. That idea is used to design groins and training walls that protect river banks. A common use in colder climates is the snow fence, which influences the deposition of snow.

6.3.3 Unsteady and Three-dimensional Separation

The stability of the flow at the point of separation is often marginal or nonexistent. A slight difference in pressure can sometimes cause a great difference in the streamline configuration. Such pressure differences can occur in an unsteady, sometimes periodic, fashion in an otherwise steady flow. One of the paradoxes of fluid flow is that a flow under steady boundary conditions can become unsteady. In flow around a cylinder the separation points oscillate rearward and forward alternately at an ill-defined range of Reynolds number from about 40 to 80. The frequency of the oscillation is one of the items in the Strouhal number

$$St = \frac{\omega D}{U} \tag{6.38}$$

in which ω is the frequency and D is the diameter of the cylinder. The Strouhal number is about 0.2 over a wide range of Reynolds numbers. The oscillation causes vortices to be shed alternately from the top and the bottom of the cylinder, which results in a pattern of vortices rotating in opposite directions behind the cylinder. This pattern is called a *Kármán vortex street* (Fig. 6.6). The movement of the separation point on each side of the cylinder causes an alternating pressure. The pressure, in turn, can cause structural vibrations transverse to the free stream. The vibrations can be destructive, especially if they happen to coincide with a natural frequency of the structure. (The coincidence of vortex shedding and the natural frequency can be induced by a vibrating structure itself. Vibrations, which tend to be in the natural frequency, may themselves induce changes in the boundary layer that cause separation and increased vibration.) Smokestacks are often protected by having devices that induce separation in a fixed point and prevent the oscillations. Because the phenomenon is Reynolds number dependent, it can happen on a variety of objects ranging from transmission

FIGURE 6.6
The Kármán vortex street (top, from G. S. Richards *Philosophical Transactions of the Royal Society of London*, Series A, 233. 1935, courtesy of the Royal Society of London; middle after Kármán, from Ludwig Prandtl, *Essentials of Fluid Dynamics*, 1952, p. 183, by permission of Blackie Academic & Professional, an imprint of Chapman & Hall; bottom courtesy of Professor S. Taneda).

wires (causing the "singing" sometimes heard in a strong wind) to large smokestacks and even buildings.

This sort of fluid-induced structural vibration should not be confused with *flutter*. Flutter is usually associated with aircraft structures, especially the wings, tail, and control surfaces, but it also occurs on other parts. Flutter is a combination of two or more modes of vibration that couple with the flow field. The flow feeds energy into the system, thus increasing the amplitude of the vibration until it is either damped by some nonlinear mechanism or there is a failure. Most airplanes are limited in maximum speed to a value safely below the *flutter speed*. There have, however, been some spectacular crashes due to flutter. Pilots in combat situations have exceeded the flutter speed in dives. The most recent flutter crash involving commercial aircraft occurred in 1960. Many suspension bridges are subject to flutter in strong winds. A graphic example of this occurrence is the failure of the Tacoma Narrows Bridge.

Vorticity theory plays essential roles in the treatment of boundary layers. Obviously, the vorticity in a thin boundary layer is high. In any sort of separation phenomenon, that vorticity is diffused into the flow and is the cause of important changes in the fluid motion.

On three-dimensional objects boundary layer flows often form complex patterns that are difficult to predict. Very small, almost insignificant, transverse pressure gradients can cause marked changes in the fluid flow near the surface where the velocities are naturally small, the inertia of the fluid is small, and the direction of flow is easily influenced. Consider, for example the flow around a pitched ellipsoid (Fig. 6.7). The normal increase in pressure on the rearward part of the upper surface might cause separation on a two-dimensional object. On the ellipsoid it causes separation, but the separation takes on a different aspect as the streamlines tend to diverge and the flow near the top of the ellipsoid is swept toward the sides. The same phenomenon occurs on swept aircraft wings, which have a pressure gradient toward the ends. The undesirable effects of the spanwise flow is combatted by the construction of a boundary layer fence (Fig. 6.8).

These transverse velocities are closely related to *secondary currents* that occur commonly in a wide variety of situations. Consider the flow in the pipe bend of Fig. 6.9. The pressure must be greater on the outside of the curve. In general the increase in pressure is just sufficient to provide the inward acceleration so that the streamlines can follow the wall curvature; that is, the pressure gradient balances the centrifugal force. On the sides of the conduit, however, the velocity is retarded, leading to less centrifugal force. Since the pressure gradient is undiminished in those areas, it causes a flow toward the center of curvature. Conservation of mass dictates a return

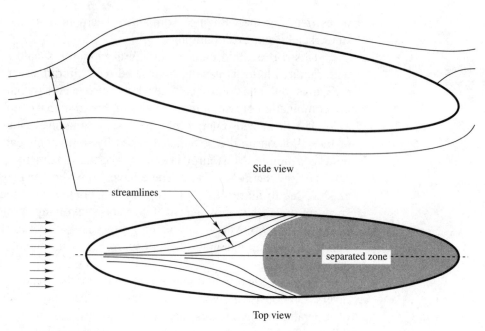

Side view

streamlines

separated zone

Top view

FIGURE 6.7
Separated flow about an ellipsoid at an angle of attack.

flow in the central part of the conduit. Such flows occur in all types of pipes and conduits. (There is another type of secondary flow—"weak" secondary flow—that occurs in straight conduits of noncircular cross section when the flow is turbulent. It should not be confused with the flow described above, which can occur in both laminar and turbulent flow. See Sec. 7.6.) Secondary flows have a marked influence on the distribution of shear on the boundaries. In rivers secondary flows cause the bottom current to flow inward in the bends. Because the majority of the sediment is next to the bed, it is deposited on the inside of bends whereas the outer banks tend to be eroded. The result is that the river tends to "meander," progressively removing material from the outside of bends and depositing it in the inside. Over time secondary currents have a great influence on the geological development of rivers. The time scale for meandering is sufficiently short (sometimes as short as a single storm event) that it has major consequences for construction projects—we would like the river to continue to flow under the bridge that was constructed just last year.

Vortex motion within the fluid can cause separation at points where it is entirely unexpected. Although the automobile and aircraft industries, among others, have many years of experience in drag reduction programs and have developed a vast store of knowledge and experience, wind tunnel

FIGURE 6.8
Lear jet. (Courtesy of Learjet, Inc.)

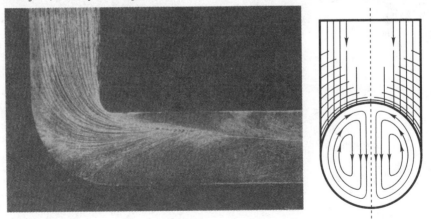

FIGURE 6.9
Secondary flow in a curved conduit (from Ludwig Prandtl, *Essentials of Fluid Dynamics*, 1952, pp. 146, 147, by permission of Blackie Academic & Professional, an imprint of Chapman & Hall).

and flight testing still remain the primary modes for determining efficient shapes. The calculations and the experience of the engineers can go far in designing efficient shapes, but the actual flow often contains surprises.

6.4   BOUNDARY LAYER THICKNESS

In deriving the boundary layer equations, a thickness parameter was defined as the point where the velocity near the wall achieves 99 percent (or some other large percentage) of the potential flow velocity when calculated with the same geometry. The velocity approaches the potential flow velocity asymptotically, but there is no clear point where the boundary layer ends and frictionless flow begins. There are, however, some unique definitions.

Consider the flow on a flat plate as shown in Fig. 6.10. The plate is assumed of infinitesimal thickness and aligned parallel to the oncoming stream so that it would have no disturbance at all in a frictionless flow. Due to the no-slip condition in viscous flow and the retarding of velocity near the plate, the streamlines are displaced outward. The amount of displacement is called the *displacement thickness*, δ_d, and is a function of distance along the boundary layer. Consider the streamline shown in Fig. 6.10. The quantity of flow between this streamline and the plate is a constant regardless of position along the plate, giving the equation

$$\int_0^{y_s+\delta_d} u_x \, dy = \text{constant} = U y_s \tag{6.39}$$

At the leading edge of the plate the boundary layer has no thickness, $\delta_d = 0$,

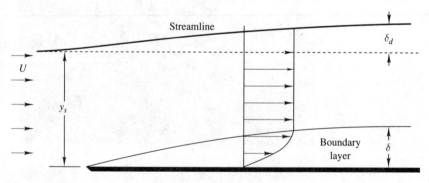

FIGURE 6.10
Boundary layer on a flat plate.

and $u_x = U$. Equal integrals are subtracted from each side of (6.39) to give

$$\int_{y_s}^{y_s+\delta_d} u_x \, dy = U\delta_d = \int_0^{y_s} (U - u_x)dy \qquad (6.40)$$

Since y_s is an arbitrary quantity, it could be infinity. Solving for δ_d

$$\delta_d = \int_0^\infty \left(1 - \frac{u_x}{U}\right) dy \qquad (6.41)$$

Because δ_d is the displacement of the streamlines, it could be considered as an added thickness of the body. That is, the potential flow computation would be more accurate if it considers the flow about a body that is fatter by the amount of δ_d. Some computer programs that solve for flow around wings iterate on the boundary layer, first computing the potential flow for the solid profile, then computing the displacement thickness, then recomputing the potential flow around a wing with the thickness increased by the displacement thickness until a converged solution is reached.

The velocity profile and the displacement thickness are shown in Fig. 6.11. The figure indicates that the amount of flow between the line a distance δ_d from the wall and the wall is equal to the amount of flow that is missing due to the viscous action from the distance δ_d to infinity.

The *momentum thickness* δ_m is defined in a similar manner as

$$\delta_m = \int_0^\infty \frac{u_x}{U} \left(1 - \frac{u_x}{U}\right) dy \qquad (6.42)$$

The momentum thickness represents the displacement of free stream momentum transport and is closely related to the skin friction drag on an

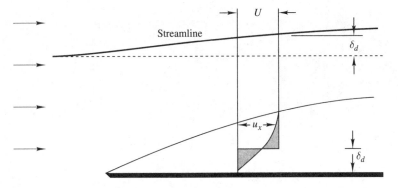

FIGURE 6.11
The velocity profile and displacement thickness.

object. From a simple control volume analysis the drag is

$$D = \rho \int_0^\infty u_x(U - u_x)\,dy = \rho U^2 \delta_m \tag{6.43}$$

The drag is also the integral of the wall shear

$$D = \int_0^x \tau_0\,dx \qquad \text{or} \qquad \tau_0 = \frac{dD}{dx} \tag{6.44}$$

Using (6.43) and (6.44)

$$\tau_0 = \rho U^2 \frac{d\delta_m}{dx} \tag{6.45}$$

The *energy thickness* (or *energy dissipation thickness*) δ_e is used to represent the loss of kinetic energy due to viscosity. The loss of kinetic energy in an element of distance, Δy, normal to the boundary is $\rho u_x(U^2 - u_x^2)\Delta y/2$. The loss of the transport of kinetic energy is equated with the equivalent transport in the free stream

$$\frac{1}{2}\rho U^3 \delta_e = \frac{\rho}{2} \int_0^\infty u_x(U^2 - u_x^2)\,dy \tag{6.46}$$

so that the definition of energy thickness is

$$\delta_e = \int_0^\infty \frac{u_x}{U}\left(1 - \frac{u_x^2}{U^2}\right)dy \tag{6.47}$$

The relative magnitudes of the displacement thickness, momentum thickness, and energy thickness are shown in Fig. 6.12 for a usual shape of the boundary layer velocity profile. If the velocity profile is taken as parabolic (which approximates some boundary layers but is not exact) with

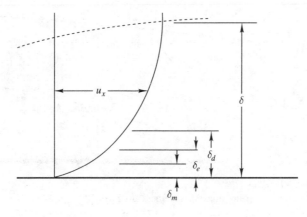

FIGURE 6.12
Comparison of the boundary layer thickness, displacement thickness, momentum thickness, and energy thickness.

the equation

$$\frac{u_x}{U} = 2\frac{y}{\delta} - \frac{y^2}{\delta^2} \tag{6.48}$$

then $u_x = U$ at $y = \delta$ and

$$\delta_d = \frac{1}{3}\delta \qquad \delta_m = \frac{2}{15}\delta \qquad \delta_e = \frac{22}{105}\delta \tag{6.49}$$

6.5 SOLUTION FOR THE BOUNDARY LAYER ON A FLAT PLATE

The simplest solution for boundary layer flow is for flow over the flat plate of Fig. 6.10. It is particularly simple because the pressure gradient is everywhere zero as shown from the trivial potential flow solution. The boundary layer equations for the flat plate are

$$\frac{\partial u_x}{\partial x} + \frac{\partial u_y}{\partial y} = 0 \tag{6.50}$$

$$u_x\frac{\partial u_x}{\partial x} + u_y\frac{\partial u_x}{\partial y} = \frac{\mu}{\rho}\frac{\partial^2 u_x}{\partial y^2} \tag{6.51}$$

and the boundary conditions are

$$u_x = 0 \quad \text{and} \quad u_y = 0 \quad \text{at} \quad y = 0 \qquad u_x = U \quad \text{at} \quad y = \infty \tag{6.52}$$

The boundary layer over a flat plate becomes turbulent at the displacement thickness Reynolds number of about 600 ($\boldsymbol{R}_{\delta_d} = \rho U \delta_d / \mu \approx 600$). The equations should not be applied for higher Reynolds numbers. Note that in individual cases the Reynolds number for turbulent flow may be quite different than the "rule of thumb."

6.5.1 The Blasius Equation

The continuity equation (6.50) is satisfied immediately by the introduction of the stream function. Then (6.51) becomes

$$\frac{\partial \Psi}{\partial y}\frac{\partial^2 \Psi}{\partial x \partial y} - \frac{\partial \Psi}{\partial x}\frac{\partial^2 \Psi}{\partial y^2} = -\frac{\mu}{\rho}\frac{\partial^3 \Psi}{\partial y^3} \tag{6.53}$$

with the boundary conditions

$$\frac{\partial \Psi}{\partial x} = 0 \quad \text{and} \quad \frac{\partial \Psi}{\partial y} = 0 \quad \text{at} \quad y = 0 \qquad \frac{\partial \Psi}{\partial y} = -U \quad \text{at} \quad y = \infty \tag{6.54}$$

The first of the conditions implies that $\Psi = $ constant on the boundary.

If we can guess how Ψ varies with x, (6.53) can be reduced to an ordinary differential equation. Balancing the dimensions helps the process. Dimensionless variables are defined as

$$\eta_* = y \sqrt{\frac{U\rho}{\mu x}} \qquad \Psi_*(\eta) = -\sqrt{\frac{\rho}{U\mu x}} \Psi \qquad (6.55)$$

in which Ψ_* is a dimensionless stream function and is a function of η_*. Substitution into (6.53) produces the *Blasius equation*

$$2\frac{d^3\Psi}{d\eta_*^3} + \Psi_* \frac{d^2\Psi_*}{d\eta_*^2} = 0 \qquad (6.56)$$

which was first solved by Blasius (1908); the solution was improved by Howarth (1938). The boundary conditions in terms of the new variables are

$$\Psi_* = 0 \quad \text{and} \quad \frac{d\Psi_*}{d\eta_*} = 0 \quad \text{at} \quad \eta_* = 0 \qquad \frac{d\Psi_*}{d\eta_*} = 1 \quad \text{at} \quad \eta_* = \infty$$
$$(6.57)$$

The differential equation can be solved numerically, but the boundary condition at infinity is awkward to impose. That condition can be replaced by a condition at $\eta_* = 0$ by a transformation. Let

$$\Psi_* = \alpha \, g(h) \qquad h = \beta \eta_* \qquad (6.58)$$

Assume that at $h = 0$, $d^2\Psi/d\eta_*^2 = \gamma$, where γ is a yet undetermined number, then

$$\frac{d^2g}{dh^2} = \frac{\gamma}{\alpha\beta^2} \qquad (6.59)$$

A convenient choice of α and β is such that $\gamma/\alpha\beta^2 = 1$. Then the boundary conditions become

$$g(0) = 0 \qquad \frac{d}{dh}g(0) = 0 \qquad \frac{d^2}{dh^2}g(0) = 1 \qquad (6.60)$$

which places all of the conditions at $h = 0$. We also choose $\beta = \alpha$ so the equation in g becomes

$$2\frac{d^3g}{dh^3} + g\frac{d^2g}{dh^2} = 0 \qquad (6.61)$$

An additional equation for the constants γ, β, and α stems from the solution of (6.61) under the conditions of (6.60). That solution leads to a value for $dg(\infty)/dh$. From (6.57) and the relationship $\alpha = \beta$, $dg(\infty)/dh = 1/\alpha^2$ determines α and β. Then $\gamma = \alpha^3$.

Using numerical methods with the conditions (6.60), it is not difficult to solve (6.61) on a digital computer. A Runge-Kutta equation solver from one of the libraries (for example, the International Mathematical and Statistical Library, IMSL) can easily produce a solution. A tabulated solution is given in Schlichting (1960). A plot of that solution from Schlichting is shown in Fig. 6.13.

6.5.2 Flat Plate Results

The experimental points on Fig. 6.13 were taken from measurements by Nikuradse (1933). The five different Reynolds numbers show excellent agreement with the theory. This result should not be translated to another problem because even a small pressure gradient can cause changes. Also, we must remember that the result applies to the laminar boundary layer, not to turbulent cases. The Reynolds number (based on distance along the plate) is

$$R_x = \frac{U \rho x}{\mu} \tag{6.62}$$

Laminar flow exists for $R_x < 300{,}000$, approximately. The critical value for the Reynolds number is influenced by such factors as the free stream

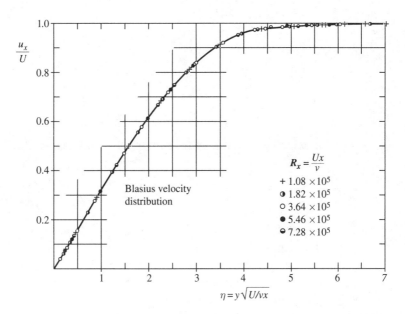

FIGURE 6.13
Solution of the Blasius equation (after Schlichting, 1960).

turbulence and the plate roughness. It could reach ten times the above value for a smooth plate and a completely nonturbulent free stream. Note that the Reynolds number increases with length along the plate so almost any flow will become turbulent if the plate is long enough.

Using the Blasius solution, the boundary layer thicknesses are

$$\delta = \frac{4.91x}{\sqrt{R_x}} \qquad \delta_d = \frac{1.72x}{\sqrt{R_x}} \qquad \delta_m = \frac{0.664x}{\sqrt{R_x}} \tag{6.63}$$

in which δ is taken at $u_x/U = 0.99$. If we consider the flow of water (kinematic viscosity $\nu = \mu/\rho = 10^{-6} m^2/s$) with a velocity of 1 m/s, the thicknesses at 1 m from the leading edge of the plate are $\delta = 0.5$ cm, $\delta_d = 0.2$ cm, and $\delta_m = 0.1$ cm. In air these quantities would be about 3.9 times as much. In either case the boundary layer is very thin.

The drag due to skin friction is defined up to the distance x as described by the first of (6.44) and τ_0 is the shear on the plate given by

$$\tau_0 = \mu \left(\frac{\partial u_x}{\partial y}\right)_{y=0} = \sqrt{\frac{U^3 \mu \rho}{x}} \Psi_*''(0) = 0.332 \sqrt{\frac{U^3 \mu \rho}{x}} \tag{6.64}$$

For water flow at 1 m/s the drag on the plate at 1 m is about 330 N/m^2 (about 0.6 lbs/ft^2).

Drag is usually expressed as a coefficient. The *skin friction coefficient* is

$$c_f = \frac{\tau_0}{\frac{1}{2}\rho U^2} \tag{6.65}$$

and the overall *skin friction coefficient* is

$$C_f = \frac{1}{\frac{1}{2}\rho U^2 L} \int_0^L \tau_0 \, dx \tag{6.66}$$

in which L is the length of the plate. For Blasius flow on one side of a flat plate

$$c_f = \frac{0.664}{\sqrt{R_x}} \qquad C_f = \frac{1.328}{\sqrt{R_L}} \tag{6.67}$$

where the L subscript on the Reynolds number indicates that it is based on the length of the plate.

From (6.63) the boundary layer thickness grows as the square root of distance along the plate. As the boundary layer thickens, the velocity gradients and the local shear diminish, as is stated in (6.64) where τ_0 varies as $1/\sqrt{x}$. From (6.64) the local shear varies as $U^{3/2}$, which may be an unexpected result for laminar flow.

The boundary layer equations do not necessarily apply to the leading edge of the plate. At that point the theory indicates that the velocity changes suddenly from U to zero. In deriving the boundary layer equations, we assumed that $\partial^2 u_x / \partial x^2$ is small, which is obviously not true near the leading edge.

6.6 SIMILARITY SOLUTIONS

The Blasius solution assumed that the velocity profiles at all distances along the plate were similar, that is,

$$\frac{u_x}{\tilde{u}_x} = \frac{d}{d\eta} f(\eta) \qquad \eta = \frac{y}{\delta} \tag{6.68}$$

in which f is a type of stream function. This sort of analysis can be performed for any type of problem where the potential flow velocity can be written

$$\tilde{u}_x = Cx^m \tag{6.69}$$

in which m is a constant. The $m = 0$ solution is the Blasius case. In the flow over a wedge as shown in Fig. 6.14

$$m = \frac{\theta}{\pi - \theta} \qquad \theta \le \pi \tag{6.70}$$

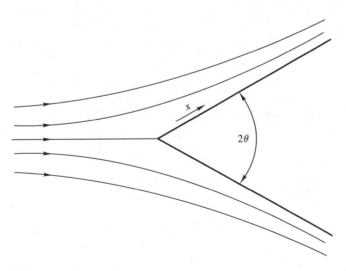

FIGURE 6.14
Flow over a wedge.

For $\theta = \pi/2$, the wedge becomes a flat plate and the flow is the stagnation flow described in Sec. 6.1.1. If $\theta > \pi/2$ ($m > 1$), the wedge becomes a corner; θ is limited to π. The ordinary differential equation (the *Falkner-Skan equation*) that results is

$$\frac{d^3 f}{d\eta^3} + \frac{m+1}{2} f \frac{d^2 f}{d\eta^2} - m \left(\frac{df}{d\eta}\right)^2 + m = 0 \qquad (6.71)$$

with the boundary conditions

$$f(0) = 0 \qquad \frac{d}{d\eta} f(0) = 0 \qquad \frac{d}{d\eta} f(\infty) = 1 \qquad (6.72)$$

The Blasius equation (6.56) is a special case of (6.71) with $\theta = m = 0$. Solutions for various values of m are shown in Fig. 6.2.

6.7   THE KÁRMÁN MOMENTUM EQUATION

The basis for some of the boundary layer calculations is not a direct solution of the differential equations, but a solution of the integral of those equations. The method is particularly flexible and admits a variety of other approximations. It is used for calculations concerning the turbulent boundary layer. For that reason it is included herein.

The integral of (6.25) with the pressure and elevation terms replaced by the Euler equation for potential flow is

$$\int_0^\infty \left(u_x \frac{\partial u_x}{\partial x} + u_y \frac{\partial u_x}{\partial y}\right) dy = \int_0^\infty \left(\tilde{u}_x \frac{d\tilde{u}_x}{dx} + \frac{\mu}{\rho} \frac{\partial^2 u_x}{\partial y^2}\right) dy \qquad (6.73)$$

The last term is the shear stress divided by ρ so that

$$\int_0^\infty \left(u_x \frac{\partial u_x}{\partial x} + u_y \frac{\partial u_x}{\partial y} - \tilde{u}_x \frac{d\tilde{u}_x}{dx}\right) dy = -\frac{\tau_0}{\rho} \qquad (6.74)$$

From the equation of continuity (6.24)

$$u_y = -\int_0^y \frac{\partial u_x}{\partial x} dy \qquad (6.75)$$

Using (6.75) in the second term of (6.74) and integrating that term by parts gives

$$-\int_0^\infty \frac{\partial u_x}{\partial y} \left(\int_0^y \frac{\partial u_x}{\partial x} dy\right) dy = -u_x \int_0^y \frac{\partial u_x}{\partial x} dy \Big|_{y=0}^{y=\infty} + \int_0^\infty u_x \frac{\partial u_x}{\partial x} dy \qquad (6.76)$$

The u_x immediately on the right of the equals sign is zero at $y = 0$ and is the potential flow velocity at $y = \infty$. Substitution of (6.76) into (6.74)

produces

$$\int_0^\infty \left(2u_x \frac{\partial u_x}{\partial x} - \tilde{u}_x \frac{\partial u_x}{\partial x} - \tilde{u}_x \frac{d\tilde{u}_x}{dx} \right) dy = -\frac{\tau_0}{\rho} \tag{6.77}$$

Rearrangement yields

$$\frac{\partial}{\partial x} \int_0^\infty u_x (\tilde{u}_x - u_x) dy + \frac{d\tilde{u}_x}{dx} \int_0^\infty (\tilde{u}_x - u_x) dy = \frac{\tau_0}{\rho} \tag{6.78}$$

From (6.41) and (6.42) the integrals in (6.78) represent the momentum thickness and the displacement thickness, respectively. Using these quantities, the equation becomes

$$\frac{d}{dx} \left(\tilde{u}_x^2 \delta_m \right) + \tilde{u}_x \delta_d \frac{d\tilde{u}_x}{dx} = \frac{\tau_0}{\rho} \tag{6.79}$$

Equation (6.79) is called the *Kármán momentum equation* (von Kármán, 1921; Pohlhausen, 1921) for the incompressible boundary layer. Even though we began with the laminar boundary layer equations, (6.79) should hold for nonlaminar flows if τ_0 is represented properly.

6.8   THE KÁRMÁN-POHLHAUSEN METHOD

The Kármán momentum equation was used by Pohlhausen to solve the laminar boundary layer problem by assuming that the velocity profile could be expressed as a fourth-order polynomial. The method is now seldom used for the laminar boundary layer since the computer can solve the more exact boundary layer equations. It is, however, used in more complex cases (turbulent flows) where the laminar equations do not apply.

Defining again $\eta = y/\delta$, the velocity profile is expressed as

$$\frac{u_x}{\tilde{u}_x} = a\eta + b\eta^2 + c\eta^3 + d\eta^4 \tag{6.80}$$

where a, b, c, and d are undetermined coefficients. The coefficients are determined from the conditions

$$\frac{\partial^2 u_x}{\partial y^2} = -\frac{\rho}{\mu} \tilde{u}_x \frac{d\tilde{u}_x}{dx} \quad \text{at} \quad y = 0$$

$$\tag{6.81}$$

$$u_x = \tilde{u}_x \quad \text{and} \quad \frac{\partial u_x}{\partial y} = 0 \quad \text{and} \quad \frac{\partial^2 u_x}{\partial y^2} = 0 \quad \text{at} \quad y = \delta$$

The condition that $u_x = 0$ at $y = 0$ is not listed since it is automatically satisfied by the form of the series. The other condition at $y = 0$ comes directly from the boundary layer equations. Pohlhausen further introduced the dimensionless parameter

$$\lambda = \frac{\rho \delta^2}{\mu} \frac{d\tilde{u}_x}{dx} = -\frac{\partial p}{\partial x} \frac{\delta^2}{\mu U}$$ (6.82)

which can be interpreted as the ratio of pressure forces to viscous forces. A positive value of λ indicates a favorable pressure gradient whereas a negative value denotes an adverse pressure gradient. On the flat plate $\lambda = 0$. The parameters of (6.80) are found to be

$$a = 2 + \frac{\lambda}{6} \qquad b = -\frac{\lambda}{2} \qquad c = 2 - \frac{\lambda}{2} \qquad d = 1 - \frac{\lambda}{6}$$ (6.83)

Equation (6.80) defines similar velocity profiles if the parameters are constants. In general, however, $\lambda = \lambda(x)$ indicating that a, b, c, and d vary with distance. In that case λ plays the part of a *shape parameter* that determines how the velocity profile changes along the boundary layer. The displacement thickness, momentum thickness, and wall shear stress are easily computed in terms of λ as

$$\frac{\delta_d}{\delta} = \frac{3}{10} - \frac{\lambda}{120} \qquad \frac{\delta_m}{\delta} = \frac{37}{315} - \frac{\lambda}{945} - \frac{\lambda^2}{9072} \qquad \frac{\tau_0 \delta}{\mu \tilde{u}_x} = 2 + \frac{\lambda}{6}$$ (6.84)

These quantities plus δ from (6.82) are substituted into the Kármán momentum equation (6.79) that then becomes a single equation for the shape parameter λ. The problem is then reduced to one ordinary differential equation in one unknown. Holstein and Bohlen (1940) have presented an efficient method of hand calculation (Schlichting, 1960).

Testing the Pohlhausen method for the flat plate solution, the Kármán equation becomes

$$\frac{d}{dx}(U^2 \delta_m) = \frac{\tau_0}{\rho}$$ (6.85)

After the appropriate substitutions, the relationships

$$\delta = \frac{5.84x}{\sqrt{R_x}} \qquad \delta_d = \frac{1.75x}{\sqrt{R_x}} \qquad \delta_m = \frac{0.685x}{\sqrt{R_x}}$$ (6.86)

are easily found, and can be compared to the exact quantities of (6.63). The local and overall skin friction coefficients are

$$c_f = \frac{0.685}{\sqrt{R_x}} \qquad C_f = \frac{1.370}{\sqrt{R_L}}$$ (6.87)

which can be compared to (6.67).

For flow around blunt objects the calculation begins at the stagnation point. The stagnation point corresponds to a value of $\lambda = 7.052$ and higher values of λ would represent corner flow.

A separation point occurs for a value of $\lambda = -12$. The calculation cannot be extended for lower values of λ. Experiment has shown the Pohlhausen method to be accurate for accelerated flow, but not for retarded flow; hence, calculations at negative values of λ may be suspect. The computed location of the separation point is no better than approximate.

6.9   INTEGRATION FOR TWO-DIMENSIONAL BOUNDARY LAYERS

Numerical methods can be used to compute the boundary layer for arbitrary pressure gradients. The relevant equations are (6.24) and (6.25) with the pressure and elevation terms replaced by the potential flow velocities

$$\frac{\partial u_x}{\partial x} + \frac{\partial u_x}{\partial y} = 0 \tag{6.88}$$

$$\frac{\partial u_x}{\partial t} + u_x \frac{\partial u_x}{\partial x} + u_y \frac{\partial u_x}{\partial y} = \tilde{u}_x \frac{\partial \tilde{u}_x}{\partial x} + \frac{\mu}{\rho} \frac{\partial^2 u_x}{\partial y^2} \tag{6.89}$$

and with the boundary conditions

$$u_x = 0 \quad \text{and} \quad u_y = 0 \quad \text{at} \quad y = 0 \qquad u_x = \tilde{u}_x(x) \quad \text{at} \quad y = \infty \tag{6.90}$$

6.9.1   The Dimensionless Equations

The x and y coordinates are replaced by dimensionless variables

$$\xi = \frac{x}{L} \qquad \eta = \frac{y}{x} \sqrt{\frac{\mu}{\tilde{u}_x \rho x}} \tag{6.91}$$

The term under the square root is a form of $\boldsymbol{R}_x$ but uses the local potential flow velocity instead of the free stream velocity. The η-coordinate is sometimes defined in terms of the object Reynolds number and the free stream velocity as

$$\eta = \frac{y}{L} \sqrt{\frac{L\tilde{u}_x}{xU}} \boldsymbol{R} \tag{6.92}$$

in which $\boldsymbol{R} = U\rho L/\mu$. The two definitions are completely equivalent.

Nondimensional velocities are introduced as

$$u_x^*(\xi, \eta) = \frac{u_x}{\tilde{u}_x} \qquad u_\eta^*(\xi, \eta) = \frac{u_y}{\tilde{u}_x} \sqrt{\frac{\tilde{u}_x \rho x}{\mu}} \tag{6.93}$$

Substitution into the equations of continuity and motion gives

$$\xi \frac{\partial u_x^*}{\partial \xi} + \frac{\eta}{2} \left(\frac{x}{\tilde{u}_x} \frac{d\tilde{u}_x}{dx} - 1 \right) \frac{\partial u_x^*}{\partial \eta} + \frac{\partial u_\eta^*}{\partial \eta} = -\frac{x}{\tilde{u}_x} \frac{d\tilde{u}_x}{dx} u_x^* \tag{6.94}$$

$$u_x^* \xi \frac{\partial u_x^*}{\partial \xi} + \left[u_\eta^* + \frac{\eta u_x^*}{2} \left(\frac{x}{\tilde{u}_x} \frac{d\tilde{u}_x}{dx} - 1 \right) \right] \frac{\partial u_x^*}{\partial \eta}$$

$$= \left(1 - u_x^{*2} \right) \frac{x}{\tilde{u}_x} \frac{d\tilde{u}_x}{dx} + \frac{\partial^2 u_x^*}{\partial \eta^2} \tag{6.95}$$

The boundary conditions are

$$u_x^*(\xi, 0) = 0 \qquad u_\eta^*(\xi, 0) = 0 \qquad u_x^*(0, \eta) = \frac{u_x}{\tilde{u}_x} \tag{6.96}$$

The last condition of (6.96) allows the boundary layer to begin at an arbitrary position. Actually, the last of (6.90) gives

$$u_x^*(\xi, \infty) = 1 \tag{6.97}$$

On the leading edge of the flat plate, $x = 0$ and $\eta \to \infty$.

6.9.2   The Starting Solution

The correct boundary conditions at the upstream part of the boundary layer computation are not always easy to find. There are really only two cases: (1) the boundary layer begins at the sharp edge of a body that is in the form of a wedge or (2) the boundary layer begins at the stagnation point of a bluff object. The case where the object has a cusp (the top and bottom contours are tangent at the leading edge) is a special case of (1) in which the flat plate solution applies. In either of these cases the similarity solution provides a starting point. That is, the stagnation flow or wedge flow is applied very near the stagnation point or the point of the wedge, respectively. That flow can be continued a short distance downstream to provide the starting solution for the boundary layer over the actual object.

   The potential flow solution is a critical part of the boundary layer solution. In some cases the potential velocities should be recalculated with a revised object shape in which the displacement thickness is added to the solid boundary for the purpose of finding the potential flow solution. However, if the boundary layer is thin, that addition will not add much to the solution; if the boundary layer is thick, the approximation is not valid.

6.10   WAKES AND JETS

Free shear flows (also called *boundary-free shear flows*) are those in which two layers of flow with different velocities are in close proximity without being bounded by a wall. Boundary layer theory can be used in some highly restricted cases to compute the velocity distributions, including the entrainment of fluids in jets. Unfortunately, laminar flow seldom applies to such cases because the flow is highly unstable and quickly breaks up into turbulence. Most of the important cases from an engineering point of view occur in air or water where any sort of shear flow is almost certainly turbulent. Some of the principles and methods used to calculate laminar flow apply, however, to the turbulent case.

6.10.1 The Wake Behind a Plate

Figure 6.15 shows the boundary layer over a flat plate (exaggerated in thickness, as usual) with the region of disturbed flow behind it. The drag on the plate can be calculated using (6.64) in (6.44) or using the overall skin friction coefficient (6.66). Taking into account the drag on both sides of the plate

$$D = \frac{1.328\rho U^2 L}{\sqrt{R_L}} \tag{6.98}$$

in which D is the drag per unit width of plate and R_L is the Reynolds number based on plate length. Using the control volume indicated by the box shown by the dashed line of Fig. 6.15

$$D = \rho \int_{-\infty}^{\infty} u_x (U - u_x) dy \tag{6.99}$$

where the integral is taken over any section behind the plate. Whatever the velocity distribution behind the plate, it must satisfy the two preceding equations.

The viscous flow solution comes at the expense of additional approximation. The *velocity defect* (difference between the free stream velocity

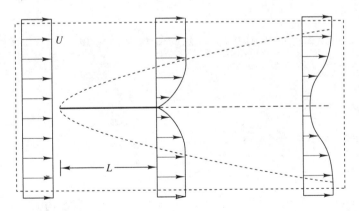

FIGURE 6.15
Flow over a flat plate with a wake behind the plate.

and the velocity in the wake) is

$$u = U - u_x \qquad (6.100)$$

We now assume that $u \ll U$, an assumption that is obviously invalid at the rear of the plate where $u = U$. The results of this approximation can be used to derive the properties of the wake far downstream from the plate ($x > 3L$), but not in the vicinity of the plate. Using (6.100) in the boundary layer equation produces

$$U \frac{\partial u}{\partial x} = \frac{\mu}{\rho} \frac{\partial^2 u}{\partial y^2} \qquad (6.101)$$

where the equation has been linearized by neglecting the vertical velocity and the product of u and $\partial u / \partial x$. The boundary conditions are

$$u = 0 \quad \text{at} \quad y = \infty \qquad \text{and} \qquad \frac{\partial u}{\partial y} = 0 \quad \text{at} \quad y = 0 \qquad (6.102)$$

The solution to the differential equation is

$$u = \frac{uC}{\sqrt{\frac{x}{L}}} e^{-Uy^2 \rho / (4\mu x)} \qquad (6.103)$$

in which C is an undetermined constant. Using (6.103) in (6.99) produces

$$D = 2\sqrt{\pi} C \rho U \sqrt{\frac{\mu L}{\rho U}} \qquad (6.104)$$

Equating (6.98) and (6.104) gives $2C\sqrt{\pi} = 1.328$. The velocity is

$$u_x = U \left(1 - \frac{0.664}{\sqrt{x}} \sqrt{\frac{L}{x}} e^{-Uy^2 \rho / (4\mu x)} \right) \qquad (6.105)$$

A plot of (6.105), in Fig. 6.16, shows that the velocity profile contains an

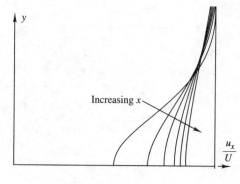

Increasing x

$\frac{u_x}{U}$

FIGURE 6.16
Velocity profiles downstream from an object.

inflection point, indicating that it is unstable for most cases. Even if the flow over the plate is laminar, the wake is likely to be turbulent except at very low Reynolds numbers. Free shear flows of any sort show similar behavior.

Although the problem was one of the wake behind a flat plate, the wake behind any object should exhibit similar behavior at large distances from the object with the very serious (and usually unrealistic) restriction that the flow is laminar.

6.10.2 A Two-dimensional Jet

Consider a jet that issues from a slit as shown in Fig. 6.17. As in the case of a wake, the jet quickly becomes turbulent, but the method of analyzing either a laminar or turbulent jet is similar. We assume that there is no pressure gradient in the x-direction and thus the momentum in the jet is constant

$$M = \rho \int_{-\infty}^{\infty} u_x^2 \, dy \tag{6.106}$$

Again, the boundary layer equations are transformed into an ordinary differential equation. The equation of continuity is satisfied by the introduction

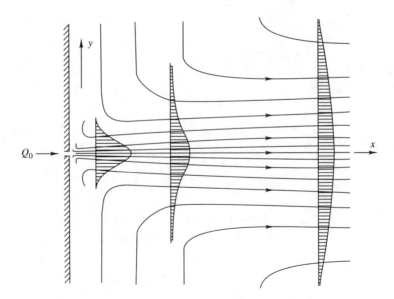

FIGURE 6.17
Two-dimensional jet flow (from Herman Schlicting, *Boundary Layer Theory*, p. 165, ©1960, reproduced by permission of McGraw-Hill, Inc.).

of a stream function. The two new variables are

$$\eta = \frac{1}{\nu^{1/3}} \frac{y}{x^{1/3}} \qquad \Psi = \nu^{1/3} x^{1/3} f(\eta) \qquad (6.107)$$

After substitution into the boundary layer equations with the pressure gradient set equal to zero, there results

$$\frac{d^3 f}{d\eta^3} + f \frac{d^2 f}{d\eta^2} + \left(\frac{df}{d\eta}\right)^2 = 0 \qquad (6.108)$$

The boundary conditions are

$$f = 0 \quad \text{and} \quad \frac{d^2 f}{d\eta^2} = 0 \quad \text{at} \quad \eta = 0 \qquad \frac{df}{d\eta} = 0 \quad \text{at} \quad \eta = \infty \quad (6.109)$$

The solution is

$$f = C \frac{1 - e^{-C\eta}}{1 + e^{-C\eta}} \qquad (6.110)$$

in which C is an arbitrary constant. The other constants have been chosen to satisfy the boundary conditions.

The constant C can be chosen with reference to the momentum in the jet. First, the velocity in the x-direction is

$$u_x = \frac{1}{6} C^2 x^{-1/3} \left(1 - \tanh^2 \frac{C\eta}{2}\right) \qquad (6.111)$$

From (6.106)

$$M = \frac{1}{6} \rho C^3 \sqrt{\frac{\mu}{\rho}} \int_{-\infty}^{\infty} \left(1 - \tanh^2 \frac{C\eta}{2}\right)^2 d\eta = \frac{2}{9} \rho C^3 \sqrt{\frac{\mu}{\rho}} \qquad (6.112)$$

Given the quantity of flow leaving the slit, the momentum is known. Assuming a uniform velocity distribution at the exit

$$M_0^2 = \frac{\rho Q_0^2}{b} \qquad (6.113)$$

where Q_0 is the flow rate per unit length of slot (L^2/T) and b is the width of the slot. Equating (6.113) and (6.112)

$$C = \left(\frac{9}{2} \frac{Q_0^2}{b} \sqrt{\frac{\rho}{\mu}}\right)^{1/3} \qquad (6.114)$$

The primary results are the distribution of velocity in the x-direction and the flow rate in the jet as it disperses in the fluid. The former is given in

(6.111); the latter is

$$Q = \int_{-\infty}^{\infty} u_x \, dy = 2C \sqrt{\frac{\mu}{\rho}} x^{1/3} \tag{6.115}$$

The jet entrains fluid from the surroundings and increases its flow rate as the downstream distance to the one-third power. The width of the jet also increases in the downstream direction.

6.10.3 The Axisymmetric Jet

A more common case is the jet issuing from a round hole. Again, the momentum is taken as constant

$$M = 2\pi\rho \int_0^{\infty} u_x^2 R \, dR \tag{6.116}$$

The boundary layer equations, written in cylindrical coordinates without a pressure gradient, are

$$\frac{\partial u_x}{\partial x} + \frac{\partial u_R}{\partial R} + \frac{u_R}{u_x} = 0 \tag{6.117}$$

$$u_x \frac{\partial u_x}{\partial x} + u_R \frac{\partial u_x}{\partial R} = \frac{\mu}{\rho R} \frac{\partial}{\partial R} \left(R \frac{\partial u_x}{\partial R} \right) \tag{6.118}$$

The boundary conditions are

$$u_R = 0 \quad \text{and} \quad \frac{\partial u_x}{\partial R} = 0 \quad \text{at} \quad R = 0 \qquad u_x = 0 \quad \text{at} \quad R = \infty \tag{6.119}$$

Applying the relationships

$$\eta = \frac{R}{x} \qquad u_x = \frac{\mu}{x\rho\eta} \frac{d}{d\eta} f(\eta) \qquad u_R = \frac{\mu}{\rho x} \left(\frac{d}{d\eta} f(\eta) - \frac{f(\eta)}{\eta} \right) \tag{6.120}$$

leads to an ordinary differential equation

$$\frac{d^3 f}{d\eta^3} + \frac{f-1}{\eta} \frac{d^2 f}{d\eta^2} + \frac{1-f}{\eta^2} \frac{df}{d\eta} + \frac{1}{\eta} \left(\frac{df}{d\eta} \right)^2 = 0 \tag{6.121}$$

The solution is

$$f = \frac{C\eta^2}{1 + 0.25 C\eta^2} \qquad u_x = \frac{2\mu C}{\rho x \left(1 + \dfrac{C}{4} \dfrac{R^2}{x^2} \right)^2} \tag{6.122}$$

Taking the momentum of the emerging jet as $M = \rho Q_0^2 / 2\pi R_0^2$, where Q_0 and R_0 are the flow rate at the exit of the jet and the radius of the hole, the constant becomes

$$C = \frac{3\rho^2 Q_0^2}{32\pi^2 \mu^2 R_0^2} \tag{6.123}$$

Actually, for developed laminar flow in a pipe, the momentum would be 4/3 of the above amount with a consequent change in C. The flow rate in the jet is given by

$$Q = 8 \frac{\pi \mu x}{\rho} \qquad (6.124)$$

indicating that it increases linearly in the downstream direction. The flow rate is independent of the constant C and, thus, independent of the momentum in the jet. More surprising is the result that the flow in the jet is independent of the flow from the orifice. A jet of small velocity spreads rapidly, carrying with it a comparatively large amount of ambient fluid; a jet of large velocity spreads slowly, carrying along the same amount of ambient fluid as the small jet.

Table 7.1 in Chap. 7, which treats turbulent flow, summarizes those results and compares them with turbulent flow cases.

PROBLEMS

6.1. Using the Kármán-Pohlhausen method, the boundary layer profile can be assumed as any polynomial. The number of boundary conditions that can be applied must be adjusted to the number of terms in the polynomial. Find (*i*) the boundary layer thickness, (*ii*) the displacement thickness, (*iii*) the momentum thickness, (*iv*) the energy thickness, and (*v*) the skin friction coefficient for
  (*a*) A linear profile $u_x/U = a\eta$; $u_x = U$ at $y = \delta$.
  (*b*) A quadratic profile $u_x/U = a\eta + b\eta^2$; $u_x = U$ and $\partial u_x/\partial y = 0$ at $y = \delta$.
  (*c*) A cubic profile $u_x/U = a\eta + b\eta^2 + c\eta^3$; $u_x = U$ and $\partial u_x/\partial y = 0$ at $y = \delta$; $\partial^2 u_x/\partial y^2 = 0$ at $y = 0$.

6.2. Instead of a polynomial, solve Prob. 6.1 using $u_x/U = \sin(\pi\eta/2)$, where $\eta = y/\delta$, and compare to the Blasius profile.

6.3. For a flat plate moving through still (*i*) air or (*ii*) water, compute the maximum length of the plate if the boundary layer is to remain laminar. Give this length as a function of the speed of the plate. Find also the drag on the plate as a function of this length.

6.4. A relationship between the Strouhal number and the Reynolds number describes the frequency at which vortices are shed from a circular cylinder. This relationship is

$$St = 0.198 \left(1 - \frac{19.7}{R}\right) \qquad 250 < R < 2 \times 10^5$$

Find the frequency of vortex shedding from a wire of 2 mm diameter in a wind of 30 m/s. (Use $\nu = 1.46 \times 10^{-5}$ m^2/s)

6.5. Fig. 6.18 shows shear vs. rate of strain for a Newtonian fluid and an ideal plastic (a non-Newtonian fluid). Whereas the Newtonian fluid flows for any shear, no matter how small, the ideal plastic does not move until a threshold shear τ_0 is reached. Newton's law of viscosity is replaced by

$$\frac{du}{dy} = 0 \quad \text{for} \quad \tau \leq \tau_0$$

$$\mu_p \frac{du}{dy} = \tau - \tau_0 \quad \text{for} \quad \tau \geq \tau_0$$

Consider the flow of an ideal plastic between flat plates (similar to Poiseuille flow).

(*a*) Sketch the shear stress diagram between the plates. Assume that the maximum shear stress is greater than the threshold.

(*b*) Find an expression for the velocity distribution as a function of the pressure gradient, dp/dx.

(*c*) Find an expression for the flow rate Q as a function of the pressure gradient.

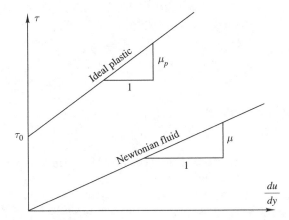

FIGURE 6.18

6.6. Consider the problem of uniform flow down an inclined plane as shown in Fig. 6.19. (Uniform means that all x-derivatives, except pressure, are zero.) For laminar flow derive

(*a*) An equation that expresses the shear distribution throughout the fluid.

(*b*) The velocity distribution in the fluid.

(*c*) The rotation of a fluid particle anywhere in the fluid.

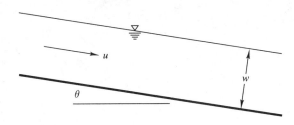

FIGURE 6.19

6.7. Figure 6.20 shows a piston and a cylinder that are parts of an air compressor. The velocity of the piston is

$$u_p = U_p \sin \omega t$$

To analyze the motion of the gas in the cylinder, it is necessary to solve the Navier-Stokes equations.

(*a*) Write the boundary conditions for each of the numbered surfaces in the diagram. Be explicit.

(*b*) During the compression cycle, what is the sign (positive or negative) of the energy terms

$$p \frac{\partial u_i}{\partial x_i} \quad \text{and} \quad \tau_{ij} \frac{\partial u_j}{\partial x_i}$$

Explain the answer. Do you expect internal energy to increase or decrease?

(*c*) Answer part (*b*) for the intake cycle.

(*d*) Look at the corner between surfaces ① and ② (the corner between the piston and the cylinder wall) and refer to part (*a*). Explain the conflict of boundary conditions at this point.

(*e*) If you were to build a model of this process, what similarity criteria would you use? What criteria would be neglected? Explain.

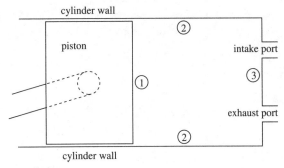

FIGURE 6.20

6.8. A cylindrical pipe of radius R_2 has a solid rod of radius R_1 at its center (Fig. 6.21). The pipe is filled with a fluid of viscosity μ. Assume that the pipe is very long so that the end effects can be neglected.

(a) Calculate the velocity distribution in the pipe if the rod is rotated with angular velocity ω. What can you say about the stability of the flow?

(b) Calculate the velocity distribution in the pipe if the rod is fixed but the pipe is rotated with angular velocity ω. What can you say about the stability of the flow?

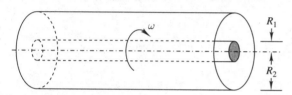

FIGURE 6.21

6.9. A large, cylindrical disc of radius R_0 undergoes axial oscillations (that is, rotates about its own axis) of small amplitude θ_0 where $v_\theta = \theta_0 e^{i\sigma t}$ in a viscous fluid. To the first approximation

$$v_\theta = R\sigma\theta_0 e^{-x\sqrt{\sigma/2\nu}} \sin\left(\sigma t - x\sqrt{\sigma/2\nu}\right)$$

in which R is the distance from the center of the disc, σ is the frequency of the oscillations, ν is the kinematic viscosity, x is the distance from the surface of the plate, and t is time.

(a) Show that the above equation is correct.

(b) Find the moment exerted on one side of the plate by the fluid.

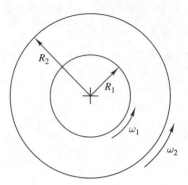

FIGURE 6.22

6.10. A viscous fluid is contained between two round cylinders as shown in Fig. 6.22.

(a) Give an equation for the velocity of the fluid, u_θ, as a function of the radius, R.

(b) Even though this is viscous flow, the flow becomes irrotational for a certain relationship between ω_1 and ω_2. Find this relationship. The condition for irrotational flow in cylindrical coordinates is

$$\frac{\partial}{\partial R}(Rv_\theta) - \frac{\partial v_R}{\partial \theta} = 0$$

(c) For the condition of irrotational flow, find the pressure as a function of R using the Navier-Stokes equation in the R-direction and check your results using Bernoulli's equation.

CHAPTER 7

TURBULENT FLOW

The majority of flows that are of engineering interest are turbulent. They are, unfortunately, the most difficult to analyze. In principle, and largely in fact, we can solve most laminar flow problems with the computing power that is available today. The same statement cannot be made of turbulent flows; the opposite is true and there seems to be little hope of achieving that goal in the near future.

Professor D. Allan Bromley (1986) in a review of the state of physics asserts

> One of the most intractable problems in all of physics has been that of turbulence; anyone who has watched the flow of water around an obstacle or watched the smoke rise from a lighted cigarette has observed the transition from ordered laminar flow into ever more complex eddies and finally into completely chaotic motion. Such motion, at the interface between ordered and chaotic, has defied numerical analysis and yet has enormous importance; it is responsible for the drag on ships and planes moving through fluids, it is responsible for most of the noise from jet aircraft, and it is responsible for the ultimate damage to human heart valves, buffeted over periods of years in turbulent blood flow.

Despite the negative tone of Bromley's assessment, a world without turbulence is unimaginable. The phenomenon is fundamental to life itself.

Since it increases diffusion by a couple of orders of magnitude, it would be impossible for air-breathing animals to obtain sufficient oxygen without turbulence. All other natural processes would change so drastically that the face of the earth would be unrecognizable by humans. Perhaps life could evolve in a nonturbulent environment, but even that supposition is far from obvious.

Air and water are the two most common fluids. The natural flows of the atmosphere (wind) and earth take place at highly turbulent Reynolds numbers. The approximation of potential flow can often yield some solutions. Laminar flow is rare in the above-ground environment and its study contributes mainly to a general understanding of the principles of fluid mechanics, many of which translate into the analysis of turbulent flows.

Exactly what is meant by the word "turbulence" must remain somewhat intuitive. Even after a study of the phenomenon, a concise definition is difficult to formulate. Hinze (1959) writes, "Turbulent fluid motion is an irregular condition of flow in which the various quantities show a random variation with time and space coordinates, so that statistically distinct average values can be discerned." The random feature is one of the primary items that makes the study of turbulence difficult. The other feature, not mentioned in the above definition, is that turbulence is strongly three-dimensional. When combined with the fact that the motion takes place over a large range of scales (say, 0.1 mm to several kilometers in the case of geophysical motions), these characteristics create almost insurmountable difficulty.

Turbulence is a phenomenon whose physics are known at a molecular or microscopic level but not at the "macroscopic" level necessary for computation. Despite huge resources and a large amount of research activity spanning more than 30 years, the numerical weather forecasting problem remains largely unsolved even though early optimism predicted that long-term numerical forecasts were just over the horizon. Current models for weather forecasting use about an eighty-kilometer grid spacing. If all the turbulent processes were to be included in the model, the grid spacing would have to be about 1 mm. Since all parts of the earth interact, the grid would have to cover the globe, requiring some 10^{18} grid points. Such a calculation is far beyond the capacity of present-day computers or any foreseeable computer. Research efforts are underway to invent models that can be accurate at much larger grid scales. Current models can compute accurately at large grid scales for a short time but eventually (say about two weeks in real time), the interaction of the large scale phenomena with the small scale phenomena renders the calculation worthless. (There are, of course, other obstacles besides raw computing power that stand in the way of long-range forecasts. Obtaining data for initial or corrective conditions on a small scale may be an even more insurmountable problem.)

The "butterfly effect" is a tongue-in-cheek description of the process of atmospheric disturbance. The story is that a butterfly decides to fly from the limb of a tree to a nearby flower in, say, Beijing, thereby creating an atmospheric disturbance. The disturbance is magnified by nonlinear atmospheric processes, eventually causing thunderstorms and tornados in some distant location like St. Louis.

The potential economic benefits of a long-range weather forecast are tremendous. Perhaps agriculture would benefit most. Farmers could decide when to plant, when to fertilize, and when to harvest. They could choose what crops to plant and use a variety that is resistant to either drought or to wet conditions. Crop failure, wasted energy, and associated economic resources would become rare. A host of other fields would benefit: transportation; construction; control of damage due to floods, hurricanes, and tornados; water supply and waste disposal are obvious examples. Of course such forecasts would be important to Jane while she is deciding whether to hold her wedding in the church or in the garden.

One of the misconceptions regarding turbulence is that some fluids are turbulent. It is true that many fluids, especially air and water, have low viscosities so that when they move, the movement tends to be turbulent. It is, however, the *flow* that is turbulent. This apparently self-evident statement is important in that the properties or parameters of turbulence are sometimes attributed to the fluid, whereas the fluid has no such property.

7.1 STABILITY

Laminar flow in a pipe occurs at low Reynolds numbers, generally less than about 2100. If we had no experimental evidence, there is nothing obvious about the equations of motion that would lead us to believe that laminar flow does not occur for all Reynolds numbers. In fact, in solving the laminar flow equations, the question must be asked "Is the solution stable?" By that question we mean: If there is some disturbance, that may or probably does occur in nature, does that disturbance grow or is it damped? If it grows, does it lead to a disorderly flow? If the answer to these questions is "yes," then a solution that assumes laminar flow is probably not characteristic of natural flows.

The classical Reynolds experiment provides a well-known answer to the previous questions. Fig. 7.1 shows a schematic of the Reynolds apparatus. The valve is arranged so that the flow rate can be carefully varied. A small stream of dye is injected into the flow to make it visible. At low Reynolds numbers the stream of dye is coherent. As the Reynolds number

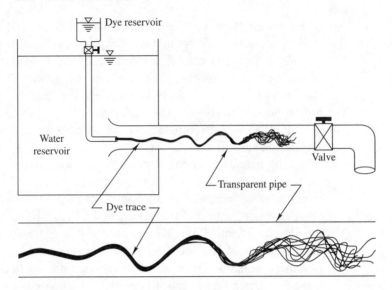

FIGURE 7.1
The Reynolds apparatus. The dye in the tube shows laminar flow and the breakup into turbulence.

increases, the dye stream begins to waver; a disturbance is clearly present. For still higher Reynolds numbers the dye breaks up into chunks and swirls are visible. Further increases in flow show that the swirls become more violent and finally the dye mixes with surrounding fluid so completely that it is no longer distinguishable.

7.1.1 Interfacial Stability—Kelvin-Helmholtz Instability

Consider the two-dimensional flow of a stratified fluid as shown in Fig. 7.2. Both fluids will be considered as inviscid and the flows are governed by

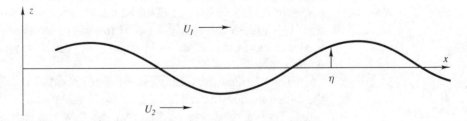

FIGURE 7.2
The interface between two fluids

Laplace's equation. For both fluids

$$\nabla^2 \Phi = 0 \tag{7.1}$$

$$\frac{\partial \eta}{\partial t} = \frac{\partial \Phi}{\partial x} \frac{\partial \eta}{\partial x} - \frac{\partial \Phi}{\partial z} \qquad \text{on} \quad z = \eta \tag{7.2}$$

$$-\frac{\partial \Phi}{\partial t} + \frac{1}{2} \left[\left(\frac{\partial \Phi}{\partial x} \right)^2 + \left(\frac{\partial \Phi}{\partial z} \right)^2 \right] + g\eta = -\frac{p}{\rho} + B \qquad \text{on} \quad z = \eta \tag{7.3}$$

in which η is the height of the interface measured from the equilibrium surface and B is the Bernoulli constant. (See also Sec. 10.5 in which surface tension is included.) The potentials consist of a parallel flow plus a small perturbation due to any change in the interface from horizontal

$$\Phi_1 = -U_1 x + \phi_1 \qquad \Phi_2 = -U_2 x + \phi_2 \tag{7.4}$$

where ϕ is the perturbation, which also satisfies Laplace's equation.

The potentials are substituted into the free surface conditions and those conditions are linearized by assuming that the perturbation potential and the deviation from the equilibrium surface ($z = 0$) are small and by neglecting products of the perturbation quantities. The result is applied to the equilibrium surface

$$\frac{\partial \eta}{\partial t} = -U \frac{\partial \eta}{\partial x} - \frac{\partial \phi}{\partial z} \qquad \text{on} \quad z = 0 \tag{7.5}$$

$$-\frac{\partial \phi}{\partial t} + \frac{1}{2} U^2 - U \frac{\partial \phi}{\partial x} + g\eta = -\frac{p}{\rho} + B \qquad \text{on} \quad z = 0 \tag{7.6}$$

The linearized boundary conditions apply to both fluids and the quantities can take on either the subscript 1 or 2, which designate the upper and lower fluids, respectively. In the case of the dynamic equation the pressure at the interface is the same whether viewed from the lower fluid or the upper fluid. The unperturbed flow ($\phi = \eta = 0$) gives the Bernoulli constant

$$B_1 = \frac{1}{2} U_1^2 + \frac{p_0}{\rho_1} \qquad B_2 = \frac{1}{2} U_2^2 + \frac{p_0}{\rho_2} \tag{7.7}$$

in which p_0 is the pressure at the interface under the unperturbed conditions. Equating the pressure from the two equations produces

$$\rho_2 \left(-\frac{\partial \phi_1}{\partial t} - U_1 \frac{\partial \phi_1}{\partial x} + g\eta_1 \right) = \rho_1 \left(-\frac{\partial \phi_2}{\partial t} - U_2 \frac{\partial \phi_2}{\partial x} + g\eta_2 \right)$$

$$\text{on} \quad z = 0 \tag{7.8}$$

At this point the equations are linear, meaning that the solutions can be superposed to form new solutions. These new solutions can be made up of harmonic terms as in a Fourier series. Each term can have a distinct wave length and frequency, but the wave length and frequency are related through the equations. One such term in each of the variables can be represented by

$$\phi_1 = \tilde{\phi}_1(z)\, e^{-i(kx-\sigma t)} \tag{7.9}$$

$$\phi_2 = \tilde{\phi}_2(z)\, e^{-i(kx-\sigma t)} \tag{7.10}$$

$$\eta = \eta_0 e^{i(kx-\sigma t)} \quad \text{on} \quad z = 0 \tag{7.11}$$

in which k is taken as a real constant but σ can be complex and is written

$$\sigma = \sigma_r + i\sigma_i \tag{7.12}$$

where $i = \sqrt{-1}$ in the expression but i designates the imaginary part when used as a subscript. If the amplitude functions $\tilde{\phi}_1$ and $\tilde{\phi}_2$ in (7.9) and (7.10) are to satisfy Laplace's equation, they must be of the form

$$\tilde{\phi}_1 = A_1 e^{az} + C_1 e^{-az} \qquad \tilde{\phi}_2 = A_2 e^{az} + C_2 e^{-az} \tag{7.13}$$

From the physical problem $\phi_1 \to 0$ as $z \to \infty$ and $\phi_2 \to 0$ as $z \to -\infty$, leading to $A_1 = C_2 = 0$. Substitution of (7.13) into (7.9) and (7.10) and substituting the resulting expressions into Laplace's equation gives

$$a = k \tag{7.14}$$

Further substitution into the combined dynamic boundary condition (7.8) and the kinematic conditions (7.5) written for each fluid yields the equations

$$C_1(i\rho_1\sigma - i\rho_1 U_1 k) + A_2(-i\rho_2\sigma + i\rho_2 U_2 k) - \eta_0(\rho_2 g - \rho_1 g) = 0$$
$$C_1 \qquad\qquad\qquad\qquad\qquad\qquad\qquad\qquad + \eta_0(i\sigma - iU_1 k) = 0$$
$$A_2(-k) \qquad\qquad\qquad\qquad\qquad\qquad + \eta_0(i\sigma - iU_2 k) = 0 \tag{7.15}$$

Equations (7.15) could be considered three equations in the amplitudes C_1, A_2, and η_0. Because the equations are homogeneous in these quantities, a trivial solution (zero for all the amplitudes) is avoided only if the coefficient determinant is equal to zero. Setting the coefficient determinant to zero yields a relationship between σ, k, ρ_1, ρ_2, U_1, and U_2. The result, solved for the wave speed, is

$$\frac{\sigma}{k} = \frac{\rho_1 U_1 + \rho_2 U_2}{\rho_1 + \rho_2} \pm \sqrt{\frac{g}{k}\frac{\rho_2 - \rho_1}{\rho_1 + \rho_2} - \frac{\rho_1 \rho_2}{(\rho_1 + \rho_2)^2}(U_2 - U_1)^2} \qquad (7.16)$$

which is a dispersion equation for waves on moving fluids of different densities.

From (7.9)–(7.11) the system is unstable if the imaginary part of σ, σ_i, is positive. Because the square root in (7.16) has both plus and minus signs, some σ_i will be positive unless the term in brackets is positive, in which case $\sigma_i = 0$. The stability condition is

$$(U_2 - U_1)^2 \leq \frac{gL}{2\pi}\frac{\rho_2^2 - \rho_1^2}{\rho_1 \rho_2} \qquad (7.17)$$

where the wave number has been replaced by the wave length, $L = 2\pi/k$. The surprising conclusion is that the interface is unstable for any difference in velocities, $U_2 - U_1$, for sufficiently short wave lengths L. Even for a light breeze blowing over water, the water surface would appear ruffled due to the instability. (That conclusion is modified by surface tension, which acts most strongly on short waves—see Sec. 10.5.)

This type of instability is called *Kelvin-Helmholtz instability*. It is common in the atmosphere and in liquids (Fig. 7.3). The analysis indicates that all shear flows are to some extent (at least for some sufficiently short wave lengths) unstable in the absence of viscosity. Viscosity modifies that conclusion in the direction of stability. The analysis has assumed an abrupt interface with a discontinuity in velocities. If viscosity is included, the discontinuity in velocities is not possible but the shear layer will extend over a finite thickness instead. Yet instability is possible and common, resulting in turbulence.

If $\rho_1 > \rho_2$ with zero velocity difference, the fluid is unstable simply because the heavier fluid is on top. This buoyancy instability is also common and a producer of turbulence. (Then, how can one turn a small glass of water upside down with the water remaining in the glass?)

If viscosity is considered, the analysis is much more complex. The instability can form a cusp, in which case it becomes a *Holmboe instability* (Holmboe, 1962; Smyth and Peltier, 1991). The distinction between these types of instability is governed by the velocity and density gradients (the Richardson number) and the ratio of diffusion of salt or heat to the diffusion of momentum (the Prandtl number or Schmidt number). The Holmboe instability is much more likely to occur in the ocean than in the atmosphere.

FIGURE 7.3
The Kelvin-Helmholtz instability (top and middle photograph courtesy Caltech). In the lower picture waves are marked by clouds (courtesy of Dr. Robert Spigel, University of Canterbury, New Zealand).

7.1.2 The Orr-Sommerfeld Equation

A two-dimensional analysis of stability is carried out by substituting the stream function into the equations of motion. Even though the analysis is restricted to two dimensions, it appears to yield valid predictions for the onset of instability and breakup into turbulence. According to some experiments, a two-dimensional instability is a primary trigger. Lin (1955) has indicated that a two-dimensional disturbance is more unstable than a three-dimensional disturbance. The initial two-dimensional instability causes secondary motions that are three-dimensional and often become unstable.

The pressure can be eliminated from the equations of motion by cross differentiation. The equation for the x-direction is differentiated with respect to y and the equation for the y-direction is differentiated with respect to x. After subtracting to eliminate the cross derivative of the pressure,

$$
\rho \frac{\partial}{\partial t} \left(\frac{\partial u_x}{\partial y} - \frac{\partial u_y}{\partial x} \right) + \rho u_x \left(\frac{\partial^2 u_x}{\partial x \partial y} - \frac{\partial^2 u_y}{\partial x^2} \right) + \rho u_y \left(\frac{\partial^2 u_x}{\partial y^2} - \frac{\partial^2 u_y}{\partial x \partial y} \right)
$$

$$
+ \rho \left(\frac{\partial u_x}{\partial y} \frac{\partial u_x}{\partial x} + \frac{\partial u_x}{\partial y} \frac{\partial u_y}{\partial y} - \frac{\partial u_x}{\partial x} \frac{\partial u_y}{\partial x} - \frac{\partial u_y}{\partial x} \frac{\partial u_y}{\partial y} \right) \tag{7.18}
$$

$$
= \mu \left(\frac{\partial^3 u_x}{\partial^2 x \partial y} + \frac{\partial^3 u_x}{\partial y^3} - \frac{\partial^3 u_y}{\partial x^3} - \frac{\partial^3 u_y}{\partial y^2 \partial x} \right)
$$

Using the stream function in (7.18) results in

$$
\rho \frac{\partial}{\partial t} (\nabla^2 \Psi) - \rho \frac{\partial \Psi}{\partial y} \frac{\partial}{\partial x} (\nabla^2 \Psi) + \rho \frac{\partial \Psi}{\partial x} \frac{\partial}{\partial y} (\nabla^2 \Psi) = \mu \nabla^2 (\nabla^2 \Psi) \tag{7.19}
$$

We now assume that the flow can be divided into a primary flow plus a perturbation that is small compared to the primary flow. The stream function becomes

$$
\Psi = \Psi_0 + \psi \tag{7.20}
$$

in which Ψ_0 represents the main flow and ψ represents the perturbation. Of course the primary flow, represented by Ψ_0, must satisfy (7.19). After (7.20) is substituted into (7.19), and (7.19) with ψ replaced by Ψ_0 is subtracted from the result

$$
\rho \frac{\partial}{\partial t} (\nabla^2 \psi) - \rho \left[\frac{\partial \Psi_0}{\partial y} \frac{\partial}{\partial x} (\nabla^2 \psi) + \frac{\partial \psi}{\partial y} \frac{\partial}{\partial x} (\nabla^2 \Psi_0) + \frac{\partial \psi}{\partial y} \frac{\partial}{\partial x} (\nabla^2 \psi) \right]
$$

$$
+ \rho \left[\frac{\partial \Psi_0}{\partial x} \frac{\partial}{\partial y} (\nabla^2 \psi) + \frac{\partial \psi}{\partial x} \frac{\partial}{\partial y} (\nabla^2 \Psi_0) + \frac{\partial \psi}{\partial x} \frac{\partial}{\partial y} (\nabla^2 \Psi_0) \right] \tag{7.21}
$$

$$
= \mu \nabla^2 (\nabla^2 \psi)
$$

Products of ψ and its derivatives are eliminated by assuming that the disturbance is small compared with the primary flow. The result is the elimination of the nonlinear terms in ψ

$$\rho\left[\frac{\partial}{\partial t}(\nabla^2\psi) - \frac{\partial\Psi_0}{\partial y}\frac{\partial}{\partial x}(\nabla^2\psi) + \frac{\partial\Psi_0}{\partial x}\frac{\partial}{\partial y}(\nabla^2\psi) - \frac{\partial\psi}{\partial y}\frac{\partial}{\partial x}(\nabla^2\Psi_0)\right.$$

$$\left. +\frac{\partial\psi}{\partial x}\frac{\partial}{\partial y}(\nabla^2\Psi_0)\right] = \mu\nabla^2(\nabla^2\psi) \tag{7.22}$$

The equation is made dimensionless by the introduction of the relationships

$$x_* = \frac{x}{L} \quad y_* = \frac{y}{L} \quad t_* = t\frac{U}{L} \quad u_x^* = \frac{u_x}{U}$$

$$u_y^* = \frac{u_y}{U} \quad \Psi_* = \frac{\Psi_0}{UL} \quad \psi_* = \frac{\psi}{UL} \quad \nabla_*^2 = L^2\nabla^2 \tag{7.23}$$

The dimensionless equation is

$$\frac{\partial}{\partial t_*}\left(\nabla_*^2\psi_*\right) - \frac{\partial\Psi_*}{\partial y_*}\frac{\partial}{\partial x_*}\left(\nabla_*^2\psi_*\right) + \frac{\partial\Psi_*}{\partial x_*}\frac{\partial}{\partial y_*}\left(\nabla_*^2\psi_*\right)$$

$$- \frac{\partial\psi_*}{\partial y_*}\frac{\partial}{\partial x_*}\left(\nabla_*^2\Psi_*\right) + \frac{\partial\psi_*}{\partial x_*}\frac{\partial}{\partial y_*}\left(\nabla_*^2\Psi_*\right) = \frac{1}{R}\nabla_*^2\left(\nabla_*^2\psi_*\right) \tag{7.24}$$

Following the same technique of the analysis of the Kelvin-Helmholtz instability, we assume

$$\psi_* = \phi(y_*)e^{ik_*(x_* - c_*t_*)} \tag{7.25}$$

where k_* is a dimensionless wave number and c_* is a dimensionless wave speed that may be complex, $c_* = c_r^* + ic_i^*$. The stability depends on the sign of c_i^* with a positive value indicating instability. The Orr-Sommerfeld equation is written for the amplitude function ϕ. Using $U = -\partial\Psi/\partial y$ and with comparable dimensionless quantities

$$\left(U^* - c_*\right)\left(\frac{d^2\phi}{dy_*^2} - k_*^2\phi\right) - \frac{d^2U^*}{dy_*^2}\phi = -\frac{i}{k_*R}\left(\frac{d^4\phi}{dy_*^4} - 2k_*^2\frac{d^2\phi}{dy_*^2} + k_*^4\phi\right) \tag{7.26}$$

The boundary conditions are that the amplitude of the disturbance and its derivative are zero on the boundaries. Arbitrarily locating the boundaries at $y_* = 0$ and $y_* = 1$,

$$\phi(0) = \phi(1) = \frac{d}{dy_*}\phi(0) = \frac{d}{dy_*}\phi(1) = 0 \tag{7.27}$$

Somewhat different conditions would apply to boundary layer flow. Equation (7.26) stems from the work of Orr (1907) and Sommerfeld (1908).

Solutions to (7.26) are extraordinarily difficult even though the equation is linear. Most attempts have been to find curves where $c_i^* = 0$, the curves of neutral stability. Such curves give a relationship between the wave number k_* of a disturbance and the Reynolds number R. Fig. 7.4 shows such a curve for plane Poiseuille flow (flow between stationary flat plates; see Sec. 5.1.1). The Reynolds number in Fig. 7.4 is based on the channel half-width. The *critical Reynolds number* for this flow is the value of the minimum of the neutral stability curve, in this case 5300. At Reynolds numbers above the critical, the flow may be turbulent if disturbances of the proper wave length are present. In most flows there are disturbances present at many wave lengths so the critical Reynolds number is the most important.

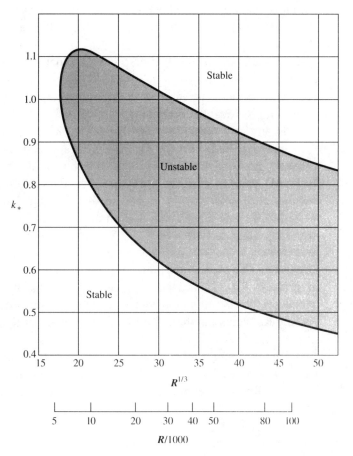

FIGURE 7.4
Stability diagram for Poiseuille flow (adapted from Shen, 1954).

7.1.3 Comments on Stability

Disturbances that cause most engineering flows to become turbulent are everywhere. They may be caused by pipe or channel roughness, by the vibrations of a pump, by wind, etc. According to the Orr-Sommerfeld equation, laminar flow is possible at high Reynolds numbers if the wave numbers of the unstable region can be avoided. Laboratory experiments have demonstrated that phenomenon. Laminar pipe flows have been produced at Reynolds numbers of 50,000 and higher whereas the flow usually becomes turbulent at a Reynolds number of approximately 2100. The experiment has to be conducted with great care. The pipe must be very smooth, the entrance region must be constructed with care (in fact, a properly shaped entrance region is the key to damping turbulent fluctuations that may be present in the fluid as it enters the pipe), the flow should be not too turbulent before entrance, and the entire apparatus should be free of vibration. At such high Reynolds numbers a slight tap on the pipe will produce instant turbulence. Because resistance to flow is much less if the flow is laminar, great economic benefit would result if commercial pipelines could carry liquids and gases at laminar flows and at reasonable flow rates at the same time. That goal is so difficult to achieve that no commercial exploitation of laminar flow at high Reynolds numbers has occurred.

In any one experiment the critical Reynolds number may differ from the minimum found on the stability diagram. Boundary layers with zero pressure gradient tend to become unstable at a displacement thickness Reynolds number of about 600. However, the disturbance that causes turbulence may grow slowly at first so that the actual turbulent flow occurs at higher Reynolds numbers. The curve of neutral stability for the Blasius boundary layer is shown in Fig. 7.5 where the dimensionless frequency of disturbance is plotted against the Reynolds number based on displacement thickness. The critical Reynolds number, based on displacement thickness, is given as 420. The transition from marginal instability to turbulence depends on the amplification factor, which is a function of the disturbance and the Reynolds number. Amplification curves for the flat plate are shown in Fig. 7.6.

There is much more to the theory of hydrodynamic stability. Stability is influenced by stratification and, in an analogous process, rotation. The curvature of streamlines changes the stability characteristics. Consider the flow between two rotating cylinders (Fig. 7.7). If the outer cylinder is rotating and the inner cylinder is stationary, the velocity increases outward in a near-linear manner. The centrifugal force on the fluid particles tends to move them outward and that force is counteracted by the pressure gradient acting inward. Since the outer fluid particles have greater velocity,

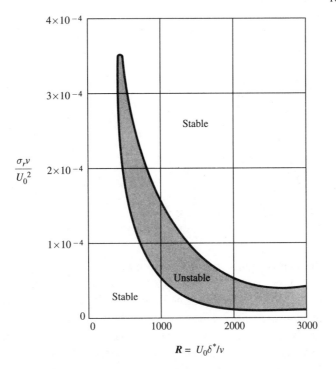

FIGURE 7.5
Stability diagram for the Blasius boundary layer (adapted from Shen, 1954).

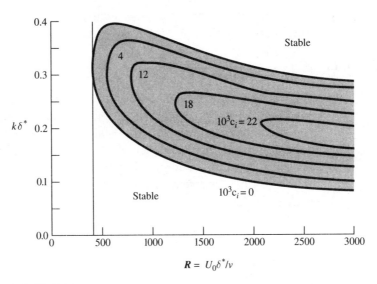

FIGURE 7.6
Amplification curves for the boundary layer on a flat plate (adapted from Shen, 1954).

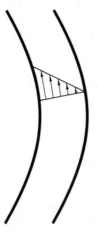

FIGURE 7.7
Flow between two rotating cylinders.

the centrifugal force on those particles is greater and they tend to stay in the outer part of the motion, a tendency that leads to stability. If the inner cylinder is rotating and the outer cylinder is stationary (as shown), the effect is opposite and the flow tends toward instability. The same considerations hold for concave and convex walls. The fluid following a convex wall—analogous to the stationary inner cylinder—tends to be more stable than the fluid following a concave wall.

Heat transfer from a wall to the fluid has stability consequences due to changes in viscosity and buoyancy. Compressibility of the fluid in the absence of heat transfer appears to have little effect at low Mach numbers. Boundary layer control measures such as suction along a porous boundary affect stability. (Suction greatly increases stability.)

7.2 THE MEAN FLOW EQUATIONS

There is no such thing as a steady, turbulent flow. Even for apparently steady initial and boundary conditions, turbulence exhibits unsteady fluctuations. The fluctuations are three-dimensional. In the flow of water—in rivers, for example—these fluctuations are visible in the form of swirls and eddies. In the atmosphere the wind clearly comes in gusts. For the solution to most fluid mechanics problems we are interested in the mean flow, some time average of the velocity and pressure.

7.2.1 Time Averages

The velocity and pressure can be divided into average quantities plus fluctuating quantities

$$u_x = \bar{u}_x + u'_x \qquad u_y = \bar{u}_y + u'_y \qquad p = \bar{p} + p' \qquad (7.28)$$

in which the overbar refers to time-averaged quantities and the prime refers to fluctuating quantities. The time average of an arbitrary quantity f, either velocity or pressure, at time t_0 is defined as

$$\bar{f} = \frac{1}{2T} \int_{t_0-T}^{t_0+T} f \, dt \tag{7.29}$$

The limits of integration must be taken sufficiently large so that the average quantity is independent of those limits. In an unsteady problem (*i.e.*, unsteady in the sense that the time-averaged velocities vary with time), the limits must be large compared to the time scale of the unsteady (turbulent) fluctuations but small compared to the time scale of the mean motions. In a steady problem the limits can be infinite, $T \to \infty$. The difference between an unsteady problem and a steady problem is shown in Fig. 7.8. In the unsteady case T must be large compared to the time scale of the turbulence but small compared to time scale of the basic flow.

By definition the time average of the fluctuating quantities is zero,

$$\frac{1}{2T} \int_{t_0-T}^{t_0+T} f' dt = 0 \tag{7.30}$$

For unsteady flow we should be able to define a value of Δt that is large compared to the time scale of the turbulent fluctuations so that

$$\int_{t_0-T}^{t_0+T} \bar{u}_x \, dt = \int_{t_0-T-\Delta t}^{t_0+T+\Delta t} \bar{u}_x \, dt \tag{7.31}$$

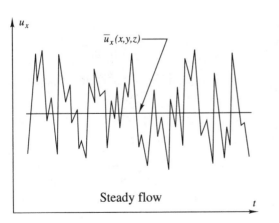

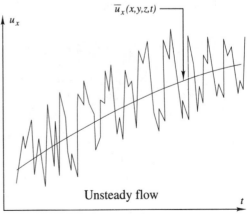

FIGURE 7.8
Turbulent fluctuations in "steady" flow and unsteady flow.

Several other properties of the averages are

$$\bar{\bar{f}} = \bar{f} \qquad \overline{(f+g)} = \bar{f} + \bar{g} \qquad \overline{f'f'} > 0 \quad \text{unless } f' = 0$$

$$\overline{f'g'} = 0 \quad \text{only if } f' \text{ and } g' \text{ are independent (uncorrelated)}$$

$$\overline{\frac{\partial f}{\partial s}} = \frac{\partial \bar{f}}{\partial s} \qquad \overline{\int f \, ds} = \int \bar{f} \, ds$$

7.2.2 The Equations of Motion

Only a constant density fluid is considered in this section; otherwise, the density must be divided into average and fluctuating quantities. The equation of continuity is

$$\frac{\partial u_i}{\partial x_i} = \frac{\partial}{\partial x_i} \left(\bar{u}_i + u_i' \right) = 0 \tag{7.32}$$

We take the average of (7.32), but the average of the fluctuating term is zero so

$$\frac{\partial \bar{u}_i}{\partial x_i} = 0 \tag{7.33}$$

Continuity of the average velocity is the same as the continuity of the total velocity. Also the fluctuating velocities are continuous

$$\frac{\partial u_i'}{\partial x_i} = 0 \tag{7.34}$$

Substitution of the fluctuating quantities into the steady flow equations of motion yields

$$\left(\bar{u}_j + u_j' \right) \frac{\partial}{\partial x_j} \left(\bar{u}_i + u_i' \right) = -\frac{1}{\rho} \frac{\partial}{\partial x_i} \left(\bar{p} + p' \right) + \frac{\mu}{\rho} \frac{\partial^2}{\partial x_j \partial x_j} \left(\bar{u}_i + u_i' \right) \tag{7.35}$$

Applying the integral of (7.30) gives an equation in the fluctuating and average velocities

$$\bar{u}_j \frac{\partial \bar{u}_i}{\partial x_j} + \overline{u_j' \frac{\partial u_i'}{\partial x_j}} = -\frac{1}{\rho} \frac{\partial \bar{p}}{\partial x_i} + \frac{\mu}{\rho} \frac{\partial^2 \bar{u}_i}{\partial x_j \partial x_j} \tag{7.36}$$

The fluctuating quantities occur only in the nonlinear terms. In the second term both of the fluctuating velocities can be moved inside the derivative sign since the extra term is zero from (7.34). The equation is often rearranged as

$$\rho \bar{u}_j \frac{\partial \bar{u}_i}{\partial x_j} = -\frac{\partial \bar{p}}{\partial x_i} + \frac{\partial}{\partial x_j} \left(\mu \frac{\partial \bar{u}_i}{\partial x_j} - \rho \overline{u_i' u_j'} \right) \tag{7.37}$$

Written in this way the last term is interpreted as an additional stress term, in addition to the viscous stress. The turbulence stresses are called *Reynolds stresses*. The stress tensor is

$$\tau_{ij} = -\mu \left(\bar{u}_{i,j} + \bar{u}_{j,i} \right) + \rho \overline{u_i' u_j'} \tag{7.38}$$

The last term really represents a momentum transport. The $u_i' u_j'$ could be the j-momentum transported in the i-direction by u_i'. The term $\overline{u_i' u_j'}$ is called the *velocity correlation* of the velocity in the i-direction with the velocity in the j-direction. The correlation coefficient is

$$c_{ij} = \frac{\overline{u_i' u_j'}}{\sqrt{\overline{u_i'^2} \ \overline{u_j'^2}}} \tag{7.39}$$

Obviously, if $i = j$ the correlation is perfect and $c_{ij} = 1$. In a shear flow the correlation is mostly negative if $i \neq j$. Consider the situation of Fig. 7.9 in which $\partial \bar{u}_x / \partial y > 0$. A positive fluctuation of a fluid particle in the y-direction would, on the average, find a velocity that is greater and would retard that velocity, causing the x-fluctuation to be negative. Conversely, a negative fluctuation in the y-direction would find, on the average, a velocity that is less and would transport positive momentum to those particles. Thus, the Reynolds stress is mostly of the same sign as the viscous stresses.

Six new terms appear in (7.37). (Because the Reynolds stress term is symmetric, the nine-term tensor has only six independent terms.) There are now more unknowns than equations since the correlations must be treated as unknown. We could write more equations in $\vec{u}'$, but doing so would lead

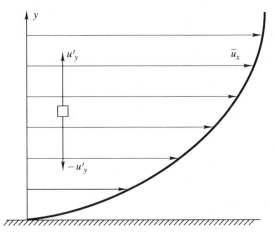

FIGURE 7.9
Fluctuating velocity in a shear flow.

to higher order correlations such as

$$\overline{u_i'u_i'u_j'} \quad \text{no sum on } i \quad \text{or} \quad \overline{p'u_i'}$$

Again, equations can be written for these terms, but they, in turn, would involve correlations of even higher order. Consequently, the system always has more unknowns than equations, a situation known as the *closure problem*, which comes about from the introduction of the average and fluctuating velocities into the Navier-Stokes equations.

Turbulent flow solutions contain some assumption on how the Reynolds stresses vary. This assumption can be at the basic level or it can be with the higher order correlations. Traditionally, researchers have tried to guess a relationship between Reynolds stresses and average velocities. The alternative is to attempt a detailed description of the fine structure of the turbulence, apparently a hopeless task in most practical problems at this time.

7.2.3 Vortex Stretching

The velocity distribution of either laminar or turbulent flow causes rotation in the fluid. The vortices, or *eddies*, result in high local velocity gradients that—through viscosity—cause a transfer of mechanical energy to thermal energy. The axis of vorticity can be in any direction, but according to Tennekes and Lumley (1972), the axes most effective at the conversion of energy are those that are roughly aligned with the velocity gradient (see

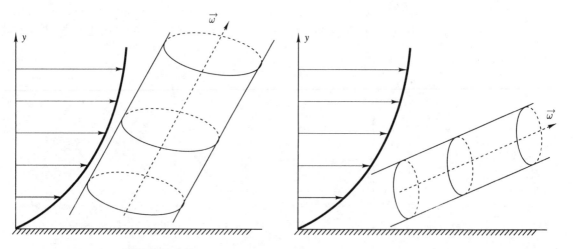

FIGURE 7.10
The velocity distribution converts a slowly rotating, fat vortex into a high-speed, thin vortex by stretching.

Fig. 7.10). A rotational axis aligned in the direction of the velocity gradient becomes tilted since the upper part is in a zone of higher velocity than the lower part. It acquires a perfect position to be stretched by the differing fluid velocities. Since the vortex is being stretched, the law of conservation of angular momentum dictates that the rate of rotation must increase. Consider the analogy of the child's toy that consists of a string with a weight (say, a button) in the middle. The weight is set to rotate in a large arc. Then as the child pulls on the string, the radius of rotation decreases, but the rate of rotation increases. A graphic example of vortex stretching is the occurrence of tornados. The worst tornados occur in cases of strong winds aloft with a lifting mechanism (moisture and heat) plus some rotation to get the process started.

The rotation in a vortex causes a good correlation between the fluctuating velocity components and results in high Reynolds stresses. This picture also provides a mechanism by which the larger whirls become smaller and more dissipative. Because the smaller vortex structure is dissipative, it requires a constant source of energy from the larger structures and ultimately from the primary flow.

To see more exactly how vortex stretching affects the fluid, consider the vorticity transport Eq. (1.134) that is repeated here

$$\frac{D\omega_i}{Dt} = \omega_j \frac{\partial u_i}{\partial x_j} + \frac{\mu}{\rho} \frac{\partial^2 \omega_i}{\partial x_j \partial x_j} \tag{7.40}$$

From (1.54) and (1.55)

$$\omega_j \frac{\partial u_i}{\partial x_j} = \omega_j \left(d_{ij} + \omega_{ij} \right) \tag{7.41}$$

Using the identity (A.57) on the last term of (7.41), it is

$$\boldsymbol{\omega} \cdot \vec{\omega} = \vec{\omega} \times \vec{\omega} = 0 \tag{7.42}$$

and the vorticity transport equation becomes

$$\frac{D\omega_i}{Dt} = \omega_j d_{ij} + \frac{\mu}{\rho} \frac{\partial^2 \omega_i}{\partial x_j \partial x_j} \tag{7.43}$$

The first term on the right is the production of vorticity by the product of the vorticity and the rate of strain. As stated in (1.136), this term disappears in two-dimensional flow.

Consider a flow in which $u_x = u_x(x)$ and the only component of vorticity is aligned with the x-axis, ω_x. Then the x-component of the term $\omega_j d_{ij}$ is

$$\omega_x d_{xx} = \frac{1}{2} \frac{\partial u_x}{\partial x} \omega_x \tag{7.44}$$

An increase in velocity in the x-direction, a positive $\partial u_x / \partial x$, means that the x-vorticity is amplified, as shown from (7.43) using (7.44).

The fluctuating terms can contribute to the vorticity of the overall flow. The rate of strain and vorticity are divided into averaged and fluctuating components

$$d_{ij} = \bar{d}_{ij} + d'_{ij} \qquad \omega_{ij} = \bar{\omega}_{ij} + \omega'_{ij} \tag{7.45}$$

with the fluctuating part defined in the obvious manner in terms of the fluctuating velocities. Using (7.45) in (7.43) and taking a time average produces

$$\frac{D\bar{\omega}_i}{Dt} = \bar{\omega}_j \bar{d}_{ij} + \frac{\mu}{\rho} \frac{\partial^2 \bar{\omega}_i}{\partial x_j \partial x_j} - \overline{u'_j \frac{\partial \omega'_i}{\partial x_j}} + \overline{\omega'_j d'_{ij}} \tag{7.46}$$

where the third term on the right comes from the substantial derivative. The last two terms show that the fluctuating quantities can contribute to the time-averaged vorticity.

7.2.4 Relative Size of the Reynolds Stress

Consider friction in pipe flow. The Darcy-Weisbach equation is

$$h_L = f \frac{L}{D} \frac{U^2}{2g} \tag{7.47}$$

in which h_L is the head loss in a length L, D is the pipe diameter and U is the average velocity. The friction-factor diagram (for uniform "sand grain" roughness), which gives the Darcy friction factor f in terms of the Reynolds number and the pipe roughness, is shown in Fig. 7.11. A similar quantity, the Fanning friction factor, is one-fourth of f. For rough pipes f becomes constant with the Reynolds number. For smooth pipes Blasius found that

$$f = 0.316 R^{-1/4} \qquad 3000 < R < 100,000 \tag{7.48}$$

which contrasts to the laminar flow case where $f = 64/R$. Clearly, the resistance in turbulent flow is much higher than in laminar flow for high Reynolds numbers. The same is apparent in flow around a sphere where in Stokes flow the drag coefficient is $C_D = 24/R$ whereas it becomes (almost) independent of the Reynolds number for highly turbulent flow.

Measurements in turbulent flow confirm that the Reynolds stresses tend to be much larger than the viscous stresses. Fig. 7.12 shows the Reynolds stress together with the total stress in a channel. The difference between these two curves is the viscous stress. The fluctuating velocity correlation c_{yx} is also shown on the figure. For most of the channel the viscous

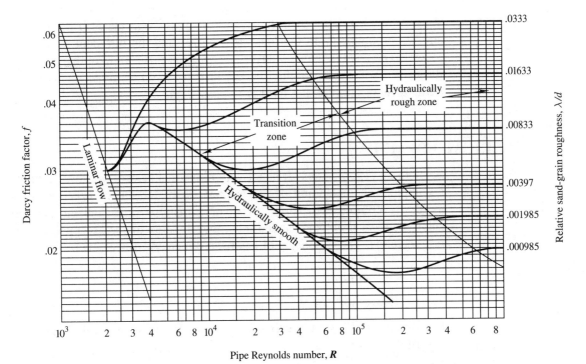

FIGURE 7.11
The Darcy friction factor for uniform sand-grain roughness.

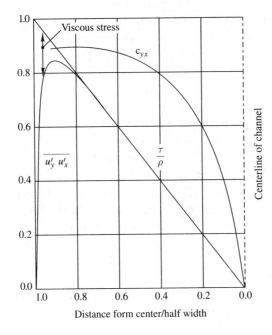

FIGURE 7.12
Shear stress, Reynolds stress, and velocity correlation for flow between plates normalized on the maximum shear.

stress is much smaller than the Reynolds stress. Near the wall the viscous stresses predominate in the *laminar sublayer* and the turbulent stress falls to zero. The term laminar sublayer is somewhat misleading in that a strong, unsteady vorticity in the flow occurs in that region. A more modern, and perhaps more appropriate, term is *viscous sublayer*.

7.3 EMPIRICAL FORMULAS

Schlichting (1960) states "It is not very likely that science will ever achieve a complete understanding of the mechanism of turbulence because of its extremely complicated nature." That pessimistic outlook and the need to solve practical problems has led most investigators to seek empirical solution methods. The primary tool has been dimensional analysis.

7.3.1 Eddy Viscosity

Boussinesq (1877) simply used the analogy with Newton's law of viscosity to write

$$\tau_{yx}^{t} = -\mu_t \frac{d\bar{u}_x}{dy} \tag{7.49}$$

in which τ_{yx}^{t} is the turbulent shear and μ_t is the *eddy viscosity*. If μ_t were a property of the fluid, a constant, then turbulent flows could by solved with the same ease of laminar flow. Even though μ_t is actually a strong property of the flow, and of position in the flow, (7.49) has served as a useful model in some cases.

The eddy viscosity formulation comes about from an analogy with molecular motion. If the movement of molecules is taken as the mechanism for momentum transport in laminar flow, then (7.49) is analogous with the molecules replaced by "fluid particles." Such an analogy is immediately suspect because in a resting fluid (*i.e.*, one in which there is no mean or fluctuating motion, which can be achieved by filling a tank and permitting it to sit without disturbance for a long period of time), there remains considerable molecular motion. On the other hand in a shear flow where there are large fluctuations of velocity, the molecular activity is almost unchanged. The molecular activity is unaffected by the state of the flow (or else μ would not be a fluid property), but turbulent activity is dependent on the flow field. If we were to form a correlation coefficient for molecular velocities analogous to (7.39), we would find that it is very much smaller than the c_{ij}.

7.3.2 Prandtl's Mixing Length

Prandtl (1925) attempted to relate the eddy viscosity to the velocity gradient and a *mixing length*, ξ, in the form

$$v_t = \xi^2 \left| \frac{d\bar{u}_x}{dy} \right| \qquad v_t = \frac{\mu_t}{\rho} \qquad (7.50)$$

Mixing length theory extends the analogy with molecular transport of momentum. The mixing length is analogous to the mean free path of molecular motion and, hence, to some turbulent scale. The hope is that the mixing length can be related to the scale of a problem. Obviously, the mixing length changes with position in the flow. For example, immediately next to a smooth wall the cross-flow mixing must tend to zero. That tendency, however, gives some hope that the relationship between the mixing length and the flow can be estimated, at least in an approximate manner. In fact some calculations have been successful using the mixing length concept. It remains, however, an empirical concept and does little to explain the fundamentals of turbulent flow.

There are some obvious difficulties with the mixing length theory. One is that the eddy viscosity is zero at points where $d\bar{u}_x/dy$ is zero—in the center of a pipe, for example. Experiment would show that the velocity fluctuations do not disappear in such locations. Most such objections can be answered by additional complication and additional empirical "constants" that must be found by experiment. Some of these additions may aid in engineering calculation of a particular problem and, in this way, have some merit. They do not substitute for a fundamental understanding of turbulence.

7.3.3 The von Kármán Similarity Theory

Von Kármán (1930) assumed that the velocity fluctuations at any point in the flow field could be related to those at any other point through transformations of time and length scales. Von Kármán's argument involves the equations of motion, but for the purposes herein dimensional reasoning alone is used. The simplest scale length that contains only the time-averaged velocities in a shear flow is

$$l_0 = \frac{\dfrac{\partial \bar{u}_x}{\partial y}}{\dfrac{\partial^2 \bar{u}_x}{\partial y^2}} \qquad (7.51)$$

Von Kármán assumed that the mixing length is proportional to this length scale

$$\xi = \kappa l_0 \qquad (7.52)$$

in which κ is a "universal constant." In fact numerous experiments have given an approximate value of $\kappa = 0.4$. The Reynolds stress is then

$$\tau_{yx}^t = -\rho \overline{u_y' u_x'} = -\rho \kappa^2 \frac{\left| \left(\dfrac{d\bar{u}_x}{dy} \right)^3 \right|}{\left(\dfrac{d^2\bar{u}_x}{dy^2} \right)^2} \frac{d\bar{u}_x}{dy} \qquad (7.53)$$

For multidimensional flows the Kármán similarity formulation must be generalized. It is

$$\tau_{ij}^t = -\rho \kappa^2 \left| \frac{\left(2\bar{d}_{mn} \bar{d}_{nm} \right)^{3/2}}{\bar{\Omega}_{pq} \bar{\Omega}_{qp}} \right| \bar{d}_{ij} \qquad (7.54)$$

in which

$$\bar{\Omega}_{ij} = \frac{\partial \bar{\omega}_i}{\partial x_j} + \frac{\partial \bar{\omega}_j}{\partial x_i} \qquad (7.55)$$

and where $\bar{\bar{\omega}}$ is the vorticity vector of the averaged flow. For flow in a round pipe

$$\tau_{Rx}^t = -\rho \kappa^2 \left| \frac{\left(\dfrac{d\bar{u}_x}{dR} \right)^3}{\left(\dfrac{d^2\bar{u}_x}{dR^2} - \dfrac{1}{R} \dfrac{d\bar{u}_x}{dR} \right)^2} \right| \frac{d\bar{u}_x}{dR} \qquad (7.56)$$

Example 7.1   Turbulent Poiseuille flow.   In the steady, uniform flow between two flat plates the equation of motion gives

$$\frac{\partial \tau}{\partial y} = \frac{\partial p}{\partial x}$$

Since the pressure gradient is a constant,

$$\tau = \tau_\omega \frac{2y}{d} \quad \text{for } y \geq 0 \qquad \tau = -\tau_\omega \frac{2y}{d} \quad \text{for } y \leq 0$$

where τ_ω is the wall shear and y is measured from the centerline between the plates. Using the Kármán similarity formulation in the top half of the channel

$$\tau_\omega \frac{2y}{d} = -\rho \kappa \frac{\left| \left(\dfrac{d\bar{u}_x}{dy} \right)^3 \right|}{\left(\dfrac{d^2\bar{u}_x}{dy^2} \right)^2} \frac{d\bar{u}_x}{dy}$$

The solution in the top half of the flow is

$$\bar{u}_x = \bar{u}_m + \frac{1}{\kappa}\sqrt{\frac{\tau_w}{\rho}}\left[\ln\left(1 - \sqrt{\frac{2y}{d}}\right) + \sqrt{\frac{2y}{d}}\right] \qquad (7.57)$$

where $\bar{u}_m$ is the maximum velocity, which occurs at the centerline, where $y = 0$.

Two obvious defects appear in the equation for the velocity. First, there is a discontinuity in $d\bar{u}_x/dx$ in the center of the channel, at $y = 0$, if the velocity equation is reflected symmetrically about the centerline. Such a discontinuity in the slope of the velocity curve is contrary to experiment—and to that sometimes unreliable human prejudice, intuition. It comes about because the eddy viscosity is zero in the center if the velocity gradient is zero.

The second difficulty occurs near the wall, at $y = d/2$, where the equation shows that the velocity goes to negative infinity. That anomaly comes about from the neglect of the viscous terms, which predominate very close to the wall. In spite of these difficulties, the velocity profile over a large portion of the channel is remarkably close to realistic.

We now repeat the calculation using Prandtl's mixing length. Prandtl assumed that the mixing length is proportional to the distance from the wall. He also reasoned from (7.50) that

$$\tau = \rho\kappa^2 y^2 \left|\frac{d\bar{u}}{dy}\right|\frac{d\bar{u}}{dy}$$

Now Prandtl assumed that $\tau = \tau_w$, the shear stress at the wall, everywhere! (Why? We can only say that the results resemble reality in spite of the approximation.) The mixing length was taken proportional to the distance from the wall

$$l_0 = \kappa\left(\frac{d}{2} - y\right)$$

which gives

$$\frac{d\bar{u}}{dy} = \frac{\sqrt{\frac{\tau_w}{\rho}}}{\kappa\left(\frac{d}{2} - y\right)}$$

After integrating

$$\bar{u}_x = \frac{1}{\kappa}\sqrt{\frac{\tau_w}{\rho}}\ln\left(\frac{d}{2} - y\right)$$

Since at $y = 0$, $\bar{u}_x = \bar{u}_{\max}$,

$$\bar{u}_x = \bar{u}_{\max} - \frac{1}{\kappa}\sqrt{\frac{\tau_w}{\rho}}\ln\frac{\frac{d}{2}}{\frac{d}{2} - y} \qquad (7.58)$$

The curves of (7.57) and (7.58) are remarkably close.

Historical Note: Theodore von Kármán (1881–1963) received his first degree in mechanical engineering from the Royal Polytechnic Institute of Budapest. His professional life coincided with the great advances in aerodynamics and the modern development of the fluid mechanics. He was lucky enough to study at Göttingen where he received his doctorate and then remained on the faculty. Von Kármán was the first director of the Polytechnic Institute of Aachen. His latter years were spent at the California Institute of Technology. He was a brilliant engineer who made maximum use of mathematics in a wide variety of mechanics problems. The Kármán vortex street was one of his first insights in fluid mechanics. That discovery served him well in the United States. The Tacoma Narrows Bridge failure occurred when he was at Cal Tech. The bridge authority considered the occurrence such a rare phenomenon that they proposed to build the bridge again using exactly the same design. Von Kármán sent them a telegram that stated "If the bridge is rebuilt in the same way it will fail in the same way." (However, the failure was actually due to flutter, not simply vortex shedding.) His autobiography, *The Wind and Beyond*, not only recounts many humorous incidents in his life but is also a fascinating history of modern mechanics.

7.3.4 Deissler's Formula

Deissler (1955) proposed the following empirical formula for the velocity distribution near a wall:

$$\tau_{yx}^t = -\rho n^2 \bar{u}_x y \left[1 - \exp\left(-\frac{\rho n^2 \bar{u}_x y}{\mu} \right) \right] \frac{d\bar{u}_x}{dy} \tag{7.59}$$

in which n is a constant to be determined from empirical data. Experiments have shown n to be about 0.124 in smooth pipes. In (7.59) y is the distance from the wall; thus, the turbulent shear goes to zero as the wall is approached.

Example 7.2   Velocity distribution in a pipe.   To find the velocity distribution in a round pipe (Fig. 7.13) the shear stress is taken as the laminar stress plus the turbulent stress from (7.59) in the zone near the wall. For the outer solution, nearer the center of the pipe, we will follow the recommendation of Deissler and use

$$\frac{\bar{u}_x}{u_*} - 12.85 = \frac{1}{0.36} \ln \frac{\rho y u_*}{26\mu} \tag{7.60}$$

in which y is the distance from the pipe wall. Eq. (7.60) is in the form of (7.58) with specific choices made for the constants. The constant κ is chosen to be 0.36 (instead of the recommended value of 0.40). In place of choosing a centerline velocity, the constants 12.85 and 26 are used. The notation $u_* = \sqrt{\tau_w/\rho}$ is common and is called the *friction velocity*.

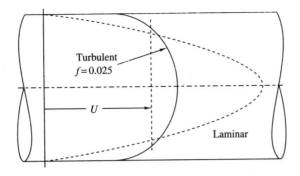

FIGURE 7.13

Laminar and turbulent flow in a pipe.

For the part of the flow near the wall the laminar shear is added to Deissler's equation

$$\tau_{Rx} = -\mu \frac{d\bar{u}_x}{dR}$$

$$- \rho n^2 \bar{u}_x (R_0 - R) \left[1 - \exp\left(-\frac{\rho n^2 \bar{u}_x}{\mu} (R_0 - R) \right) \right] \frac{d\bar{u}_x}{dR} \qquad (7.61)$$

in which R_0 is the pipe radius. The shear stress in the pipe very near the wall is written

$$\tau_{Rx} = \tau_w \left(1 - \frac{R_0 - R}{R_0} \right) \approx \tau_w \qquad (7.62)$$

The approximation of (7.62) is that near the wall $R \approx R_0$. That same approximation cannot be used in (7.61) since the distance $R_0 - R$ is a vital part of the equation.

Equation (7.61) can be integrated numerically and matched to (7.60) for the velocity in the pipe cross section. The result is shown in Fig. 7.14. The result agrees well with experiment, but this was the sort of experiment from which the empirical relationships were derived.

7.4 JETS AND WAKES

As noted in the chapter on laminar flow, the flow in jets and wakes is nearly always turbulent. The absence of the boundary makes these flows somewhat simpler than turbulent flows that are constrained by walls. In turbulent jets and wakes the viscosity of the fluid plays no role other than the final dissipation of the turbulent energy. The velocity profile and entrainment into the jet or wake can be calculated without regard to the viscosity.

A jet is created by fluid issuing from a slit (two-dimensional), orifice (axisymmetric), or some other shape. A wake is the region of disturbed flow behind an object in otherwise undisturbed flow. A wake, like a jet,

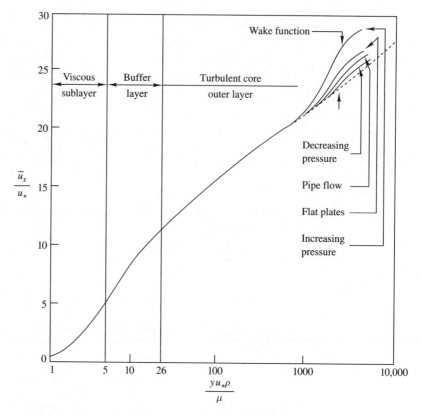

FIGURE 7.14
Velocity distribution in a pipe.

can be two-dimensional, axisymmetric, or arbitrary. A closely related flow is the shear flow that results from the boundary between two flows having different free stream velocities. The latter is sometimes called a half-jet, although some aspects of a shear flow are different from those of a jet. The assumption is usually made that the characteristics of these flows are similar with distance along the axis of the flow. A similar assumption is that the turbulence is *self-preserving*, meaning that the turbulent structure, as expressed by a time and length scale, remains similar along the axis of the flow. Only two-dimensional and axisymmetric flows are considered herein. Also, all flows are steady from the point of view of the time-averaged velocities.

7.4.1 Length and Time Scales for Wakes

The primary assumption on scales is that changes occur slowly in the down-stream direction compared to the cross-stream direction, $\partial/\partial x \ll \partial/\partial y$. This assumption is essentially the one made in boundary layer flow. There are two length scales for the problem, the width b of the wake (suitably defined, such as the distance between the points where the *velocity defect*, $U - \bar{u}_x$, falls below some percentage of the free stream velocity) and an ill-defined scale change in the downstream direction, L. The assumption, better expressed as the derivative, is that the former scale is small compared to the latter ($b \ll L$). The time scale is defined by the velocity. Two characteristic velocities are taken, the free stream velocity, and the free stream velocity minus the centerline velocity (the centerline velocity defect) in the wake, $W = U - \bar{u}_c$. Both velocities, U and W, are used in scaling the equations. Both are velocities in the direction of the primary flow, and U is used only in those terms that have an undifferentiated $\bar{u}_x$.

As always, the terms of the continuity equation must be of the same order of magnitude for the equation to balance; thus,

$$\frac{\partial \bar{u}_x}{\partial x} = -\frac{\partial \bar{u}_y}{\partial y} = \mathrm{O}\left(\frac{W}{L}\right) = \mathrm{O}\left(\frac{V}{b}\right) \tag{7.63}$$

where V is a characteristic velocity in the y-direction. The streamline and cross-stream velocities are assumed to be $W = \mathrm{O}(VL/b)$. If $L \gg b$, then $W \gg V$.

The equations of motion are

$$\bar{u}_j \frac{\partial \bar{u}_i}{\partial x_j} = -\frac{1}{\rho}\frac{\partial \bar{p}}{\partial x_i} - \frac{\partial}{\partial x_j}\left(\overline{u_i' u_j'}\right) \tag{7.64}$$

in which the viscous terms have been omitted. In this form the equations are independent of the Reynolds number. The terms on the left scale as

$$\bar{u}_x \frac{\partial \bar{u}_x}{\partial x} = \mathrm{O}\left(\frac{UW}{L}\right) \qquad \bar{u}_y \frac{\partial \bar{u}_x}{\partial y} = \mathrm{O}\left(\frac{VW}{b}\right)$$

$$\bar{u}_x \frac{\partial \bar{u}_y}{\partial x} = \mathrm{O}\left(\frac{UV}{L}\right) \qquad \bar{u}_y \frac{\partial \bar{u}_y}{\partial y} = \mathrm{O}\left(\frac{V^2}{b}\right) \tag{7.65}$$

At this point there must be some assumption regarding the Reynolds stresses. First, we assume that the correlation coefficients are of order one. Then take $u' = \mathrm{O}(\hat{u})$ where u' is either of the fluctuating quantities and $\hat{u}$ is defined later. Using these results in the equations for the x-direction and y-direction results in

$$\bar{u}_x \frac{\partial \bar{u}_x}{\partial x} \ + \ \bar{u}_y \frac{\partial \bar{u}_x}{\partial y} \ = \ -\frac{1}{\rho}\frac{\partial \bar{p}}{\partial x} \ - \ \frac{\partial \overline{u_x'^2}}{\partial x} \ - \ \frac{\partial \overline{u_x' u_y'}}{\partial y} \tag{7.66}$$

$$\frac{UW}{L} \qquad\qquad \frac{VW}{b} \qquad\qquad\qquad \frac{\hat{u}^2}{L} \qquad \frac{\hat{u}^2}{b}$$

$$\bar{u}_x \frac{\partial \bar{u}_y}{\partial x} \;+\; \bar{u}_y \frac{\partial \bar{u}_y}{\partial y} \;=\; -\frac{1}{\rho}\frac{\partial \bar{p}}{\partial y} \;-\; \frac{\partial \overline{u'_x u'_y}}{\partial x} \;-\; \frac{\partial \overline{u'^2_y}}{\partial y} \qquad (7.67)$$

$$\underset{\dfrac{UV}{L}}{} \qquad \underset{\dfrac{V^2}{b}}{} \qquad\qquad \underset{\dfrac{\hat{u}^2}{L}}{} \qquad \underset{\dfrac{\hat{u}^2}{b}}{}$$

The pressure terms are left without an indication of their size at this point. In (7.66) the last term is larger than the next-to-last term. If at least one of the turbulence terms is to be retained, the last term must be at least as large as the first term, which gives an indication of the size of the turbulence fluctuations, $\hat{u}^2 \geq O(UWb/L)$. If that is the case, however, the last term of (7.67) is larger than any of the other three terms that have been sized because $W \gg V$. If the cross-stream equation is to balance, the last term can be offset only by the cross-stream pressure gradient

$$-\frac{1}{\rho}\frac{\partial \bar{p}}{\partial y} = \frac{\partial \overline{u'^2_y}}{\partial y} \qquad \text{or} \qquad \frac{p_0 - \bar{p}}{\rho} = \overline{u'^2_y} \qquad (7.68)$$

where p_0 is the free stream pressure.

To scale the equation for the x-direction, we must have an indication of the streamline pressure gradient. Equation (7.68) gives

$$-\frac{1}{\rho}\frac{\partial \bar{p}}{\partial x} = \frac{\partial \overline{u'^2_y}}{\partial x} = O\left(\frac{\hat{u}^2}{L}\right) \qquad (7.69)$$

indicating that the pressure term is small compared to the last term. If the turbulence term is to be retained, the first and last terms must be of the same order. The implied restriction is that the equations will apply far downstream where the velocity defect in the wake is small. The second term is written $VW/b = O(W^2/L)$ from the results of the continuity equation. Far downstream in the wake $W \ll U$ so that the second term can be neglected in comparison with the first.

The replacement of $\bar{u}_x$ by U is a consistent approximation, leading to the x-momentum equation

$$U\frac{\partial \bar{u}_x}{\partial x} = -\frac{\partial \overline{u'_x u'_y}}{\partial y} \qquad (7.70)$$

7.4.2 Length and Time Scales for Jets

In the case of the jet there is no free stream velocity to form a time scale. The U is taken as the centerline velocity in the jet and $W = U$. Therefore, the first two terms of (7.66) must be retained in the equation

$$\bar{u}_x \frac{\partial \bar{u}_x}{\partial x} + \bar{u}_y \frac{\partial \bar{u}_x}{\partial y} = -\frac{\partial \overline{u'_x u'_y}}{\partial x} \qquad (7.71)$$

7.4.3 Similarity Hypothesis in Wakes

In the wake we assume that the velocity defect divided by the centerline velocity defect is a function only of the ratio of the y-coordinate and the width of the wake

$$\frac{W}{W_c} = f\left(\frac{y}{b}\right) \qquad W = U - \bar{u}_x \tag{7.72}$$

where W is the velocity defect and W_c is the maximum velocity defect (located at the centerline). The width b is to be defined later. Also the Reynolds stress is assumed to be similar, leading to

$$-\overline{u'_x u'_y} = W_c^2 g\left(\frac{y}{b}\right) \tag{7.73}$$

Defining $\eta = y/b$ and differentiating (7.72) with respect to x

$$\frac{\partial \bar{u}_x}{\partial x} = -\frac{dW_c}{dx}f + W_c \frac{\eta}{b}\frac{db}{dx}\frac{df}{d\eta} \tag{7.74}$$

Using (7.73) and (7.74) in (7.70) produces

$$\frac{dg}{d\eta} = \frac{U}{W_c}\frac{db}{dx}\eta\frac{df}{d\eta} - \frac{Ub}{W_c^2}\frac{dW_c}{dx}f \tag{7.75}$$

The similarity of the velocity profiles requires that the coefficients of f and $\eta df/d\eta$ be constants, meaning that the right side of (7.70) must be a function of y/b

$$\frac{1}{W_c^2}\frac{db}{dx} = \text{constant} \qquad \frac{b}{W_c^2}\frac{dW_c}{dx} = \text{constant} \tag{7.76}$$

The solution to (7.76) is $b = Ax^n$ and $W_c = Bx^{n-1}$ where A and B are constants.

Another constraint on the velocity profile is the constancy of momentum. The drag on an object is derived in Sec. 6.10.1 as [see (6.99)]

$$D = \rho \int_{-\infty}^{\infty} \bar{u}_x(U - \bar{u}_x)dy \tag{7.77}$$

where the integral can be taken at any section downstream of the object; thus, the integral is not a function of x. Using (7.72) in (7.77)

$$\frac{D}{\rho} = UbW_c \int_{-\infty}^{\infty} f\, d\eta - bW_c^2 \int_{-\infty}^{\infty} f^2 d\eta \tag{7.78}$$

Since terms of the order of W_c/U have already been neglected, the last term of (7.78) can be neglected. Because U and the integral are independent of x, bW_c must be independent of x if the first integral on the right of (7.78) is to be constant; therefore,

$$bW_c = ABx^{2n-1} \qquad n = \frac{1}{2} \tag{7.79}$$

so that

$$b = A\sqrt{x} \qquad W_c = \frac{B}{\sqrt{x}} \tag{7.80}$$

The behavior of the centerline velocity defect is the same as in the laminar wake as expressed by (6.105).

The velocity profile is found by making the additional assumption of a constant eddy viscosity, $\nu^t = \mu^t/\rho$. Then a *turbulent Reynolds number* is defined as

$$R_t = \frac{bW_c}{\nu^t} \tag{7.81}$$

Because $bW_c = $ constant, the turbulent Reynolds number does not change with distance. Experiment has found this quantity to be $R_t \approx 12.5$. Using the definitions of f and g, the definition of the eddy viscosity, (7.72), and (7.73), the turbulent Reynolds number is

$$R_t = -\frac{g}{f} \tag{7.82}$$

Substituting (7.80) into (7.75) and replacing g by (7.82) gives

$$\frac{dg}{d\eta} = \frac{1}{2}\frac{UA}{B}\left(\eta\frac{df}{d\eta} + f\right) = -\frac{1}{R_t}\frac{d^2 f}{d\eta^2} \tag{7.83}$$

The solution to (7.83) is

$$f = C \exp\left(-\frac{R_t}{2}\frac{UA}{B}\eta^2\right) \tag{7.84}$$

We can still choose the width scale of the jet and the C in the above equation. Following Tennekes and Lumley (1972) $C = 1$ and $R_t U A/B = 1$ so that

$$\bar{u}_x = U - (U - \bar{u}_c)e^{-(y^2/b^2)} \tag{7.85}$$

where $y = b$ at the point $W \approx 0.6W_c$.

Experiment (Townsend, 1956) shows that the velocity profile is in good agreement with (7.85) for $\eta < 1.3$. On the outer edges of the wake the velocity defect is larger than predicted. Tennekes and Lumley (1972) indicate that the error is due to *intermittency*—near the edges of the wake the turbulence is intermittent. A point encounters only turbulent flow if it is close to the centerline (point A, Fig. 7.15), and encounters only laminar flow if it is far from the centerline (point D), but intermediate points are in the turbulent zone part of the time and the potential flow zone part of the time (points B and C). The effective viscosity in the intermittent region is somewhat less than the eddy viscosity v^t. In view of the approximations, the velocity profile is remarkably accurate; the calculation error is no more than 5 percent of the centerline velocity over the entire profile (Tennekes and Lumley, 1972).

With reference to Fig. 7.15, note that the maximum long term average velocity (averaging time long compared to the time scale of the major fluctuations) occurs on the centerline, but the maximum short term average velocity (long compared to the turbulent time scale but short compared to the time scale of the major eddies) will be located somewhere off the centerline and can be much larger than the long term average. The approximations have limited the applicability of results to points far downstream in the wakes. The wake behind a cylinder agrees with the values of W and b beyond about 80 diameters.

7.4.4 The Axisymmetric Wake

To prepare for the calculation of the axisymmetric wake, an expression for Reynolds stresses in axisymmetric flow is needed. We will consider only the equation of motion in the R-direction

$$u_R \frac{\partial u_R}{\partial R} + \frac{u_\theta}{R} \frac{\partial u_R}{\partial \theta} - \frac{u_\theta^2}{R} + u_x \frac{\partial u_R}{\partial x} = -\frac{1}{\rho} \frac{\partial p}{\partial R} + v \left\{ \frac{\partial}{\partial R} \left[\frac{1}{R} \frac{\partial}{\partial R} (R u_R) \right] \right.$$

$$\left. + \frac{1}{R^2} \frac{\partial^2 u_R}{\partial \theta^2} - \frac{2}{R^2} \frac{\partial u_\theta}{\partial \theta} + \frac{\partial^2 u_R}{\partial R^2} \right\}$$

$$(7.86)$$

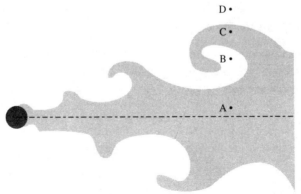

FIGURE 7.15
The wake behind an object.

Using the equation of continuity

$$\frac{1}{R}\frac{\partial}{\partial R}(Ru_R) + \frac{1}{R}\frac{\partial u_\theta}{\partial R} + \frac{\partial u_x}{\partial x} = 0 \tag{7.87}$$

the left side of (7.86) can be written

$$\frac{1}{R}\frac{\partial}{\partial R}(Ru_R^2) + \frac{1}{R}\frac{\partial}{\partial \theta}(u_R u_\theta) - \frac{u_\theta^2}{R} + \frac{\partial}{\partial x}(u_R u_x)$$

The axisymmetric Reynolds stresses are derived in an obvious manner.

The development for wakes follows that in Sec. 7.4.1 and 7.4.3. The equation corresponding to (7.70) is

$$U\frac{\partial \bar{u}_x}{\partial x} = -\frac{1}{R}\frac{\partial}{\partial R}\left(R\overline{u'_R u'_x}\right) \tag{7.88}$$

Equations (7.72) and (7.73) become

$$\frac{W}{W_c} = f(\eta) \qquad -R\overline{u'_x u'_R} = W_c^2 g(\eta) \qquad \eta = \frac{R}{b} \tag{7.89}$$

Equations (7.75) and (7.76) are the same in the axisymmetric analysis and the same conclusions follow, $b = Ax^n$, $W_c = Bx^{n-1}$. The drag formula becomes

$$D = 2\pi\rho \int_0^\infty \bar{u}_x(U - \bar{u}_x)R\,dR \approx Ub^2 W_c \int_0^\infty f\eta\,d\eta \tag{7.90}$$

with the approximation that $W_c \ll U$. The requirement that $b^2 W_c$ is not a function of x is $n = 1/3$. Thus the velocity defect varies as $x^{-2/3}$ and the jet width as $x^{1/3}$. The major difference in the plane and axisymmetric wakes is that the Reynolds number, $R = Ub/\nu$, is not a constant in the axisymmetric case but decreases in the downstream direction. When R becomes of order one, the flow will become laminar, but that is likely to happen a very large distance downstream of the disturbing object.

7.4.5 The Plane Jet

In the jet it is not the velocity defect that becomes small in the downstream direction but the velocity itself. The velocity is written in terms of the centerline velocity as

$$\bar{u}_x = \bar{u}_c f(\eta) \qquad \eta = \frac{y}{b} \tag{7.91}$$

in which $\bar{u}_c$ is the centerline velocity. From the equation of continuity

$$\bar{u}_y = -\int_0^y \frac{\partial \bar{u}_x}{\partial x}\,dy = -b\int_0^\eta \left(\frac{d\bar{u}_c}{dx}f - \frac{\bar{u}_c}{b}\frac{db}{dx}\eta\frac{df}{d\eta}\right)d\eta \tag{7.92}$$

The Reynolds stress is written similarly to (7.73)

$$-\overline{u'_x u'_y} = \bar{u}_c^2 g(\eta) \tag{7.93}$$

The f and g are now used in (7.71)

$$\frac{bf^2}{\bar{u}_c}\frac{d\bar{u}_c}{dx} - \frac{db}{dx}\eta f\frac{df}{d\eta} - \frac{b}{\bar{u}_c}\frac{df}{d\eta}\frac{d\bar{u}_c}{dx}\int_0^\eta f\,d\eta + \frac{db}{dx}\frac{df}{d\eta}\int_0^\eta \eta\frac{df}{d\eta}d\eta = \frac{dg}{d\eta} \tag{7.94}$$

Similarity requires that

$$\frac{b}{\bar{u}_c}\frac{d\bar{u}_c}{dx} = \text{constant} \qquad \frac{db}{dx} = \text{constant} \tag{7.95}$$

The second equation of (7.95) implies that the jet width grows linearly with distance downstream. The first equation is satisfied by $\bar{u}_c = Ax^n$ where n can be any power. The constancy of momentum can be used to determine the proper value of n.

The plane jet is assumed to emerge from a slot as shown in Fig. 7.16. Assuming a uniform velocity distribution in the slot, the momentum is

$$\rho U_0^2 b_0 = \rho\int_{-\infty}^\infty \bar{u}_x^2\,dy = \rho\bar{u}_c^2 b\int_{-\infty}^\infty f^2 d\eta \tag{7.96}$$

in which U_0 and b_0 are the velocity and height of the exit channel. In order that (7.96) be independent of x, $n = -1/2$ and the centerline velocity decreases as $1/\sqrt{x}$.

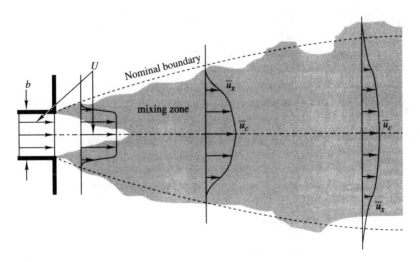

FIGURE 7.16
Jet emerging from a slot.

Using the eddy viscosity assumption, (7.94) becomes

$$\frac{d^2 f}{d\eta^2} = \frac{B\bar{u}_c b}{\nu^t}\left(-\frac{f^2}{2} - \eta f\frac{df}{d\eta} + \frac{1}{2}\frac{df}{d\eta}\int_0^\eta f\,d\eta + \frac{df}{d\eta}\int_0^\eta \eta\frac{df}{d\eta}d\eta\right)$$
(7.97)

The last three terms in the parentheses reduce to the negative of the third term, yielding

$$\frac{d^2 f}{d\eta^2} = -\frac{B}{2}R_t\left(f + \frac{df}{d\eta}\int_0^\eta f\,d\eta\right)$$
(7.98)

The width of the jet is defined by taking $BR_t/2 = 1$. Then the solution to (7.98) is

$$f = \frac{1}{\cosh^2\dfrac{\eta}{\sqrt{2}}}$$
(7.99)

The similarity assumption is confirmed to downstream distances greater than about five times channel height, as shown by experiment. Equation (7.99) fits the velocity profiles well up to about $\eta = 1.5$. At the edge of the jet, where $\eta > 1.5$, the predicted values of the velocity are too high.

7.4.6 The Axisymmetric Jet

The axisymmetric jet issues from an orifice of diameter b_0 with a velocity U_0. The algebraic development is basically the same as for the plane jet. Although the equation corresponding to (7.94) is slightly different, the conditions expressed by (7.95) are the same. The momentum is

$$\frac{\pi}{4}\rho U_0^2 b_0^2 = 2\pi\rho\int_0^\infty R\bar{u}_c^2\,dR = 2\pi\rho\bar{u}_c^2 b^2\int_0^\infty \eta f^2 d\eta$$
(7.100)

which leads to the conclusion that $\bar{u}_c$ is proportional to $1/x$ and that the width is proportional to x. As in the case of the plane wake, R is a constant. The velocity profiles appear to be similar after about eight jet diameters downstream.

7.4.7 Laminar and Turbulent Jets and Wakes

Table 7.1 gives the width of the flow and the centerline velocity as a function of a power of x for both laminar and turbulent flows.

7.5 WALL TURBULENCE

The turbulence near a wall is more complex than that in free shear flows due to the fact that there are more length scales. Some of those flows were described in Sec. 7.3 where the various empirical formulas were presented.

TABLE 7.1
Power laws for velocity and width of jets and wakes

	Laminar flow		Turbulent flow	
	Width	Velocity	Width	Velocity
Plane wake	$x^{1/2}$	$x^{-1/2}$	$x^{1/2}$	$x^{-1/2}$
Circular wake	$x^{1/2}$	x^{-1}	$x^{1/3}$	$x^{-2/3}$
Plane jet	$x^{2/3}$	$x^{-1/3}$	x	$x^{-1/2}$
Circular jet	x	x^{-1}	x	x^{-1}

7.5.1 Smooth Walled Channels and Pipes

The established flow between parallel walls is one of the simplest of those dominated by wall turbulence. For moderate and high Reynolds numbers the radius of curvature of the wall in pipe flow is large compared to the thicknesses of the sublayers that are described below, and thus the description for flow between plates can be used for pipe flow also. Assuming a smooth wall, the flow can be divided into four regions as shown in Fig. 7.17. The *viscous sublayer* is necessary near the wall to fulfill the no-slip condition. In the *outer core* (or *outer layer*) and the *inertial sublayer* the Reynolds stresses dominate the viscous stresses. The *buffer layer* is a turbulent region that provides the transition between the core and the viscous sublayer.

The assumption that the flow is established (i.e., uniform; derivatives in the x-direction, except for pressure, are zero) and steady makes the inertial terms in the equations of motion disappear

$$\bar{u}_i \frac{\partial \bar{u}_j}{\partial x_i} = 0 \tag{7.101}$$

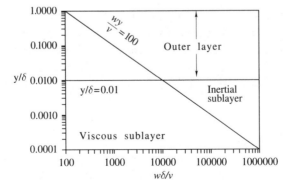

FIGURE 7.17
Zones of wall turbulence where w is a characteristic level of turbulent velocity fluctuations (after H. Tennekes and J. Lumley, *A First Course in Turbulence*, The MIT Press, p. 148, ©1972 by the Massachusetts Institute of Technology).

The pressure gradient is constant and a known function of the wall shear

$$\frac{d\bar{p}}{dx} = -\frac{2\tau_w}{d} = -\frac{2\rho u_*^2}{d} \tag{7.102}$$

in which $u_* = \sqrt{\tau_w/\rho}$ has been defined as the friction velocity.

The equations of motion become

$$0 = -\frac{1}{\rho}\frac{\partial\bar{p}}{\partial x} - \frac{\partial}{\partial y}\left(\overline{u'_x u'_y}\right) + \nu\frac{\partial^2\bar{u}_x}{\partial y^2} \tag{7.103}$$

$$0 = -\frac{1}{\rho}\frac{\partial\bar{p}}{\partial y} - \frac{\partial\overline{u'^2_y}}{\partial y} \tag{7.104}$$

Both equations can immediately be integrated with respect to y

$$\frac{y}{\rho}\frac{\partial\bar{p}}{\partial x} + \overline{u'_x u'_y} = \nu\frac{\partial\bar{u}_x}{\partial y} - \frac{\tau_w}{\rho} \tag{7.105}$$

$$\overline{u'^2_y} = -\frac{\bar{p} - p_w}{\rho} \tag{7.106}$$

in which p_w is the pressure at the wall. Although $\partial\bar{p}/\partial x$ has been assumed to be constant and independent of y, (7.106) indicates that $\bar{p}$ is a function of distance from the wall due to the *turbulent pressure* generated by the fluctuating velocity u'_y. Clearly, (7.105) reduces to Newton's law of viscosity at the wall where $y = 0$ and the turbulent fluctuations disappear. (We have now defined pressure three ways: the thermodynamic pressure; the mean pressure, $p + \tau_{ii}/3$; and the turbulent pressure.)

Three layers can be identified. In the viscous sublayer the distance from the wall is very small and can be neglected, the fluctuating velocity terms are small, and (7.105) reduces to

$$u_x = \bar{u}_x = \frac{\tau_w y}{\nu\rho} = u_*^2\frac{y}{\nu} \tag{7.107}$$

indicating that the velocity profile is linear with distance from the wall. The extent of the viscous sublayer is limited to $y_+ = u_* y/\nu < 10$.

In the inertial sublayer (also called the *turbulent sublayer* or *turbulent surface layer*) the turbulence stress predominates over the viscous stress. However, y is still very small so that (7.105) becomes

$$\frac{\partial\bar{u}_x}{\partial y} = \frac{\overline{u'_x u'_y}}{\nu} \tag{7.108}$$

The logarithmic velocity distribution results if the right side of (7.108) is taken as a constant and the eddy viscosity is proportional to the distance from the wall so that

$$y\frac{\partial \bar{u}_x}{\partial y} = \text{constant} \tag{7.109}$$

where the constant is determined below. Integrating (7.109) yields

$$\frac{\bar{u}_x}{u_*} = \frac{1}{\kappa}\ln y_+ + 5 \qquad y_+ > 30 \tag{7.110}$$

in which κ is the von Kármán constant and the constant of (7.109) has been selected so that (7.110) fits the experimental data. The constant of integration is to be determined so that the velocity profile is continuous from the viscous sublayer to the inertial sublayer.

In the zone called the buffer layer, neither the viscous stress nor the turbulent stress predominates. In engineering calculations the buffer layer is often ignored by connecting the viscous sublayer to the inertial sublayer at about $y_+ = 11$.

As noted in Ex. 7.1, there is still an inconsistency near the centerline where the logarithmic laws would intersect. Experimental data have lead to an equation for the inner core

$$\frac{\bar{u}_x}{u_*} = 1 - 2.5\ln\frac{y}{R_0} + W \qquad W = \frac{1}{2}\left[\sin\pi\left(\frac{y}{R_0} - \frac{1}{2}\right) + 1\right] \tag{7.111}$$

in which R_0 is the radius of the pipe (or distance to the centerline between plates). The function W is called the *wake function*. The entire curve of the velocity distribution is shown in Fig. 7.14 in Sec. 7.3.

7.5.2 Rough Walls

If the wall is *rough* the concept of the viscous sublayer must be modified. "Rough" means that the projections from the mean position of the wall are not small compared to the thickness of the viscous sublayer. The definition of rough refers not only to the characteristics of the wall but also to the thickness of the viscous sublayer, which becomes thinner as the Reynolds number becomes larger. The wall roughness adds a characteristic length λ to those used for the smooth wall.

The velocity near the wall is poorly defined because of the random geometry along the wall. The no-slip condition still applies to the solid boundary, but the position of the boundary is defined only from a statistical point of view where λ is something like the mean roughness height or the root-mean-square of the roughness height. The *roughness Reynolds number* is defined as

$$\boldsymbol{R}_\lambda = \frac{\lambda u_*}{\nu} \tag{7.112}$$

If $R_\lambda < 5$, the wall can be considered smooth from the point of view of the overall friction. In such a case (7.110) applies. For R_λ large the *rough wall* velocity profile is described as

$$\frac{\bar{u}_x}{u_*} = \frac{1}{\kappa} \ln \frac{y}{\lambda} + \text{constant} \qquad (7.113)$$

The constant is poorly defined since the position at $y = 0$ is poorly defined. In practice the constant is often taken as zero with the value of λ adjusted so that the velocity profile matches the data.

There is a difference between uniform roughness and the type of random roughness that results from manufacturing processes. The friction factor for pipes with uniform roughness is shown in Fig. 7.11 in Sec. 7.2. That diagram can be compared to Fig. 7.18, which is used for practical calculation. In either case the friction becomes independent of the Reynolds number after a certain point, the boundary of the *wholly rough zone*. From

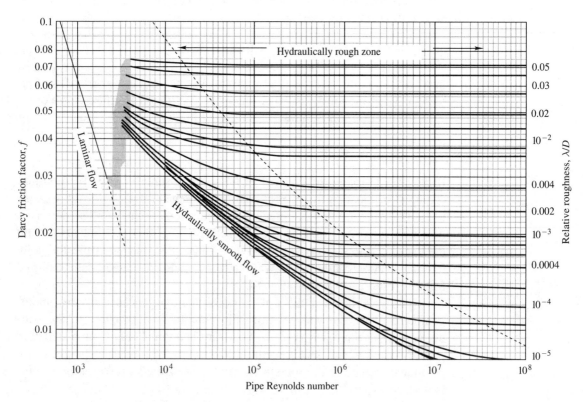

FIGURE 7.18
Friction factors for commercial pipes (Moody, 1944).

the practical point of view, roughness is *defined* for pipe flow by locating the friction factor on this diagram—an inverse calculation.

> **Historical Note:** At the beginning of the twentieth century, pipe friction was a vital issue that needed to be clearly defined for design purposes. The definitive measurements were made by Augustus V. Saph and Ernest W. Schoder (1879–1968) in a joint Ph.D. thesis at Cornell University (Saph and Schoder, 1902) under the supervision of Professor G. S. Williams (of the Hazen-Williams equation). These were meticulous measurements made in smooth (glass and brass) and rough (galvanized) pipes. The empirical equations that they derived set the standard for the measurement of pipe friction. Schoder later became Professor of Experimental Hydraulics at Cornell and co-authored the well known book *Hydraulics* by Schoder and Dawson (1934).
>
> Unfortunately, Saph and Schoder did not put their equations in dimensionless form and did not recognize the role of the Reynolds number (Buckingham published his pi-theorem in 1914). Thus the analysis was left to Paul Richard Heinrich Blasius (1883–1970) in Berlin in 1911. Blasius made a Reynolds number diagram very much like Fig. 7.18 based on the Saph and Schoder data plus some of his own data. The Blasius experiments used both water and air and varied the temperature, thus recognizing the effect of viscosity and how that effect could be consolidated in the Reynolds number. The Reynolds number diagram is now known as either the Stanton diagram, named after Thomas Edward Stanton (1865–1931), a colleague and student of Osborne Reynolds at Owens College, or as the Stanton-Moody diagram, named after Lewis Ferry Moody (1880–1953) (Moody, 1944).

7.5.3 Flat Plates and Pipes

Many of the results for external boundary layer flow have come directly from measurements on pipe flow. Although most of the equations of the preceding section can be justified by dimensional reasoning and order of magnitude studies, many of the practical turbulence formulas are purely empirical. The velocity profile on a plate or in a pipe looks as though it can be represented by an exponential of the type

$$\frac{\bar{u}_x}{U} = \left(\frac{y}{\delta}\right)^{1/n} \tag{7.114}$$

in which U is the free stream velocity in the case of the plate or the centerline velocity in the case of the pipe, and δ is the boundary layer thickness on a plate or the radius of a pipe. The exponent n is a function of the Reynolds number, increasing with increasing Reynolds number, but a common value is $n = 7$. Using (7.114) produces (Schlichting, 1960)

$$\frac{\bar{u}_x}{u_*} = 8.74 \left(\frac{y u_*}{\nu}\right)^{1/7} \tag{7.115}$$

and the shear stress is

$$\frac{\tau_0}{\rho U^2} = 0.0225 \, \mathbf{R}_\delta{}^{-1/4} \tag{7.116}$$

in which $\mathbf{R}_\delta$ is the Reynolds number based on δ.

In the case of the flat plate the displacement and momentum thicknesses [see (6.41) and (6.42)] are

$$\delta_d = \frac{1}{8}\delta \qquad \delta_m = \frac{7}{72}\delta \tag{7.117}$$

The rough plate has one fundamental difference with the rough pipe. Assuming that the roughness λ remains the same along the plate, the relative roughness, λ/δ, decreases along the plate due to the thickening of the boundary layer.

7.5.4 Boundary Layers

Practical boundary layer calculations for turbulent flow contain a large degree of empiricism. There are many methods (Cebeci and Smith, 1974). The earliest methods, described by Schlichting (1960), are designed for hand calculation but still figure prominently in many engineering calculations (Kline, et al., 1968). Since that time several machine techniques have been developed that attempt to solve a differential equation with some assumptions (Young, 1989; Bradshaw, 1972). The latter methods are rather detailed, but can probably yield more accurate solutions than the older techniques. In this section we deal lightly with only the former type of method in two dimensions.

The integral equation methods are designed to solve the Kármán momentum equation (6.79), which is repeated here

$$\frac{d}{dx}\left(\tilde{u}_x^2 \delta_m\right) + \tilde{u}_x \delta_d \frac{d\tilde{u}_x}{dx} = \frac{\tau_0}{\rho} \tag{7.118}$$

After introducing a *shape factor* as the ratio of the displacement and momentum thicknesses,

$$H = \frac{\delta_d}{\delta_m} \tag{7.119}$$

the momentum equation is rewritten as

$$\frac{d\delta_m}{dx} + \frac{\delta_m}{\tilde{u}_x}(H + 2)\frac{d\tilde{u}_x}{dx} = \frac{\tau_0}{\rho \tilde{u}_x^2} \tag{7.120}$$

In (7.120) the displacement thickness has been eliminated in favor of the shape factor H leaving the unknowns δ_m, H, and τ_0. The potential flow velocity distribution $\tilde{u}_x$ is assumed known. No approximations have been made to this point. The remainder of the calculation depends on the assumptions that are to be made.

Following Schlichting (1960) the shear stress is assumed to be of the same form as that of a flat plate [see (7.116)]

$$\frac{\tau_0}{\rho \tilde{u}^2} = \frac{\alpha}{\left(\dfrac{\tilde{u}_x \delta_d}{\nu}\right)^{1/n}} \tag{7.121}$$

in which the constant free stream velocity U has been replaced by the variable potential flow velocity $\tilde{u}_x$. The "constants" α and n are taken to be $n = 4$, $\alpha = 0.0128$ or $n = 6$, $\alpha = 0.0065$. It makes little difference to the final calculation which pair of values are chosen. In addition, the ratio of the displacement thickness to the momentum thickness is taken as a constant, $H = 1.4$. Using (7.121) in (7.120) and integrating produces

$$\delta_m \left(\frac{\tilde{u}_x \delta_m}{\nu}\right)^{1/n} = \tilde{u}_x^{-b}\left(C + a\int_{x_t}^{x} \tilde{u}_x^b dx\right) \tag{7.122}$$

in which x_t is the point of transition between the laminar and the turbulent boundary layers, C is a constant that is to be determined from the condition of the laminar boundary layer at the point of transition, and

$$a = \frac{n+1}{n}\alpha \qquad b = \frac{n+1}{n}(H+2) - \frac{1}{n} \tag{7.123}$$

Given the potential flow velocity, (7.122) can easily be solved for the momentum thickness.

7.6   NONCIRCULAR CONDUITS

7.6.1   Friction

The resistance to turbulent flow in noncircular conduits is more difficult to characterize than that in round pipes. There are two reasons. First, conduits of arbitrary cross-section cannot be described with a single parameter such as a pipe diameter. Thus, a plot of friction factor vs. Reynolds number, corresponding to Fig. 7.18, is not adequate. Second, there is the matter of secondary currents, treated in the next section, which redistribute the shears and have an effect on boundary layer thicknesses.

In an attempt to describe all shapes of conduits the *hydraulic radius* is defined as

$$R_h = \frac{A}{P} \qquad (7.124)$$

in which A is the area of the conduit and P is the part of its boundary that is in contact with the fluid—the entire perimeter except in the case of a free surface. The Darcy-Weisbach equation is written

$$h_L = f \frac{L}{4R_h} \frac{U^2}{2g} \qquad (7.125)$$

which reduces to the normal pipe formula in the case of circular conduits. Friction factors obtained experimentally and plotted as in Fig. 7.18 show a much greater scatter than similar data for round pipes.

7.6.2 Secondary Currents

If a conduit is not axially symmetric, a flow is generated at right angles to the primary current. Such flows are a direct result of the Reynolds stresses. If the streamlines were always aligned with the direction of the conduit, there would be no vorticity in the axial direction. To show that secondary currents can exist in straight conduits, we show that the vorticity vector in the direction of the flow is not zero, or at least cannot remain zero, under turbulent conditions. Taking the x-axis in the direction of flow, we assume that $\bar{\omega}_x = 0$ and ask if $\partial \bar{\omega}_x / \partial t = 0$. If so, the flow can exist without axial vorticity.

The equation of vorticity in turbulent flow is (7.46). First, we need to show that flow in noncircular conduits can exist without vorticity in the absence of turbulence. Equation (7.46) without turbulence is

$$\frac{\partial \bar{\omega}_i}{\partial t} + \bar{u}_j \frac{\partial \bar{\omega}_i}{\partial x_j} = \bar{\omega}_j \bar{d}_{ij} + \frac{\mu}{\rho} \frac{\partial^2 \bar{\omega}_i}{\partial x_j \partial x_j} \qquad (7.126)$$

Taking the i-direction as the x-direction and assuming that $\bar{\omega}_x$ is zero everywhere at time $t = 0$, its space derivatives must also be zero so the last term of (7.126) disappears as do all the terms on the left except the time derivative. Then

$$\frac{\partial \bar{\omega}_x}{\partial t} = \bar{\omega}_j \bar{d}_{xj} = \frac{1}{2}\left(\frac{\partial \bar{u}_y}{\partial z} - \frac{\partial \bar{u}_z}{\partial y}\right)\frac{\partial \bar{u}_x}{\partial x} + \frac{1}{4}\left(\frac{\partial \bar{u}_z}{\partial x} - \frac{\partial \bar{u}_x}{\partial z}\right)\left(\frac{\partial \bar{u}_x}{\partial y} + \frac{\partial \bar{u}_y}{\partial x}\right)$$

$$+ \frac{1}{4}\left(\frac{\partial \bar{u}_x}{\partial y} - \frac{\partial \bar{u}_y}{\partial x}\right)\left(\frac{\partial \bar{u}_x}{\partial z} + \frac{\partial \bar{u}_z}{\partial x}\right) \qquad (7.127)$$

Assuming that all x-derivatives disappear and that the velocity is in the x-direction at the initial time, the right side of (7.127) is zero. Thus, turbulent-free flow can exist in the noncircular conduit without secondary currents.

The remaining part of the vorticity equation is

$$\frac{\partial \bar{\omega}_i}{\partial t} = -\overline{u'_j \frac{\partial \omega'_i}{\partial x_j}} + \overline{\omega'_j d'_{ij}} \tag{7.128}$$

The first term on the right, after multiplying by 2, is

$$-2\overline{u_j \frac{\partial \omega'_x}{\partial x_j}} = -\overline{u'_x \frac{\partial^2 u'_z}{\partial x \partial y}} + \overline{u'_x \frac{\partial^2 u'_y}{\partial x \partial z}} - \overline{u'_y \frac{\partial^2 u'_z}{\partial y^2}} + \overline{u'_y \frac{\partial^2 u'_y}{\partial y \partial z}} - \overline{u'_z \frac{\partial u'_z}{\partial y \partial z}} + \overline{u'_z \frac{\partial^2 u'_y}{\partial z^2}} \tag{7.129}$$

We now apply the assumption of uniform flow; the x-derivative of any averaged quantity is zero, $\partial(\cdots)/\partial x = 0$. Also, the x-derivative of a fluctuating quantity is eliminated by the equation of continuity (7.34)

$$\frac{\partial u'_x}{\partial x} = -\frac{\partial u'_y}{\partial y} - \frac{\partial u'_z}{\partial z} \tag{7.130}$$

Thus,

$$\overline{u'_x \frac{\partial^2 u'_z}{\partial x \partial y}} = \frac{\partial}{\partial x}\overline{u'_x \frac{\partial u'_z}{\partial y}} - \overline{\frac{\partial u'_x}{\partial x}\frac{\partial u'_z}{\partial y}} = \overline{\frac{\partial u'_y}{\partial y}\frac{\partial u'_z}{\partial y}} + \overline{\frac{\partial u'_z}{\partial z}\frac{\partial u'_z}{\partial y}}$$

$$\overline{u'_z \frac{\partial^2 u'_y}{\partial x \partial z}} = \frac{\partial}{\partial x}\overline{u'_x \frac{\partial u'_y}{\partial z}} - \overline{\frac{\partial u'_x}{\partial x}\frac{\partial u'_y}{\partial z}} = \overline{\frac{\partial u'_y}{\partial y}\frac{\partial u'_y}{\partial z}} + \overline{\frac{\partial u'_z}{\partial z}\frac{\partial u'_y}{\partial z}} \tag{7.131}$$

Using (7.131) in (7.129)

$$-2\overline{u'_j \frac{\partial \omega'_x}{\partial x_j}} = -\overline{\frac{\partial u'_y}{\partial y}\frac{\partial u'_z}{\partial y}} - \overline{\frac{\partial u'_z}{\partial z}\frac{\partial u'_z}{\partial y}} + \overline{\frac{\partial u'_y}{\partial y}\frac{\partial u'_y}{\partial z}} + \overline{\frac{\partial u'_z}{\partial z}\frac{\partial u'_y}{\partial z}} - \overline{u'_y \frac{\partial^2 u'_z}{\partial y^2}}$$

$$+ \overline{u'_y \frac{\partial^2 u'_y}{\partial y \partial z}} - \overline{u'_z \frac{\partial u'_z}{\partial y \partial z}} + \overline{u'_z \frac{\partial^2 u'_y}{\partial z^2}} \tag{7.132}$$

The second term on the right of (7.128), after multiplying by 2, is

$$2\overline{\omega'_j d'_{xj}} = \overline{\frac{\partial u'_x}{\partial x}\frac{\partial u'_z}{\partial y}} - \overline{\frac{\partial u'_x}{\partial x}\frac{\partial u'_y}{\partial y}} + \overline{\frac{\partial u'_x}{\partial z}\frac{\partial u'_y}{\partial x}} - \overline{\frac{\partial u'_z}{\partial x}\frac{\partial u'_x}{\partial y}} \tag{7.133}$$

Using (7.130) to eliminate the x-derivatives of u'_x and combining with

(7.132) gives

$$2\frac{\partial \bar{\omega}_x}{\partial t} = -2\overline{\frac{\partial u'_y}{\partial y}\frac{\partial u'_z}{\partial y}} + 2\overline{\frac{\partial u'_z}{\partial z}\frac{\partial u'_y}{\partial z}} - \overline{\frac{\partial u'_z}{\partial z}\frac{\partial u'_z}{\partial y}} + \overline{\frac{\partial u'_y}{\partial y}\frac{\partial u'_y}{\partial z}}$$

$$- \overline{u'_y\frac{\partial^2 u'_z}{\partial y^2}} + \overline{u'_y\frac{\partial^2 u'_y}{\partial y\partial z}} - \overline{u'_z\frac{\partial^2 u'_z}{\partial y\partial z}} + \overline{u'_z\frac{\partial u'_y}{\partial z^2}} \qquad (7.134)$$

$$- \overline{\frac{\partial u'_z}{\partial z}\frac{\partial u'_z}{\partial y}} + \overline{\frac{\partial u'_y}{\partial y}\frac{\partial u'_y}{\partial z}} + \overline{\frac{\partial u'_x}{\partial z}\frac{\partial u'_y}{\partial x}} - \overline{\frac{\partial u'_z}{\partial x}\frac{\partial u'_x}{\partial y}}$$

The derivatives of u'_x are eliminated by

$$\overline{\frac{\partial u'_x}{\partial z}\frac{\partial u'_y}{\partial x}} = \frac{\partial}{\partial x}\overline{u'_y\frac{\partial u'_x}{\partial z}} - \overline{u'_y\frac{\partial^2 u'_x}{\partial x\partial z}} = \overline{u'_y\frac{\partial^2 u'_y}{\partial y\partial z}} + \overline{u'_y\frac{\partial u'_z}{\partial z^2}}$$

$$\overline{\frac{\partial u'_x}{\partial y}\frac{\partial u'_z}{\partial x}} = \frac{\partial}{\partial x}\overline{u'_z\frac{\partial u'_x}{\partial y}} - \overline{u'_z\frac{\partial^2 u'_x}{\partial x\partial y}} = \overline{u'_z\frac{\partial^2 u'_y}{\partial y^2}} + \overline{u'_z\frac{\partial^2 u'_z}{\partial y\partial z}} \qquad (7.135)$$

where, again, the x-derivatives of averaged quantities have been set to zero. Using (7.135) in (7.134) produces

$$2\frac{\partial \bar{\omega}_x}{\partial t} = -2\overline{\frac{\partial u'_y}{\partial y}\frac{\partial u'_z}{\partial y}} + 2\overline{\frac{\partial u'_z}{\partial z}\frac{\partial u'_y}{\partial z}} - \overline{\frac{\partial u'_z}{\partial z}\frac{\partial u'_z}{\partial y}} + \overline{\frac{\partial u'_y}{\partial y}\frac{\partial u'_y}{\partial z}} - \overline{u'_y\frac{\partial u'_z}{\partial y^2}}$$

$$+ 2\overline{u'_y\frac{\partial^2 u'_y}{\partial y\partial z}} - 2\overline{u'_z\frac{\partial^2 u'_z}{\partial y\partial z}} + \overline{u'_z\frac{\partial^2 u'_y}{\partial z^2}} - \overline{\frac{\partial u'_z}{\partial z}\frac{\partial u'_z}{\partial y}}$$

$$+ \overline{\frac{\partial u'_y}{\partial y}\frac{\partial u'_z}{\partial z}} + \overline{u'_y\frac{\partial^2 u'_z}{\partial z^2}} - \overline{u'_z\frac{\partial^2 u'_y}{\partial y^2}}$$

$$(7.136)$$

This is the result of Einstein and Li (1958) who simplify the expression to

$$2\frac{\partial \bar{\omega}_x}{\partial t} = \frac{\partial^2}{\partial y\partial z}\left(\overline{u'^2_y} - \overline{u'^2_z}\right) + \left(\frac{\partial^2}{\partial z^2} - \frac{\partial^2}{\partial y^2}\right)\overline{u'_y u'_z} \qquad (7.137)$$

Consider flow in the x-direction over a flat plate that is parallel to the xy-plane. Because the flow is two-dimensional, the derivatives with respect to y are zero, eliminating the first and last terms on the right of (7.137). The quantity $\overline{u'_y u'_z}$ can be considered the Reynolds stress for flow in the y-direction. Since there is no flow in the y-direction, that correlation is zero. Thus, the time derivative of the vorticity is zero. However, (7.137) does not answer the question as to whether such a flow is stable. In fact a three-dimensional structure is observed in many plane flows where some sort of transverse current develops.

The same considerations hold for flow in a round pipe (see Prob. 7.3). The axial symmetry for flow in a pipe means that all the terms on the right of (7.137) vanish if the flow is initially axially symmetric. In fact no experiment has detected ordered secondary currents in pipes (with the exception of cases where the axial symmetry was spoiled by different roughnesses around the walls).

Flows in noncircular conduits have secondary currents. Consider the corner shown as Fig. 7.19. At a position nearer the vertical wall than the horizontal wall (point A), the value of $\overline{u_z'^2}$ is greater than $\overline{u_y'^2}$. The x- and y-derivatives are positive. The $\overline{u_y'u_z'}$ correlation is small. Thus the time derivative of the vorticity is negative from the first term of (7.137). At point B, on the other hand, $\overline{u_y'^2}$ is greater than $\overline{u_z'^2}$ and the time derivative of the vorticity is positive. The result is a secondary current pattern that is shown in the upper part of Fig. 7.20. The flow toward the corner carries momentum into the corner and tends to make the shear distribution on the walls more uniform—although it still goes to zero at the corner. Lines of equal primary velocity $\bar{u}_x$, *isovels*, are shown in the right side of Fig. 7.20. The secondary currents also have a consequence in heat transfer, as many radiators use tubes that are not circular. Obviously, a rigorous analysis of resistance to flow would have to take into account an analysis of the secondary currents—a difficult task.

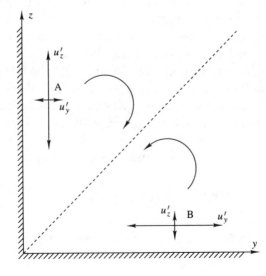

FIGURE 7.19
Turbulent velocities and direction of rotation in a corner.

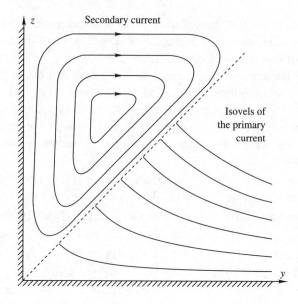

FIGURE 7.20
Streamlines of the secondary current and lines of equal velocity of the primary current.

7.7 TURBULENT ENERGY

Every student of turbulence learns the four line verse,

Big whirls have little whirls
That feed their velocity.
Little whirls have lesser whirls
And so on to viscosity.

It expresses the fact that large-scale turbulence tends to break down into smaller and smaller eddies. Finally, when the eddies are sufficiently small, the viscous conversion of mechanical energy into thermal energy eliminates the motion. The process that creates the turbulence feeds energy into the large eddies and the large eddies pass this energy to smaller eddies until viscosity damps the process at a very small scale. The Navier-Stokes equations describe the entire process and, if the resolution of a numerical calculation could be made fine enough to encompass the smallest eddy scale with a sufficient number of grid points and at the same time the solution region made large enough to include the largest eddies, a numerical solution to those equations would solve turbulent flow problems.

7.7.1 The Energy Spectrum

Fig. 7.21 shows the turbulent energy as a function of the reciprocal of the eddy size (the horizontal axis is the wave number, $k_e = 2\pi/L_e$, in which the eddy size is L_e, the wave length). The decomposition of energy (or

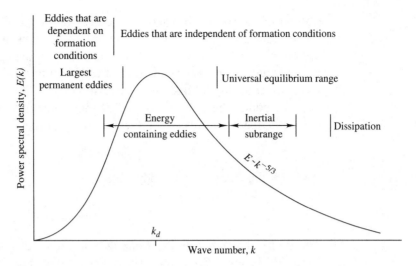

FIGURE 7.21
The power spectral density as a function of wave number.

any other measurement) into waves (eddies) of different lengths is known as *spectral analysis*. Much of what is known about turbulence comes from spectral analysis. Although there is a distinction between waves and eddies (an eddy can contain many wave lengths), we shall not make that difference herein.

The process of passing energy from one wave length to another is nonlinear and is driven by vortex stretching, as that is the only mechanism for one eddy size to feed another eddy size. The majority of the process is inviscid as the action of viscosity becomes important only toward the high wave number (small wave length) end of the spectrum.

The largest eddies are created by the flow process itself and are very much dependent on the conditions of the flow. They are relatively long lived and contain on the order of 20 percent of the turbulent energy. The Prandtl mixing length is of the order of the size of the larger eddies but always smaller than the physical dimensions of the flow, smaller than the diameter of a pipe, for example. At somewhat higher wave numbers the creation process becomes less important because the medium-sized eddies are daughters of the larger eddies. Although the largest eddies are anisotropic, those that are independent of the formation process tend more toward isotropy. These medium-sized eddies contain the majority of the turbulent energy.

The small eddies tend to be isotropic even when the larger eddies are strongly directionally dependent. As the size becomes smaller, the energy contained in the eddies rapidly diminishes. The *equilibrium range* is that portion of the spectrum where the eddies are independent of the formation process but draw their energy from the larger eddies, and at the same time are not affected by viscosity. Thus, the eddies in the equilibrium range act as a conduit for the transformation of turbulent energy from the production mechanism to the dissipation mechanism.

To define the process somewhat more quantitatively, a frequency f_e can be defined in terms of a typical velocity in the eddy u_e as

$$f_e = u_e k_e \qquad (7.138)$$

and with the characteristic eddy time

$$t_e = \frac{L_e}{u_e} \qquad L_e = \frac{2\pi}{k_e} \qquad (7.139)$$

The dissipation range is the *Kolmogoroff micro scale* L_K where the *eddy Reynolds number*, $\boldsymbol{R}_e = u_e L_K / \nu$, is unity. At that scale the final decay of the mechanical energy takes place. In pipe flow, for example, the largest eddies are the size of the radius of the pipe, the Prandtl mixing length is about 16 percent of the radius, the energy containing eddies are of size 0.1 $R_0 \boldsymbol{R}^{-0.13}$ (where R_0 is the pipe radius and $\boldsymbol{R}$ is the pipe Reynolds number based on diameter), the energy-dissipating eddies are of size 40 $R_0 \boldsymbol{R}^{-0.78}$, and the Kolmogoroff scale is 8 $R_0 \boldsymbol{R}^{-0.78}$.

7.7.2 The Energy Cascade

The dissipation rate in turbulence is given by

$$\epsilon = -\frac{1}{2} \frac{\partial \overline{u_i' u_i'}}{\partial t} \qquad (7.140)$$

which represents the decay of the correlation of the fluctuating velocities after the mechanism of turbulence creation has been removed. Although (7.140) is apparently not derivable, it has been checked by numerous experiments. The important consideration is that (7.140) does not contain viscosity. The conduit of turbulent energy from the large to the small scales passes energy that is roughly proportional to the square of the fluctuating velocity regardless of the viscosity. The dissipation at the small eddy size of the spectrum, the large wave numbers, is automatically adjusted to the rate of energy that passes down the spectrum.

The turbulent kinetic energy per unit of mass for all the eddies is defined as

$$\int_0^\infty E(k)\,dk = \frac{1}{2}\overline{u_i' u_i'} \qquad (7.141)$$

A characteristic velocity is $u_e = \sqrt{k_e E}$. Kolmogoroff (1941) found that for the central part of the spectrum, E obeys the relationship

$$E(k) \sim k^{-5/3} \qquad (7.142)$$

The latter proportionality is an important result in that when modeling turbulent flows the model should reproduce that statistical behavior.

The mechanical energy equation for turbulence is

$$\frac{D}{Dt}\overline{u_i'u_i'} = -\frac{\partial}{\partial x_i}\left[\overline{u_i'\left(\frac{p'}{\rho} + u_j'u_j'\right)}\right] - \overline{u_i'u_j'}\frac{\partial u_j'}{\partial x_i}$$

Change | Advection of total | Production
in kinetic | energy by | of energy
energy/mass | turbulence | by turbulent
 | | motion

$$+\nu\frac{\partial}{\partial x_i}\left[\overline{u_j'\left(\frac{\partial u_i'}{\partial x_j} + \frac{\partial u_j'}{\partial x_i}\right)}\right] - \nu\overline{\left(\frac{\partial u_i'}{\partial x_j} + \frac{\partial u_j'}{\partial x_i}\right)\frac{\partial u_j'}{\partial x_i}} \qquad (7.143)$$

Work done by | Dissipation (ϵ)
viscous shear |

PROBLEMS

7.1. The diameter of the earth is approximately 12,900 km. The highest mountain on the earth is about 8.9 km above sea level, but a tall mountain in most parts of the world is 3.7 km high. Thus, the ratio of mountain height to diameter is 2.9×10^{-4}. Using this same ratio, compute the height of a ridge on a billiard ball (about 4 cm diameter) that would appear mountainous to the boundary layer for flow around the ball. For flow around a sphere (the ball) what is the thickness of the boundary layer at mid diameter? What is a comparable thickness of the earth's boundary layer?

7.2. A common approximation for the turbulent boundary layer is

$$\frac{\bar{u}_x}{U} = \left(\frac{y}{\delta}\right)^{1/7}$$

(a) Compute the displacement and momen-tum thicknesses in terms of the bound-ary layer thickness, δ.

(b) An experimental equation for the shear stress in a turbulent boundary layer is

$$\tau_0 = 0.0228\rho U^2 R_\delta^{-1/4} \qquad R_\delta = \frac{\delta U}{\nu}$$

Use the momentum equation for drag

$$D = \rho \int_0^\infty u_x(U - u_x)dy = \rho U^2 \delta_m$$

$$= \int_0^x \tau_0\, dx$$

to derive an expression of the boundary layer thickness as a function of x.

7.3. Transform (7.137) into cylindrical coordinates ($y = R\cos\theta$, $z = R\sin\theta$). Use the result to argue that the time derivative of vorticity in a round pipe is zero and thus there are no secondary currents.

CHAPTER
8

SHALLOW
WATER
FLOW

In Chap. 5 we defined free surface flow as the case where the size and shape of the solution region were part of the solution. That definition explicitly excluded the type of "free surface flow" of this chapter. Shallow water theory makes the "dimensional approximation"; it reduces the basic problem to one or two dimensions that are approximately horizontal. The equations of motion are integrated in the vertical from a lower boundary to the free surface, eliminating the free surface as a boundary of the solution region. Since the variable part of the boundary of the solution domain is eliminated, the domain becomes fixed. Thus, there are two simplifications: reducing the dimensions and fixing the solution domain. These approximations change the nature of the equations, the types (or properties) of solutions, and the methods of solution. In the subsequent parts of this chapter we frequently use the term "free surface," by which we mean the limits of the depth of the fluid. That term does not imply a free surface problem.

Especially in environmental fluid mechanics, there are many, many flows that are included in the category of shallow water theory. As ridiculous as it may seem, the oceans can be considered shallow in the solution of many problems. The criterion is that the wave length must be long compared to the depth. That is not true, of course, for the sorts of waves that are most visible on the surface of the oceans, but there are others—tides, tsunamis—that fulfill that condition. In fact the oceans are at most a few kilometers deep whereas their breadth is hundreds of kilometers.

In many respects shallow water problems are closer to hydraulics than to hydrodynamics by virtue of the fact that the frictional aspects are almost always taken as empirical. The applications are universally in turbulent flow. (At least we know of no reasonable application in laminar flow with the possible exception of runoff in a thin sheet, which we do not consider a practical problem.) The empirical formulations avoid the complications of turbulence at the price of universality and some doubts when new problems are solved.

8.1 THE SHALLOW WATER EQUATIONS

Like the derivation of the equations of basic hydrodynamics, the equations of shallow water can be obtained in a number of different ways. Perhaps the most basic is to begin with the hydrodynamic equations because that method displays most graphically the assumptions and approximations. We need both the integral and the differential equations. In all cases, however, the density is taken as a constant. Constant density does exclude some phenomena that belong in the category of shallow water theory such as certain atmospheric flows and some occurrences in water where the flow is continuously stratified.

8.1.1 Conservation Equations and Boundary Conditions

The relevant equations come directly from Chap. 1. The integral that must be satisfied to conserve mass is

$$\int_{CS} (\vec{u} \cdot \vec{n}) \, dA = 0 \tag{8.1}$$

The corresponding differential equation is

$$\frac{\partial u_i}{\partial x_i} = 0 \tag{8.2}$$

Conservation of momentum gives

$$\vec{F} = \int_{CV} \frac{\partial}{\partial t} (\rho \vec{u}) \, dV + \int_{CS} \rho \vec{u} \, (\vec{u} \cdot \vec{n}) \, dA \tag{8.3}$$

$$\rho \frac{\partial u_i}{\partial t} + \rho u_j \frac{\partial u_i}{\partial x_j} = -\frac{\partial p}{\partial x_i} - \rho g \frac{\partial h}{\partial x_i} - \frac{\partial \tau_{ji}}{\partial x_j} \tag{8.4}$$

To these equations we add the following boundary conditions. First, the pressure on the free surface is taken as zero

$$p = 0 \quad \text{on} \quad z = \eta + H \tag{8.5}$$

where $\eta + H$ is the elevation of the free surface (Fig. 8.1). On the free surface we—again—use the condition that particles on the surface remain on the surface. An equation defining the free surface is

$$S(x, y, z, t) = \eta(x, y, t) + H - z = 0 \tag{8.6}$$

The substantial derivative of (8.6) is

$$\frac{DS}{Dt} = \frac{\partial \eta}{\partial t} + \tilde{u}_x \frac{\partial}{\partial x}(\eta + H) + \tilde{u}_y \frac{\partial}{\partial y}(\eta + H) - \tilde{u}_z = 0 \tag{8.7}$$

$$\text{on} \quad z = \eta + H$$

where the tilde indicates that these quantities are to be evaluated at $z = \eta + H$. On the solid bottom where $z = H(x, y)$, the boundary is not a function of time and the previous equation can be rewritten

$$\underline{u}_x \frac{\partial H}{\partial x} + \underline{u}_y \frac{\partial H}{\partial y} = \underline{u}_z \quad \text{on} \quad z = H \tag{8.8}$$

in which the underline denotes that the velocity is evaluated at the bottom.

Thus far no approximations have been made; the problem has only been specialized somewhat. The flow is contained between a solid bottom and a free surface. The elevation of the free surface, $\eta(x, y, t) + H(x, y)$, remains an unknown.

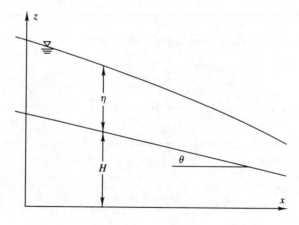

FIGURE 8.1
Definitions for the horizontal coordinate system.

8.1.2 Conservation of Mass

Formally integrating (8.2) over the depth gives

$$\int_{H}^{\eta+H} \left(\frac{\partial u_x}{\partial x} + \frac{\partial u_y}{\partial y} + \frac{\partial u_z}{\partial z} \right) dz = 0 \tag{8.9}$$

We need to reverse the order of integration and differentiation. Leibnitz' theorem is

$$\frac{\partial}{\partial t} \int_{a(y,t)}^{b(y,t)} f(x, y, t) dx = \int_{a(y,t)}^{b(y,t)} \frac{\partial f}{\partial t} dx - f(a, y, t) \frac{\partial a}{\partial t} + f(b, y, t) \frac{\partial b}{\partial t} \tag{8.10}$$

The first term of (8.9) becomes

$$\int_{H}^{\eta+H} \frac{\partial u_x}{\partial x} dz = \frac{\partial}{\partial x} \int_{H}^{\eta+H} u_x dz - \tilde{u}_x \frac{\partial}{\partial x}(\eta + H) + \underline{u}_x \frac{\partial H}{\partial x} \tag{8.11}$$

If the no-slip condition is to be applied—it usually is not applied—the latter velocity, $\underline{u}_x$, is zero. Similarly in the y-direction

$$\int_{H}^{\eta+H} \frac{\partial u_y}{\partial y} dz = \frac{\partial}{\partial y} \int_{H}^{\eta+H} u_y dz - \tilde{u}_y \frac{\partial}{\partial y}(\eta + H) + \underline{u}_y \frac{\partial H}{\partial y} \tag{8.12}$$

To remove the integrals, an average velocity over the depth is defined as

$$\bar{u}_x = \frac{1}{\eta} \int_{H}^{\eta+H} u_x dz \qquad \bar{u}_y = \frac{1}{\eta} \int_{H}^{\eta+H} u_y dz \tag{8.13}$$

Using (8.11), (8.12), and (8.13) in (8.9)

$$\frac{\partial}{\partial x} \left(\bar{u}_x \eta \right) + \frac{\partial}{\partial y} \left(\bar{u}_y \eta \right) - \tilde{u}_x \frac{\partial}{\partial x}(\eta + H)$$

$$+ \underline{u}_x \frac{\partial H}{\partial x} - \tilde{u}_y \frac{\partial}{\partial y}(\eta + H) + \underline{u}_y \frac{\partial H}{\partial y} + \tilde{u}_z - \underline{u}_z = 0 \tag{8.14}$$

Applying the boundary conditions (8.7) and (8.8)

$$\frac{\partial \eta}{\partial t} + \frac{\partial}{\partial x} \left(\bar{u}_x \eta \right) + \frac{\partial}{\partial y} \left(\bar{u}_y \eta \right) = 0 \tag{8.15}$$

8.1.3 Conservation of Momentum

Equations (8.4) are integrated in the vertical. For the present the shear terms are neglected; they will be represented later by "friction" terms that are evaluated empirically. In the x-direction

$$\int_H^{H+\eta} \left[\frac{\partial u_x}{\partial t} + \frac{\partial}{\partial x} \left(u_x^2 \right) + \frac{\partial}{\partial y} \left(u_x u_y \right) + \frac{\partial}{\partial z} \left(u_x u_z \right) + \frac{1}{\rho} \frac{\partial \hat{p}}{\partial x} \right] dz = 0$$

$$\hat{p} = p + \rho g h$$

(8.16)

where the equation of conservation of mass has been used. The time integral is

$$\int_H^{H+\eta} \frac{\partial u_x}{\partial t} dz = \frac{\partial}{\partial t} \int_H^{H+\eta} u_x dz - \tilde{u}_x \frac{\partial \eta}{\partial t} = \frac{\partial}{\partial t} \left(\bar{u}_x \eta \right) - \tilde{u}_x \frac{\partial \eta}{\partial t} \quad (8.17)$$

The first space integral is

$$\int_H^{H+\eta} \frac{\partial}{\partial x} \left(u_x^2 \right) dz = \frac{\partial}{\partial x} \int_H^{H+\eta} u_x^2 \, dz - \tilde{u}_x^2 \frac{\partial (H + \eta)}{\partial x} + \underline{u}_x^2 \frac{\partial H}{\partial x} \quad (8.18)$$

To avoid the appearance of the integral in the equation, the *momentum correction factor* is defined as

$$\beta_{xx} = \frac{\int_H^{H+\eta} u_x^2 \, dz}{\bar{u}_x^2 \eta} \quad (8.19)$$

so that

$$\int_H^{H+\eta} \frac{\partial}{\partial x} \left(u_x^2 \right) dz = \frac{\partial}{\partial x} \left(\beta_{xx} \bar{u}_x^2 \eta \right) - \tilde{u}_x^2 \frac{\partial (H + \eta)}{\partial x} + \underline{u}_x^2 \frac{\partial H}{\partial x} \quad (8.20)$$

The y-derivative is

$$\int_H^{H+\eta} \frac{\partial}{\partial y} \left(u_x u_y \right) dz = \frac{\partial}{\partial y} \left(\beta_{xy} \bar{u}_x \bar{u}_y \eta \right) - \tilde{u}_x \tilde{u}_y \frac{\partial (H + \eta)}{\partial y} + \underline{u}_x \underline{u}_y \frac{\partial H}{\partial y}$$

(8.21)

with the definition

$$\beta_{xy} = \frac{\int_H^{H+\eta} u_x u_y dz}{\bar{u}_x \bar{u}_y \eta} \quad (8.22)$$

The vertical derivative is

$$\int_H^{H+\eta} \frac{\partial}{\partial z} \left(u_x u_z \right) dz = \tilde{u}_x \tilde{u}_z - \underline{u}_x \underline{u}_z \quad (8.23)$$

When (8.20), (8.21), and (8.23) are added, the terms in the surface and bottom velocities disappear by virtue of (8.7) and (8.8). The pressure term is left for further development. A similar derivation is made for the y-equation where analogous values of the correction factor are defined.

Before integrating the vertical momentum equation, we make the primary approximation of shallow water theory—the approximation that defines shallow water theory. A small parameter is

$$\epsilon = \frac{D}{L} \ll 1 \tag{8.24}$$

in which D is defined as a vertical dimension—say, the water depth—and L is a horizontal dimension—say, a wave length. The equations that follow depend on the ratio D/L being small. Parallel to that definition, the characteristic velocities U and W are defined in the horizontal and vertical directions, respectively. Using the dimensionless variables

$$x^* = \frac{x}{L} \qquad y^* = \frac{y}{L} \qquad z^* = \frac{z}{D} \qquad u_x^* = \frac{u_x}{U} \qquad u_y^* = \frac{u_y}{U} \qquad u_z^* = \frac{u_z}{W}$$

$$t^* = \frac{Ut}{L} \qquad p^* = \frac{p}{\rho U^2} \qquad \eta^* = \frac{\eta}{D} \tag{8.25}$$

the dimensionless equation of continuity is

$$\frac{\partial u_x^*}{\partial x^*} + \frac{\partial u_y^*}{\partial y^*} + \frac{LW}{DU}\frac{\partial u_z^*}{\partial z^*} = 0 \tag{8.26}$$

The ratio LW/DU must be of order one or smaller for the equation to balance in all cases. Thus, we set

$$\frac{W}{U} = \frac{D}{L} \tag{8.27}$$

The usual definition is

$$U = \sqrt{gD} \qquad W = U\frac{D}{L} \tag{8.28}$$

The vertical momentum equation is

$$\epsilon^2 \left(\frac{\partial u_z^*}{\partial t^*} + u_x^*\frac{\partial u_z^*}{\partial u_x^*} + u_y^*\frac{\partial u_z^*}{\partial y^*} + u_z^*\frac{\partial u_z^*}{\partial z^*} \right) = -\frac{\partial p^*}{\partial z^*} - 1 \tag{8.29}$$

Neglecting terms of the order of ϵ^2, integrating, and returning to dimensional variables

$$p = \rho g(\eta + H - z) \tag{8.30}$$

which is the equation of hydrostatic pressure where the constant of integration has been chosen to satisfy the condition that the pressure is zero on the free surface.

We are now in a position to integrate the pressure derivatives in the x- and y-directions.

$$\int_H^{\eta+H} \frac{\partial p}{\partial x}dz = \rho g \int_H^{\eta+H} \frac{\partial}{\partial x}(\eta + H - z)dz = \rho g \eta \left(\frac{\partial \eta}{\partial x} + \frac{\partial H}{\partial x} \right) \tag{8.31}$$

A common assumption is that the xy "plane" lies in the bed, even when the bed is sloped or uneven, which is equivalent to neglecting the cosine of the bed angle with the horizontal, an assumption that $\cos\theta_x \approx 1$ and $\cos\theta_y \approx 1$, where θ_x and θ_y are the angles the bed makes with the horizontal. This assumption means that the free surface is at $z = \eta/\cos\theta \approx \eta$ and distance measurements made along the bed of the channel are equivalent to horizontal distance measurements. The assumption that the bed is nearly horizontal is our first approximation. The derivatives of the bottom elevation, $\partial H/\partial x$ and $\partial H/\partial y$, are written in terms of bottom slope

$$S_{0x} = \sin\theta_x = -\frac{\partial H}{\partial x} \qquad S_{0y} = \sin\theta_y = -\frac{\partial H}{\partial y} \qquad (8.32)$$

so that H remains the actual bottom elevation but will disappear from the equations. S_{0x} is often interpreted as the slope of the bed, which implies that it is $\tan\theta_x$ instead of $\sin\theta_x$, but the slopes are usually small so that there is no practical difference in these definitions. Figure 8.2 shows the coordinate system with the angle of the bed, θ, greatly exaggerated.

With similar development in the y-direction the "horizontal" equations of conservation of momentum become

$$\frac{\partial}{\partial t}(\eta\bar{u}_x) + \frac{\partial}{\partial x}\left(\beta_{xx}\eta\bar{u}_x^2\right) + \frac{\partial}{\partial y}\left(\beta_{xy}\eta\bar{u}_x\bar{u}_y\right) + g\eta\frac{\partial\eta}{\partial x} = g\eta S_{0x} \qquad (8.33)$$

$$\frac{\partial}{\partial t}(\eta\bar{u}_y) + \frac{\partial}{\partial x}\left(\beta_{yx}\eta\bar{u}_x\bar{u}_y\right) + \frac{\partial}{\partial y}\left(\beta_{yy}\eta\bar{u}_y^2\right) + g\eta\frac{\partial\eta}{\partial y} = g\eta S_{0y} \qquad (8.34)$$

There is no assumption that the velocities are not functions of z (constant in the vertical), but the β terms are usually taken as unity, which is equivalent. The primary and most limiting assumption is that ϵ is small.

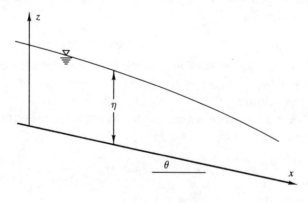

FIGURE 8.2
The coordinate system in the plane of the bed.

Clearly, waves of one meter length on a stream of one meter depth lead to $\epsilon = 1$, not a small number. On the other hand, tsunamis on the ocean may have a wave length of 200 kilometers on an ocean one kilometer deep, leading to $\epsilon = 1/20$, which is sufficiently small for accurate calculations ($\epsilon^2 = 1/400$, a *very* small number). In these examples the ocean is shallow whereas the stream is deep.

8.1.4   The Compressible Flow Analogy

The Euler equations for two-dimensional compressible flow are

$$\frac{\partial \rho}{\partial t} + \frac{\partial}{\partial x_i}(\rho u_i) = 0 \tag{8.35}$$

$$\frac{\partial u_i}{\partial t} + u_j \frac{\partial u_i}{\partial x_j} + \frac{1}{\rho}\frac{\partial p}{\partial x_i} = 0 \tag{8.36}$$

To these is added an equation of state, that of polytropic flow

$$p = K\rho^\gamma \tag{8.37}$$

where K is a constant and $\gamma = c_p/c_v$ is the ratio of specific heats (c_p is the specific heat at constant pressure and c_v is the specific heat at constant volume). For air $\gamma = 1.405$. If $\gamma = 2$ (no gas has this value), then

$$\frac{\partial p}{\partial x_i} = K\gamma\rho^{\gamma-1}\frac{\partial \rho}{\partial x_i} = 2K\rho\frac{\partial \rho}{\partial x_i} \tag{8.38}$$

Substituting (8.38) into (8.36) produces the shallow water equations with ρ identified as η and $2K$ replacing g.

If the equations are written in terms of wave velocities, there is no need to take a special value of γ. The speed of sound in a gas is given by

$$c = \sqrt{\frac{dp}{d\rho}} \tag{8.39}$$

The speed of a wave in shallow water (which is derived and used frequently in subsequent sections) is

$$c = \sqrt{g\eta} \tag{8.40}$$

In compressible flow and shallow water, respectively,

$$\frac{1}{\rho}\frac{\partial p}{\partial x_i} = \frac{1}{\rho}\frac{dp}{d\rho}\frac{\partial \rho}{\partial x_i} = \frac{c^2}{\rho}\frac{\partial \rho}{\partial x_i} \qquad g\frac{\partial \eta}{\partial x_i} = \frac{g\eta}{\eta}\frac{\partial \eta}{\partial x_i} = \frac{c^2}{\eta}\frac{\partial \eta}{\partial x_i} \tag{8.41}$$

The equations of motion are

$$\frac{\partial u_i}{\partial t} + u_j \frac{\partial u_i}{\partial x_j} + \frac{c^2}{\rho}\frac{\partial \rho}{\partial x_i} = 0 \qquad \frac{\partial u_i}{\partial t} + u_j \frac{\partial u_i}{\partial x_j} + \frac{c^2}{\eta}\frac{\partial \eta}{\partial x_i} = 0 \qquad (8.42)$$

In the early days of high-speed aerodynamics, the analogy was used to produce laboratory results in flumes using water that could not then be done in wind tunnels. The flow of information was later reversed when military spending for aerodynamics produced a bountiful supply of mathematical advances that could also be applied to shallow water theory. Although some basic laboratory results can be translated, those phenomena that are not a direct result of the equations—and the approximations contained therein—are basically different. In modern times, there is little incentive to use experiments in one medium to study problems in the other. If the phenomenon is predictable from the equations, it can be calculated; if not, it can't be translated from the experiment.

8.1.5 Linearization

The mathematics of linear equations is easier than that of nonlinear equations and often the linear equations can show some of the properties of the nonlinear equations. We assume

$$u_i = u_i^0 + \hat{u}_i \qquad \eta = \eta^0 + \hat{\eta} \qquad (8.43)$$

in which $\hat{u}_i$ and $\hat{\eta}$ represent small departures of velocity and depth from a steady, uniform flow. We assume that $\hat{\eta}/\eta^0$ is small and products of $\hat{u}$ and $\hat{\eta}$ and their derivatives can be neglected. For no slope, no friction, and uniform velocity, the equations then become

$$\frac{\partial \hat{\eta}}{\partial t} + \eta^0 \frac{\partial \hat{u}_i}{\partial x_i} + u_i^0 \frac{\partial \hat{\eta}}{\partial x_i} = 0 \qquad \frac{\partial \hat{u}_i}{\partial t} + u_j^0 \frac{\partial \hat{u}_i}{\partial x_j} + g\frac{\partial \hat{\eta}}{\partial x_i} = 0 \qquad (8.44)$$

Because there is no friction in the equations, the uniform velocity u_i^0 can be taken as zero without loss of generality (i.e., a translation of the coordinate system). Then combining the two expressions of (8.44) produces the *wave equation*

$$\frac{\partial^2 \hat{u}_i}{\partial x_j \partial x_j} = \frac{1}{c_0^2}\frac{\partial^2 \hat{u}_i}{\partial t^2} \qquad \frac{\partial^2 \hat{\eta}}{\partial x_j \partial x_j} = \frac{1}{c_0^2}\frac{\partial \hat{\eta}}{\partial t^2} \qquad (8.45)$$

in which $c_0 = \sqrt{g\eta^0}$ is the wave speed of the base flow. Limiting the problem to one space dimension, the solution to (8.45) is

$$\hat{u} = f_1(x - c_0 t) + F_1(x + c_0 t)$$
$$\hat{\eta} = f_2(x - c_0 t) + F_2(x + c_0 t) \qquad (8.46)$$

where f_1, F_1, f_2 and F_2 are general functions. These functions are related in order to satisfy (8.44)

$$f_2 = \frac{c_0}{g} f_1 \qquad F_2 = -\frac{c_0}{g} F_1 \qquad (8.47)$$

Thus,

$$\hat{u} = f(x - c_0 t) + F(x + c_0 t)$$

$$\hat{\eta} = \frac{c_0}{g} [f(x - c_0 t) - F(x + c_0 t)] \qquad (8.48)$$

Along a line in the xt plane that is given by the equation $x - c_0 t =$ constant, the function f is constant; along $x + c_0 t =$ constant, F is constant. Thus f can be considered a disturbance (wave) that travels in the positive x-direction with speed c_0; F is a disturbance traveling in the negative x-direction. The lines in the xt plane along which these disturbances travel (i.e., $x \pm c_0 t =$ constant) are called *characteristics*. Since c_0 has been assumed constant, the characteristics are straight lines. Also, the disturbances f and F remain unmodified (except by boundary conditions) throughout the solution. Consider a case (Fig. 8.3) in which $\hat{u}$ and $\hat{\eta}$ are given as initial conditions everywhere on the x-axis. Writing (8.48) at point A and eliminating F, then writing these equations at point B and eliminating f produces

$$\hat{u}_A + \frac{g\hat{\eta}_A}{c_0} = 2f \qquad \hat{u}_B - \frac{g\hat{\eta}_B}{c_0} = 2F \qquad (8.49)$$

Since f is constant along $x = c_0 t +$ constant (the line C$^+$) and F is constant along $x = -c_0 t +$ constant (the line C$^-$)

$$\hat{u}_A + \frac{g\hat{\eta}_A}{c_0} = \hat{u}_P + \frac{g\hat{\eta}_P}{c_0} \qquad \hat{u}_B - \frac{g\hat{\eta}_B}{c_0} = \hat{u}_P - \frac{g\hat{\eta}_P}{c_0} \qquad (8.50)$$

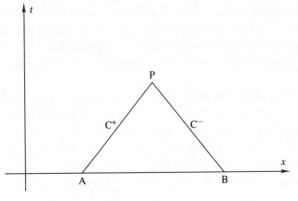

FIGURE 8.3
Characteristics from points A and B to point P.

Solving for $\hat{u}_P$ and $\hat{\eta}_P$

$$\hat{u}_P = \frac{1}{2}\left(\hat{u}_A + \hat{u}_B + \frac{g\hat{\eta}_A}{c_0} - \frac{g\hat{\eta}_B}{c_0}\right)$$

$$\hat{\eta}_P = \frac{c_0}{2g}\left(\hat{u}_A - \hat{u}_B + \frac{g\hat{\eta}_A}{c_0} + \frac{g\hat{\eta}_B}{c_0}\right)$$

(8.51)

At every point in the xt plane a *forward characteristic* and a *backward characteristic* intersect. Each characteristic carries information about the variables. In combination that information is sufficient to find the solution at the intersection. At boundary points there must be a boundary condition that substitutes for the information along the missing forward characteristic (upstream boundary) or along the missing backward characteristic (downstream boundary). With two initial conditions (each variable prescribed on the *initial line*) and two boundary conditions (one variable on the upstream boundary and one variable on the downstream boundary) the solution can be found anywhere in the xt plane.

In the following section analogous behavior is presented for the full, nonlinear, shallow water equations. In the full equations the wave speed is not constant but depends on the depth, η, itself a part of the solution. Since the depth is not, in general, a constant, the speed of the disturbance is not constant and the characteristics are not straight lines.

8.2   THE THEORY OF CHARACTERISTICS

The theory of characteristics is of paramount importance in the treatment of the shallow water equations, those of gas dynamics, and many other equations of mathematical physics. It is helpful in the solution of problems and in the physical interpretation of associated phenomena. Aside from the applications presented in this chapter, it is a guide to the numerical solutions of the equations and in some cases the numerical method is a direct consequence of the theory of characteristics. The theory is developed from a general point of view and then connected to the shallow water equations. In this way the various forms of the shallow water equations can be treated and their solutions compared to similar equations of mathematical physics.

Consider the following set of *quasi-linear* partial differential equations. The term quasi-linear refers to the fact that the equations are linear in the derivatives of the dependent variables, but in general are nonlinear. The equations are not completely general in that they contain only two independent variables (x and y), but they contain an arbitrary number of dependent variables. A parallel development of the theory is possible for more than two independent variables but is longer. The equations are

$$a_{ij}\frac{\partial u_j}{\partial x} + b_{ij}\frac{\partial u_j}{\partial y} + d_i = 0 \qquad i = 1, 2, \ldots, n \qquad (8.52)$$

in which the repeated index is summed over dependent variables $u_1, u_2, \ldots,$ u_n. The a_{ij}, b_{ij}, and d_i are, in general, functions of the dependent variables but not of their derivatives. Even though only the first derivatives appear in the equations, they can represent, in combination, higher order equations. For example, Laplace's equation,

$$\frac{\partial^2 u}{\partial x^2} + \frac{\partial^2 u}{\partial y^2} = 0 \qquad (8.53)$$

is written as

$$\frac{\partial u_1}{\partial x} + \frac{\partial u_2}{\partial y} = 0 \qquad \frac{\partial u_1}{\partial y} - \frac{\partial u_2}{\partial x} = 0 \qquad (8.54)$$

by taking $u_1 = \partial u/\partial x$ and $u_2 = \partial u/\partial y$. Thus, (8.54) is equivalent to (8.53).

8.2.1 Directional Derivatives

The object of the development is to write the set (8.52) in the form

$$A_{ij}\left(\lambda\frac{\partial}{\partial x} + \frac{\partial}{\partial y}\right)u_j + D_i = 0 \qquad i = 1, 2, \ldots, n \qquad (8.55)$$

where A_{ij}, λ, and D_i are functions of a_{ij}, b_{ij}, and d_i. The quantity in parentheses is a differential operator that "operates" on the dependent variables u_i. Equation (8.55) is similar to an ordinary differential equation in that it differentiates the dependent variable in a single direction given by

$$\frac{du_i}{ds} = \left[\left(\lambda\frac{\partial}{\partial x} + \frac{\partial}{\partial y}\right)u_i\right]\frac{dy}{ds} \qquad (8.56)$$

where s is distance along some curve in the xy plane as shown in Fig. 8.4.

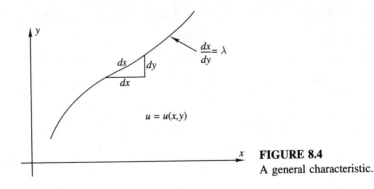

FIGURE 8.4

A general characteristic.

The total derivative of u_i along a curve in the xy plane is

$$\frac{du_i}{ds} = \frac{\partial u_i}{\partial x}\frac{dx}{ds} + \frac{\partial u_i}{\partial y}\frac{dy}{ds} \tag{8.57}$$

Assume that the equation of the curve is given by

$$\frac{dx}{dy} = \lambda(x, y) \tag{8.58}$$

Then

$$\frac{du_i}{ds} = \frac{\partial u_i}{\partial x}\lambda\frac{dy}{ds} + \frac{\partial u_i}{\partial y}\frac{dy}{ds} = \frac{dy}{ds}\left(\lambda\frac{\partial}{\partial x} + \frac{\partial}{\partial y}\right)u_i \qquad \text{or}$$

$$\frac{du_i}{dy} = \left(\lambda\frac{\partial}{\partial x} + \frac{\partial}{\partial y}\right)u_i \tag{8.59}$$

The distinction between partial and total derivatives is important to recognize. The total derivative must be defined as to how it is to be taken, in this case along the curve $dx/dy = \lambda$. The partial derivative implies that all independent variables except that in the derivative remain constant, so that, for example, $\partial u_i/\partial x$ implies the change of u_i along the line $y = 0$. Taken alone the symbol du_i/dx does not specify how the derivative is to be taken and must be defined such as in (8.59). The derivative along a curve is called a *directional derivative*. Partial derivatives are special cases of directional derivatives in which the direction is specified by holding all but one of the independent variables constant.

8.2.2 Transformation of the Quasi-linear Equations

Equations (8.52) are multiplied by a transformation matrix, t_{ij},

$$t_{ij}a_{jk}\frac{\partial u_k}{\partial x} + t_{ij}b_{jk}\frac{\partial u_k}{\partial y} + t_{ik}d_k = 0 \tag{8.60}$$

The t_{ij} may be functions of the dependent and independent variables but not of the derivatives of the dependent variables. Comparing (8.60) with (8.55)

$$A_{ik}\lambda = t_{ij}a_{jk} \qquad A_{ik} = t_{ij}b_{jk} \tag{8.61}$$

Eliminating A_{ik} produces

$$t_{ij}a_{jk} = \lambda t_{ij}b_{jk} \tag{8.62}$$

Using (8.62) to eliminate the a_{ij} in (8.60) gives

$$\lambda t_{ij}b_{jk}\frac{\partial u_k}{\partial x} + t_{ij}b_{jk}\frac{\partial u_k}{\partial y} + t_{ik}d_k = 0 \tag{8.63}$$

which is of the desired form of (8.55)

$$t_{ij}b_{jk}\left(\lambda\frac{\partial}{\partial x} + \frac{\partial}{\partial y}\right)u_k + t_{ik}d_k = 0 \qquad (8.64)$$

The matter of determining the t_{ij} still remains. Obviously, the t_{ij} must satisfy (8.62), which may be considered as n^2 equations in the n^2+1 values of t_{ij} and λ. Actually, only n of these equations are distinct since we could set $t_{ij} = t_{kj}$. Since the equations are homogeneous in t_{ij}, the determinant of the coefficient matrix must be zero to avoid the trivial solution that all the t_{ij} are not zero, thus

$$\begin{vmatrix} a_{11} - \lambda b_{11} & a_{12} - \lambda b_{12} & \cdots & a_{1n} - \lambda b_{1n} \\ a_{21} - \lambda b_{21} & a_{22} - \lambda b_{22} & \cdots & a_{2n} - \lambda b_{2n} \\ \vdots & \vdots & \ddots & \vdots \\ a_{n1} - \lambda b_{n1} & a_{n2} - \lambda b_{n2} & \cdots & a_{nn} - \lambda b_{nn} \end{vmatrix} = 0 \qquad (8.65)$$

The determinant forms an n-degree polynomial for λ. This polynomial is key to the system of equations. If there are n real and distinct roots, λ_i, of the polynomial, then the original set of differential equations is *hyperbolic*; if all the roots are imaginary, the equations are *elliptic*. An intermediate case is that of *parabolic* equations. These types of partial differential equations have very different properties and different boundary conditions must be applied to obtain a well-posed system.

8.2.3 Examples of Equations

Consider first Laplace's Equation (8.53) and the equivalent pair of first-order equations (8.54). The a_{ij} are the coefficients of the x-derivatives and form the matrix

$$[a] = \begin{bmatrix} \text{first eq, first variable } (a_{11}) & \text{first eq, second variable } (a_{12}) \\ \text{second eq, first variable } (a_{21}) & \text{second eq, second variable } (a_{22}) \end{bmatrix}$$

The b_{ij} are the same except they are the coefficients of the y-derivatives. For (8.54)

$$[a] = \begin{bmatrix} 1 & 0 \\ 0 & -1 \end{bmatrix} \qquad [b] = \begin{bmatrix} 0 & 1 \\ 1 & 0 \end{bmatrix} \qquad (8.66)$$

The determinant corresponding to (8.65) is

$$\det\left(a_{ij} - \lambda b_{ij}\right) = \begin{vmatrix} 1 & -\lambda \\ -\lambda & -1 \end{vmatrix} = -\lambda^2 - 1 \qquad (8.67)$$

Setting the determinant equal to zero yields

$$\lambda = \pm\sqrt{-1} = \pm i \qquad (8.68)$$

Because all the λ are imaginary, the original equation—or set of equations—is elliptic and doesn't possess characteristics.

As the second example consider the heat equation

$$\frac{\partial^2 u}{\partial x^2} - \frac{\partial u}{\partial t} = 0 \tag{8.69}$$

Defining $u_1 = \partial u/\partial x$ and $u_2 = u$ the equivalent set of first-order equations is

$$\frac{\partial u_1}{\partial x} - \frac{\partial u_2}{\partial t} = 0 \qquad u_1 - \frac{\partial u_2}{\partial x} = 0 \tag{8.70}$$

Then

$$[a] = \begin{bmatrix} 1 & 0 \\ 0 & -1 \end{bmatrix} \qquad [b] = \begin{bmatrix} 0 & -1 \\ 0 & 0 \end{bmatrix} \qquad \det\left(a_{ij} - \lambda b_{ij}\right) = \begin{bmatrix} 1 & \lambda \\ 0 & -1 \end{bmatrix} \tag{8.71}$$

There is no finite value of λ that makes the determinant equal to zero. The heat equation is parabolic.

The wave equation—a one-dimensional version of (8.45)—is

$$\frac{\partial^2 u}{\partial x^2} - \frac{1}{c^2}\frac{\partial^2 u}{\partial t^2} = 0 \tag{8.72}$$

with c a constant. Defining $u_1 = \partial u/\partial x$ and $u_2 = \partial u/\partial t$ the equivalent first-order set is

$$\frac{\partial u_1}{\partial x} - \frac{1}{c^2}\frac{\partial u_2}{\partial t} = 0 \qquad \frac{\partial u_1}{\partial t} - \frac{\partial u_2}{\partial x} = 0 \tag{8.73}$$

Then

$$[a] = \begin{bmatrix} 1 & 0 \\ 0 & -1 \end{bmatrix} \qquad [b] = \begin{bmatrix} 0 & -\dfrac{1}{c^2} \\ 1 & 0 \end{bmatrix}$$

$$\det\left(a_{ij} - \lambda b_{ij}\right) = \begin{vmatrix} 1 & \dfrac{\lambda}{c^2} \\ -\lambda & -1 \end{vmatrix} = -1 + \frac{\lambda^2}{c^2} \tag{8.74}$$

leading to the solution

$$\lambda = \pm c \tag{8.75}$$

Because there are two real and distinct roots, the wave equation is hyperbolic. There are two directional derivatives of the form (8.58) with y replaced by t. The equations of the characteristics are

$$\frac{dx}{dt} = \pm c \qquad \text{or} \qquad x = \pm ct + \text{constant} \tag{8.76}$$

For the two values of λ, let $\lambda_1 = c$ and $\lambda_2 = -c$. In matrix notation (8.62) for the t_{ij} becomes

$$\begin{bmatrix} t_{11} & t_{12} \\ t_{21} & t_{22} \end{bmatrix} \begin{bmatrix} 1 & 0 \\ 0 & -1 \end{bmatrix} = \begin{bmatrix} c & 0 \\ 0 & -c \end{bmatrix} \begin{bmatrix} t_{11} & t_{12} \\ t_{21} & t_{22} \end{bmatrix} \begin{bmatrix} 0 & -\dfrac{1}{c^2} \\ 1 & 0 \end{bmatrix} \qquad (8.77)$$

which leads to the equations

$$t_{11} = ct_{12} \qquad t_{12} = \frac{t_{11}}{c} \qquad t_{21} = -ct_{22} \qquad t_{22} = -\frac{t_{21}}{c} \qquad (8.78)$$

The t_{ij} are not unique, but their ratios are

$$\frac{t_{11}}{t_{12}} = c \qquad \frac{t_{21}}{t_{22}} = -c \qquad (8.79)$$

and an acceptable transformation matrix is

$$[t] = \begin{bmatrix} c & 1 \\ -c & 1 \end{bmatrix} \qquad (8.80)$$

Knowing the t_{ij}, we are ready to write the normal form of the equations such as (8.64), which is

$$\frac{1}{c}\left(c\frac{\partial}{\partial x} + \frac{\partial}{\partial t} \right) u_2 - \left(c\frac{\partial}{\partial x} + \frac{\partial}{\partial t} \right) u_1 = 0$$

$$\frac{1}{c}\left(-c\frac{\partial}{\partial x} + \frac{\partial}{\partial t} \right) u_2 + \left(-c\frac{\partial}{\partial x} + \frac{\partial}{\partial t} \right) u_1 = 0 \qquad (8.81)$$

Since c is constant in the linear wave equation, (8.81) can be written

$$\left(c\frac{\partial}{\partial x} + \frac{\partial}{\partial t} \right)(u_2 - cu_1) = 0 \qquad \left(-c\frac{\partial}{\partial x} + \frac{\partial}{\partial t} \right)(u_2 + cu_1) = 0 \quad (8.82)$$

The first of (8.82) indicates that $u_2 - cu_1$ is constant along the line $dx/dt = c$ (or $x = ct + \text{constant}$); the second indicates that $u_2 + cu_1$ is constant along $dx/dt = -c$. These quantities, $u_2 - cu_1$ and $u_2 + cu_1$, are called *Riemann invariants* for the original set of equations.

Compare this solution with the known general solution for the wave equation as in (8.48). From the definitions of u_1 and u_2

$$u_1 = \frac{\partial u}{\partial x} = f' + F' \qquad u_2 = \frac{\partial u}{\partial t} = -cf' + cF' \qquad (8.83)$$

so that

$$u_2 - cu_1 = -2cf' \qquad u_2 + cu_1 = 2cF' \qquad (8.84)$$

Since F and F' do not change along the lines $x + ct = \text{constant}$ and f and f' do not change along $x - ct = \text{constant}$ (Fig. 8.5), the results of the

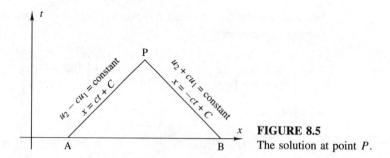

FIGURE 8.5
The solution at point P.

theory of characteristics are verified by the general solution. Unfortunately, we do not know the general solution to most partial differential equations and, therefore, must resort to some other process such as the theory of characteristics.

8.2.4 Boundary Data

The solution to the wave equation indicates the type of boundary data that is necessary. To find the solution at point P of Fig. 8.5, we draw the characteristics from point P to the x-axis at points A and B. Then from (8.82)

$$(u_2 - cu_1)_P = (u_2 - cu_1)_A \qquad (u_2 + cu_1)_P = (u_2 + cu_1)_B \qquad (8.85)$$

Given u_1 and u_2 on the x-axis, the solution at P is given directly by (8.85). At point Q on the upstream boundary (Fig. 8.6) or point Q$'$ on the downstream boundary, the boundary condition replaces the equation along one of the characteristics.

The rule is that one item of data must be given along a line for each characteristic that stems from that line. In the example of the wave

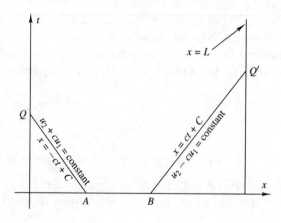

FIGURE 8.6
Solution for the boundary points.

equation, two items of data are required on the initial line and one item of data on each of the boundary lines. With some other equations a boundary will act as an initial line (all of the characteristics stem from the line) or a boundary may have no characteristics originating at it, in which case it needs no conditions.

In a general hyperbolic equation the characteristics are not straight lines; the slope, dx/dt, of the characteristics is a function of the dependent variables, which change along the characteristics. In those cases the slope of the characteristics is part of the problem and not known a priori. The influence of the boundary data—or any disturbance in the plane of the solution—is defined by the pattern of the characteristics. A small disturbance at point Q (Fig. 8.7) influences the solution only in the *range of influence* of the point. An observer at point R would remain ignorant of the disturbance.

The data along a portion of an initial line AB (Fig. 8.8) will determine the complete solution within the *zone of determinacy* of AB. On the other hand, the portion AB of the initial line is called the *domain of dependence* of point R. The solution at point P is influenced by the information on the line AB, but not determined since information outside of AB also has an influence.

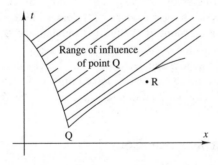

FIGURE 8.7
Definition of the range of influence.

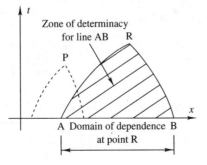

FIGURE 8.8
Definitions of the zone of determinacy and the domain of dependence.

8.3 UNSTEADY, ONE-DIMENSIONAL FLOW IN RECTANGULAR CHANNELS

Although most rivers, canals, and waterways are of arbitrary cross-section, the properties of the shallow water equations are easier to develop for rectangular channels. A later section will treat the more general case. The equations are

$$\frac{\partial \eta}{\partial t} + u\frac{\partial \eta}{\partial x} + \eta\frac{\partial u}{\partial x} = 0 \qquad (8.86)$$

$$\frac{\partial u}{\partial t} + u\frac{\partial u}{\partial x} + g\frac{\partial \eta}{\partial x} = g\left(S_{0x} - S_{fx}\right) \qquad (8.87)$$

The *friction slope*, S_{fx}, has been included in lieu of the viscous terms; it usually takes a form such as

$$S_{fx} = C\frac{u^r}{\eta^s} \qquad (8.88)$$

in which C, r and s depend on the particular friction law that is chosen. We have also assumed that $\beta = 1$ so that $u = \bar{u}$.

> **Historical Note:** Equations (8.86) and (8.87) are often known as the Saint Venant equations. Jean-Claude Barré de Saint Venant (1797–1886) wrote about the hydraulics of pipe and channel flow. Some of his work was published after his death—for example, his defense of the principle that the drag on an object is the same if the fluid is moving and the object fixed or vice versa, contrary to the experiments of Pierre Louis Georges Du Buat (1734–1809). Saint Venant recognized the role of turbulence in fluid friction. He worked in wave mechanics and adopted the word "celerity" for the speed of a wave in order to avoid confusing wave velocity with fluid velocity.

8.3.1 The Characteristic Equations

Using (8.86) and (8.87) with $[a]$ the coefficients of the x-derivatives and $[b]$ as the coefficients of the t-derivatives

$$[a] = \begin{bmatrix} \eta & u \\ u & g \end{bmatrix} \qquad [b] = \begin{bmatrix} 0 & 1 \\ 1 & 0 \end{bmatrix}$$

$$\det\left(a_{ij} - \lambda b_{ij}\right) = \begin{vmatrix} \eta & u - \lambda \\ u - \lambda & g \end{vmatrix} = \eta g - (u - \lambda)^2 \qquad (8.89)$$

After solving for λ, we find that the characteristic directions are

$$\left(\frac{dx}{dt}\right)_1 = u + \sqrt{g\eta} \qquad \left(\frac{dx}{dt}\right)_2 = u - \sqrt{g\eta} \qquad (8.90)$$

Clearly, the wave speed is given by $c = \sqrt{g\eta}$ and the disturbance speed is found by adding or subtracting the velocity. Assuming a positive velocity, the first characteristic will always have a positive slope—disturbances travel in the downstream direction at a speed greater than the velocity. The second characteristic may have either a positive or negative slope depending on the relative magnitude of u and c. If $u < c$ the flow is *subcritical*, if $u > c$ it is *supercritical*, and the intermediate case, $u = c$, is *critical flow*. In supercritical flow a small disturbance is felt only downstream of the point of the disturbance. For subcritical flow the disturbance travels both upstream and downstream and eventually influences the flow in the entire channel. The characteristics for these two cases are shown in Fig. 8.9. The depth at which the wave speed is equal to the velocity, $u = c$, is called *critical depth*.

Equations (8.62) become

$$\begin{bmatrix} t_{11} & t_{12} \\ t_{21} & t_{22} \end{bmatrix} \begin{bmatrix} \eta & u \\ u & g \end{bmatrix} = \begin{bmatrix} u+c & 0 \\ 0 & u-c \end{bmatrix} \begin{bmatrix} t_{11} & t_{12} \\ t_{21} & t_{22} \end{bmatrix} \begin{bmatrix} 0 & 1 \\ 1 & 0 \end{bmatrix} \quad (8.91)$$

A transformation matrix that satisfies these equations is

$$[t] = \begin{bmatrix} c & \eta \\ -c & \eta \end{bmatrix} \quad (8.92)$$

The normal form of the shallow water equations is

$$\eta \left[(u+c)\frac{\partial}{\partial x} + \frac{\partial}{\partial t} \right] u + c \left[(u+c)\frac{\partial}{\partial x} + \frac{\partial}{\partial t} \right] \eta + \eta \left[-g\left(S_{0x} - S_{fx}\right) \right] = 0$$

$$\eta \left[(u-c)\frac{\partial}{\partial x} + \frac{\partial}{\partial t} \right] u - c \left[(u-c)\frac{\partial}{\partial x} + \frac{\partial}{\partial t} \right] \eta + \eta \left[-g\left(S_{0x} - S_{fx}\right) \right] = 0$$

$$(8.93)$$

As in the case of the linear wave equation, it is convenient to write the relationships so that they each contain a single operator that operates on the Riemann invariants. We divide through by η, obtaining

$$\left[(u \pm c)\frac{\partial}{\partial x} + \frac{\partial}{\partial t} \right] u \pm \frac{c}{\eta} \left[(u \pm c)\frac{\partial}{\partial x} + \frac{\partial}{\partial t} \right] \eta - g\left(S_{0x} - S_{fx}\right) = 0 \quad (8.94)$$

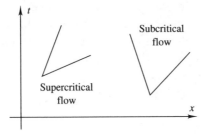

FIGURE 8.9

Characteristics for subcritical and supercritical flows.

The second term has the form

$$\frac{c}{\eta}\frac{d\eta}{dt} = \frac{c}{g\eta}\frac{d}{dt}(g\eta) = \frac{c}{c^2}\frac{d}{dt}c^2 = 2\frac{dc}{dt} \tag{8.95}$$

The normal form becomes

$$\left[(u \pm c)\frac{\partial}{\partial x} + \frac{\partial}{\partial t}\right](u \pm 2c) = g\left(S_{0x} - S_{fx}\right) \tag{8.96}$$

The quantities $u \pm 2c$ are not really invariant along their respective characteristics since the right sides of the two equations (8.96) are not zero.

The process of finding the normal form could have been shortened if we had been able to divine an integrating factor. Equation (8.86) is multiplied by c and (8.87) by η to obtain

$$c\frac{\partial\eta}{\partial t} + cu\frac{\partial\eta}{\partial x} + c\eta\frac{\partial u}{\partial x} = 0 \qquad \eta\frac{\partial u}{\partial t} + \eta u\frac{\partial u}{\partial x} + g\eta\frac{\partial\eta}{\partial x} = g\eta\left(S_{0x} - S_{fx}\right) \tag{8.97}$$

If these equations are first added then subtracted, the two equations that result are (8.93). Note that in order to add these equations, the terms of each must have the same dimensions, an aid in determining the integrating factor.

8.3.2 The Equation Without Slope and Friction

If $S_{0x} = S_{fx} = 0$ (or if $S_{0x} \equiv S_{fx}$), the Riemann invariants are invariant; $u + 2c = $ constant and $u - 2c = $ constant along their respective characteristics, $dx/dt = u \pm c$. The solution plane (the xt plane) can be divided into several regions. A *uniform region* is one in which both sets of characteristics are straight. In that case the equations of the characteristics give

$$u + c = \text{constant} \qquad u - c = \text{constant} \tag{8.98}$$

leading to the trivial solution that u and c—and hence η—are everywhere constant in the region.

In a *simple wave* one family of characteristics is straight and the other is curved. For these simplified equations, if one set of characteristics is straight, then the neighboring characteristics of the same family are also straight. Consider the situation in Fig. 8.10 in which the characteristic A_0B_0 is specified as straight. Because A_0A and B_0B are backward characteristics, from (8.96) without slope and friction

$$u_A - 2c_A = u_{A_0} - 2c_{A_0} \qquad u_B - 2c_B = u_{B_0} + 2c_{B_0} \tag{8.99}$$

The first of equations (8.90) and (8.96) indicates that along the forward

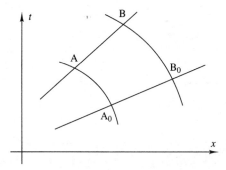

FIGURE 8.10
Characteristics for a simple wave.

characteristics

$$\frac{dx}{dt} = \text{constant} = u_{A_0} + c_{A_0} = u_{B_0} + c_{B_0} \qquad u_{A_0} + 2c_{A_0} = u_{B_0} + 2c_{B_0}$$
(8.100)

and along AB

$$u_A + 2c_A = u_B + 2c_B \tag{8.101}$$

From (8.100) $u_{A_0} = u_{B_0}$ and $c_{A_0} = c_{B_0}$. Using these relationships in (8.99) gives

$$u_A - 2c_A = u_{B_0} - 2c_{B_0} = u_B - 2c_B \tag{8.102}$$

Taken together, (8.101) and (8.102) indicate that $u_A = u_B$ and $c_A = c_B$, showing that the characteristic AB is also straight and the dependent variables are constant along it. Note, however, that the dependent variables on AB are not necessarily the same as those on A_0B_0 and AB does not necessarily have the same slope as A_0B_0.

A *complex region* is one where both sets of characteristics are curved. No simple equations apply in the complex region.

8.3.3   Dam Breaking

The failure of a dam in an ideal channel is presented as an illustration of the above points and as a minor application. Consider the situation depicted in the top part of Fig. 8.11. To begin we consider the highly unrealistic case of a moveable dam, one that can slide along the stream at a prescribed speed. The position of the dam with respect to time is shown on the xt plane in the center part of Fig. 8.11. It first starts slowly and accelerates to time t_1 and then remains at a constant speed. The speed of the dam is designated by u_D. The first movement of the dam sends a disturbance into the reservoir and that disturbance travels with speed $c_0 = \sqrt{g\eta_0}$ where η_0 is the depth of the water behind the dam. The *initial characteristic* (a backward characteristic)

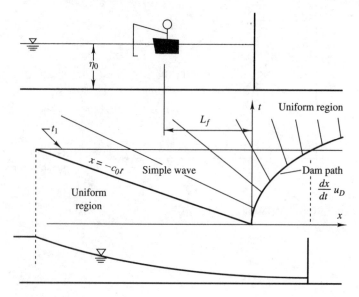

FIGURE 8.11
A simple wave.

is given by the equation $x = -c_0 t$. Consider a happy fisherman sitting in a rowboat at a distance L_f behind the dam. The disturbance caused by the movement of the dam does not reach the boat until time $t_f = L_f/c_0$; up to that time the boat has been floating in the uniform region.

Because only a simple wave can border a uniform region (except at a point), the boat now enters the simple wave—or simple region—in which one of the two families of characteristics is straight and the other is curved. Along the curved characteristics—which are forward characteristics (not shown in Fig. 8.11) and originate in the uniform region—we know that

$$u + 2c = 2c_0 \qquad (8.103)$$

At the position of the dam, the velocity of the water is known to be u_D so, using (8.103), the wave speed is $c = c_0 - u_D/2$. The slope of the backward (straight) characteristics on the dam path is

$$\frac{dx}{dt} = u - c = \frac{3}{2}u_D - c_0 \qquad (8.104)$$

Since both u and c are constant along the straight characteristics, the motion in the simple region can be described. If the dam reaches a constant velocity, another uniform region will appear with constant water speed (equal to the speed of the dam) and constant depth. The last part of Fig. 8.11 shows the water surface after some time t_1.

A special case occurs if the dam is accelerated instantly to its final velocity. All of the characteristics that form the simple wave then intersect at the origin (Fig. 8.12). The simple wave is then called a *centered simple wave*. From the geometry of Fig. 8.12, the slope of the characteristics in the simple wave is

$$\frac{dx}{dt} = \frac{x}{t} = \frac{3}{2}u - c_0 \tag{8.105}$$

Since $c = c_0 - u/2$, the solution in the simple wave is

$$u = \frac{2}{3}\left(\frac{x}{t} + c_0\right) \qquad c = -\frac{1}{3}\frac{x}{t} + \frac{2}{3}c_0 \tag{8.106}$$

The simple wave is limited by the initial characteristic $x = -c_0 t$ and the final characteristic

$$x = \left(\frac{3}{2}u_D - c_0\right)t \tag{8.107}$$

in which u_D is the final speed of the dam.

There is a limit to the effective speed of the dam. In the simple wave $c = c_0 - u_D/2$, but c must remain positive; thus the limiting speed is

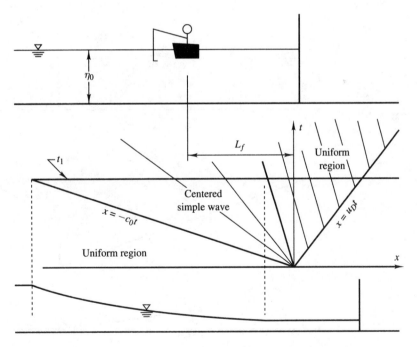

FIGURE 8.12
The centered simple wave and two uniform regions.

$u_L = 2c_0$. If the dam moves at a speed faster than the limiting velocity, it outruns the water, and a dewatered zone appears (Fig. 8.13). The water depth becomes zero at the line $x = 2c_0t + $ constant. Of course the speed of the dam does not influence the problem if it is greater or equal to the limit. The dam breaking problem is a special case where the dam moves downstream at a speed greater than $2c_0$ from the time $t = 0$ (Fig. 8.14).

Aside from the assumptions that the flow occurs in an idealized channel of zero slope and friction, the approximation of hydrostatic pressure is violated during the first instances of the disappearance of the dam. For example, at the time of removal of the dam the pressure on the vertical wall of water is zero, certainly a large departure from hydrostatic. Nevertheless, these equations have been used for practical calculations. During the Second World War there was a question of the consequences of bombing dams in the United States and Europe. Solutions of this type were used to solve the problem. In later years several computer programs have been written to solve the dam break problem, but these programs are either based on the theory of characteristics or have some of the aspects of stability and accuracy defined by the theory of characteristics.

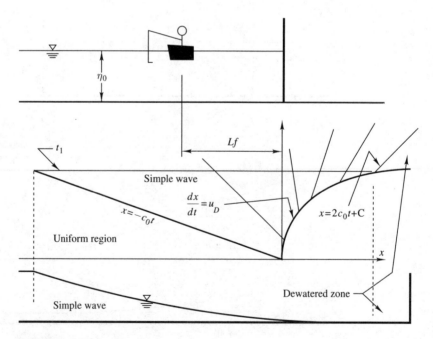

FIGURE 8.13
Case where the dam outruns the water.

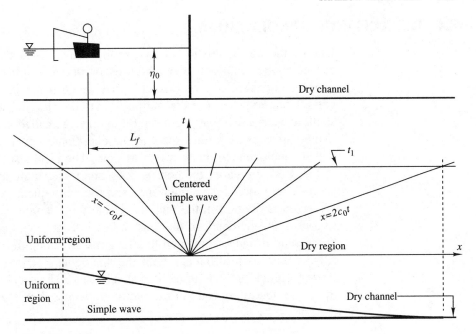

FIGURE 8.14
The dam break problem without tailwater.

Before leaving this problem, note that in the solution to the dam break problem the t-axis (the line $x = 0$ of Fig. 8.14) is one of the backward characteristics. From (8.106)

$$u(0, t) = \frac{2}{3}c_0 \qquad c(0, t) = \frac{2}{3}c_0 \qquad (8.108)$$

At the dam site $u = c$, the flow passes through critical depth, and the depth and velocity are constant with time.

We return to the plight of the fisherman who was initially a distance L_f behind the dam when it disappeared. Remaining unaware of events until time $t_f = L_f/c_0$, he then becomes paralyzed with fear and can only drift in the boat with the current. In the simple wave the velocity of the boat is described by (8.106) and its position is given by

$$\frac{dx_f}{dt} = u = \frac{2}{3}\left(\frac{x_f}{t} + c_0\right) \qquad (8.109)$$

Integrating and using the condition that $x_f = -L_f$ at time t_f,

$$x_f = 2c_0 t - 3L_f^{1/3} c_0^{2/3} t^{2/3} \qquad t \geq \frac{L_f}{c_0} \qquad (8.110)$$

8.4 DISCONTINUOUS SOLUTIONS

One of the characteristics of hyperbolic equations is that, unlike elliptic and parabolic equations, they admit discontinuous solutions, that is, the dependent variables may be discontinuous. Even if the initial and boundary conditions are smooth, a discontinuity may develop in the solution. In the shallow water equations these discontinuities are called *hydraulic jumps* or, if they are moving with respect to the coordinate system, *hydraulic bores*, and are analogous to shock waves in a gas. The ideal discontinuity is shown in Fig. 8.15. In reality the bore is not a sharp discontinuity due to the necessary curvature of the streamlines and the consequent lack of hydrostatic pressure distribution in the vicinity. Usually the length of the bore is three to four times the depth of the fluid, a large distance compared to the depth but a small distance compared to a length scale along the waterway. In that respect there is a difference with the analogous compressible flow where the streamlines are not curved in a shock wave. The shock wave has a finite thickness due to nonequilibrium thermodynamics, but is *much* thinner than the bore.

8.4.1 The Basic Relationships in One Dimension

In Fig. 8.15a the bore is traveling with velocity u_J; in Fig. 8.15b the coordinate system moves with the bore so that the discontinuity is stationary. In either case the fact that friction is neglected is a good approximation over the length of the jump or bore. The integral equations (8.1) and (8.3), written

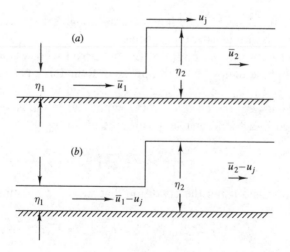

FIGURE 8.15
The hydraulic bore (*a*) and hydraulic jump (*b*).

for steady state, are the key to the jump relations. Conservation of mass applied to the situation in Fig. 8.15*b* immediately gives

$$\bar{u}_{r_1}\eta_1 = \bar{u}_{r_2}\eta_2 \qquad (8.111)$$

where $\bar{u}_r = \bar{u} - u_J$ is the relative velocity averaged over the depth. In applying (8.3) the only forces on the control volume are the hydrostatic forces before and after the jump. After integrating

$$\frac{g}{2}\left(\eta_1^2 - \eta_2^2\right) = \bar{u}_{r_1}\eta_1\left(\beta_{xx_2}\bar{u}_{r_2} - \beta_{xx_1}\bar{u}_{r_1}\right) = \bar{u}_{r_2}\eta_2\left(\beta_{xx_2}\bar{u}_{r_2} - \beta_{xx_1}\bar{u}_{r_1}\right) \qquad (8.112)$$

where β_{xx} is given by (8.19).

The energy equation also plays a part in the calculation. Unlike the gas dynamics problem, however, we are interested only in the mechanical energy—the kinetic and potential energies. The rate of change of mechanical energy in the control volume is (assuming uniform velocities in the cross sections)

$$\frac{dE_m}{dt} = \left(\frac{\bar{u}_{r_1}^2}{2} + g\eta_2\right)\rho\bar{u}_{r_2}\eta_2 - \left(\frac{\bar{u}_{r_1}^2}{2} + g\eta_1\right)\rho\bar{u}_{r_1}\eta_1 \qquad (8.113)$$

As in any liquid flow, there is no mechanism to change thermal energy into mechanical energy; therefore, the mechanical energy at section 1 (in Fig. 8.15) must be greater that at section 2, signifying that $dE_m/dt \leq 0$. After some manipulation using (8.111), (8.112), and (8.113)

$$\frac{dE_m}{dt} = \frac{\rho g\bar{u}_{r_1}}{4\eta_2}(\eta_1 - \eta_2)^3 \leq 0 \qquad (8.114)$$

or $\eta_2 \geq \eta_1$ as indicated in the figure. The energy relationship—which is really an entropy requirement—precludes the existence of a "hydraulic drop." The discontinuity can occur only with a net flux of mass from the low side to the high side. That fact can also be shown by the theory of characteristics where a "jump" can maintain itself, but if a "drop" is postulated, it will be smeared over the flow and will not remain a discontinuity.

8.4.2 The Equations of Bore Movement

The equations that are used in calculations all stem from (8.111) and (8.112) together with the definition of the relative velocities. The particular form of any of the equations can be obtained by algebraic manipulation. The height and velocity ratios of the jump—taking $\beta_{xx} = 1$—are

$$\frac{\eta_2}{\eta_1} = \frac{1}{2}\left(\sqrt{1 + \frac{8\bar{u}_{r_1}^2}{g\eta_1}} - 1\right) \qquad \frac{\bar{u}_{r_2}}{\bar{u}_{r_1}} = \frac{\eta_1}{\eta_2} \qquad (8.115)$$

Note that the relative upstream Froude number is $F_1 = \bar{u}_{r_1}/\sqrt{g\eta_1}$. F_1 must be greater than or equal to one in order that the inequality of (8.114) is not violated. As F_1 approaches one, the height of the jump goes to zero and the speed of the jump relative to the fluid approaches the wave speed, $c = \sqrt{g\eta}$.

Dealing with the moving bore is not always such a simple matter because if the bore speed is not known, the relative velocities on each side are not known. Relative to the velocity in the shallower fluid the bore speed and velocity in the deeper fluid are

$$\bar{u}_1 - u_J = \sqrt{\frac{g\eta_2}{2}\left(1 + \frac{\eta_2}{\eta_1}\right)} \qquad \bar{u}_2 - \bar{u}_1 = (u_J - \bar{u}_1)\left(1 - \frac{\eta_1}{\eta_2}\right) \quad (8.116)$$

A problem occurs when the upstream and downstream velocities, (i.e., $\bar{u}_1$ and $\bar{u}_2$), are known and η_1 is known leaving η_2 and the relative bore velocity, $u_J - \bar{u}_1$, as unknowns. Eliminating η_2 from (8.116)

$$(u_J - \bar{u}_1)(u_J - \bar{u}_2) = \frac{g\eta_1}{2}\left(1 + \frac{u_J - \bar{u}_1}{u_J - \bar{u}_2}\right) \quad (8.117)$$

The difficulty arises from the algebraic complexity of (8.117) as an equation for $u_J - \bar{u}_1$. Solving for $u_J - \bar{u}_1$

$$u_J - \bar{u}_1 = \frac{g\eta_1}{2} \frac{u_J - \bar{u}_2}{(u_J - \bar{u}_2)^2 - \frac{g\eta_1}{2}} \quad (8.118)$$

Equation (8.118) is plotted together with $u_J - \bar{u}_2$ in Fig. 8.16. The circles represent possible points of solution. From (8.117) u_J and $u_J - \bar{u}_2$ must have the same sign; therefore, the solution point is in the first or third quadrant, eliminating the centermost point. For the case illustrated in Fig. 8.17a, both $\bar{u}_2 - \bar{u}_1$ and $u_J - \bar{u}_1$ are positive, leading to a solution in the first quadrant. If $\eta_2 > \eta_1$, that will always be the case. Figure 8.17b illustrates the contrary case where $u_J - \bar{u}_1$ and $\bar{u}_2 - \bar{u}_1$ are of opposite sign.

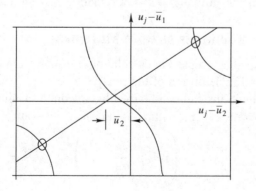

FIGURE 8.16
The bore relationships.

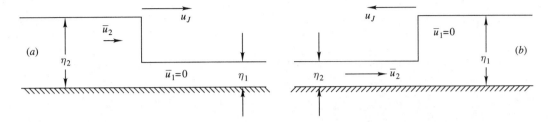

FIGURE 8.17
Bore progressing into still water (*a*) and bore progressing into oncoming flow (*b*).

The choice of roots between the first and third quadrants must be such that the energy relationship is satisfied; that is, the flux passes from the low side to the high side of the bore.

Two additional equations for the relative velocities are

$$\bar{u}_1 - u_J = \sqrt{g\frac{\eta_2}{\eta_1}\frac{\eta_1 + \eta_2}{2}} \qquad \bar{u}_2 - u_J = \sqrt{g\frac{\eta_1}{\eta_2}\frac{\eta_1 + \eta_2}{2}} \qquad (8.119)$$

Note that $\bar{u}_1 - u_J$ is larger than the local wave speed, $c_1 = \sqrt{g\eta_1}$, and $\bar{u}_2 - u_J$ is smaller than c_2. Small disturbances—those that travel with the velocity c—that are in the shallower water are swept into the jump; they cannot move upstream. Those that are in the deeper water can move in both directions and they can affect the jump even though they are downstream. This fact has far ranging consequences; it is significant in the next section.

8.4.3 Dam Break Revisited

Section 8.3.3 began with a moveable dam that slid downstream. Suppose, however, that Superman arrived and pushed the dam in the upstream direction, into the reservoir. The path of the dam is depicted in Fig. 8.18. The characteristics that stem from the dam path are still straight but they are sloped so that they tend to cross. The dependent variables u and c are constant along the straight characteristics, but different characteristics carry different values of these variables. If the characteristics cross, the solution at the crossing point becomes double valued and there is a discontinuity in the dependent variables. It is not difficult to show that such an intersection occurs from the time the dam begins to move, but instead consider the case where the dam is accelerated instantly to its final velocity as shown in Fig. 8.19. The straight characteristics are shown in two zones, both uniform regions. In the still water the slope of the characteristics is $dx/dt = c_0$, whereas it is $dx/dt = u_D + c$ in the zone that borders the dam. The bore that is formed is shown in Fig. 8.19. The velocity is known upstream of

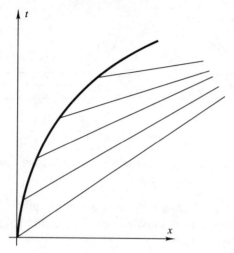

FIGURE 8.18
Dam moving into the reservoir.

the bore and the depth known downstream of the bore. The equations of the last section can be used to solve for the unknown upstream depth and the speed of the bore.

We are now prepared to consider the dam break problem with tailwater as shown in Fig. 8.20. The solution is as shown in the figure. The first two regions are just as in the former case. As shown in Fig. 8.20, the uniform region (1) of undisturbed water is under the initial characteristic given by $x = -c_1 t$ and the simple wave (2) borders the uniform region. But the simple wave ends in another uniform region (3), which in turn ends with a bore moving into the undisturbed water downstream of the dam (4). Along the forward characteristics that are born in region (1)

$$u + 2c = 2c_1 \qquad (8.120)$$

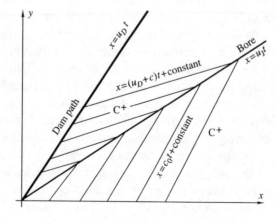

FIGURE 8.19
The bore created by a dam moving into the reservoir with constant speed.

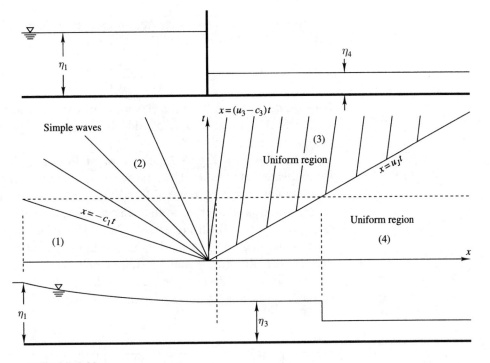

FIGURE 8.20
Dam break with tailwater.

In the centered simple wave of region (2)

$$\frac{dx}{dt} = \frac{x}{t} = u_2 - c_2 \tag{8.121}$$

Combining (8.120) and (8.121) gives the solution in the simple wave

$$u_2 = \frac{2}{3}\left(c_1 + \frac{x}{t}\right) \qquad c_2 = \frac{2}{3}c_1 - \frac{1}{3}\frac{x}{t} \tag{8.122}$$

The equation for the bore comes from (8.111). Using the current notation

$$\frac{c_3^2}{c_4^2} = \frac{\eta_3}{\eta_4} = \frac{1}{2}\left(\sqrt{1 + 8\frac{u_J^2}{c_4^2}} - 1\right) \qquad \frac{\eta_3}{\eta_4} = \frac{u_J}{u_J - u_3} \tag{8.123}$$

A useful combination is

$$\frac{u_3}{c_4} = \frac{u_J}{c_4}\frac{\sqrt{1 + 8\dfrac{u_J^2}{c_4^2}} - 3}{\sqrt{1 + 8\dfrac{u_J^2}{c_4^2}} - 1} \tag{8.124}$$

Equations (8.123) and twice the square root of (8.122) are plotted in Fig. 8.21. The third curve of Fig. 8.21 is the sum of the other two

$$\frac{u_3}{c_4} + 2\frac{c_3}{c_4} = 2\frac{c_1}{c_4} \tag{8.125}$$

Now c_1/c_4 is known from the elevations of the headwater and the tailwater. Figure 8.21 then gives u_J/c_4 and also leads to $2c_3/c_4$ and u_3/c_4, thus completing the solution.

As shown in Fig. 8.22, the flow at the dam site is unaffected by the tailwater. However, if the tailwater is higher such that $u_3 < c_3$, the characteristic dividing zones (2) and (3) would have a negative slope, the t-axis would no longer be in the centered simple wave, and (8.123) would not apply at $x = 0$. In either case the velocity and depth are constant in time at the dam site. In the latter case the depth at the dam site will be greater than the $4\eta_1/9$ that we found previously and the flow rate will be less. The flow rate at the dam site is unaffected by the tailwater until the tailwater depth is 13.84 percent of the reservoir depth (Fig. 8.22).

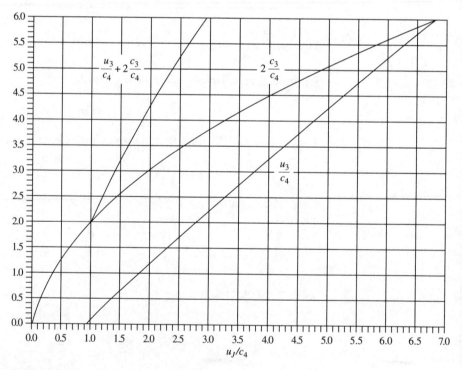

FIGURE 8.21
The bore equations for the dam break problem with tailwater.

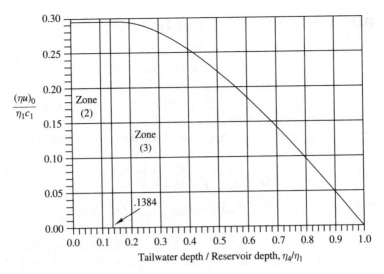

FIGURE 8.22
Flow rate at the dam site.

The bore height is zero for no tailwater and zero for a tailwater depth equal to the reservoir depth. It reaches a maximum at $\eta_4/\eta_1 = 0.176$ (Fig. 8.23).

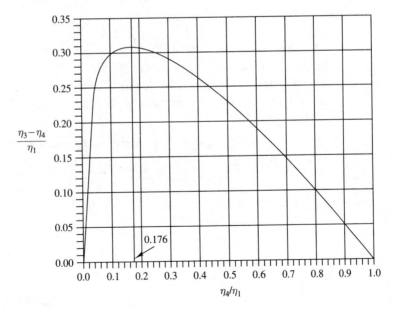

FIGURE 8.23
Height of the bore.

8.5 ONE-DIMENSIONAL FLOW IN CHANNELS

Much of the use of the shallow water equations is concerned with one-dimensional flow in watercourses that may be of arbitrary shape. Such calculations are usually done numerically, but this section prepares some of the equations. The problem is one of approximating a basically three-dimensional solution in one dimension. It would be neither economical nor necessarily more accurate to solve the equations in two or three dimensions.

8.5.1 The Basic Equations

Consider the channel cross-section of Fig. 8.24. The width of the channel at any point is $w(x, z)$. Since the basic equations have already been integrated in the vertical, we need only to integrate across the channel with the appropriate averages. At the beginning it is assumed that $\bar{u}_y \equiv 0$; all the flow is in the x-direction.

Integrating (8.15)

$$\int_{w_L}^{w_R} \left[\frac{\partial \eta}{\partial t} + \frac{\partial}{\partial x} (\bar{u}_x \eta) \right] dy = 0 \tag{8.126}$$

The order of integration and differentiation can be interchanged in both the case of differentiation with x and with t since $\eta(w_R) = \eta(w_L) = 0$. The cross-sectional area can be described in two ways,

$$A = \int_{w_L}^{w_R} \eta \, dy \qquad A = \int_0^{\tilde{\eta}} w(x, z) \, dz \tag{8.127}$$

in which $w = w_L - w_R$ is the width of the channel as a function of height from the bottom and length downstream and $\tilde{\eta}$ is the maximum depth (actually, the z-coordinate of the free surface). The average velocity in the channel is

$$\bar{\bar{u}} = \frac{1}{A} \int_A u_x dA = \int_{w_L}^{w_R} \bar{u}_x \eta \, dy \tag{8.128}$$

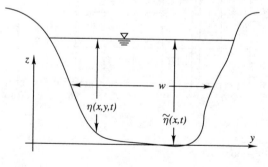

FIGURE 8.24
An irregular channel cross-section.

The double bar over u indicates that it has been averaged twice (vertically and in the y-direction) and the subscript has been dropped since the only velocity is in the x-direction. Using these definitions, the equation of conservation of mass becomes

$$\frac{\partial A}{\partial t} + \frac{\partial}{\partial x}\left(A\bar{\bar{u}}\right) = 0 \tag{8.129}$$

We need to express the derivatives in terms of the variables $\tilde{\eta}$ and $\bar{\bar{u}}$. The time and space derivatives are

$$\frac{\partial A}{\partial t} = \int_0^{\tilde{\eta}} \frac{\partial}{\partial t} w(x,z)\,dz + \tilde{w}\frac{\partial\tilde{\eta}}{\partial t} \qquad \frac{\partial A}{\partial x} = \int_0^{\tilde{\eta}} \frac{\partial w}{\partial x}\,dz + \tilde{w}\frac{\partial\tilde{\eta}}{\partial x} \tag{8.130}$$

in which $\tilde{w} = w(x,\tilde{\eta})$ is the width at the water surface. Since the banks are fixed, the derivative of $w(x,z)$ with respect to time is zero. If the channel is *prismatic*

$$\frac{\partial w}{\partial x} = 0 \tag{8.131}$$

meaning that the geometry does not change in the downstream direction. The integral in the second of (8.130) is often written

$$\int_0^{\tilde{\eta}} \frac{\partial w}{\partial x}\,dz = \left(\frac{\partial A}{\partial x}\right)_{\tilde{\eta}} \tag{8.132}$$

indicating that the change of area is to be taken with $\tilde{\eta}$ held constant, disregarding changes in area due to changes in depth. In the subsequent equations, we assume that the channel is prismatic. Conservation of mass is expressed

$$\tilde{w}\frac{\partial\tilde{\eta}}{\partial t} + \bar{\bar{u}}\tilde{w}\frac{\partial\tilde{\eta}}{\partial x} + A\frac{\partial\bar{\bar{u}}}{\partial x} = 0 \tag{8.133}$$

The relevant equation of motion is (8.33) with the friction term (friction slope) added. The separate integrals over the width are

$$\int_{w_L}^{w_R} \frac{\partial}{\partial t}(\eta\bar{u}_x)\,dy = \frac{\partial}{\partial t}(A\bar{\bar{u}}) = A\frac{\partial\bar{\bar{u}}}{\partial t} + \bar{\bar{u}}\frac{\partial A}{\partial t} = A\frac{\partial\bar{\bar{u}}}{\partial t} + \bar{\bar{u}}\tilde{w}\frac{\partial\tilde{\eta}}{\partial t} \tag{8.134}$$

$$\int_{w_L}^{w_R} \frac{\partial}{\partial x}\left(\beta_{xx}\eta\bar{u}_x^2\right)dy = \frac{\partial}{\partial x}\left(\beta A\bar{\bar{u}}^2\right)$$
$$\beta = \frac{1}{A\bar{\bar{u}}^2}\int_{w_L}^{w_R} \beta_{xx}\eta\bar{u}_x^2\,dy = \frac{1}{A\bar{\bar{u}}^2}\int_A u_x^2\,dA \tag{8.135}$$

$$\int_{w_L}^{w_R} g\eta\frac{\partial\eta}{\partial x}\,dy = gA\frac{\partial\tilde{\eta}}{\partial x} \tag{8.136}$$

$$\int_{w_L}^{w_R} g\eta \left(S_{0x} - S_{fx}\right) dy = gA \left(S_{0x} - S_{fx}\right) \qquad (8.137)$$

In writing (8.136) we assumed that the water surface slope does not change in the cross-stream direction. In (8.137) the integrated friction slope is to be expressed as

$$S_{fx} = C\frac{\bar{\bar{u}}^r}{R_h^s} \qquad R_h = \frac{A}{P} \qquad (8.138)$$

where R_h is the *hydraulic radius*, the area divided by the wetted perimeter, P. Although the friction slope has undergone an integration and uses the doubly averaged velocity, any errors are absorbed in the coefficient. Expressing the equation of motion with derivatives of depth and average velocity gives

$$\bar{\bar{u}}\tilde{w}\frac{\partial\tilde{\eta}}{\partial t} + A\frac{\partial\bar{\bar{u}}}{\partial t} + \left(\beta\bar{\bar{u}}^2\tilde{w} + gA\right)\frac{\partial\tilde{\eta}}{\partial x} + 2\beta\bar{\bar{u}}A\frac{\partial\bar{\bar{u}}}{\partial x} = gA(S_{0x} - S_{fx}) \quad (8.139)$$

In reaching this equation we have assumed that:

1. There is no y-velocity; $\bar{u}_y \equiv 0$.
2. There is no transverse slope of the water surface.
3. The channel is prismatic.
4. The derivative $d\beta_{xx}/dx$ has been neglected.

   The sort of averaging that has preceded must be done with some caution. For example, a point disturbance violates the first two assumptions since it travels in the cross-stream direction. Boundary friction could constitute such a disturbance. In a well-defined channel the averages may produce an acceptable solution, but if a large flood plane is attached to the channel, the approximations may not be satisfactory. On the other hand, the one-dimensional calculations might be more accurate than an attempt at multidimensional analysis.

   Carrying β in the equations is usually not justified. In fact $\partial\beta/\partial x$ and a transverse surface slope may become important where boundary layer changes occur, in transitions for example. In many of the same situations the assumption of hydrostatic pressure may cause an equally important error. Since these factors are not taken into account, to worry about velocity distribution corrections in most calculations is somewhat academic. The errors made in open channel calculations stem primarily from the lack of knowledge of friction, secondarily from the violation of hydrostatic pressure and transverse surface slope assumptions, and only after an accounting for these factors is the velocity distribution important. Although pathological cases no doubt exist, the β factor is seldom important and does not lead to a better understanding of open channel phenomena.

Writing the $[a]$ (x-derivative) and $[b]$ (t-derivative) matrices as in (8.52) gives

$$[a] = \begin{bmatrix} \bar{\bar{u}}\tilde{w} & A \\ \beta\bar{\bar{u}}^2\tilde{w} + gA & 2\beta\bar{\bar{u}}A \end{bmatrix} \qquad [b] = \begin{bmatrix} \tilde{w} & 0 \\ \bar{\bar{u}}\tilde{w} & A \end{bmatrix} \qquad (8.140)$$

Setting the determinant $||a_{ij} - \lambda b_{ij}||$ equal to zero and solving for λ

$$\lambda = \frac{dx}{dt} = \beta\bar{\bar{u}} \pm \sqrt{\frac{gA}{\tilde{w}} + \beta\bar{\bar{u}}^2(\beta - 1)} \qquad (8.141)$$

where dx/dt is the slope of the characteristics. Unlike the previous cases separation of λ into velocity and wave speed is impossible. However, a solution for a *critical velocity* by setting dx/dt equal to zero is possible. Selecting the minus sign, and solving for u

$$\bar{\bar{u}}_c = \sqrt{\frac{gA}{\tilde{w}\beta}} \qquad (8.142)$$

The wave speed is defined as the critical velocity, $c = \sqrt{gA/\tilde{w}\beta}$. In that case the wave speed reduces to $c = \sqrt{g\eta}$ for a rectangular channel with $\beta = 1$.

8.5.2 Steady Flow

Removing the time derivatives from (8.133) and (8.139), eliminating $\partial\bar{\bar{u}}/\partial x$, and solving for the slope of the free surface gives

$$\frac{d\tilde{\eta}}{dx} = \frac{S_{0x} - S_{fx}}{1 - \dfrac{\beta\bar{\bar{u}}^2}{\dfrac{gA}{\tilde{w}}}} \qquad (8.143)$$

Clearly, the wave speed appears in the denominator. The Froude number can be defined as

$$F = \frac{\bar{\bar{u}}}{\sqrt{\dfrac{gA}{\tilde{w}}}} \qquad \text{or} \qquad F = \frac{\bar{\bar{u}}}{\sqrt{\dfrac{gA}{\tilde{w}}}}\sqrt{\beta} \qquad (8.144)$$

There is an equivalent derivation that is worth mentioning because it indicates an error that is frequently made. Conservation of mass and momentum is written

$$\frac{d}{dx}\left(A\bar{\bar{u}}\right) = 0 \qquad \beta\bar{\bar{u}}\frac{d\bar{\bar{u}}}{dx} + g\frac{d\tilde{\eta}}{dx} = g\left(S_{0x} - S_{fx}\right) \qquad (8.145)$$

The first of (8.145) is

$$A\bar{\bar{u}} = Q = \text{constant} \tag{8.146}$$

where Q is the volumetric rate of flow. Assuming that the derivative of β is zero, the equation of motion yields

$$\frac{d}{dx}\left(\frac{\beta\bar{\bar{u}}^2}{2} + g\tilde{\eta}\right) = g\left(S_{0x} - S_{fx}\right) \tag{8.147}$$

Because the quantity on the left is interpreted as the rate of change of kinetic and potential energy along the channel, the factor β sometimes is incorrectly replaced by the energy correction factor, α. [See (1.127)]

8.5.3 Unsteady Flow

In this section β is taken as one and (8.129) and (8.139) apply. After writing the equations in normal form

$$A\left[(\bar{\bar{u}} + c)\frac{\partial}{\partial x} + \frac{\partial}{\partial t}\right]\bar{\bar{u}} + c\tilde{w}\left[(\bar{\bar{u}} + c)\frac{\partial}{\partial x} + \frac{\partial}{\partial t}\right]\tilde{\eta}$$

$$= (\bar{\bar{u}} - c)\left(\frac{\partial A}{\partial x}\right)_{\tilde{\eta}}\bar{\bar{u}} + Ag(S_{0x} - S_{fx})$$

$$A\left[(\bar{\bar{u}} - c)\frac{\partial}{\partial x} + \frac{\partial}{\partial t}\right]\bar{\bar{u}} - c\tilde{w}\left[(\bar{\bar{u}} - c)\frac{\partial}{\partial x} + \frac{\partial}{\partial t}\right]\tilde{\eta}$$

$$= (\bar{\bar{u}} + c)\left(\frac{\partial A}{\partial x}\right)_{\tilde{\eta}}\bar{\bar{u}} + \left(S_{0x} - S_{fx}\right)$$

$$\tag{8.148}$$

After dividing by A the second term is in the form $(c\tilde{w}/A)\, d\tilde{\eta}/dt$, which we would like to write as, say, $d\omega/dt$, where ω is some function. Therefore,

$$\frac{c\tilde{w}}{A}\frac{d\tilde{\eta}}{dt} = \frac{d\omega}{dt} \quad \text{or} \quad \omega = \int_0^{\tilde{\eta}}\sqrt{\frac{g\tilde{w}}{A}}\,dz \tag{8.149}$$

where ω is called a "stage variable" (sometimes called the "Escoffier stage variable," Escoffier and Boyd, 1962). The stage variable is a function of depth and channel geometry. For any given cross-section it can be plotted—or tabulated—against depth. Using (8.149) the characteristic equations (8.148) become

$$\left[(\bar{\bar{u}} + c)\frac{\partial}{\partial x} + \frac{\partial}{\partial t}\right](\bar{\bar{u}} + \omega) = -\frac{c}{A}\left(\frac{\partial A}{\partial x}\right)_{\tilde{\eta}}\bar{\bar{u}} + g(S_{0x} - S_{fx})$$

$$\left[(\bar{\bar{u}} - c)\frac{\partial}{\partial x} + \frac{\partial}{\partial t}\right](\bar{\bar{u}} - \omega) = \frac{c}{A}\left(\frac{\partial A}{\partial x}\right)_{\tilde{\eta}}\bar{\bar{u}} + g\left(S_{0x} - S_{fx}\right)$$

$$\tag{8.150}$$

If the right sides of (8.150) are taken as zero, the quantities $\bar{\bar{u}} \pm \omega$ become Riemann invariants and the problem possesses simple wave solutions.

The proper boundary conditions depend on whether the flow is subcritical or supercritical. In supercritical flow both of the conditions must be placed on the upstream boundary whereas in subcritical flow there is one condition on each boundary. Stage-discharge relationships do not make good boundary conditions as they contain hysteresis in the absence of a nearby *control*—a section with a Froude number greater than or equal to one. The initial conditions require that both velocity and depth (stage) are known at all points at the initial time. The initial conditions present a practical problem since velocity is almost never known and stage is known only at a very few discrete points. The remedy is to begin the calculation at some time prior to the times of interest and allow the computation to converge to the correct curve.

8.5.4 Friction

Unfortunately, a correct calculation using the unsteady flow equations depends on the choice of the correct frictional law. Manning's equation,

$$S_{fx} = n^2 R_h^{-4/3} \bar{\bar{u}}^2 \quad \text{in SI units}$$

$$S_{fx} = \frac{n^2}{(1.486)^2} R_h^{-4/3} \bar{\bar{u}}^2 \quad \text{in English units} \tag{8.151}$$

has been used most commonly. The value of n is almost universally assumed to be the same in unsteady, nonuniform flow as it is in the situation for which Manning's equation was originally intended to apply—steady, uniform flow. A few experiments such as those of Daily and Jordaan (1956) have tended to bear out that assumption.

Generally, Manning's n is believed to decrease with increasing Reynolds number as indicated by Chow (1959) and Woo and Brater (1962). The measurement of n can be difficult. Fig. 8.25 shows the results obtained by Baltzer and Lai (1968). Baltzer and Lai indicate that the curve is the result of a "typical least-squares plot" but they do not say what sort of curve they tried to fit to the data. The large scatter for small Reynolds number is to be expected since low flow regions are subject to high errors with any known computational method.

Figure 8.26 shows the error that might be expected when using an incorrect value of n. Unfortunately, if there is an error of 20 percent in n, the calculations for the flow quantities probably differ by a like percentage. The photographs and tables in Chow (1959) are helpful in determining

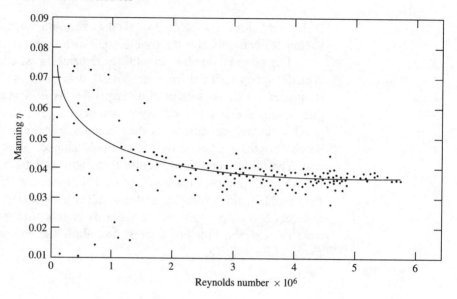

FIGURE 8.25
Plot of Manning's n (after R. A. Baltzer and C. Lai, "Computer simulation of unsteady flows in waterways," *Journal of the Hydraulics Division*, ASCE, vol. 94, no. HY4, July 1968, p. 1098, reproduced by permission of ASCE).

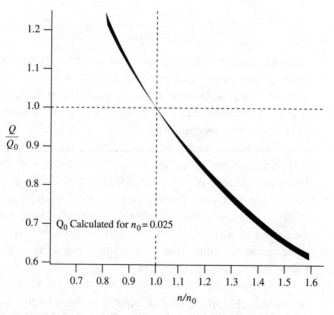

FIGURE 8.26
Flow errors as a function of n (after Liggett and Graf, 1966).

friction factors. In the end, however, it is necessary to "calibrate" any mathematical model against field data before confidence can be placed in the computations.

8.6 STABILITY IN ONE-DIMENSIONAL FLOW

As with many equations, the shallow water equations should be investigated for stability. In fact we will find that—especially in steep channels—an otherwise steady, smooth flow sometimes breaks up into an unsteady, pulsating flow. The stability criteria were first derived by Escoffier and Boyd (1962) and were later generalized somewhat by Dracos and Glenne (1967). As in all stability analyses, the idea is to ask if a disturbance grows or is damped. The growing wave will eventually break and become a hydraulic bore.

8.6.1 Stability Criteria for Uniform, Steady Flow

We begin with (8.150) but with a prismatic channel. After assuming a disturbance, $\Delta\omega$ and $\Delta\bar{\bar{u}}$

$$\left[(\bar{\bar{u}} + c)\frac{\partial}{\partial x} + \frac{\partial}{\partial t}\right](\bar{\bar{u}} + \Delta\bar{\bar{u}} + \omega + \Delta\omega) = g\left(S_{0x} - S_{fx} - \Delta S_{fx}\right)$$

$$\left[(\bar{\bar{u}} - c)\frac{\partial}{\partial x} + \frac{\partial}{\partial t}\right](\bar{\bar{u}} + \Delta\bar{\bar{u}} - \omega - \Delta\omega) = g\left(S_{0x} - S_{fx} - \Delta S_{fx}\right)$$

$$(8.152)$$

Subtracting the original equations from (8.152) produces

$$\frac{d}{dt_+}\left(\Delta\bar{\bar{u}} + \Delta\omega\right) = -g\Delta S_{fx} \qquad \frac{d}{dt_-}\left(\Delta\bar{\bar{u}} - \Delta\omega\right) = -g\Delta S_{fx} \quad (8.153)$$

where d/dt_+ and d/dt_- are the directional derivatives along the forward and backward characteristics, respectively. The friction slope is taken as (8.138). Substituting the disturbance $\Delta\bar{\bar{u}}$ and $\Delta\tilde{\eta}$

$$S_{fx} + \Delta S_{fx} = C\frac{(\bar{\bar{u}} + \Delta\bar{\bar{u}})^r}{\left(R_h + \dfrac{dR_h}{d\tilde{\eta}}\Delta\tilde{\eta}\right)^s} \qquad (8.154)$$

For small $\Delta\bar{\bar{u}}$ and $\Delta\tilde{\eta}$

$$S_{fx} + \Delta S_{fx} = C\frac{\bar{\bar{u}}^r}{R_h^s}\left(1 - s\frac{R_h'}{R_h}\Delta\tilde{\eta} + \frac{r}{\bar{\bar{u}}}\Delta\bar{\bar{u}}\right)$$

$$\Delta S_{fx} = C\frac{\bar{\bar{u}}^r}{R_h^s}\left(\frac{r}{\bar{\bar{u}}}\Delta\bar{\bar{u}} - s\frac{R_h'}{R_h}\Delta\tilde{\eta}\right)$$

$$(8.155)$$

$\Delta\omega$ is easily expressed in terms of $\Delta\tilde{\eta}$ using the definition (8.149)

$$\Delta\omega = \sqrt{\frac{g\tilde{\omega}}{A}}\,\Delta\tilde{\eta} \tag{8.156}$$

Consider the forward characteristic AB as shown in Fig. 8.27 and a disturbance along that characteristic. Along the backward characteristic B′B according to (8.153)

$$\Delta\bar{\bar{u}} - \Delta\omega = -g\int_{B'}^{B} \Delta S_{fx}\,dt \tag{8.157}$$

But ΔS_{fx} is zero until the point B is reached. That is, B′ can be moved arbitrarily close to B without changing the right side of (8.157) and, since ΔS_{fx} is always finite, the value of the integral must be zero, or

$$\Delta\bar{\bar{u}} = \Delta\omega \tag{8.158}$$

Using (8.155), (8.156), and (8.158) in the first of (8.153) yields

$$\frac{d}{dt_+}\Delta\bar{\bar{u}} = -\frac{g}{2}C\frac{\bar{\bar{u}}^r}{R_h^s}\left(1 - \frac{s}{r}\frac{R_h'}{R_h}\sqrt{\frac{A}{g\tilde{\omega}}}\bar{\bar{u}}\right)\frac{r}{\bar{\bar{u}}}\Delta\bar{\bar{u}} \tag{8.159}$$

Equation (8.159) is of the form that is needed to investigate stability,

$$\frac{d}{dt_+}\left|\Delta\bar{\bar{u}}\right| = f\left(\bar{\bar{u}}, \tilde{\eta}\right)\left|\Delta\bar{\bar{u}}\right| \tag{8.160}$$

If $f\left(\bar{\bar{u}}, \tilde{\eta}\right)$ is positive, a nonzero $\Delta\bar{\bar{u}}$ will make $d\left|\Delta\bar{\bar{u}}\right|/dt_+$ positive, indicating further growth in $\Delta\bar{\bar{u}}$—an unstable condition. On the other hand a negative $f\left(\bar{\bar{u}}, \tilde{\eta}\right)$ leads to a negative derivative and a decrease in the magnitude of $\Delta\bar{\bar{u}}$—a stable condition. Assuming that $\bar{\bar{u}}$ is positive, the condition

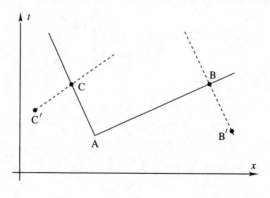

FIGURE 8.27
Characteristics.

for stability along the forward characteristic is

$$\frac{s}{r}\frac{R'_h}{R_h}\sqrt{\frac{A}{g\tilde{w}}}\,\bar{\bar{u}} \le 1 \quad \text{or} \quad \frac{s}{r}\frac{R'_h}{R_h}\frac{A}{\tilde{w}}F \le 1 \tag{8.161}$$

For the backward characteristic the equation corresponding to (8.159) is

$$\frac{d}{dt_-}\Delta\bar{\bar{u}} = -\frac{g}{2}C\frac{\bar{\bar{u}}^r}{R_h^s}\left(1 + \frac{s}{r}\frac{R'_h}{R_h}\sqrt{\frac{A}{g\tilde{w}}}\,\bar{\bar{u}}\right)\frac{r}{\bar{\bar{u}}}\Delta\bar{\bar{u}} \tag{8.162}$$

which leads to

$$\frac{s}{r}\frac{R'_h}{R_h}\sqrt{\frac{A}{g\tilde{w}}}\,\bar{\bar{u}} \ge -1 \quad \text{or} \quad \frac{s}{r}\frac{R'_h}{R_h}\frac{A}{\tilde{w}}F \ge -1 \tag{8.163}$$

Combining the two criteria

$$-1 \le \frac{s}{r}\frac{R'_h}{R_h}\frac{A}{\tilde{w}}F \le 1 \tag{8.164}$$

The *Vedernikov number* (Vedernikov, 1945, 1946; Powell, 1948) is defined as

$$V_e = \frac{s}{r}F\frac{A}{\tilde{w}}\frac{R'_h}{R_h} \tag{8.165}$$

In terms of P, the wetted perimeter, the Vedernikov number becomes

$$V_e = \left(1 - R_h\frac{dP}{dA}\right)\frac{s}{r}F \tag{8.166}$$

8.6.2 Channels of Various Shapes

Any channel will be unstable for a sufficiently high Froude number unless the Vedernikov number is zero for all Froude numbers, or

$$R_h\frac{dP}{dA} = 1 \quad \text{or} \quad \frac{dP}{dA} = \frac{P}{A} \tag{8.167}$$

Integrating (8.167) and solving for the width as a function of depth indicates that an absolutely stable channel would have rapidly diverging sides. Such channels have not been designed.

For rectangular channels the relationships are

$$A = w_0\tilde{\eta} \qquad P = w_0 + 2\tilde{\eta} \qquad \frac{dP}{dA} = \frac{2}{w_0}$$

$$R_h = \frac{w_0\tilde{\eta}}{w_0 + 2\tilde{\eta}} \qquad V_e = \frac{s}{r}F\frac{w_0}{w_0 + 2\tilde{\eta}}$$

(8.168)

in which w_0 is the bottom width (in this case equal to $\tilde{w}$). The Vedernikov number is always positive and thus only the forward type of instability can exist. Setting $V_e = 1$, the condition for stable flow is

$$F \le \frac{r}{s}\frac{w_0 + 2\tilde{\eta}}{w_0}$$

(8.169)

If Manning's equation is used, the exponents are $r = 2$ and $s = 4/3$ so that

$$F \le \frac{3}{2} + 3\frac{\tilde{\eta}}{w_0} \qquad \text{stability condition for Manning friction}$$

(8.170)

For the Chezy equation, $r = 2$ and $s = 1$ leading to

$$F \le 2 + 4\frac{\tilde{\eta}}{w_0} \qquad \text{stability condition for Chezy friction}$$

(8.171)

The rather large difference in the stability criteria with different frictional laws points out the approximate nature of the analysis. We have not really asked which *flows* are stable but instead have asked when the *solutions to the equations* are stable. When performing numerical calculations, the latter question is as important as the first because we need to know if a computational instability is due to the numerical method or is inherent in the equations.

Flow over a plane is a special case of the rectangular channel. The hydraulic radius is

$$R_h = \lim_{w_0 \to \infty} \frac{w_0\tilde{\eta}}{w_0 + 2\tilde{\eta}} = \tilde{\eta}$$

(8.172)

and $dP/dA = 0$. Thus, the stability condition is

$$F \le \frac{r}{s}$$

(8.173)

If Manning's equation is used the Froude number must be less than 1.5.

The Vedernikov criterion is really a statement of roughness and slope, as is shown for the case of flow over a plane. For uniform flow with $r = 2$ and $s = 1$

$$S_{0x} = S_{fx} = C\frac{\bar{\bar{u}}_0^2}{\bar{\eta}_0} = gCF^2$$

(8.174)

From (8.173) $F^2 \le 4$, which means that for stability

$$4gC \ge S_{0x} \tag{8.175}$$

Thus, instabilities occur more readily in smooth channels—or steep channels—than in rough channels. The same statement can be made for any shape.

For a trapezoidal channel with equal side slopes α (Fig. 8.28)

$$\tilde{w} = w_0 + 2\alpha\tilde{\eta} \qquad A = w_0\tilde{\eta} + \frac{\tilde{\eta}}{2}(\tilde{w} - w_0) = \frac{\tilde{\eta}}{2}(w_0 + \tilde{w}) = w_0\tilde{\eta} + \alpha\tilde{\eta}^2$$

$$P = w_0 + 2\tilde{\eta}\sqrt{\alpha^2 + 1} \qquad \frac{dP}{dA} = 2\frac{\sqrt{\alpha^2 + 1}}{w_0 + 2\alpha\tilde{\eta}} \qquad R_h = \frac{w_0\tilde{\eta} + \alpha\tilde{\eta}^2}{w_0 + 2\tilde{\eta}\sqrt{\alpha^2 + 1}}$$

$$V_e = \frac{s}{r}F\left[1 - 2\frac{(w_0\tilde{\eta} + \alpha\tilde{\eta}^2)\sqrt{1 + \alpha^2}}{\left(w_0 + 2\tilde{\eta}\sqrt{1 + \alpha^2}\right)(w_0 + 2\alpha\tilde{\eta})}\right]$$

$$\tag{8.176}$$

First, we seek a stable depth, that is, a depth where $V_e = 0$. Setting the term in brackets in the last of (8.176) equal to zero and solving for $\tilde{\eta}$ gives

$$\tilde{\eta} = \frac{w_0}{2\sqrt{\alpha^2 + 1}}\left(\sqrt{1 - \frac{2\sqrt{\alpha^2 + 1}}{\alpha}} - 1\right) \tag{8.177}$$

The quantity under the square root is negative unless α is negative, that is, unless the sides slope inward, indicating that there is no absolutely stable depth for a channel such as that pictured in Fig. 8.28, but instead the sides must slope inward. If the depth conforms to (8.177), there is no instability for any Froude number and a small departure from this depth would require a large Froude number for instability. For positive α, the last of (8.176) defines the Vedernikov number that must be less than unity for stability.

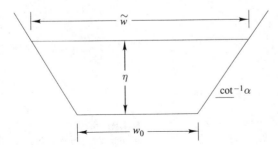

FIGURE 8.28
A trapezoidal channel.

The triangular channel is a special case of the trapezoidal channel with $w_0 = 0$ and $\alpha > 0$. The Vedernikov number is

$$V_e = \frac{1}{2}\frac{s}{r}F \tag{8.178}$$

which indicates that the channel is stable for $F < 3$ (Manning) or $F < 4$ (Chezy).

The circular channel running partly full (Fig. 8.29) is an especially interesting case and one of the most common types of open channel flow. Often the depth is replaced by the angle θ. Then

$$A = \frac{r_c^2}{2}(2\theta - \sin 2\theta) \qquad P = 2r_c\theta \qquad \frac{dP}{dA} = \frac{2}{r_c(1 - \cos 2\theta)}$$

$$R_h = \frac{r_c}{2}\left(1 - \frac{\sin 2\theta}{2\theta}\right) \qquad V_e = \frac{s}{r}F\frac{\sin 2\theta - 2\theta\cos 2\theta}{2\theta(1 - \cos 2\theta)} \tag{8.179}$$

Again, we can look for a stable depth by setting the Vedernikov number to zero. The solution is $\theta = 128.7°$ for absolute stability. A complete stability diagram is shown in Fig. 8.30.

The capacity curve for the circular conduit indicates that its maximum capacity occurs before it fills, but experimentally its maximum capacity occurs when it is full and is less than the theoretical, nearly-full capacity. The stability diagram is independent of the frictional coefficient and simply assumes a law of the form (8.138). If, indeed, the friction depends on the velocity and hydraulic radius to simple positive powers, the sign of the Vedernikov number can become negative, leading to the backward type of instability. Apparently, the backward instability causes a nearly full channel to fill and reduces the practical capacity to that of the full pipe. In all the foregoing, the particular values of the frictional constants do not enter the argument.

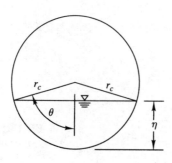

FIGURE 8.29
Circular channel running part full.

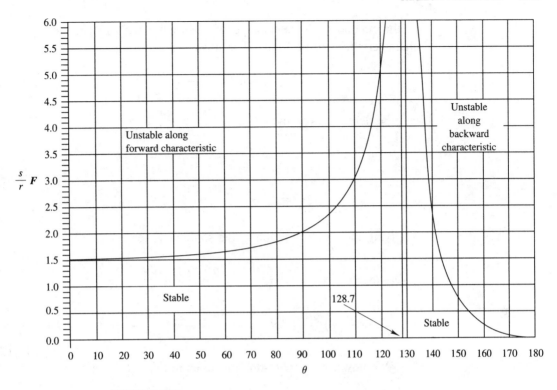

FIGURE 8.30
Stability diagram for the circular channel (see Liggett, 1975).

8.6.3 Distance Required for Bore Formation

Consider a channel in which the flow is uniform and steady but the Vedernikov number is in the unstable range. After a minor disturbance (always present) a wave begins to form, but it takes some time and distance for the instability to become apparent or for the wave to break. Before that time or distance the channel may end or conditions may change, in which case it would not be obvious to an observer that the flow was ever unstable. Our purpose in this section is to derive the distance necessary for the formation of bores or for the breaking of waves after the initial instability.

Stoker (1957) defines the following transformation:

$$\xi = x \qquad \tau = \left(\bar{\bar{u}}_0 + c_0\right)t - x \qquad (8.180)$$

in which the subscript refers to the undisturbed flow. The transformation defines constant τ along a characteristic of the uniform flow. The space derivative remains unchanged so that we can easily define a change of a variable with respect to distance. However, the derivative $\partial/\partial x$ implies that t is to be held constant whereas $\partial/\partial \xi$ implies that τ is to be held constant.

In the second case the change of a variable with respect to ξ along a characteristic is to be found. The inverse transformation is

$$x = \xi \qquad t = \frac{\tau + \xi}{\bar{\bar{u}}_0 + c_0} \tag{8.181}$$

and the derivatives transform as

$$\frac{\partial}{\partial x} = \frac{\partial \xi}{\partial x}\frac{\partial}{\partial \xi} + \frac{\partial \tau}{\partial x}\frac{\partial}{\partial \tau} = \frac{\partial}{\partial \xi} - \frac{\partial}{\partial \tau}$$

$$\frac{\partial}{\partial t} = \frac{\partial \xi}{\partial t}\frac{\partial}{\partial \xi} + \frac{\partial \tau}{\partial t}\frac{\partial}{\partial \tau} = \left(\bar{\bar{u}}_0 + c_0\right)\frac{\partial}{\partial \tau} \tag{8.182}$$

Applying the transformation to the shallow water equations

$$\tilde{w}\left(\bar{\bar{u}}_0 + c_0\right)\frac{\partial \tilde{\eta}}{\partial \tau} + \bar{\bar{u}}\tilde{w}\left(\frac{\partial \tilde{\eta}}{\partial \xi} - \frac{\partial \tilde{\eta}}{\partial \tau}\right) + A\left(\frac{\partial \bar{\bar{u}}}{\partial \xi} - \frac{\partial \bar{\bar{u}}}{\partial \tau}\right) = 0$$

$$\left(\bar{\bar{u}} + c_0\right)\frac{\partial \bar{\bar{u}}}{\partial \tau} + \bar{\bar{u}}\left(\frac{\partial \bar{\bar{u}}}{\partial \xi} - \frac{\partial \bar{\bar{u}}}{\partial \tau}\right) + g\left(\frac{\partial \tilde{\eta}}{\partial \xi} - \frac{\partial \tilde{\eta}}{\partial \tau}\right) = g(S_{0x} - S_{fx}) \tag{8.183}$$

We now form a perturbation series in the unknowns

$$\bar{\bar{u}} = \bar{\bar{u}}_0 + \bar{\bar{u}}_1(\xi)\tau + \bar{\bar{u}}_2(\xi)\tau^2 + \cdots$$

$$\tilde{\eta} = \tilde{\eta}_0 + \tilde{\eta}_1(\xi)\tau + \tilde{\eta}_2(\xi)\tau^2 + \cdots \tag{8.184}$$

in which $\bar{\bar{u}}_1, \bar{\bar{u}}_2, \ldots, \tilde{\eta}_1, \tilde{\eta}_2, \ldots$ are functions of ξ but not of τ. Our purpose is to look at the solution in the neighborhood of the characteristic $\tau = 0$. The series (8.184) applies for small τ. Equations (8.184) will be substituted into (8.183) and equivalent powers of τ are set equal. First, however, consider the friction law (8.138). The hydraulic radius can be expanded as

$$R_h = R_{h0} + \frac{dR_h}{d\tilde{\eta}}\Delta\tilde{\eta} + \frac{1}{2}\frac{d^2R_h}{d\tilde{\eta}^2}(\Delta\tilde{\eta})^2 + \cdots \tag{8.185}$$

and

$$\frac{1}{R_h^s} = \frac{1}{R_{h0}^s}\left\{1 - s\frac{R_h'}{R_{h0}}\Delta\tilde{\eta} + \left[\frac{s^2+s}{2}\left(\frac{R_h'}{R_{h0}}\right)^2 - \frac{s}{2}\frac{R_h''}{R_{h0}}\right](\Delta\tilde{\eta})^2 + \cdots\right\} \tag{8.186}$$

$\Delta\tilde{\eta}$ is defined as

$$\Delta\tilde{\eta} = \tilde{\eta} - \tilde{\eta}_0 = \tilde{\eta}_1\tau + \tilde{\eta}_2^2\tau^2 + \cdots \tag{8.187}$$

so that

$$\frac{1}{R_h^s} = \frac{1}{R_{h0}^s}\left\{1 - s\frac{R_h'}{R_{h0}}\tilde{\eta}_1\tau\right.$$

$$\left. + \left[\left(\frac{s^2+s}{2}\left(\frac{R_h'}{R_{h0}}\right)^2 - \frac{s}{2}\frac{R_h''}{R_{h0}}\right)\tilde{\eta}_1^2 - s\frac{R_h'}{R_{h0}}\tilde{\eta}_2\right]\tau^2 + \cdots\right\} \qquad (8.188)$$

The velocity expansion is

$$\bar{\bar{u}}^r = \bar{\bar{u}}_0^r\left[1 + r\frac{\bar{\bar{u}}_1}{\bar{\bar{u}}_0}\tau + \frac{r(r-1)}{2}\left(\frac{\bar{\bar{u}}_2}{\bar{\bar{u}}_0}\right)^2\tau^2 + \cdots\right] \qquad (8.189)$$

Finally,

$$S_{0x} - S_{fx} = C\frac{\bar{\bar{u}}_0^r}{R_{h0}^s}\left[\left(r\frac{\bar{\bar{u}}_1}{\bar{\bar{u}}_0} - s\frac{R_h'}{R_{h0}}\tilde{\eta}_1\right)\tau + \cdots\right] \qquad (8.190)$$

where we have assumed uniform flow in using $S_{0x} = S_{fx0}$. Also we need

$$\tilde{w} = \tilde{w}_0 + \tilde{w}'\tilde{\eta}_1\tau + \cdots \qquad A = A_0 + \tilde{w}_0\tilde{\eta}_1\tau + \cdots \qquad (8.191)$$

The perturbation equations are now substituted into (8.183) with the derivatives of $\bar{\bar{u}}_0$ and $\tilde{\eta}_0$ equal to zero, which conforms to uniform flow. Only those terms that are first order in τ are considered,

$$\left(\tilde{w}_0 + \tilde{w}'\tilde{\eta}_1\tau\right)\left(\bar{\bar{u}}_0 + c_0\right)\left(\tilde{\eta}_1 + 2\tilde{\eta}_2\tau\right)$$

$$+ \left(\bar{\bar{u}}_0 + \bar{\bar{u}}_1\tau\right)\left(\tilde{w}_0 + \tilde{w}'\tilde{\eta}_1\tau\right)\left(\frac{\partial\tilde{\eta}_1}{\partial\xi}\tau - \tilde{\eta}_1 - 2\tilde{\eta}_2\tau\right) \qquad (8.192)$$

$$+ \left(A_0 + \tilde{w}_0\tilde{\eta}_1\tau\right)\left(\frac{\partial\bar{\bar{u}}_1}{\partial\xi}\tau - \bar{\bar{u}}_1 - 2\bar{\bar{u}}_2\tau\right) = 0$$

$$\left(\bar{\bar{u}}_0 + c_0\right)\left(\bar{\bar{u}}_1 + 2\bar{\bar{u}}_2\tau\right) + \left(\bar{\bar{u}}_0 + \bar{\bar{u}}_1\tau\right)\left(\frac{\partial\bar{\bar{u}}_1}{\partial\xi}\tau - \bar{\bar{u}}_1 - 2\bar{\bar{u}}_2\tau\right)$$

$$+ g\left(\frac{\partial\tilde{\eta}_1}{\partial\xi}\tau - \tilde{\eta}_1 - 2\tilde{\eta}_2\tau\right) = gC\frac{\bar{\bar{u}}_0^r}{R_{h0}^s}\left(r\frac{\bar{\bar{u}}_1}{\bar{\bar{u}}_0} - s\frac{R_h'}{R_{h0}}\tilde{\eta}_1\right)\tau \qquad (8.193)$$

Equating terms without τ (i.e., τ^0) gives

$$\tilde{w}_0 c_0 \tilde{\eta}_1 = A\bar{\bar{u}}_1 \qquad c_0\bar{\bar{u}}_1 = g\tilde{\eta}_1 \qquad (8.194)$$

Equations (8.194) express the same result as (8.158), which can be shown for the special case of a rectangular channel by expanding the wave velocity

$$c = \sqrt{g\tilde{\eta}} = \sqrt{g\tilde{\eta}_0}\sqrt{1 + \frac{\tilde{\eta}_1}{\tilde{\eta}_0} + \frac{\tilde{\eta}_2}{\tilde{\eta}_0}\tau^2}$$

$$= \sqrt{g\tilde{\eta}_0}\left[1 + \frac{1}{2}\frac{\tilde{\eta}_1}{\tilde{\eta}_0}\tau - \frac{1}{4}\left(\frac{\tilde{\eta}_1}{\tilde{\eta}_0}\tau\right)^2 + \cdots\right] = c_0 + c_1\tau + c_2\tau^2 + \cdots$$

$$(8.195)$$

For the rectangular channel (8.194) gives

$$\bar{\bar{u}}_1 = \sqrt{g\tilde{\eta}_0}\frac{\tilde{\eta}_1}{\tilde{\eta}_0} \tag{8.196}$$

which is $\bar{\bar{u}}_1 = 2c_1$ using the definition of c_1 from (8.195). As in (8.158), $\omega = 2c_1$ for a rectangular channel.

Next, the terms of first order in τ of (8.192) and (8.193) are equated to obtain

$$2(1 + F_0)\frac{\partial\tilde{\eta}_1}{\partial\xi} - \left(3\frac{g}{c_0^2} - \frac{w'}{w_0}\right)\tilde{\eta}_1^2 + gC\frac{\bar{\bar{u}}_0^r}{R_{h0}^s}\frac{r}{c_0\bar{\bar{u}}_0}(1 - V_e)\tilde{\eta}_1 = 0 \quad (8.197)$$

To shorten the expressions, let

$$\varXi = \frac{3\dfrac{g}{c_0^2} - \dfrac{w'}{w_0}}{2(1 + F_0)} \qquad \varLambda = gC\frac{\bar{\bar{u}}_0^r\dfrac{r}{c_0\bar{\bar{u}}_0}(1 - V_e)}{2R_{h0}^s(1 + F_0)} \tag{8.198}$$

Then the solution to (8.197) is

$$\tilde{\eta}_1 = \frac{\kappa}{\kappa\dfrac{\varXi}{\varLambda} + e^{\varLambda\xi}} \tag{8.199}$$

in which κ is a constant of integration.

From the expansion of (8.184), $\tilde{\eta}_1$ is $\partial\tilde{\eta}/\partial\tau$ for $\tau = 0$. Thus, $\tilde{\eta}_1$ is the rate of rise of the water surface. Setting $\xi = 0$ in (8.199), we obtain the constant of integration

$$\kappa = \frac{\left(\dfrac{\partial\tilde{\eta}}{\partial\tau}\right)_0}{1 - \dfrac{\varXi}{\varLambda}\left(\dfrac{\partial\tilde{\eta}}{\partial\tau}\right)_0} \tag{8.200}$$

If $(\partial\tilde{\eta}/\partial\tau)_0$ is small, (8.199) becomes approximately

$$\tilde{\eta}_1 = \kappa e^{-\varLambda\xi} \tag{8.201}$$

In order that $\tilde{\eta}_1$ not grow exponentially, $\varLambda$ must be greater than zero, leading to the stability criterion that $V_e < 1$.

A measure of the distance that an instability requires in order to cause a substantial presence is the length between the disturbance and the point where the wave breaks. To find approximately the point where the wave breaks, we set the denominator of (8.199) equal to zero and use (8.200) to obtain

$$\xi_b = \frac{1}{\Lambda} \ln \frac{-\frac{\varXi}{\Lambda}\left(\frac{\partial \tilde{\eta}}{\partial \tau}\right)_0}{1 - \frac{\varXi}{\lambda}\left(\frac{\partial \tilde{\eta}}{\partial \tau}\right)_0} \tag{8.202}$$

Remember that $\Lambda < 0$ for unstable flow and that we have assumed that $V_e > 1$.

Although the distance ξ_b depends on $(\partial \tilde{\eta}/\partial \tau)_0$, it is insensitive to $(\partial \tilde{\eta}/\partial \tau)_0$. The logarithm, being the opposite of the exponential, varies slowly. For a small initial disturbance

$$\xi_b = -\frac{1}{\Lambda} \ln \epsilon \qquad \epsilon = -\frac{\frac{\varXi}{\Lambda}\left(\frac{\partial \tilde{\eta}}{\partial \tau}\right)_0}{1 - \frac{\varXi}{\Lambda}\left(\frac{\partial \tilde{\eta}}{\partial \tau}\right)_0} \tag{8.203}$$

Thus, we can choose a small value of ϵ and the value of ξ_b then depends on $-1/\Lambda \sim 1/(V_e - 1)$. Figure 8.31 shows such a curve (after Montouri, 1963; Montouri's curve is not exactly that of (8.203) because the approximation that he used was different, but it is close enough for all practical purposes). The experimental data agree as well as could be expected with the theoretical results. This curve can be used to determine if a breaking wave is likely to occur in any given case. Of course, the curve should not be applied to flows that vary considerably from uniform flow.

8.6.4 Breaking Wave Due to Rising Water

Combining expressions (8.199) and (8.200)

$$\tilde{\eta}_1 = \frac{\left(\frac{\partial \tilde{\eta}}{\partial \tau}\right)_0}{\frac{\varXi}{\Lambda}\left(\frac{\partial \tilde{\eta}}{\partial \tau}\right)_0 + \left[1 - \frac{\varXi}{\Lambda}\left(\frac{\partial \tilde{\eta}}{\partial \tau}\right)_0\right] e^{\Lambda \xi}} \tag{8.204}$$

Assuming that $\Lambda > 0$ (i.e., the Vedernikov number is in the stable range), the flow may still become unstable if $(\partial \tilde{\eta}/\partial \tau)_0$ is large enough. If

$$1 - \frac{\varXi}{\Lambda}\left(\frac{\partial \tilde{\eta}}{\partial \tau}\right)_0 > 0 \tag{8.205}$$

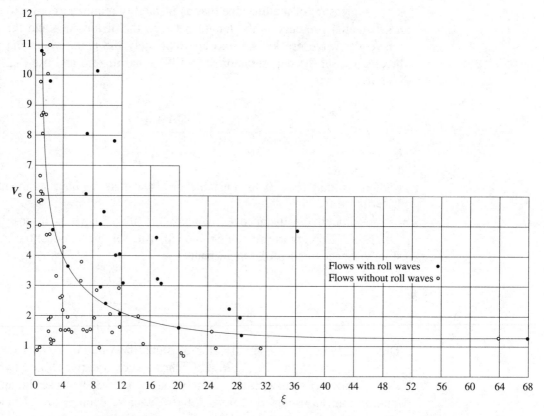

FIGURE 8.31

Distance required for bore formation (after C. Montouri, Discussion to "Stability aspects of flow in open channels," *Journal of the Hydraulics Division*, ASCE, vol. 89, no. HY4, March 1969, p. 270, reproduced by permission of ASCE).

the denominator is positive and must grow exponentially, leading to an exponential decrease in $\tilde{\eta}$. On the other hand, if

$$1 - \frac{\varXi}{\varLambda}\left(\frac{\partial \tilde{\eta}}{\partial \tau}\right)_0 < 0 \tag{8.206}$$

the denominator is positive for $\xi = 0$ but must pass through zero for some larger value of ξ. The stability condition is

$$\left(\frac{\partial \tilde{\eta}}{\partial \tau}\right)_0 < \frac{\varLambda}{\varXi} = gC\,\frac{\bar{\bar{u}}_0^r}{R_{ho}^s}\,\frac{\dfrac{r}{c_0\bar{\bar{u}}_0}(1 - V_e)}{3\dfrac{g}{c_0^2} - \dfrac{w'}{w_0}} \tag{8.207}$$

Using real time, the stability criterion in terms of the rate of rise of water at a point is

$$\left(\frac{\partial \tilde{\eta}}{\partial t}\right)_0 < \frac{\Lambda}{\Xi} = gC\frac{\bar{\bar{u}}_0^r}{R_{ho}^s}\frac{\dfrac{r}{c_0\bar{\bar{u}}_0}(1 - V_e)}{3\dfrac{g}{c_0^2} - \dfrac{w'}{w_0}}\left(\bar{\bar{u}}_0 + c_0\right) \qquad (8.208)$$

8.6.5 Roll Waves

Finally, we consider the physical manifestation of an instability. The flow may simply form a bore as in the case of rising water in an otherwise stable flow. Such bore development is not common in rivers and estuaries but does appear naturally due to rising tide in such estuaries as the Bay of Fundy and the Tsien-Tang River. The Tsien-Tang bore sometimes reaches heights as great as 6–9 meters (Stoker, 1957). Instabilities are common in very shallow flows along street surfaces, in gutters, and in pipes that are partially full.

For a high Vedernikov number, the flow may break down into a more complex arrangement known as *roll waves*. These waves consist of a series of bores separated by smooth water of variable depth as indicated in Fig. 8.32. The details of the development of roll waves are undoubtedly influenced by the depth of the otherwise steady flow. In Mayer's (1957) experiments the flow was very shallow and influenced by capillary effects.

The analysis of roll waves uses a coordinate system that is moving with the speed of the bore

$$\xi = x - u_J t \qquad (8.209)$$

The equations in a rectangular channel become

$$\left(\bar{\bar{u}} - u_J\right)\frac{d\tilde{\eta}}{d\xi} + \tilde{\eta}\frac{d\bar{\bar{u}}}{d\xi} = 0 \qquad \left(\bar{\bar{u}} - u_J\right)\frac{d\bar{\bar{u}}}{d\xi} + g\frac{d\tilde{\eta}}{d\xi} = g(S_{0x} - S_{fx}) \quad (8.210)$$

After integration, the first of (8.210) is

$$\left(\bar{\bar{u}} - u_J\right)\tilde{\eta} = \text{constant} = -\frac{Q_r}{w} \qquad (8.211)$$

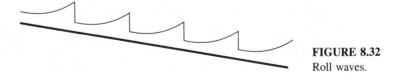

FIGURE 8.32
Roll waves.

in which Q_r is the flow rate relative to the moving coordinate system. Using (8.211) in the second of (8.210)

$$\frac{d\tilde{\eta}}{d\xi} = \frac{S_{0x} - S_{fx}}{1 - \dfrac{Q_r^2}{g\tilde{\eta}^3 w^2}} \tag{8.212}$$

which is the same as the steady flow profile equation except that Q_r is not the flow passing down the channel but the flow being overtaken by the bores.

Viewed from the moving coordinate system, the bores are stationary (hydraulic jumps) and the fluid is moving in the negative x-direction with speed $\bar{\bar{u}} = u_J$. Critical depth is

$$\eta_c = \frac{(u_J - u_c)^2}{g} \tag{8.213}$$

where u_c is the wave velocity or critical speed. The denominator of (8.212) is zero at critical depth, but since the depth must be greater than critical on the high side of the jumps and less than critical on the low side, the profile must pass smoothly—as observed in experiments—through critical depth between jumps. If the denominator passes through zero with a smooth profile, the numerator must be zero at the same point. The remainder of the analysis consists of finding the conditions so that $\partial\tilde{\eta}/\partial\xi$ is positive and finite at the point of critical depth. The derivatives of the denominator and numerator of (8.212) are

$$\frac{dD}{d\tilde{\eta}} = \frac{d}{d\tilde{\eta}}\left(1 - \frac{Q_r^2}{g\tilde{\eta}^3 w^2}\right) = 3\frac{Q_r^2}{g\tilde{\eta}^4 w^2} \tag{8.214}$$

$$\frac{dN}{d\tilde{\eta}} = \frac{d}{d\tilde{\eta}}\left[S_{0x} - C\frac{\left(u_J - \dfrac{Q_r}{\tilde{\eta}w}\right)^r}{R_h^s}\right] \tag{8.215}$$

$$= -C\frac{\left(u_J - \dfrac{Q_r}{A}\right)^r}{R_h^s}\frac{rQ_r}{\bar{\bar{u}}\tilde{\eta}A}\left(1 - \frac{r}{s}F\frac{A}{w}\frac{R_h'}{R_h}\right)$$

In (8.215) we have used $D = 0$ and $Q_r = A\sqrt{g\tilde{\eta}}$ at critical depth. In order that $dN/d\tilde{\eta} > 0$, which is the condition of a positive slope of the water surface between jumps, the opposite of the inequality (8.161) applies. Notice that the second term in the last set of parentheses of (8.215) is the Vedernikov number. Thus, the condition for the existence of periodic roll waves is identical to the condition for unstable (uniform) flow.

8.7 TWO-DIMENSIONAL, STEADY, SUPERCRITICAL FLOW

The study of supercritical flow in shallow water is very much analogous to the study of supersonic flow in a gas. The analogy was exploited by von Kármán (1938) and Preiswerk (1940). Some channels on steep slopes have been designed so that the water reaches supercritical velocities. Of course, most spillways and overflow structures are steep.

8.7.1 The Equations

Two-dimensional, steady flow uses (8.15), (8.33), and (8.34) where β is taken as unity, the time derivatives are discarded, and a friction slope is added,

$$\bar{u}_x \frac{\partial \eta}{\partial x} + \eta \frac{\partial \bar{u}_x}{\partial x} + \bar{u}_y \frac{\partial \eta}{\partial y} + \eta \frac{\partial \bar{u}_y}{\partial y} = 0 \tag{8.216}$$

$$\bar{u}_x \frac{\partial \bar{u}_x}{\partial x} + \bar{u}_y \frac{\partial \bar{u}_x}{\partial y} + g \frac{\partial \eta}{\partial x} = g(S_{0x} - S_{fx}) \tag{8.217}$$

$$\bar{u}_x \frac{\partial \bar{u}_y}{\partial x} + \bar{u}_y \frac{\partial \bar{u}_y}{\partial y} + g \frac{\partial \eta}{\partial y} = g(S_{0y} - S_{fy}) \tag{8.218}$$

To give these equations the characteristic treatment, the coefficients of the x- and y-derivatives are

$$[a] = \begin{bmatrix} \eta & 0 & \bar{u}_x \\ \bar{u}_x & 0 & g \\ 0 & \bar{u}_x & 0 \end{bmatrix} \qquad [b] = \begin{bmatrix} 0 & \eta & \bar{u}_y \\ \bar{u}_y & 0 & 0 \\ 0 & \bar{u}_y & g \end{bmatrix} \tag{8.219}$$

The coefficient determinant is

$$\det(a_{ij} - \lambda b_{ij}) = \begin{vmatrix} \eta & -\lambda \eta & \bar{u}_x - \lambda \bar{u}_y \\ \bar{u}_x - \lambda \bar{u}_y & 0 & g \\ 0 & \bar{u}_x - \lambda \bar{u}_y & -\lambda g \end{vmatrix} \tag{8.220}$$

Setting the determinant equal to zero, the roots are

$$\lambda_1 = \frac{\bar{u}_x}{\bar{u}_y} \qquad \lambda_2 = \frac{-\bar{u}_x \bar{u}_y - \sqrt{g\eta \left(\bar{u}_x^2 + \bar{u}_y^2 - g\eta\right)}}{-\bar{u}_y^2 + g\eta}$$

$$\lambda_3 = \frac{-\bar{u}_x \bar{u}_y + \sqrt{g\eta \left(\bar{u}_x^2 + \bar{u}_y^2 - g\eta\right)}}{-\bar{u}_y^2 + g\eta} \tag{8.221}$$

which form the characteristic directions as shown in Fig. 8.33. The first direction (λ_1) coincides with the streamline while the other two characteristics form an angle μ with the streamlines. These characteristics exist only in the case

$$\bar{u}_x^2 + \bar{u}_y^2 > g\eta = c^2 \qquad (8.222)$$

that is, the flow is supercritical and the equations are hyperbolic.

Consider, for example, a flow in the x-direction ($\bar{u}_y = 0$). The left- and right-going characteristics are

$$\frac{dx}{dy} = \pm\frac{\sqrt{\bar{u}_x^2 - c^2}}{c} \qquad (8.223)$$

or, inverting the derivative,

$$\frac{dy}{dx} = \pm\frac{c}{\sqrt{\bar{u}_x^2 - c^2}} = \pm\frac{1}{\sqrt{F^2 - 1}} = \pm\tan\mu \qquad (8.224)$$

in which F is the local Froude number. The later part of (8.224) also defines the angle μ (called the *Mach angle* in the analogous supersonic flow but unnamed in shallow water theory) as

$$\mu = \arctan\frac{1}{\sqrt{F^2 - 1}} = \frac{c}{\sqrt{\bar{u}_x^2 + \bar{u}_y^2 - c^2}} \qquad (8.225)$$

The Mach angle defines the limit of spreading of small disturbances. Referring to Fig. 8.34, a fluid particle that is in point A at time t_0 will travel a distance $\sqrt{\bar{u}_x^2 + \bar{u}_y^2}\,\Delta t$ in the time Δt. If a disturbance is created at point A and time t_0, that disturbance would spread with velocity c relative to the fluid in all directions. After time Δt, the disturbance would be felt within a circle of radius $c\Delta t$ about point B. The lines tangent to this circle and originating at point A form the Mach lines or characteristics from point A. Locations outside of the characteristics will never be affected by the small disturbance at A.

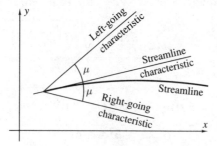

FIGURE 8.33
Characteristic directions for two-dimensional, supercritical flow.

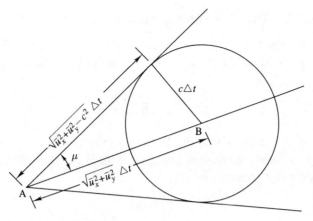

FIGURE 8.34
Spreading of small disturbances.

A suitable transformation matrix is

$$[t] = \begin{bmatrix} 0 & \lambda_1 & 1 \\ \lambda_2\bar{u}_y - \bar{u}_x & \eta & -\lambda_2\eta \\ \lambda_3\bar{u}_y - \bar{u}_x & \eta & -\lambda_3\eta \end{bmatrix} \qquad (8.226)$$

The equations in normal form are

$$\bar{u}_x\left(\lambda_1\frac{\partial}{\partial x} + \frac{\partial}{\partial y}\right)\bar{u}_x + \bar{u}_y\left(\lambda_1\frac{\partial}{\partial x} + \frac{\partial}{\partial y}\right)\bar{u}_y + g\left(\lambda_1\frac{\partial}{\partial x} + \frac{\partial}{\partial y}\right)\eta$$

$$= g[\lambda_1(S_{0x} - S_{fx}) - S_{fx}] \qquad (8.227)$$

$$\bar{u}_y\eta\left(\lambda_2\frac{\partial}{\partial x} + \frac{\partial}{\partial y}\right)\bar{u}_x - \bar{u}_x\eta\left(\lambda_2\frac{\partial}{\partial x} + \frac{\partial}{\partial y}\right)\bar{u}_y$$

$$- [\lambda_2 g\eta + \bar{u}_y(\bar{u}_x - \lambda_2\bar{u}_y)]\left(\lambda_2\frac{\partial}{\partial x} + \frac{\partial}{\partial y}\right)\eta = g\eta(S_{0x} - S_{fx} + \lambda_2 S_{fy})$$

$$(8.228)$$

$$\bar{u}_y\eta\left(\lambda_3\frac{\partial}{\partial x} + \frac{\partial}{\partial y}\right)\bar{u}_x - \bar{u}_x\eta\left(\lambda_3\frac{\partial}{\partial x} + \frac{\partial}{\partial y}\right)\bar{u}_y$$

$$- [\lambda_3 g\eta + \bar{u}_y(\bar{u}_x - \lambda_3\bar{u}_y)]\left(\lambda_3\frac{\partial}{\partial x} + \frac{\partial}{\partial y}\right)\eta = g\eta(S_{0x} - S_{fx} + \lambda_3 S_{fy})$$

$$(8.229)$$

The differential operators of (8.227), (8.228), and (8.229) differentiate the dependent variables along the streamline, the right-going characteristic, and the left-going characteristic, respectively.

8.7.2 The Frictionless Solution

The equations are somewhat simpler if slope and friction are neglected $(S_{0x} = S_{0y} = S_{fx} = S_{fy} = 0)$. In that case (8.227) becomes

$$\left(\lambda_1 \frac{\partial}{\partial x} + \frac{\partial}{\partial y}\right)\left(\frac{\bar{u}_x^2 + \bar{u}_y^2}{2} + g\eta\right) = 0 \tag{8.230}$$

which indicates that the Bernoulli sum is constant along a streamline. If, in addition, the flow is irrotational, then the Bernoulli constant is the same for every streamline

$$\frac{\bar{u}_x^2 + \bar{u}_y^2}{2} + g\eta = \text{constant} \tag{8.231}$$

Solving for η and taking the derivative in an arbitrary direction, say σ, gives

$$\frac{d\eta}{d\sigma} = -\frac{1}{g}\left(\bar{u}_x \frac{d\bar{u}_x}{d\sigma} + \bar{u}_y \frac{d\bar{u}_y}{d\sigma}\right) \tag{8.232}$$

Using (8.232) in (8.228) and (8.229)—with zero right-hand sides—to eliminate the derivatives of η produces

$$\left(c\bar{u}_y + \bar{u}_x\sqrt{\bar{u}_x^2 + \bar{u}_y^2 - c^2}\right)\left(\lambda_2 \frac{\partial}{\partial x} + \frac{\partial}{\partial y}\right)\bar{u}_x$$

$$-\left(c\bar{u}_x - \bar{u}_y\sqrt{\bar{u}_x^2 + \bar{u}_y^2 - c^2}\right)\left(\lambda_2 \frac{\partial}{\partial x} + \frac{\partial}{\partial y}\right)\bar{u}_y = 0 \tag{8.233}$$

$$\left(c\bar{u}_y - \bar{u}_x\sqrt{\bar{u}_x^2 + \bar{u}_y^2 - c^2}\right)\left(\lambda_3 \frac{\partial}{\partial x} + \frac{\partial}{\partial y}\right)\bar{u}_x$$

$$-\left(c\bar{u}_x + \bar{u}_y\sqrt{\bar{u}_x^2 + \bar{u}_y^2 - c^2}\right)\left(\lambda_3 \frac{\partial}{\partial x} + \frac{\partial}{\partial y}\right)\bar{u}_y = 0 \tag{8.234}$$

Equations (8.233) and (8.234) can be reduced to invariant form by defining

$$\theta = \arctan \frac{\bar{u}_y}{\bar{u}_x} \qquad \nu = \int_0^{\sqrt{\bar{u}_x^2 + \bar{u}_y^2}} \frac{\cot \mu}{\sigma} d\sigma \tag{8.235}$$

where θ is the angle the streamline makes with the x-axis and ν is called the *Prandtl-Meyer function* in the analogous gas dynamics problem. ν is a function of the Froude number. The functions θ and ν become new dependent variables replacing $\bar{u}_x$ and $\bar{u}_y$. The resulting equations are

$$\left(\lambda_2 \frac{\partial}{\partial x} + \frac{\partial}{\partial y}\right)(\nu + \theta) = 0 \qquad \left(\lambda_3 \frac{\partial}{\partial x} + \frac{\partial}{\partial y}\right)(\nu - \theta) = 0 \quad (8.236)$$

Thus $\nu + \theta$ is constant along the right-going characteristic while $\nu - \theta$ is constant along the left-going characteristic.

The function ν can be expressed in terms of the local Froude number. Using the total velocity as $\bar{u} = \sqrt{\bar{u}_x^2 + \bar{u}_y^2}$,

$$\nu = \int \sqrt{F^2 - 1} \, \frac{d\bar{u}}{\bar{u}} \qquad (8.237)$$

Since $\bar{u} = cF$

$$\frac{d\bar{u}}{\bar{u}} = \frac{dc}{c} + \frac{dF}{F} \qquad (8.238)$$

The wave speed is expressed in term(s) of the Froude number by using the Bernoulli equation written as $\bar{u}^2/2 + c^2 = c_0^2$

$$\frac{dc}{c} = -\frac{F^2}{F^2 + 2} \frac{dF}{F} \qquad (8.239)$$

Finally, using these quantities in (8.237) yields

$$\nu = \int \frac{2\sqrt{F^2 - 1}}{F^2 + 2} \frac{dF}{F} = \sqrt{3} \arctan \sqrt{\frac{F^2 - 1}{3}} - \arctan \sqrt{F^2 - 1} \quad (8.240)$$

The constant of integration has been chosen so that $\nu = 0$ at $F = 1$. For problem solving, F is most conveniently expressed as a function of ν, which can be obtained in the form of a plot, in a table, or from solving (8.240) as a transcendental equation.

8.7.3 Simple Waves

Consider Fig. 8.35, which shows a flow initially parallel to the x-axis and tangent to a wall. At the point PC (point of curvature) the wall curves away from the flow and the wall again becomes straight at the point PT (point of tangency). We seek simple wave solutions with straight left-going characteristics as shown. Since we postulate that the left-going characteristics are straight, and since $\nu - \theta = $ constant along these characteristics, both ν and θ are individually constant along the left-going characteristics. All right-going characteristics originate in the uniform zone to the left so that

$$\nu + \theta = \nu_1 \qquad (8.241)$$

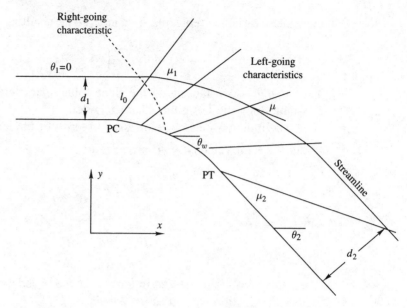

FIGURE 8.35
Flow around a curve.

holds for the entire flow. The velocity must be tangent to the wall at all points, so for a point on the wall

$$\nu = \nu_1 - \theta_w \tag{8.242}$$

where θ_w is the value of θ on the wall. As shown in the figure, θ decreases in the downstream direction and thus ν increases.

These equations are sufficient to solve the problem. For any point on the wall, ν can be found from (8.242); then F is found from (8.240). The straight characteristics can be plotted using μ as in (8.225) and θ_w. Since the dependent variables are constant along the straight characteristics, the problem is solved.

To plot the streamlines around the curve, consider first the case where the approach Froude number is unity. Using the subscript c to denote critical condition, conservation of mass gives (see Fig. 8.36 for notation)

$$Q = \bar{u}_c d_c \eta_c = \bar{u} l \eta \sin \mu \tag{8.243}$$

From the Bernoulli equation

$$\frac{\bar{u}^2}{2} + c^2 = \frac{\bar{u}_c^2}{2} + c_c^2 = \frac{3}{2} c_c^2 \tag{8.244}$$

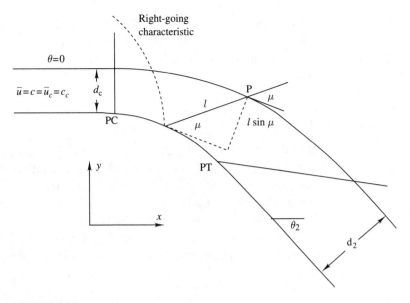

FIGURE 8.36
Flow around a curve beginning with critical flow.

which leads to

$$\frac{c_c^2}{c^2} = \frac{F+2}{3} \qquad \frac{\bar{u}_c^2}{\bar{u}^2} = \frac{F^2+2}{3F^2} \tag{8.245}$$

Using (8.243) and (8.245) with $c^2 = g\eta$ gives

$$\frac{l}{d_c} = \sqrt{\frac{F^2+2}{3F^2}} \frac{F^2+2}{3} \frac{1}{\sin\mu} = \left(\frac{F^2+2}{3}\right)^{3/2} \tag{8.246}$$

in which F is the Froude number at point P of Fig. 8.36. If the curve does not begin with critical conditions, the ratio of length l_0—along the characteristic from the wall to the streamline (Fig. 8.35)—to the variable length l is

$$\frac{l}{l_0} = \frac{l/d_c}{l_0/d_c} = \left(\frac{F^2+2}{F_0^2+2}\right)^{3/2} \tag{8.247}$$

in which F_0 is the Froude number of the approaching flow. Streamlines can now be plotted after a number of characteristics are drawn and the ratio l/l_0 is calculated.

In principle, a supercritical channel could be designed by replacing one of the outer streamlines by a wall. In that way the flow would pass through the curve smoothly without waves due to the curvature. In practice there is the difficulty that the designed geometry would be valid for only one approach Froude number and flow at other Froude numbers would have waves.

8.8 DIAGONAL JUMPS

If in Fig. 8.35 the wall curves into the flow, the characteristics will intersect instead of forming a fan; there will be a discontinuous solution just as in the analogous case of one-dimensional, unsteady flow. Consider the case where the wall bends into the flow suddenly as in Fig. 8.37. A diagonal jump will begin at the corner as shown.

To analyze the flow, consider $\bar{u}_n$ as the component of velocity normal to the jump and $\bar{u}_s$ as the component parallel to the jump. From conservation of mass

$$\bar{u}_{1s} = \bar{u}_{2s} = \bar{u}_s \qquad \bar{u}_{1n}\eta_1 = \bar{u}_{2n}\eta_2 \qquad (8.248)$$

Momentum is automatically conserved in the direction parallel to the jump since forces on the end areas of the control volume (Fig. 8.37) along the jump cancel each other. In the direction normal to the jump

$$\frac{\rho g}{2}\left(\eta_1^2 - \eta_2^2\right) = \rho\left(-\bar{u}_{1n}^2\eta_1 + \bar{u}_{2n}^2\eta_2\right) \qquad (8.249)$$

From the geometry of the figure

$$\bar{u}_{1n} = \bar{u}_1 \sin\beta \qquad \bar{u}_{2s} = \bar{u}_2 \sin(\beta - \theta) \qquad \bar{u}_{1n} = \bar{u}_s \tan\beta$$

$$\bar{u}_{2n} = \bar{u}_s \tan(\beta - \theta) \qquad (8.250)$$

After algebraic manipulation, these equations yield the following useful

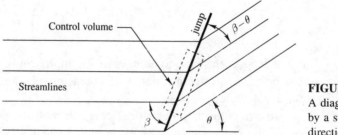

Control volume

jump

$\beta - \theta$

Streamlines

β

θ

FIGURE 8.37
A diagonal jump formed by a sudden change in direction of the flow.

relationships:

$$\frac{\eta_2}{\eta_1} = \frac{1}{2}\left(\sqrt{1 + 8F_1^2\sin^2\beta} - 1\right) \qquad \frac{\bar{u}_{1n}}{\bar{u}_{2n}} = \frac{\eta_2}{\eta_1} = \frac{\tan\beta}{\tan(\beta-\theta)}$$

$$F_2 = F_1\frac{\sin\beta}{\sin(\beta-\theta)}\left(\frac{\tan(\beta-\theta)}{\tan\beta}\right)^{3/2}$$

$$\tan\theta = \frac{\tan\beta\left(\sqrt{1 + 8F_1^2\sin^2\beta} - 3\right)}{2\tan^2\beta + \sqrt{1 + 8F_1^2\sin^2\beta} - 1} \qquad (8.251)$$

The last of (8.251) is shown in graphical form in Fig. 8.38.

From Fig. 8.38 we see that for any value of $\theta < \theta_{max}$ there are two possible solutions, each having different values of β. The larger value of β gives *strong jumps* with the flow downstream from the jump being sub-critical; the smaller value leads to *weak jumps* where the downstream flow is supercritical with the exception of the small region of β values between the dashed line ($F_2 = 1$) and the maximum of the curves. For weak jumps, the wall deflection angle dictates the angle θ. For, strong jumps the sub-critical downstream flow is dependent on conditions further downstream;

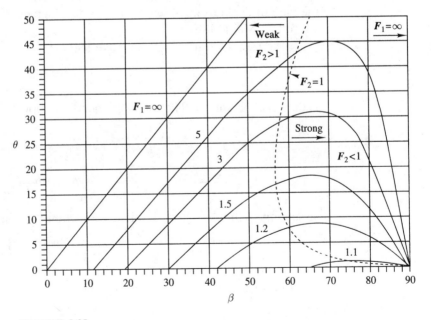

FIGURE 8.38
The diagonal jump relationships.

disturbances can be propagated upstream and thus conditions further down-stream have an effect on the jump. In the latter case the upstream Froude number and wall angle may not uniquely determine the solution.

Consider the case where $\eta_2/\eta_1 \to 1$, that is, the limit of a disappearing jump. The first of (8.251) leads to

$$\sin \beta = \frac{1}{F} \tag{8.252}$$

which indicates that in the limit of a small jump, the jump line coincides with a characteristic; that is, $\beta = \mu$ [see (8.224)].

8.9 BOUSSINESQ EQUATIONS

Shallow water theory is defined by the hydrostatic condition. If the vertical pressure distribution departs from hydrostatic—if there is significant vertical acceleration—calculations using shallow water theory lead to error. The Boussinesq equations form a first-order correction to classical shallow water theory in which the pressure is not far from hydrostatic but can depart somewhat. Although the result will not predict the short waves that are used by surfers, it does form a correction and will lead to the solitary wave. Stoker (1957) develops shallow water theory in a systematic way where the equations are written in terms of powers of a small parameter, the ratio of depth to wave length. In our development of the approximation, we follow Mei (1989).

The ratio of water depth to wave length,

$$\varepsilon = \frac{\eta_0}{L} \tag{8.253}$$

is the small parameter. The following nondimensional variables are defined:

$$x_* = \frac{x}{L} \qquad y_* = \frac{y}{L} \qquad z_* = \frac{z}{\eta_0}$$

$$t_* = t\frac{\sqrt{g\eta_0}}{L} \qquad \eta_* = \frac{\eta}{\eta_0} \qquad \Phi^* = \Phi\frac{1}{L\sqrt{g\eta_0}} \tag{8.254}$$

Assuming frictionless flow, Laplace's equation is valid

$$\varepsilon^2 \left(\frac{\partial^2 \Phi^*}{\partial x_*^2} + \frac{\partial^2 \Phi^*}{\partial y_*^2} \right) + \frac{\partial^2 \Phi^*}{\partial z_*^2} = 0 \tag{8.255}$$

The free surface condition (8.7) becomes

$$\varepsilon^2 \left(\frac{\partial \eta_*}{\partial t_*} - \frac{\partial \Phi^*}{\partial x_*}\frac{\partial \eta_*}{\partial x_*} - \frac{\partial \Phi^*}{\partial y_*}\frac{\partial \eta_*}{\partial y_*} \right) = -\frac{\partial \Phi_*}{\partial z_*} \qquad \text{on} \quad z_* = \eta_* \tag{8.256}$$

The pressure condition (the Bernoulli relation) is

$$\varepsilon^2 \left(-\frac{\partial \Phi^*}{\partial t_*} + \eta^* \right) + \frac{1}{2} \left\{ \varepsilon^2 \left[\left(\frac{\partial \Phi^*}{\partial x_*} \right)^2 + \left(\frac{\partial \Phi^*}{\partial y_*} \right)^2 \right] + \left(\frac{\partial \Phi^*}{\partial z_*} \right)^2 \right\} = 0$$

on $\quad z_* = \eta_*$

(8.257)

and finally we assume a flat bottom

$$\frac{\partial \Phi^*}{\partial z_*} = 0 \qquad \text{on} \quad z_* = 0 \qquad\qquad (8.258)$$

The potential is expanded into a series in the vertical coordinate so that

$$\Phi^*(x_*, y_*, z_*, t_*) = \sum_{n=0}^{\infty} z_*^n \Phi_n(x_*, y_*, t_*) \qquad (8.259)$$

The horizontal derivatives of the potential are

$$\frac{\partial \Phi^*}{\partial x_i^*} = \sum_{n=0}^{\infty} z_*^n \frac{\partial \Phi_n}{\partial x_i^*} \quad i = 1, 2 \qquad \frac{\partial^2 \Phi^*}{\partial x_i^* \partial x_i^*} = \sum_{n=0}^{\infty} z_*^n \frac{\partial^2 \Phi_n}{\partial x_i^* \partial x_i^*} \qquad (8.260)$$

and the vertical derivatives are

$$\frac{\partial \Phi^*}{\partial z_*} = \sum_{n=0}^{\infty} n z_*^{n-1} \Phi_n = \sum_{n=0}^{\infty} (n+1) z_*^n \Phi_{n+1}$$

$$\frac{\partial^2 \Phi}{\partial z_*^2} = \sum_{n=0}^{\infty} n(n+1) z_*^{n-1} \Phi_{n+1} = \sum_{n=0}^{\infty} (n+2)(n+1) z_*^n \Phi_{n+2}$$

(8.261)

The middle part of the latter equations is zero in the first term ($n = 0$) and thus we can substitute $n + 1$ for n.

Using these derivatives in (8.255)

$$\varepsilon^2 \left(\frac{\partial^2 \Phi^*}{\partial x_*^2} + \frac{\partial^2 \Phi^*}{\partial y_*^2} \right) + \frac{\partial^2 \Phi^*}{\partial z_*^2}$$

$$= \sum_{n=0}^{\infty} z_*^n \left[\varepsilon^2 \left(\frac{\partial^2 \Phi^*}{\partial x_*} + \frac{\partial^2 \Phi^*}{\partial y_*} \right) + (n+1)(n+2) \Phi_{n+2} \right] = 0$$

(8.262)

Since this equation must hold for any value of z_* (within the depth), the quantity in brackets must be zero, leading to a recursive formula

$$\Phi_{n+2} = -\frac{\varepsilon^2}{(n+1)(n+2)} \left(\frac{\partial^2 \Phi_n}{\partial x_*^2} + \frac{\partial^2 \Phi_n}{\partial y_*^2} \right) \qquad (8.263)$$

On the solid bottom, $z_* = 0$ and all the terms except the first of (8.259) are zero, but the first term must also be zero from (8.258); thus $\Phi_1 \equiv 0$. Using this result in the recursive formula means that all the Φ with odd subscripts are zero,

$$\Phi_1 \equiv \Phi_3 \equiv \Phi_5 \equiv \cdots \equiv 0 \tag{8.264}$$

For the evenly subscripted Φ,

$$\Phi_2 = -\frac{\varepsilon^2}{2 \cdot 1} \nabla_h^2 \Phi_0 \qquad \Phi_4 = -\frac{\varepsilon^2}{4 \cdot 3} \nabla_h^2 \Phi_2 = \frac{\varepsilon^4}{24} \nabla_h^2 \nabla_h^2 \Phi_0 \tag{8.265}$$

where ∇_h is the horizontal (dimensionless) gradient. Equation (8.259) becomes

$$\Phi^* = \Phi_0 - \frac{\varepsilon^2}{2} z_*^2 \nabla_h^2 \Phi_0 + \frac{\varepsilon^4}{24} z_*^4 \nabla_h^2 \nabla_h^2 \Phi_0 + \mathrm{O}(\varepsilon^6) \tag{8.266}$$

Using the series in the boundary conditions (8.256) and (8.257)

$$\varepsilon^2 \left[\frac{\partial \eta_*}{\partial t_*} - \nabla_h \eta_* \cdot \left(\nabla_h \Phi_0 - \frac{\varepsilon^2}{2} \eta_*^2 \nabla_h^2 \nabla_h \Phi_0 \right) \right]$$

$$= \varepsilon^2 \eta_* \nabla_h^2 \Phi_0 - \frac{\varepsilon^4}{6} \eta_*^3 \nabla_h^2 \nabla_h^2 \Phi_0 + \mathrm{O}(\varepsilon^6) \tag{8.267}$$

$$- \varepsilon^2 \left(\frac{\partial \Phi_0}{\partial t_*} - \frac{\varepsilon^2}{2} \eta_*^2 \nabla_h^2 \frac{\partial \Phi_0}{\partial t} - \eta_* \right)$$

$$+ \frac{\varepsilon^2}{2} \left[(\nabla_h \Phi_0)^2 - \varepsilon^2 \eta_*^2 \nabla_h \Phi_0 \cdot \nabla_h^2 (\nabla_h \Phi_0) \right] + \frac{\varepsilon^4}{2} \eta_*^2 \left(\nabla_h^2 \Phi_0 \right)^2 = \mathrm{O}(\varepsilon^6) \tag{8.268}$$

where η_* has replaced z_* since the equation is to be evaluated on the free surface. The zero order velocity is $\vec{u}_0 = -\nabla_h \Phi_0$. Using this quantity in (8.267)

$$\frac{\partial \eta_*}{\partial t_*} + \nabla_h \eta_* \cdot \left(\vec{u}_0 - \frac{\varepsilon^2}{2} \eta_*^2 \nabla_h^2 \vec{u}_0 \right) + \eta_* \nabla_h \cdot \vec{u}_0 - \frac{\varepsilon^2}{6} \eta_*^3 \nabla_h^2 (\nabla \cdot \vec{u}_0) = \mathrm{O}(\varepsilon^4) \tag{8.269}$$

Taking the gradient of (8.268)

$$\frac{\partial \vec{u}_0}{\partial t} + \vec{u}_0 \cdot \nabla_h \vec{u}_0 + \nabla_h \eta_*$$

$$+ \frac{\varepsilon^2}{2} \nabla_h \left\{ \eta_*^2 \left[(\nabla_h \cdot \vec{u}_0)^2 - \vec{u}_0 \cdot \nabla_h^2 \vec{u}_0 - \nabla_h \cdot \frac{\partial \vec{u}_0}{\partial t} \right] \right\} = \mathrm{O}(\varepsilon^4) \tag{8.270}$$

The actual velocity is obtained by differentiating (8.266)

$$u_x^* = -\frac{\partial \Phi^*}{\partial x_*} = u_{0x} - \frac{\varepsilon^2}{2} z_*^2 \frac{\partial}{\partial x_*} \nabla_h \cdot \vec{u}_0 + O(\varepsilon^4)$$

$$u_y^* = -\frac{\partial \Phi^*}{\partial y_*} = u_{0y} - \frac{\varepsilon^2}{2} z_*^2 \frac{\partial}{\partial y_*} \nabla_h \cdot \vec{u}_0 + O(\varepsilon^4) \qquad (8.271)$$

$$u_z^* = -\frac{\partial \Phi^*}{\partial z_*} = \varepsilon^2 z_* \nabla_h \cdot \vec{u}_0 + O(\varepsilon^4)$$

To form the vertically averaged horizontal velocity, we take the integral of the first two of these equations

$$\bar{u}_x^* = -\frac{1}{\eta_*} \int_0^{\eta_*} \frac{\partial \Phi^*}{\partial x_*} dz_* = \frac{1}{\eta_*} \int_0^{\eta_*} \left(u_{0x} - \frac{\varepsilon^2}{2} z_*^2 \frac{\partial}{\partial x_*} \nabla_h \cdot u_{0x} + O(\varepsilon^4) \right) dz_*$$

$$= u_{0x} - \frac{\varepsilon^2}{6} \eta_*^2 \nabla_h^2 u_{0x} + O(\varepsilon^4)$$

$$\bar{u}_y^* = -\frac{1}{\eta_*} \int_0^{\eta_*} \frac{\partial \Phi^*}{\partial y_*} dz_* = \frac{1}{\eta_*} \int_0^{\eta_*} \left(u_{0y} - \frac{\varepsilon^2}{2} z_*^2 \frac{\partial}{\partial y_*} \nabla_h \cdot u_{0y} + O(\varepsilon^4) \right) dz_*$$

$$= u_{0y} - \frac{\varepsilon^2}{6} \eta_*^2 \nabla_h^2 u_{0y} + O(\varepsilon^4)$$

$$(8.272)$$

The inversion of (8.272) to the stated accuracy is

$$u_{0x} = \bar{u}_x^* + \frac{\varepsilon^2}{6} \eta_*^2 \nabla_h^2 \bar{u}_x^2 + O(\varepsilon^4) \qquad u_{0y} = \bar{u}_y^* + \frac{\varepsilon^2}{6} \eta_*^2 \nabla_h^2 \bar{u}_y + O(\varepsilon^4) \quad (8.273)$$

Using (8.273) in (8.269) gives the depth-averaged equation of continuity (8.15) in dimensionless form. A depth-averaged equation of motion results from the substitution of (8.273) into (8.270),

$$\frac{\partial \bar{u}_x^*}{\partial t_*} + \bar{u}_x^* \frac{\partial \bar{u}_x^*}{\partial x_*} + \bar{u}_y^* \frac{\partial \bar{u}_x^*}{\partial y_*} + g \frac{\partial \eta_*}{\partial x_*} + \frac{\varepsilon^2}{6} \frac{\partial}{\partial t_*} \left(\eta_*^2 \nabla_h^2 \bar{u}_x^* \right)$$

$$+ \frac{\varepsilon^2}{6} \frac{\partial}{\partial x_*} \left\{ \eta_*^2 \left[3 \left(\frac{\partial \bar{u}_x^*}{\partial x_*} + \frac{\partial \bar{u}_y^*}{\partial y_*} \right)^2 - 2\bar{u}_x^* \nabla_h^2 \bar{u}_x^* - 2\bar{u}_y^* \nabla_h^2 \bar{u}_y^* \right. \qquad (8.274)$$

$$\left. - 3 \frac{\partial}{\partial t_*} \left(\frac{\partial \bar{u}_x^*}{\partial x_*} + \frac{\partial \bar{u}_y^*}{\partial y_*} \right) \right] \right\} = O(\varepsilon^4)$$

$$\frac{\partial \bar{u}_y^*}{\partial t_*} + \bar{u}_x^* \frac{\partial \bar{u}_x^*}{\partial x_*} + \bar{u}_y^* \frac{\partial \bar{u}_y^*}{\partial y_*} + g \frac{\partial \eta_*}{\partial y_*} + \frac{\varepsilon^2}{6} \frac{\partial}{\partial t_*} \left(\eta_*^2 \nabla_h^2 \bar{u}_y^* \right)$$

$$+ \frac{\varepsilon^2}{6} \frac{\partial}{\partial y_*} \left\{ \eta_*^2 \left[3 \left(\frac{\partial \bar{u}_x^*}{\partial x_*} + \frac{\partial \bar{u}_y^*}{\partial y_*} \right)^2 - 2\bar{u}_x^* \nabla_h^2 \bar{u}_x^* - 2\bar{u}_y^* \nabla_h^2 \bar{u}_y^* \right. \right. \tag{8.275}$$

$$\left. \left. - 3 \frac{\partial}{\partial t_*} \left(\frac{\partial \bar{u}_x^*}{\partial x_*} + \frac{\partial \bar{u}_y^*}{\partial y_*} \right) \right] \right\} = O(\varepsilon^4)$$

Neglecting terms of order ε^2 leads directly to the frictionless shallow water equations. In dimensional form (8.274) and (8.275) contain correction terms to the hydrostatic approximation

$$\frac{\partial \bar{u}_i}{\partial t} + \bar{u}_j \frac{\partial \bar{u}_i}{\partial x_i} + g \frac{\partial \eta}{\partial x_i} + \frac{1}{6} \left\{ \frac{\partial}{\partial t} \left(\eta^2 \frac{\partial \bar{u}_i}{\partial x_j \partial x_j} \right) \right.$$

$$\left. + \frac{\partial}{\partial x_i} \left[3\eta^2 \left(\frac{\partial \bar{u}_j}{\partial x_j} \right)^2 - 2\eta^2 \bar{u}_j \frac{\partial^2 \bar{u}_j}{\partial x_k \partial x_k} - 3\eta^2 \frac{\partial}{\partial t} \frac{\partial \bar{u}_j}{\partial x_j} \right] \right\} = 0 \tag{8.276}$$

Up to this point no restriction has been placed on the amplitude. The equation can be simplified somewhat by assuming a small amplitude. Let

$$\eta_* = 1 + \eta_*' = 1 + \delta \eta_{**} \qquad \delta = \frac{a}{\eta_0} < 1 \tag{8.277}$$

where a is the wave amplitude. We can treat the velocity the same way, $u_i = u_i^0 + u'$, but since there is no friction, no generality is lost by taking $u_i^0 = 0$ and dropping the prime. Without approximation the equation of continuity is

$$\delta \frac{\partial \eta_{**}}{\partial t_*} + \frac{\partial}{\partial x_i^*} \left[\bar{u}_i^*(1 + \delta \eta_{**}) \right]$$

$$= \delta \frac{\partial \eta_{**}}{\partial t_*} + \frac{\partial \bar{u}_i^*}{\partial x_i^*} + \delta \bar{u}_i^* \frac{\partial \eta_{**}}{\partial x_i^*} + \delta \eta_{**} \frac{\partial \bar{u}_i^*}{\partial x_i^*} = 0 \tag{8.278}$$

We conclude that $\bar{u}_i^*$ is of the order of δ and use that fact to eliminate terms in (8.274) and (8.275). In those equations terms of order ϵ^2, $\epsilon^2\delta$, and δ^2 are retained and terms of the order $\epsilon^2\delta^2$ are discarded. The result, written in dimensional variables, is

$$\frac{\partial \bar{u}_i}{\partial t} + \bar{u}_j \frac{\partial \bar{u}_i}{\partial x_j} + g \frac{\partial \eta}{\partial x_i} - \frac{\eta_0}{3} \frac{\partial^3 \bar{u}_j}{\partial x_i \partial x_j \partial t} = 0 \tag{8.279}$$

The latter expressions form the *Boussinesq equations* of shallow water theory (which should not be confused with the Boussinesq approximations in stratified flow—see Sec. 9.9—or in porous media). The pressure is given by

$$p = \rho g(\eta - z) + \rho \left(z\eta_0 + \frac{z^2}{2} \right) \frac{\partial^2 \bar{u}_i}{\partial t \partial x_i} \qquad (8.280)$$

which is no longer hydrostatic for unsteady flow. If (8.276) is used, the pressure correction is retained for steady flow.

Consider the one-dimensional, linearized equations (products of u and its derivatives neglected and the product ηu neglected). Continuity and motion give

$$\frac{\partial \eta}{\partial t} + \eta_0 \frac{\partial \bar{u}}{\partial x} = 0 \qquad \frac{\partial \bar{u}}{\partial t} + g \frac{\partial \eta}{\partial x} - \eta_0^3 \frac{\partial^3 \bar{u}}{\partial t \partial x^2} = 0 \qquad (8.281)$$

Assuming an oscillatory wave

$$\eta = ae^{i(kx - \sigma t)} \qquad \bar{u} = Ue^{i(kx - \sigma t)} \qquad (8.282)$$

Using these expressions in (8.281) and canceling the exponential terms gives

$$-i\sigma a + ik\eta_0 U = 0 \qquad -i\sigma U + ikga - \frac{\eta_0}{3}(ik)^2(-i\sigma)U = O \quad (8.283)$$

which can be considered as two equations for a and U. To avoid the trivial solution that these quantities are zero, the determinant of the coefficient matrix must be zero,

$$\begin{vmatrix} -i\sigma & ik\eta_0 \\ igk & -i\sigma \left(1 + \dfrac{k^2 \sigma^2}{3} \right) \end{vmatrix} = 0 \qquad (8.284)$$

The phase velocity is

$$c = \frac{\sigma}{k} = \frac{\sqrt{g\eta_0}}{1 + \dfrac{k^2 \eta_0^2}{3}} \qquad (8.285)$$

which shows a dependence on wave length.

Historical Note: Joseph Boussinesq (1842–1929)—born the same year as Osborne Reynolds—received a doctorate from the Faculté des Sciences in Paris in 1867, later to become a professor at the Sorbonne. He primarily studied flow in pipes and channels, but his interest in science was very broad.

Although he had a number of publications, his monumental, 700 page book *Essai sur la théorie des eaux courantes* was by far the most famous. In that book the equation for wave velocity appears as

$$c = \sqrt{g\eta_0}\left(1 + \frac{3}{4}\frac{\Delta\eta}{\eta_0} + \frac{\eta_0^2}{6\Delta\eta}\frac{d^2(\Delta\eta)}{dx^2}\right)$$

which indicates a correction for surface curvature. Boussinesq was able to derive the shape and speed of a solitary wave.

PROBLEMS

8.1. For the general approximations of (8.276) show that the pressure is given by

$$p = \rho g(\eta - z)$$

$$-\frac{\eta^2 - z^2}{2}\left[\frac{\partial^2 \bar{u}_i}{\partial t \partial x_i} + \bar{u}_i\frac{\partial^2 \bar{u}_i}{\partial x_j \partial x_j} - \left(\frac{\partial \bar{u}_i}{\partial x_i}\right)^2\right]$$

Show that this expression for pressure reduces to (8.280) for the Boussinesq approximations.

8.2. Water is two meters deep in a rectangular channel that is 200 m wide. If the rate of inflow at the upstream end is abruptly increased from zero to 28,000 m³/s, what will be the height and the speed of the resulting surge?

8.3. The equations for one-dimensional flow of gas in a pipe are

$$\frac{\partial \rho}{\partial t} + \frac{\partial}{\partial x}(u\rho) = 0$$

$$\rho\left(\frac{\partial u}{\partial t} + u\frac{\partial u}{\partial x}\right) + c^2\frac{\partial \rho}{\partial x} + f(\rho, u) = 0$$

where ρ is density, c is the wave propagation speed (the speed of sound), and u is velocity.
(a) Derive the equations of characteristics.
(b) Define subsonic and supersonic flow and state the boundary conditions that would apply in each case.

8.4. Consider the shallow water equations

$$\frac{\partial u}{\partial t} + u\frac{\partial u}{\partial x} + g\frac{\partial \eta}{\partial x} = 0$$

$$\frac{\partial \eta}{\partial t} + u\frac{\partial \eta}{\partial x} + \eta\frac{\partial u}{\partial x} = 0$$

For small amplitude waves the equations can be linearized as follows. Let

$$\eta = \eta_0 + \eta' \qquad u = u_0 + u'$$

where η_0 and u_0 are constants and it is assumed that $\eta'/\eta_0 << 1$ and $u'/u_0 << 1$. We also assume that the derivatives are small and neglect the products of small quantities. The equations then become

$$\frac{\partial u'}{\partial t} + u_0\frac{\partial u'}{\partial x} + g\frac{\partial \eta'}{\partial x} = 0$$

$$\frac{\partial \eta'}{\partial t} + u_0\frac{\partial \eta'}{\partial x} + \eta_0\frac{\partial u'}{\partial x} = 0$$

(a) Show that this pair of equations is hyperbolic.
(b) Derive the equations of characteristics.
(c) Show that if $u_0 = 0$ the equations reduce to the linear wave equation. Note that since there is no friction in the original equations, setting the velocity equal to zero is not a reduction of generality but a simple translation of coordinates.
(d) At time $t = 0$ the water in a channel is as shown in Fig. 8.39 and the velocity is everywhere zero. Using the equations of part (b), sketch the water surface after (i) one second, (ii) three seconds, and (iii) 13 seconds.

(e) At what time does the water surface come back to the initial conditions?

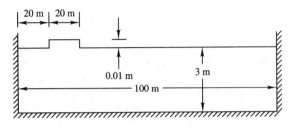

FIGURE 8.39

8.5. At time $t = 0$ the gate of Fig. 8.40 is suddenly closed. Then at time $t = 60$ s it is fully opened.

(a) At what time and distance (behind the gate) does the expansion wave (caused by the opening) overtake the bore?

(b) Sketch the situation upstream of the gate at 55 s and at 75 s. What happens after the expansion wave overtakes the bore.

(c) Will the water surface upstream of the gate return to the situation in the figure? If so, how long will it take? [Note: For the last question you may want only to set up the differential equation. It is nonlinear and must be solved numerically.]

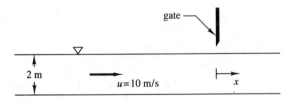

FIGURE 8.40

8.6. Consider a parabolic-shaped channel with the sides given by the equation $z = ky^2$ where $k = 4$. A dam across the channel has water 30 m deep behind it and the channel downstream is dry. The dam is suddenly destroyed.

(a) At what time is the dam failure felt by the fisherman who is 300 m behind the dam.

(b) At what speed does the toe of the water advance down the channel?

(c) What is the depth of water at the dam site after 100 s? Is this depth constant?

8.7. Electrical engineers use the *telegrapher's equation*

$$C\frac{\partial E}{\partial t} + GE + \frac{\partial I}{\partial x} = 0$$

$$L\frac{\partial I}{\partial t} + RI + \frac{\partial E}{\partial x} = 0$$

where L is the inductance of the cable, R is its resistance, C is its shunt capacity, G is its shunt conductance (loss of current divided by voltage), E is voltage, and I is current (not constant with distance). In general L, R, C, and G are functions of E and I. (The telegrapher's equation is also used in the calculation of flows in estuaries; see Ippen, 1966.)

(a) Derive the equations of characteristics.

(b) What is the speed of electricity along the cable in terms of the parameters L, R, C, and G.

(c) Will these equations give simple wave solutions? If not, what are the conditions on L, R, C, and G to obtain simple wave solutions?

8.8. A dam with headwater of 20 m and tailwater of 5 m is suddenly destroyed.

(a) Plot the water surface after one minute and after two minutes.

(b) What is the velocity of the leading bore?

(c) What is the depth and velocity of the water at the dam site?

(d) A fisherman in a boat was 20 m behind the dam at the time of failure. Where is the boat one minute and two minutes after failure?

8.9. A project to transport fresh water uses a thin-walled flexible pipe on the floor of the ocean.

Since it would be flexible and nearly neutrally buoyant (but weighted so it would not rise off the sea bed), the designer argues that the pipe wall could be very thin and would have to sustain only small pressure differences. The force in the walls is $2F = pD$ where $F = k(D - D_0)$, k is a constant, and D_0 is a reference diameter (Fig. 8.41). Assume that water is incompressible.

(a) Derive the partial differential equations that describe unsteady motion and the expansion and contraction of the walls in response to velocity changes. The dependent variables are u, the velocity in the pipe, and D. Neglect friction and assume that the sea pressure on the outside of the pipe is constant.

(b) Find the celerity of a small disturbance, c.

(c) What might be appropriate boundary conditions?

(d) If frictional losses (i.e., normal pipe friction) is taken into account, are all flows stable? If not, find a stability condition.

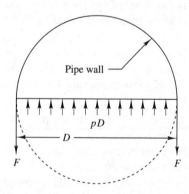

FIGURE 8.41

8.10. Transient flow in pipelines (water hammer) can cause high pressures that expand the pipe and compress the fluid. The governing equations are

$$\frac{1}{\rho A}\frac{D}{Dt}(\rho A) + \frac{\partial u}{\partial x} = 0$$

$$\frac{\partial u}{\partial t} + u\frac{\partial u}{\partial x} + \frac{1}{\rho}\frac{\partial p}{\partial x} + \frac{fu|u|}{2D} = 0$$

in which f is the Darcy friction factor and D is the diameter of the pipe. The compressibility of a liquid is often expressed as a bulk modulus,

$$K = \rho\frac{dp}{d\rho}$$

The expansion of the pipe is more complex. The change in area is given by

$$\frac{\Delta A}{A} = \frac{2}{E}(\Delta\sigma_2 - \mu\Delta\sigma_1)$$

where E is the modulus of elasticity, μ is Poisson's ratio, σ_2 is the circumferential tension (see Fig. 8.41) given by $\sigma_2 = pD/(2z)$ (where z is the thickness of the pipe wall and D is taken as a constant when differentiating σ_2), and σ_1 is the longitudinal tension given by three separate cases

1. If the pipe is anchored at only one point

$$\Delta\sigma_1 = \frac{\Delta pD}{4z}$$

2. If the pipe is constrained against longitudinal movement

$$\Delta\sigma_1 = \mu\Delta\sigma_2$$

3. If the pipe contains many expansion joints

$$\Delta\sigma_1 = 0$$

(a) Show that the speed of a wave is

$$c^2 = \frac{\dfrac{K}{\rho}}{1 + \dfrac{K\kappa D}{Ez}}$$

where $\kappa = 1 - \mu/2$, $\kappa = 1 - \mu^2$, or $\kappa = 1$ depending on the three cases of constraint.

(b) Write the equations in characteristic form.

(c) Show that the nonlinear terms, $u\partial u/\partial x$ and $u\partial\rho/\partial x$, can be neglected in the usual case that $c >> u$.

(d) Show that when the nonlinear terms (except the friction term) are neglected the characteristics are straight.

CHAPTER
9

CIRCULATION

Perhaps the most fascinating fluid flow is that of large scale natural circulation of air and water. We all live in this sort of circulation and it affects our everyday lives. These flows have all the complexities of small scale phenomena plus the complications of rotation—the rotation of the earth, which has profound effects—and stratification. Of course, there is an almost endless laundry list of additional influences: heat exchange from radiation, evaporation, condensation, and conduction; cloud physics; and air-sea interaction. Although we do not investigate these phenomena herein, complete calculations must include them. Not all are known in sufficient detail or can be computed at the range of scales that are necessary to make accurate computations, even with the most sophisticated computer that can be foreseen.

From the point of view in this chapter there are three differences—somewhat linked—with the problems of the previous chapters: (1) the scale is much larger, (2) rotation may be important, and (3) stratification may be important.

9.1 ROTATION

On the scale of local phenomena the rotation of the earth is too small to have an appreciable effect. The bathtub vortex, for example, can rotate in either direction—contrary to popular belief—or even change direction;

it is not materially affected by the rotation of the earth but instead by residual vorticity in the bath water. On the other hand, the circulation in your swimming pool could be influenced by the earth's rotation if allowed to sit undisturbed for a few days. Our initial task in analysis is to include rotation in the equations of motion. Conservation of mass remains the same as in (1.20)

$$\frac{\partial \rho}{\partial t} + \vec{\nabla} \cdot (\rho \vec{u}) = 0 \quad \text{or} \quad \frac{\partial \rho}{\partial t} + \frac{\partial}{\partial x_i}(\rho u_i) = 0 \qquad (9.1)$$

9.1.1 Particle on a Rotating Earth

To begin consideration of the equation of motion, we examine a particle moving on a spherical earth (Fig. 9.1). The particle is moving eastward with a velocity u relative to the earth and at the same time the earth is rotating eastward with the velocity ΩR_p in which Ω is the rotational velocity of the earth and R_p is the distance of the particle from the axis of rotation. The velocity of the particle with respect to an *inertial coordinate system*—a system that is not undergoing acceleration—is $u + \Omega R_p$. Since the particle must stay on the spherical earth, it experiences a centrifugal force

$$\frac{\left(\Omega R_p + u\right)^2}{R_p} M = F_c \qquad (9.2)$$

where M is the mass of the particle. The centrifugal force is directed outward and its horizontal component—the component tangent to the earth—is

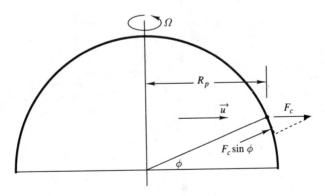

FIGURE 9.1
A particle moving in the northern hemisphere.

$F_c \sin \phi = -F_{ch}$ or

$$-F_{ch} = \left(\Omega^2 R_p + 2\Omega u + \frac{u^2}{R_p} \right) M \sin \phi \qquad (9.3)$$

The first term in parentheses represents the acceleration experienced by a particle fixed to the surface of the earth, the second term is the *Coriolis force*, and the third term often can be neglected due to the (usually) large value of R_p. Thus, (9.3) shows that there is a component of the force tangential to the earth, causing our particle to tend to turn to the right in the northern hemisphere or to the left in the southern hemisphere.

The spinning room phenomenon is an analogous example. Figure 9.2 shows a number of people sitting around a large table in a round room without windows so that they can't see that the room is rotating. Person A rolls a (frictionless) ball to person B. To A's amazement the ball does not go to B but appears to curve to the right to person C. Someone observing this game from above and not on a rotating platform would not be puzzled. The ball, in fact, did roll in a straight line, but in the time it took to cross the table, person C had rotated into the position formerly occupied by B. To those inside the room, however, there seems to be a mysterious force causing all moving objects on the table to curve to the right. Those of us on the rotating earth observe moving objects in the same way as those sitting at the table.

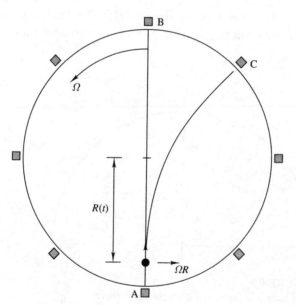

FIGURE 9.2
Apparently curved paths in a rotating room.

Figure 9.3 shows two coordinate systems. The coordinates with primes are in a rotating system and those without primes are in an inertial system. The following notation is used: $\vec{r}$ is the position vector in the inertial system, $\vec{r}'$ is the position vector in the rotating system, and $\vec{R}$ is the position vector of the orgin of the rotating system in the inertial system. Obviously,

$$\vec{r} = \vec{R} + \vec{r}' \qquad \frac{d\vec{r}}{dt} = \frac{d\vec{R}}{dt} + \frac{d\vec{r}'}{dt} \qquad (9.4)$$

and

$$\vec{r}' = x_{i'}\vec{e}_{i'} \qquad \frac{d\vec{r}'}{dt} = \frac{dx_{i'}}{dt}\vec{e}_{i'} + x_{i'}\frac{d\vec{e}_{i'}}{dt} \qquad (9.5)$$

By definition the unit vectors, $\vec{e}_{i'}$, all have magnitude one. If they always point in the same direction, that is, if the primed coordinate system does not rotate, the derivatives $d\vec{e}_{i'}/dt$ are all zero. If the primed system is rotating with angular velocity $\vec{\Omega}$ (Fig. 9.4)

$$\frac{d\vec{e}_{i'}}{dt} = \vec{\Omega} \times \vec{e}_{i'} \qquad (9.6)$$

Using (9.4)

$$\frac{d\vec{r}}{dt} = \frac{d\vec{R}}{dt} + \vec{e}_{i'}\frac{dx_{i'}}{dt} + \vec{\Omega} \times \vec{r}' \qquad (9.7)$$

The first term on the right represents the translation of the primed system with respect to the inertial system, the second term is the translation of the point P with respect to the primed system, and the third term represents the rotation of the primed system.

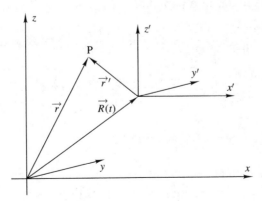

FIGURE 9.3
Inertial and rotating coordinate systems.

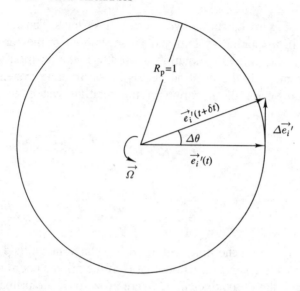

FIGURE 9.4
Demonstration of the derivative and the vector product.

We are more interested in the acceleration, so another derivative is taken

$$\frac{d^2\vec{r}}{dt^2} = \frac{d^2\vec{R}}{dt^2} + \vec{e}_{i'}\frac{d^2x_{i'}}{dt^2} + \frac{d\vec{e}_{i'}}{dt}\frac{dx_{i'}}{dt} + \frac{d\vec{\Omega}}{dt} \times \vec{r}' + \vec{\Omega} \times \frac{d\vec{r}'}{dt} \qquad (9.8)$$

Since we are going to apply the equations to a fluid, take the acceleration of a particle in the primed system as

$$\frac{D'\vec{u}'}{Dt} = \vec{e}_{i'}\frac{d^2x_{i'}}{dt^2}. \qquad (9.9)$$

Also,

$$\frac{d\vec{r}'}{dt} = \vec{u}' + \vec{\Omega} \times \vec{r}' \qquad \frac{dx_{i'}}{dt}\frac{d\vec{e}_{i'}}{dt} = \vec{\Omega} \times \vec{u}' \qquad (9.10)$$

The acceleration of a particle is

$$\frac{d^2\vec{r}}{dt^2} = \frac{d^2\vec{R}}{dt^2} + \frac{D'\vec{u}'}{Dt} + \vec{\Omega} \times \vec{u}' + \frac{d\vec{\Omega}}{dt} \times \vec{r}' + \vec{\Omega} \times (\vec{u}' + \vec{\Omega} \times \vec{r}')$$

$$= \frac{d^2\vec{R}}{dt^2} + \frac{D'\vec{u}'}{Dt} + 2\vec{\Omega} \times \vec{u}' + \vec{\Omega} \times (\vec{\Omega} \times \vec{r}') + \frac{d\vec{\Omega}}{dt} \times \vec{r}' \qquad (9.11)$$

$$\quad\ A \qquad\ B \qquad\quad C \qquad\qquad D \qquad\qquad E$$

The separate terms are as follows:

    A translational acceleration of the primed system
    B particle acceleration with respect to the primed system

C Coriolis acceleration

D centrifugal acceleration with respect to the primed system

E angular acceleration in the primed system

9.1.2 Conservation of Momentum

Since we are going to apply the equations to the rotating earth, the primed coordinate system is fixed with respect to the earth; it does not translate, it does not undergo translational acceleration, and the rotation is constant,

$$\frac{d\vec{R}}{dt} = 0 \qquad \frac{d^2\vec{R}}{dt^2} = 0 \qquad \frac{d\vec{\Omega}}{dt} = 0 \qquad (9.12)$$

The equation of conservation of momentum becomes

$$\rho\left[\frac{D'\vec{u}'}{Dt} + 2\vec{\Omega}\times\vec{u}' + \vec{\Omega}\times\left(\vec{\Omega}\times\vec{r}'\right)\right] = -\vec{\nabla}'p - \rho g'\vec{\nabla}'h + \mu\nabla'^2\vec{u}' \quad (9.13)$$

Consider the centrifugal term for a stationary particle on the earth,

$$\vec{\Omega}\times\left(\vec{\Omega}\times\vec{r}'\right) = -\Omega^2\vec{R}_p \qquad (9.14)$$

But on the earth the acceleration of gravity has been *defined* as the combination of the pull of the earth with the centrifugal acceleration taken into account. That is, we call *g* the acceleration that would be measured by dropping an object in a vacuum or by dividing weight, as measured on a common scale, by mass. In fact the earth is not a perfect sphere due to the centrifugal acceleration. Thus, if the equation of motion is to be applied on earth with the common definition of gravity, it should not contain the centrifugal acceleration so that (9.13) becomes

$$\rho\left(\frac{D'\vec{u}'}{Dt} + 2\vec{\Omega}\times\vec{u}'\right) = -\vec{\nabla}'p - \rho g\vec{\nabla}'h + \mu\nabla'^2\vec{u}' \qquad (9.15)$$

The primes are now dropped and the resulting equation applies to fluid flow on the earth in the normal way

$$\rho\left(\frac{D\vec{u}}{Dt} + 2\vec{\Omega}\times\vec{u}\right) = -\vec{\nabla}p - \rho g\vec{\nabla}h + \mu\nabla^2\vec{u} \qquad (9.16)$$

In (9.16) $\vec{\Omega}$ points to the north along the axis of the earth. The coordinate system is specialized so that *x* is east, *y* is north along the surface of the earth, and *z* points vertically upward (Fig. 9.5). The magnitude of the earth's rotation is one revolution in 23 hours, 56 minutes, 4.09 seconds—one *sidereal day*. It is not exactly 24 hours due to the movement of the earth around the sun. If the earth did not rotate, the length of a day would be one year, not infinity.

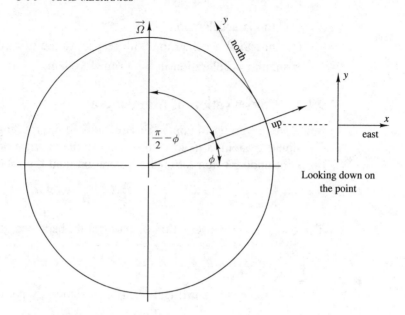

FIGURE 9.5
Coordinates from a point on the surface of the earth.

The Coriolis term is

$$\vec{\Omega} \times \vec{u} = \begin{vmatrix} \vec{e}_x & \vec{e}_y & \vec{e}_z \\ 0 & \Omega \cos \phi & \Omega \sin \phi \\ u_x & u_y & u_z \end{vmatrix} \tag{9.17}$$

$$= (\Omega u_z \cos \phi - \Omega u_y \sin \phi)\vec{e}_x + \Omega u_x \sin \phi \, \vec{e}_y - \Omega u_x \cos \phi \, \vec{e}_z$$

Written in component form, (9.16) becomes

$$\frac{Du_x}{Dt} - 2\Omega u_y \sin \phi + 2\Omega u_z \cos \phi$$

$$= -\frac{1}{\rho}\frac{\partial p}{\partial x} + \nu \left(\frac{\partial^2 u_x}{\partial x^2} + \frac{\partial^2 u_x}{\partial y^2} + \frac{\partial^2 u_x}{\partial z^2} \right)$$

$$\frac{Du_y}{Dt} + 2\Omega u_x \sin \phi = -\frac{1}{\rho}\frac{\partial p}{\partial y} + \nu \left(\frac{\partial^2 u_y}{\partial x^2} + \frac{\partial^2 u_y}{\partial y^2} + \frac{\partial^2 u_y}{\partial z^2} \right) \tag{9.18}$$

$$\frac{Du_z}{Dt} - 2\Omega u_x \cos \phi = -\frac{1}{\rho}\frac{\partial p}{\partial z} - g + \nu \left(\frac{\partial^2 u_z}{\partial x^2} + \frac{\partial^2 u_z}{\partial y^2} + \frac{\partial^2 u_z}{\partial z^2} \right)$$

The equations of motion often contain an approximation in which the vertical velocity, u_z, is considered small. First, the *Coriolis parameter* is defined as

$$f = 2\Omega \sin \phi \tag{9.19}$$

If both u_z and Du_z/Dt are neglected, the equations are

$$\frac{Du_x}{Dt} - fu_y = -\frac{1}{\rho}\frac{\partial p}{\partial x} + \nu\left(\frac{\partial^2 u_x}{\partial x^2} + \frac{\partial^2 u_x}{\partial y^2} + \frac{\partial^2 u_x}{\partial z^2}\right)$$

$$\frac{Du_y}{Dt} + fu_x = -\frac{1}{\rho}\frac{\partial p}{\partial y} + \nu\left(\frac{\partial^2 u_y}{\partial x^2} + \frac{\partial^2 u_y}{\partial y^2} + \frac{\partial^2 u_y}{\partial z^2}\right) \qquad (9.20)$$

$$0 = -\frac{1}{\rho}\frac{\partial p}{\partial z} - g$$

Written in this way, the x- and y-axes are no longer restricted to east and north but z is still vertical.

9.1.3 The Dimensionless Equations

Define nondimensional variables as

$$x_* = \frac{x}{L} \qquad y_* = \frac{y}{L} \qquad z_* = \frac{z}{D} \qquad u_x^* = \frac{u_x}{U} \qquad u_y^* = \frac{u_y}{U}$$

$$u_z^* = \frac{u_z}{W} \qquad p_* = \frac{p}{\rho U \Omega L} \qquad t_* = \Omega t \qquad (9.21)$$

where the vertical and horizonal directions are treated differently. The resulting dimensionless equations are

$$\frac{\partial u_x^*}{\partial x_*} + \frac{\partial u_y^*}{\partial y_*} + \frac{\partial u_z^*}{\partial z_*} = 0 \qquad (9.22)$$

$$\frac{\partial u_x^*}{\partial t_*} + \mathbf{R}_0\left(u_x^*\frac{\partial u_x^*}{\partial x_*} + u_y^*\frac{\partial u_x^*}{\partial y_*} + u_z^*\frac{\partial u_x^*}{\partial z_*}\right) - 2u_y^* \sin\phi + 2\frac{W}{U}u_z^* \cos\phi$$

$$= -\frac{\partial p_*}{\partial x_*} + E\left(\frac{\partial^2 u_x^*}{\partial x_*^2} + \frac{\partial^2 u_x^*}{\partial y_*^2} + \frac{\partial^2 u_x^*}{\partial z_*^2}\right) \qquad (9.23)$$

$$\frac{\partial u_y^*}{\partial t_*} + \mathbf{R}_0\left(u_x^*\frac{\partial u_y^*}{\partial x_*} + u_y^*\frac{\partial u_y^*}{\partial y_*} + u_z^*\frac{\partial u_y^*}{\partial z_*}\right) + 2u_x^* \sin\phi$$

$$= -\frac{\partial p_*}{\partial y_*} + E\left(\frac{\partial^2 u_y^*}{\partial x_*^2} + \frac{\partial^2 u_y^*}{\partial y_*^2} + \frac{\partial^2 u_y^*}{\partial z_*^2}\right) \qquad (9.24)$$

$$\frac{WD}{UL}\left[\frac{\partial u_z^*}{\partial t_*} + \mathbf{R}_0\left(u_x^*\frac{\partial u_z^*}{\partial x_*} + u_y^*\frac{\partial u_z^*}{\partial y_*} + u_z^*\frac{\partial u_z^*}{\partial z_*}\right) - 2\frac{U}{W}u_x^* \cos\phi\right]$$

$$= -\frac{\partial p_*}{\partial z_*} - \frac{gD}{U\Omega L} + \frac{WD}{UL}E\left(\frac{\partial^2 u_z^*}{\partial x_*^2} + \frac{\partial^2 u_z^*}{\partial y_*^2} + \frac{\partial^2 u_z^*}{\partial z_*^2}\right) \qquad (9.25)$$

in which WL/UD is taken as unity and the dimensionless numbers are the *Rossby number*

$$R_0 = \frac{U}{\Omega L} \tag{9.26}$$

and the *Ekman number*

$$E = \frac{\nu^t}{\Omega L^2} \tag{9.27}$$

In the Ekman number the kinematic viscosity is taken as a turbulent value. Typical values might be

$$\Omega = 7.29 \times 10^{-5}/s \qquad \nu^t = 0.05 \text{ m/s}^2 \qquad D = 2000 \text{ m} \qquad L = 10^5 \text{ m}$$

$$U \approx 1 \text{ m/s (a rather high value)} \qquad W \approx 0.02 \text{m/s}$$

$$\frac{WL}{UD} = 1 \qquad \frac{L^2}{D^2} = 10^4/2 \qquad R_0 = 1/7.29 \qquad E = \mathrm{O}(10^{-7})$$

Because the horizontal and vertical eddy viscosities are often different by several orders of magnitude, the Ekman number is divided into horizontal and vertical Ekman numbers, E_h and E_v, where the corresponding values of the eddy viscosities are used.

As usual, the dimensionless numbers give a guide as to what is important in a specific case, but they are not an infallible guide. The strongest indication we have is that the vertical friction is much more important than the horizontal friction. If the flow were laminar, as was assumed in the derivation, there would be little doubt. In turbulent flow the momentum transport may have a preferred direction; in fact $\nu^t_{\text{horizontal}} \gg \nu^t_{\text{vertical}}$. The friction term in the horizontal equations becomes

$$\nu^t_{\text{horizontal}} \left(\frac{\partial^2 u_x}{\partial x^2} + \frac{\partial^2 u_x}{\partial y^2} \right) + \nu^t_{\text{vertical}} \frac{\partial^2 u_x}{\partial z^2}$$

$$\nu^t_{\text{horizontal}} \left(\frac{\partial^2 u_y}{\partial x^2} + \frac{\partial^2 u_y}{\partial y^2} \right) + \nu^t_{\text{vertical}} \frac{\partial^2 u_y}{\partial z^2}$$

In dimensionless form

$$E \left[\frac{\nu^t_{\text{horizontal}}}{\nu^t_{\text{vertical}}} \left(\frac{\partial^2 u_x}{\partial x^2} + \frac{\partial^2 u_x}{\partial y^2} \right) + \frac{L^2}{D^2} \frac{\partial^2 u_x}{\partial z^2} \right]$$

$$E \left[\frac{\nu^t_{\text{horizontal}}}{\nu^t_{\text{vertical}}} \left(\frac{\partial^2 u_y}{\partial x^2} + \frac{\partial^2 u_y}{\partial y^2} \right) + \frac{L^2}{D^2} \frac{\partial^2 u_y}{\partial z^2} \right]$$

Thus, the horizontal friction must be of the order of four magnitudes greater than the vertical friction for the terms to be of equal importance using the example numbers above. The last expression is often used to justify neglecting the horizontal friction in favor of the vertical friction.

9.2 FRICTIONLESS FLOWS

Some of the approximations to the equations provide insight to the types of flows that may take place in the atmosphere and the ocean. In this section the Ekman number is taken as zero so that the active terms in the equations of motion are the inertial terms and the pressure gradient.

9.2.1 Geostrophic Flow

The simplest of the frictionless flows is the case where both the Rossby number and the Ekman number are zero. Then (with constant density)

$$-fu_y = -\frac{\partial}{\partial x}\left(\frac{p}{\rho}\right) \qquad fu_x = -\frac{\partial}{\partial y}\left(\frac{p}{\rho}\right) \tag{9.28}$$

where the Coriolis parameter, f, is defined in (9.19). The function $p/f\rho$ serves as a stream function and lines of constant pressure are streamlines. Given a pressure gradient, the *geostrophic velocities* are

$$u_{gx} = -\frac{\partial}{\partial y}\left(\frac{p}{\rho f}\right) \qquad u_{gy} = \frac{\partial}{\partial x}\left(\frac{p}{\rho f}\right) \tag{9.29}$$

A pressure gradient in the x-direction causes a velocity in the y-direction and a pressure gradient in the y-direction causes a velocity in the negative x-direction (Fig. 9.6). Contrary to what might be expected, the flow is at right angles to the forcing function.

An unexpected result of the equation is the *Taylor column*. Taking $p = p(x,y)$—the pressure as a function of the horizontal coordinates only—then (9.29) gives the horizontal velocities as functions of the horizontal

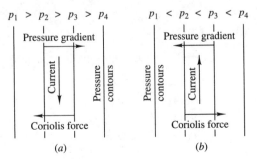

FIGURE 9.6
Geostrophic flow.

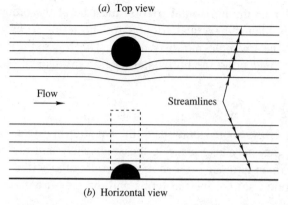

(a) Top view

Flow

Streamlines

(b) Horizontal view

FIGURE 9.7
The Taylor column.

coordinates only. Consider a bump on the sea floor as shown in Fig. 9.7. Ordinarily, the streamlines would bend to pass both over and around the bump, but due to the fact that the horizontal velocities are not functions of the vertical coordinate, the streamlines cannot bend in the vertical. Thus, the bump acts as though it were a solid column (dashed lines) obstructing the flow for its full depth. The condition that leads to the Taylor column is that the rotation is strong compared to the inertial and friction forces. This phenomenon can be easily demonstrated in a rotating container in the laboratory. The bump must extend well above the boundary layer so that the friction forces are not important.

Equations (9.29) can be written more generally by squaring and adding and taking the square root to obtain

$$\sqrt{u_{gx}^2 + u_{gy}^2} = \frac{1}{\rho f}\frac{\partial p}{\partial n} \qquad \frac{\partial p}{\partial n} = \sqrt{\left(\frac{\partial p}{\partial x}\right)^2 + \left(\frac{\partial p}{\partial y}\right)^2} \qquad (9.30)$$

If the pressure is assumed hydrostatic, a pressure gradient must be reflected in a sea surface slope, $\partial p/\partial n = \rho g\, \partial \eta/\partial n$. Taking $f = 10^{-4}$/s, $g = 9.8$ m/s^2, and a (rather rapid) velocity of 1 m/s results in a sea surface slope of about 10^{-5} or about one centimeter in one kilometer.

9.2.2 Gradient Flow

We next add the inertial term to the equations of motion

$$u_x\frac{\partial u_x}{\partial x} + u_y\frac{\partial u_x}{\partial y} - fu_y = -\frac{1}{\rho}\frac{\partial p}{\partial x}$$

$$u_x\frac{\partial u_y}{\partial y} + u_y\frac{\partial u_y}{\partial y} + fu_x = -\frac{1}{\rho}\frac{\partial p}{\partial y}$$

$$(9.31)$$

The equations are more easily solved in cylindrical coordinates. Conservation of mass is

$$\frac{1}{R}\frac{\partial}{\partial R}(Ru_R) + \frac{1}{R}\frac{\partial u_\theta}{\partial \theta} = 0 \tag{9.32}$$

and the equations of motion are

$$u_R\frac{\partial u_R}{\partial R} + \frac{u_\theta}{R}\frac{\partial u_R}{\partial \theta} - \frac{u_\theta^2}{R} - fu_\theta = -\frac{1}{\rho}\frac{\partial p}{\partial R}$$

$$u_R\frac{\partial u_\theta}{\partial R} + \frac{u_\theta}{R}\frac{\partial u_\theta}{\partial \theta} + \frac{u_\theta u_R}{R} + fu_R = -\frac{1}{\rho R}\frac{\partial p}{\partial \theta} \tag{9.33}$$

We now seek a solution where $u_R = 0$, $\partial p/\partial \theta = 0$, and $\partial u_\theta/\partial \theta = 0$. The result is

$$u_\theta = -\frac{Rf}{2} \pm \sqrt{\frac{f^2R^2}{4} + \frac{R}{\rho}\frac{\partial p}{\partial R}} \tag{9.34}$$

which is the definition of the *gradient current*. Note that u_θ can be either clockwise or counterclockwise, depending on the sign of the square root; however, u_θ is negative (in the Northern Hemisphere where $f > 0$) when $\partial p/\partial R < 0$, a high pressure area. For the high pressure case, the limit on the pressure gradient is $|\partial p/\partial R| < \rho Rf^2/4$ in order that the term under the square root is positive. The limit on the pressure gradient in atmospheric highs forms a limit on wind velocity.

If $\partial p/\partial R > 0$, a low pressure area, then u_θ tends to be positive, but can be of either sign. In fact the rotation in tornados and dust devils can be in either direction although the (Northern Hemisphere) motion of hurricanes is always counterclockwise. Unlike the limit in high pressure areas, pressure gradients can be large, thus accounting for the extreme velocities in tornados and hurricanes.

A special case of the gradient flow is that where there is no pressure gradient, $\partial p/\partial R = 0$, in which case it is called *inertial current* and

$$u_\theta = -Rf \tag{9.35}$$

The *inertial period*, or time for a particle to pass around the circumference of a circle, is

$$T = \frac{2\pi R}{|u_\theta|} = \frac{2\pi}{f} \tag{9.36}$$

Of course $u_\theta = 0$ is also a solution of the equations.

The Coriolis parameter is $f = 1.458 \times 10^{-4} \sin\phi$, so the inertial period depends on latitude. It ranges from about 12 hours at the poles to infinity at the equator. The inertial period leaves its footprints in a wide variety of

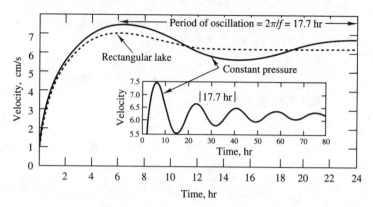

FIGURE 9.8

Inertial oscillations in an otherwise rectilinear current (after J. A. Liggett, "Unsteady circulation in shallow, homogeneous lakes," *Journal of the Hydraulics Division.* ASCE, vol. 95, No. HY4, July 1969, p. 1283, reproduced by permission of ASCE).

flows in bodies of water and the atmosphere. For example, Fig. 9.8 shows a calculation of lake current where the latitude is 42.5°. The current is driven by wind stress and starts from initial conditions of no flow. The dotted line represents the actual computation in a finite lake, but the solid line is the solution for a limitless ocean. Since friction is present, the inertial oscillation is damped.

In an oceanic current the inertial period manifests itself as indicated in Fig. 9.9. Starting at point 1 the velocity vectors change magnitude and direction as indicated by the sequence of numbers, returning to the starting point after one inertial period.

The other special case of the gradient current is *cyclostrophic motion*, where the Coriolis term is small due to small f or small R. Then

$$u_\theta = \pm\sqrt{\frac{R}{\rho}\frac{\partial p}{\partial R}} \qquad (9.37)$$

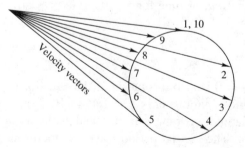

FIGURE 9.9

Wind-driven current that experiences an oscillation due to the inertial period.

TABLE 9.1
Special cases for flow

Type of flow	Operative terms
Gradient	Coriolis, pressure gradient, and inertia
Inertia	Coriolis and inertia
Geostrophic	Coriolis and pressure gradient
Cyclostrophic	Inertia and pressure gradient
Slope current	Coriolis, pressure gradient, and friction

Of course $\partial p / \partial R$ must be positive, so this type of motion can occur only in a low. This motion neglects the earth's rotation and is found in the bathtub vortex, tornados, dust devils, etc. Sometimes dust devils are observed to occur in pairs, each rotating in opposite directions.

In summary, the types of flow are given by Table 9.1. The final entry in the table is considered in Sec. 9.4.

9.3 INERTIALESS, NEARLY HORIZONTAL FLOWS

In this section we consider flows where the Rossby number is small so that the inertial terms are neglected. The force balance is between the Coriolis force and the friction. Three types of boundary conditions can be applied: (1) a shear applied at the water surface (wind-driven circulation) in an infinitely deep sea where the velocities go to zero as $z \to -\infty$; (2) a no-slip condition on the bottom of an infinitely deep sea with given velocities, u_{gx} and u_{gy}, as $z \to \infty$; and (3) a shear on the surface of an infinite sea with given velocities, u_{gx} and u_{gy}, as $z \to -\infty$. Figure 9.10 shows the surface and bottom layers separated by a region where friction is unimportant. The problem can also be solved if the surface and bottom layers merge without the presence of the geostrophic flow. In each case the equations of motion are

$$-f u_y = v^t \frac{\partial^2 u_x}{\partial z^2} \qquad f u_x = v^t \frac{\partial^2 u_y}{\partial z^2} \qquad (9.38)$$

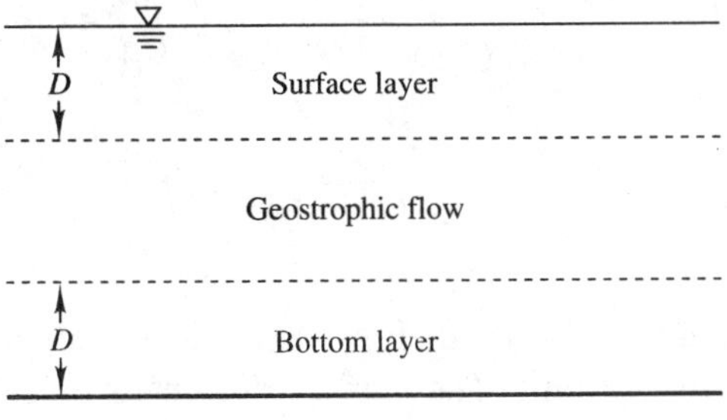

FIGURE 9.10
Surface and bottom layers divided by a central layer.

The solution is

$$u_x = C_1 e^{\sqrt{f/2v^t}\,z} \cos\left(\sqrt{\frac{f}{2v^t}}\,z + a_1\right) + C_2 e^{-\sqrt{f/2v^t}\,z} \cos\left(\sqrt{\frac{f}{2v^t}}\,z + a_2\right)$$

$$u_y = C_1 e^{\sqrt{f/2v^t}\,z} \sin\left(\sqrt{\frac{f}{2v^t}}\,z + a_1\right) - C_2 e^{-\sqrt{f/2v^t}\,z} \sin\left(\sqrt{\frac{f}{2v^t}}\,z + a_2\right)$$

(9.39)

where C_1, C_2, a_1, and a_2 are used to fit the boundary conditions.

9.3.1   Ekman Surface Current

For the first problem take $C_2 = 0$ in (9.39) so that the velocity does not go to infinity as $z \rightarrow -\infty$. Then

$$u_x = C_1 e^{\sqrt{f/2v^t}\,z} \cos\left(\sqrt{\frac{f}{2v^t}}\,z + a_1\right)$$

$$u_y = C_1 e^{\sqrt{f/2v^t}\,z} \sin\left(\sqrt{\frac{f}{2v^t}}\,z + a_1\right)$$

(9.40)

Specify $\tau_x = 0$ and $\tau_y = \mu^t du_y/dz$ on $z = 0$. Differentiating the first of (9.40) with respect to z and applying the condition that the x-shear is zero on $z = 0$ leads to

$$\cos a_1 - \sin a_1 = 0 \qquad \text{or} \qquad a_1 = \frac{\pi}{4} = 45° \qquad (9.41)$$

Differentiating the second of (9.40) with respect to z and applying the y-shear

$$C_1 \sqrt{\frac{f}{2v^t}}(\sin a_1 + \cos a_1) = \frac{\tau}{\mu^t} \qquad \text{or} \qquad C_1 = \frac{\tau}{\sqrt{\mu^t \rho f}} = U_0 \quad (9.42)$$

where U_0 is defined by the expression. Finally,

$$u_x = U_0 e^{\sqrt{f/2v^t}\,z} \cos\left(\sqrt{\frac{f}{2v^t}}\,z + \frac{\pi}{4}\right)$$

$$u_y = U_0 e^{\sqrt{f/2v^t}\,z} \sin\left(\sqrt{\frac{f}{2v^t}}\,z + \frac{\pi}{4}\right)$$

(9.43)

The result is shown in Fig. 9.11 and is called the *Ekman spiral* for the "pure drift" current. The surface current is at a 45-degree angle to the

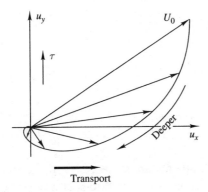

FIGURE 9.11
The Ekman spiral.

forcing of the wind. As one swims downward, the direction of the current changes so that it opposes the surface shear. The spiral terminates at the origin where the current is zero at great depth.

Notice that $\sqrt{f/2v^t}$ has the dimension of length^{-1}. When the reciprocal is multiplied by π it is the Ekman *depth of influence* and is taken as a measure of the depth of the *Ekman surface layer*,

$$D = \pi\sqrt{\frac{2v^t}{f}} \qquad (9.44)$$

where, again, the kinematic viscosity is taken as an eddy kinematic viscosity. Then equations (9.43) are written

$$u_x = U_0 e^{\pi z/D} \cos\left(\frac{z}{D} + \frac{1}{4}\right)\pi \qquad u_y = U_0 e^{\pi z/D} \sin\left(\frac{z}{D} + \frac{1}{4}\right)\pi$$
$$(9.45)$$

At $z = -D$ the current speed is $1/23$ of that at the surface; most of the effect of the surface shear has disappeared. For common values of $f = 10^{-4}$/s and $v^t = 100$ cm^2/s, $D = 44.4$ m.

The transport in the Ekman spiral is obtained by integrating (9.43),

$$T_x = \int_{-\infty}^{0} U_0 e^{\sqrt{f/2v^t}\,z} \cos\left(\sqrt{\frac{f}{2v^t}}\,z + \frac{\pi}{4}\right)dz = U_0\sqrt{\frac{v^t}{f}}$$
$$(9.46)$$
$$T_y = \int_{-\infty}^{0} U_0 e^{\sqrt{f/2v^t}\,z} \sin\left(\sqrt{\frac{f}{2v^t}}\,z + \frac{\pi}{4}\right)dz = 0$$

Remarkably, the transport is zero in the direction of the wind shear; it is at right angles to the force.

9.3.2 Ekman Bottom Current

The equations of motion are

$$
-f u_y = -f u_{gy} + v^t \frac{\partial^2 u_x}{\partial z^2}
$$

$$
f u_x = f u_{gx} + v^t \frac{\partial^2 u_y}{\partial x^2}
$$

(9.47)

where the pressure gradients are replaced by the geostrophic velocities. The solution with no-slip boundary conditions at $z = 0$ and geostrophic flow at $z = \infty$ is

$$
u_x = u_{gx} \left(1 - e^{-\sqrt{f/2v^t}\,z} \cos \sqrt{\frac{f}{2v^t}}\, z \right) - u_{gy} e^{-\sqrt{f/2v^t}\,z} \sin \sqrt{\frac{f}{2v^t}}\, z
$$

$$
u_y = u_{gx} e^{-\sqrt{f/2v^t}\,z} \sin \sqrt{\frac{f}{2v^t}}\, z + u_{gy} \left(1 - e^{-\sqrt{f/2v^t}\,z} \cos \sqrt{\frac{f}{2v^t}}\, z \right)
$$

(9.48)

The Ekman *bottom layer* thickness is defined in the same manner as the surface layer. For seas with less depth than about $2D$, the entire depth consists of the two Ekman layers and there is little geostrophic flow, a frequent case for inland waters.

The transport is calculated from

$$
T_x = \int_0^\infty (u_x - u_{gx})dz = \sqrt{\frac{v^t}{2f}}(u_{gx} - u_{gy})
$$

$$
T_y = \int_0^\infty (u_y - u_{gy})dz = \sqrt{\frac{v^t}{2f}}(u_{gx} + u_{gy})
$$

(9.49)

Consider winds in the atmosphere. The basic circulation about a high pressure area in the Northern Hemisphere is clockwise; it is counterclockwise about a low pressure area. Thus, the flows are as indicated in Fig. 9.12. The transport in the low is inward toward the center and occurs primarily near the lower boundary (the ground). Consequently, air must rise in the center. The vertical velocity brings air from the lower atmosphere to colder elevations. Since warm air can hold more moisture than cold air, condensation occurs, causing rain or snow. In a high pressure area the opposite occurs, leading to fine weather. On the other hand, high pressure tends to trap pollutants near the ground due to the descending air and light winds.

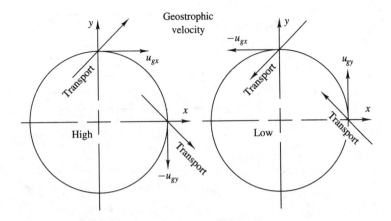

FIGURE 9.12
Circulation around highs and lows.

9.3.3   Surface Current in Geostrophic Flow

The equations are (9.47) with the boundary conditions

$$\mu^t \frac{du_x}{dz} = \tau_x \quad \text{and} \quad \mu^t \frac{du_y}{dz} = \tau_y \quad \text{at} \quad z = 0$$

$$u_x = u_{gx} \quad \text{and} \quad u_y = u_{gy} \quad \text{at} \quad z = -\infty$$

$$(9.50)$$

which represents wind stress on a sea with geostrophic flow below. The solution is

$$u_x = u_{gx}$$

$$+ \frac{1}{\mu^t}\sqrt{\frac{v^t}{2f}} e^{\sqrt{f/2v^t}\,z}\left[(\tau_y + \tau_x)\cos\sqrt{\frac{f}{2v^t}}\,z - (\tau_y - \tau_x)\sin\sqrt{\frac{f}{2v^t}}\,z\right]$$

$$u_y = u_{gy}$$

$$+ \frac{1}{\mu^t}\sqrt{\frac{v^t}{2f}} e^{\sqrt{f/2v^t}\,z}\left[(\tau_y + \tau_x)\sin\sqrt{\frac{f}{2v^t}}\,z + (\tau_y - \tau_x)\cos\sqrt{\frac{f}{2v^t}}\,z\right]$$

$$(9.51)$$

and the transport is

$$T_x = \int_{-\infty}^{0} (u_x - u_{gx})dz = \frac{\tau_y}{f\rho}$$

$$T_y = \int_{-\infty}^{0} (u_y - u_{gy})dz = -\frac{\tau_x}{f\rho}$$

$$(9.52)$$

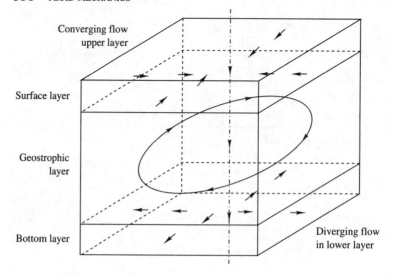

Converging flow
upper layer

Surface layer

Geostrophic
layer

Bottom layer

Diverging flow
in lower layer

FIGURE 9.13
Current system in a high pressure area in the ocean.

9.3.4   Oceanic Circulation

Figure 9.13 shows a current system under a high pressure area in the ocean. There is inward transport near the surface and outward transport near the bottom. Downwelling occurs near the center of the high and is balanced by upwelling far from the center.

In the oceans the center of circulation in an oceanic high is west of the center of the high pressure, causing "western boundary currents." Examples of such currents are the Gulf Stream in the western Atlantic and the Kuroshio current in the western Pacific. The reason for the western intensification comes from the variation of the Coriolis parameter with latitude and conservation of angular momentum. Consider a column of water at the north pole that is motionless with respect to the earth (Fig. 9.14). At the north pole the column has the same rotation as the earth. As it is moved southward, it tends to retain that rotation, so that an observer further south (at the equator in the figure) would note a rotation with respect to the earth. In a high pressure area, the column would move south on the east side of the high.

Since the rotation of the column is counter to the rotation about the high, it subtracts from the circulating current. The opposite occurs if the column is moved from the equator toward the north pole. As it is moved northward, it would tend to take on a rotation opposite to that of the earth—actually a lack of rotation as the earth is rotating under it. The northern movement on the western side of the high pressure and the apparent rotation

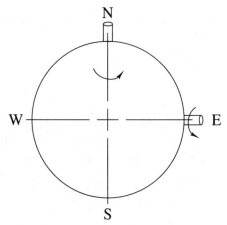

FIGURE 9.14

A column of water moved from the north pole to the equator tends to retain its rotation.

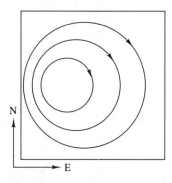

FIGURE 9.15

Crowding of streamlines toward the western coast.

add to the circulation about the high. The net result of these two factors is a decrease in current on the eastern side and an increase on the western side. Figure 9.15 shows how the streamlines would appear in a square ocean. (See also Sec. 9.9.2.)

Even the largest lakes do not show a similar western intensification because the scale is too small. In the Atlantic the Gulf Stream carries on the order of 20×10^6 m³/s to 70×10^6 m³/s. For comparison the flow of the Mississippi River averages about 0.0176×10^6 m³/s.

Historical Note: Vagn Walfrid Ekman (1874–1954) studied oceanography at the University of Uppsala (Sweden) where his thesis was on the "wind spiral." Fridtjof Nansen (1861–1930) suggested this subject after observations of floating ice that seemed to drift to the right of the wind. Although Ekman first published his results in 1902, his 1905 paper, "On the influence of the earth's rotation on ocean currents," attracted international attention. Ekman went on to study currents in the ocean, the equation of state for sea water, and stratified flow. He was also an experimentalist, developing and using current meters for oceanic measurements.

9.4 SLOPE CURRENT

Thus far we have considered only flows in deep water or the atmosphere. The *slope current* is used to satisfy the boundary conditions near a shoreline. It neglects inertia and the equations are

$$-fu_y = -\frac{1}{\rho}\frac{\partial p}{\partial x} + v^t \frac{\partial^2 u_x}{\partial z^2}$$

$$fu_x = -\frac{1}{\rho}\frac{\partial p}{\partial y} + v^t \frac{\partial^2 u_y}{\partial z^2}$$

(9.53)

The y-axis is taken along the gradient of the water surface slope (Fig. 9.16) so that

$$\frac{\partial p}{\partial x} = 0 \qquad \frac{\partial p}{\partial y} = -g\rho \tan \beta \approx -g\rho\beta$$

(9.54)

The easiest method of solution is to let

$$U_x = u_x - \frac{g\beta}{f} \qquad U_y = u_y \qquad W = U_x + iU_y$$

(9.55)

Then the resulting equations combine to

$$\frac{d^2 W}{dz^2} = \frac{f}{v^t} i W$$

(9.56)

with the solution

$$W = C_1 e^{(i+1)\sqrt{f/2v^t}\,z} + C_2 e^{(i+1)\sqrt{f/2v^t}\,z}$$

(9.57)

We will take the boundary conditions

$$\frac{du_x}{dz} = \frac{du_y}{dz} = 0 \qquad \text{at } z = 0 \qquad \text{(no wind shear)}$$

$$u_x = u_y = 0 \qquad \text{at } z = -d \qquad \text{(no slip on the bottom)}$$

(9.58)

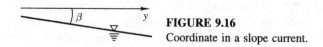

FIGURE 9.16
Coordinate in a slope current.

The velocity components are

$$u_x = -\frac{g\beta}{f}\left[\frac{\cosh\sqrt{\frac{f}{2v^t}}(d-z)\cos\sqrt{\frac{f}{2v^t}}(d+z)}{\cosh 2\sqrt{\frac{f}{2v^t}}d + \cos 2\sqrt{\frac{f}{2v^t}}d}\right.$$

$$\left. +\frac{\cosh\sqrt{\frac{f}{2v^t}}(d+z)\cos\sqrt{\frac{f}{2v^t}}(d-z)}{\cosh 2\sqrt{\frac{f}{2v^t}}d + \cos 2\sqrt{\frac{f}{2v^t}}d} - 1\right]$$

$$u_y = \frac{g\beta}{f}\left[\frac{\sinh\sqrt{\frac{f}{2v^t}}(d-z)\sin\sqrt{\frac{f}{2v^t}}(d+z)}{\cosh 2\sqrt{\frac{f}{2v^t}}d + \cos 2\sqrt{\frac{f}{2v^t}}d}\right.$$

$$\left. +\frac{\sinh\sqrt{\frac{f}{2v^t}}(d+z)\sin\sqrt{\frac{f}{2v^t}}(d-z)}{\cosh 2\sqrt{\frac{f}{2v^t}}d + \cos 2\sqrt{\frac{f}{2v^t}}d}\right]$$

(9.59)

Integrating for the volume transport

$$T_x = \int_{-d}^{0} u_x dz = \frac{Dg\beta}{2\pi f}\left(2\pi\frac{d}{D} - \frac{\sinh\frac{2\pi d}{D} + \sin\frac{2\pi d}{D}}{\cosh\frac{2\pi d}{D} + \cos\frac{2\pi d}{D}}\right)$$

$$T_y = \int_{-d}^{0} u_y dz = \frac{Dg\beta}{2\pi f}\frac{\sinh\frac{2\pi d}{D} - \sin\frac{2\pi d}{D}}{\cosh\frac{2\pi d}{D} + \cos\frac{2\pi d}{D}}$$

(9.60)

where $D = \pi\sqrt{2v^t/f}$ is the Ekman depth of influence. If the depth is large, $d/D \gg 1$,

$$T_x \approx \frac{g\beta d}{f}\left(1 - \frac{D}{2d\pi}\right) \qquad T_y \approx \frac{g\beta d}{f}\frac{D}{2d\pi}$$

(9.61)

so that T_x increases linearly with depth but T_y becomes constant for large depth. Consider a lateral boundary as shown in Fig. 9.17. The pressure gradient arranges itself so that the net transport normal to the coast is zero. We could solve for the pressure gradient by taking

$$(T_x)_{\text{shallow}} + (T_x)_{\text{deep}} = 0$$

(9.62)

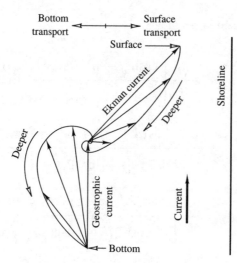

FIGURE 9.17
Slope current in deep water along a shoreline.

The unknown would be the angle the sea surface makes with horizontal, β. Of course, in a realistic situation the current is modified by the bottom topography, which often shows a great variation in the region near the coast. Further, the calculation is not realistic in deep water because we have not really assumed an impenetrable wall, but instead a line across which there is no net transport. The calculation would (unrealistically) indicate transport seaward at the bottom and transport toward the coast at the surface.

The calculation does give some qualitative results. With the current as shown in Fig. 9.17, there must be a downwelling at the coast due to the landward surface current and the seaward bottom current. If the current is from right to left facing the sea—in the opposite direction of Fig. 9.17—there would be an upwelling along the coast. The same considerations hold in shallow water except that the velocity envelope is much thinner (Fig. 9.18). Obviously the coastlines play a primary role in modifying oceanic currents.

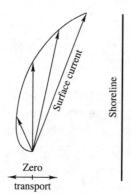

FIGURE 9.18
Ekman spiral in shallow water.

9.5   THE VERTICALLY INTEGRATED EQUATIONS

Often the dimension of a problem is reduced by integrating the equations of motion over the vertical, very much like the shallow water equations. In fact, the result is the set of shallow water equations, except that we will retain the viscous terms but express the viscosity as an eddy viscosity. See Sec. 8.1 for the detailed derivation. The equation of conservation of mass becomes

$$\int_{-d}^{\eta} \left(\frac{\partial u_x}{\partial x} + \frac{\partial u_y}{\partial y} + \frac{\partial u_z}{\partial z} \right) dz = \frac{\partial \eta}{\partial t} + \frac{\partial}{\partial x}(\bar{u}_x \eta) + \frac{\partial}{\partial y}(\bar{u}_y \eta) \qquad (9.63)$$

where $\bar{u}_x$ and $\bar{u}_y$ are defined by (8.13).

In the case of the equations of motion, we neglect the inertial terms ($R_0 \ll 1$) although that is not always a good approximation. In a numerical calculation these terms can be included if necessary. The equations are

$$\int_{-d}^{\eta} \left[f u_y - \frac{1}{\rho} \frac{\partial p}{\partial x} + v_v^t \frac{\partial^2 u_x}{\partial z^2} + v_h^t \left(\frac{\partial^2 u_x}{\partial x^2} + \frac{\partial^2 u_x}{\partial y^2} \right) \right] dz = 0$$

$$\int_{-d}^{\eta} \left[-f u_y - \frac{1}{\rho} \frac{\partial p}{\partial y} + v_v^t \frac{\partial^2 u_y}{\partial z^2} + v_h^t \left(\frac{\partial^2 u_y}{\partial x^2} + \frac{\partial^2 u_y}{\partial y^2} \right) \right] dz = 0$$

$$(9.64)$$

where a distinction has been made between the vertical kinematic eddy viscosity and the horizontal kinematic eddy viscosity. The Coriolis terms and the viscous terms form the differences between these equations and the shallow water equations. Taking the eddy viscosity as constant, the vertical viscous terms are

$$\int_{-d}^{\eta} v_v^t \frac{\partial^2 u_x}{\partial z^2} dz = \frac{\tau_{sx} - \tau_{Bx}}{\rho} \qquad \int_{-d}^{\eta} v_v^t \frac{\partial^2 u_y}{\partial z^2} dz = \frac{\tau_{sy} - \tau_{By}}{\rho} \qquad (9.65)$$

in which τ_{sx} and τ_{sy} are the surface shears in the coordinate directions and τ_{Bx} and τ_{By} are the bottom shears. The integrated equations are

$$-(\eta + d) f \bar{u}_x = -\frac{(\eta + d)}{\rho} \frac{\partial p}{\partial x} + \frac{\tau_{sx} - \tau_{Bx}}{\rho} + v_h^t \int_{-d}^{\eta} \nabla_h^2 u_x dz$$

$$(\eta + d) f \bar{u}_y = -\frac{(\eta + d)}{\rho} \frac{\partial p}{\partial y} + \frac{\tau_{sy} - \tau_{By}}{\rho} + v_h^t \int_{-d}^{\eta} \nabla_h^2 u_y dz$$

$$(9.66)$$

where ∇_h^2 is the Laplacian in the horizontal coordinates. There are a few assumptions implicit in (9.66). The fluid is assumed to be unstratified and the eddy viscosities are taken as constant. (If the viscosities are not constant, as is often the assumed situation in many numerical calculations, the eddy viscosity is placed inside the derivative, $\partial[v_v^t(\partial u_x/\partial z)]/\partial z$ and $\partial[v_h^t(\partial u_y/\partial x)]/\partial x$.) In addition, the equations are often taken over large sections of the earth so that the Coriolis parameter cannot be taken as a constant. The *beta-plane* approximation is the Coriolis term expanded in a series with the *y*-axis restricted to north,

$$f = f_0 + \beta y + \cdots = 2\Omega \sin \phi \qquad \frac{\partial f}{\partial y} \approx \beta \qquad (9.67)$$

and where the series is truncated at the first term in *y*. Other approximations sometimes include neglecting bottom friction, $\tau_{Bx} = 0$, $\tau_{By} = 0$; taking the horizontal friction as zero, $v_h^t = 0$; writing the horizontal friction in terms of the velocity, $\nabla_h^2 u_x = k u_x$ or $= k u_x^2$; neglecting bottom topography, $\partial d/\partial x = 0$, $\partial d/\partial y = 0$; and neglecting surface slope, $\partial \eta/\partial x = 0$, $\partial \eta/\partial y = 0$. The last two assumptions taken together allow the Laplacian to be taken outside of the integral so only the vertically averaged velocities appear in the equations. Sometimes the complete inertial terms are included, in which case the equations revert to the shallow water equations with the Coriolis term added.

The vertically integrated equations are often used and do give good answers in certain cases, but they can be misleading. In many situations the flow is strongly three-dimensional and no two-dimensional approximation suffices. Wind-driven circulation is an obvious example (treated subsequently), especially in limited seas or lakes.

9.6 HOMOGENEOUS LAKE CIRCULATION

The causes and effects of circulation in lakes are similar to those in oceans. The differences are primarily of scale; for example, the Coriolis parameter can be taken as constant to a high degree of accuracy in most lakes. Lakes have limited boundaries on which definite boundary conditions can (must) be applied. In both cases the primary cause of circulation is the wind stress on the surface, but in lakes the tides are usually small and tide-driven circulation is negligible. Both lakes and oceans become stratified and the stratification can make a large difference in the circulation, but that is the subject of a subsequent section. In both cases the flow is three-dimensional, but in lakes more than oceans no satisfactory two-dimensional approximation exists.

The equation of conservation of mass is written in the usual form

$$\frac{\partial u_x}{\partial x} + \frac{\partial u_y}{\partial y} + \frac{\partial u_z}{\partial z} = 0 \qquad (9.68)$$

Although the approximations made below are not always invoked, they simplify the solution and retain the essential elements. For conservation of momentum the nonlinear inertial terms are neglected (occasionally a barely palatable approximation but one that in practice leads to less error than other assumptions), the horizontal friction is neglected (but can be taken into account if necessary) and vertical eddy viscosity is assumed constant (often an unacceptable assumption). The shallow water approximation is invoked (hydrostatic pressure), which means that the depth is small compared to the horizontal extent of the lake—an assumption almost always fulfilled. The shallow water approximation greatly simplifies the solution since it precludes the short surface waves and leaves the mathematics free to solve the circulation problem. The equations of conservation of momentum are

$$\frac{\partial u_x}{\partial t} - f u_y = -\frac{1}{\rho}\frac{\partial p}{\partial x} + v_v^t \frac{\partial^2 u_x}{\partial z^2}$$

$$\frac{\partial u_y}{\partial t} + f u_x = -\frac{1}{\rho}\frac{\partial p}{\partial y} + v_v^t \frac{\partial^2 u_y}{\partial z^2} \qquad (9.69)$$

$$g = -\frac{1}{\rho}\frac{\partial p}{\partial z}$$

Since vertical momentum is neglected, vertical velocity appears only in the equation of conservation of mass and is a consequence of the horizontal motion.

The no-slip boundary conditions are

$$u_x = u_y = 0 \qquad \text{on all solid boundaries where } z = -d \qquad (9.70)$$

Because u_z does not appear in a horizontal viscous term in the equations, there is no method that a no-slip condition can be enforced on vertical boundaries. Thus, the lake is restricted to be contained between an upper surface (that is nearly flat) and a lower boundary defined by $z = -d(x, y)$. On the surface a shear is imposed

$$\rho v_v^t \frac{\partial u_x}{\partial z} = \tau_x \qquad \rho v_v^t \frac{\partial u_y}{\partial z} = \tau_y \qquad \text{on the surface} \qquad (9.71)$$

where τ_x and τ_y are prescribed surface shears. An alternate dimensionless friction parameter is defined

$$m = \sqrt{\frac{f d^2}{2 v_v^t}} \qquad (9.72)$$

where $2m^2 = f d^2 / v_v^t$ is the Taylor number and $1/2m^2$ is the Ekman number. To neglect the horizontal friction in comparison with the bottom friction, the depth must be not much greater than the Ekman depth of influence, $\pi \sqrt{2 v_v^t / f}$.

The form of the equations of motions assumes that the eddy viscosity is a constant, often a poor approximation. The actual value should decrease with depth. In some of the numerical solutions that assumption is not necessary, but if a partial analytical solution is used, it makes the equations tractable.

9.6.1   Wind Stress

The circulation in lakes is driven primarily by inflow and outflow and by wind stress. The latter is also a primary driving force in oceanic circulation. As the equations have been prescribed herein, the surface stress is a boundary condition that must be prescribed for a solution. Unfortunately, the magnitude of the shear that the wind exerts on a water surface is highly variable and poorly known. First, the wind itself may not be well known. Second, there are a great many variables that enter into a calculation. These include the *fetch* (the distance the wind has been blowing over the surface), the roughness of the surface and the roughness of the terrain upwind of the lake, the length of time the wind has been blowing (which affects the wave conditions), the turbulence in the air, the lapse rate in the atmosphere, and the relative temperature of the wind and water.

The relative temperature of the wind and water, for example, determines the stability of the air and the development of the boundary layer. If a warm wind is blowing over a cold lake, the air near the surface is cooled and the boundary layer becomes stably stratified. As a result the turbulence is damped, the velocity gradient near the surface is milder (since there is less mixing), and the shear on the water is decreased. On the other hand an unstable local condition can enhance the turbulence and can transport more momentum from the air to the water.

Nearly all the formulas for wind stress ignore most of the factors that are listed above and take the simple (empirical) form

$$\tau = C\rho_a u_a^2 \tag{9.73}$$

in which ρ_a is the density of the air and u_a is the wind velocity measured at some given height (often 3 m). C is a coefficient that must be guessed. There are a great many formulas for C (Wu, 1969). A crude calculation is

$$C = \begin{cases} 0.5\sqrt{u_a} \times 10^{-3} & 1 < u_a < 15 \text{ m/s} \\ 2.6 \times 10^{-3} & u_a \geq 15 \text{ m/s} \end{cases} \tag{9.74}$$

For the higher velocities

$$\tau = \begin{cases} 3.2 \times 10^{-6} u_a^2 & \text{kN/m}^2 & u_a \text{ in m/s} \\ 6.2 \times 10^{-6} u_a^2 & \text{lbs/ft}^2 & u_a \text{ in ft/sec} \end{cases} \tag{9.75}$$

Consider the effect on set-up of the water surface (Fig. 9.19). The wind force is balanced by the pressure gradient. Assuming hydrostatic pressure

$$\tau = \frac{\rho g}{2L}\left[(d + \delta\eta)^2 - (d - \delta\eta)^2\right] = \frac{\rho g d\,\delta\eta}{L} \qquad \delta\eta = \frac{\tau L}{\rho g d} \qquad (9.76)$$

For a strong wind of 50 km/hr, the set-up is about 7 cm in 65 km.

9.6.2 Steady State

When the time derivative is discarded, the equations of motion (9.69) can be integrated in the z-direction to obtain

$$
\begin{aligned}
u_x &= -\frac{1}{\rho f}\frac{\partial p}{\partial y} + \left(C_2 e^{mz/d} - C_4 e^{-mz/d}\right)\cos\frac{mz}{d} \\
&\quad - \left(C_1 e^{mz/d} - C_3 e^{-mz/d}\right)\sin\frac{mz}{d} \\
u_y &= \frac{1}{\rho f}\frac{\partial p}{\partial x} + \left(C_1 e^{mz/d} + C_3 e^{-mz/d}\right)\cos\frac{mz}{d} \\
&\quad + \left(C_2 e^{mz/d} + C_4 e^{-mz/d}\right)\sin\frac{mz}{d}
\end{aligned}
\qquad (9.77)
$$

where C_1, C_2, C_3, and C_4 are constants of integration but must remain functions of x and y because the integration was carried out only with respect to z. The surface boundary conditions provide a condition on the C

$$
\begin{aligned}
C_1 &= C_3 + \frac{1}{\rho\sqrt{2 f v^t}}\left(\tau_y - \tau_x\right) \\
C_2 &= C_4 + \frac{1}{\rho\sqrt{2 f v^t}}\left(\tau_y + \tau_x\right)
\end{aligned}
\qquad (9.78)
$$

where z has been taken as zero. Since the problem is steady, defining the coordinate system in the surface does not constitute an approximation.

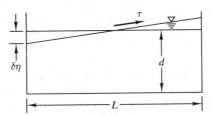

FIGURE 9.19
Wind stress and set-up.

The no-slip conditions provide two additional equations that determine the constants. The results are algebraically complex. The following are defined:

$$c_1 = \cos^2 \frac{mh}{d} \left(e^{-mh/d} + e^{mh/d}\right)^2 + \sin^2 \frac{mh}{d} \left(e^{-mh/d} - e^{mh/d}\right)^2$$

$$c_2 = \frac{d}{2m} e^{-mh/d} \left(\sin \frac{mh}{d} - \cos \frac{mh}{d}\right)$$

$$c_3 = \sin \frac{mh}{d} \left(e^{-mh/d} - e^{mh/d}\right)$$

$$c_4 = \frac{d}{2m} e^{-mh/d} \left(\sin \frac{mh}{d} + \cos \frac{mh}{d}\right) \tag{9.79}$$

$$c_5 = \cos \frac{mh}{d} \left(e^{-mh/d} + e^{mh/d}\right)$$

$$c_6 = \frac{d}{2m} e^{mh/d} \left(\sin \frac{mh}{d} + \cos \frac{mh}{d}\right)$$

$$c_7 = \frac{d}{2m} e^{mh/d} \left(\sin \frac{mh}{d} - \cos \frac{mh}{d}\right)$$

Then

$$C_3 = \frac{1}{c_1} \left[-\frac{c_5}{\rho f} \frac{\partial p}{\partial x} + \frac{c_3}{\rho f} \frac{\partial p}{\partial y} + \frac{\tau_y d}{\rho v_v^t} (c_2 c_5 - c_4 c_3) + \frac{\tau_x d}{\rho v_v^t} (c_2 c_3 + c_4 c_5) \right]$$

$$\tag{9.80}$$

$$C_4 = \frac{1}{c_1} \left[-\frac{c_3}{\rho f} \frac{\partial p}{\partial x} + \frac{c_5}{\rho f} \frac{\partial p}{\partial y} + \frac{\tau_y d}{\rho v_v^t} (c_2 c_3 + c_4 c_5) - \frac{\tau_x d}{\rho v_v^t} (c_2 c_5 + c_4 c_3) \right]$$

After solving for the horizontal velocities, the vertical velocity is

$$u_z = \int_{-d}^{z} \left(\frac{\partial u_x}{\partial x} + \frac{\partial u_y}{\partial y} \right) dz \tag{9.81}$$

The pressure remains in the equations as an unknown.

Although the velocities are three-dimensional, the hydrostatic condition makes the pressure two-dimensional. The pressure equation comes from (9.81) with the upper limit on the integral taken as zero

$$\int_{-d}^{0} \left(\frac{\partial u_x}{\partial x} + \frac{\partial u_y}{\partial y} \right) dz$$

$$= \sqrt{\frac{v_v^t}{2f}} \left\{ \left[\frac{\partial C_3}{\partial y} - \frac{\partial C_4}{\partial x} \right] \left[-e^{-mh/d} \left(\cos \frac{mh}{d} - \sin \frac{mh}{d} \right) \right. \right.$$

$$\left. + e^{mh/d} \left(\cos \frac{mh}{d} + \sin \frac{mh}{d} \right) \right]$$

$$- \left[\frac{\partial C_3}{\partial x} + \frac{\partial C_4}{\partial y} \right] \left[e^{-mh/d} \left(\sin \frac{mh}{d} + \cos \frac{mh}{d} \right) \right. \tag{9.82}$$

$$\left. \left. + e^{mh/d} \left(\sin \frac{mh}{d} - \cos \frac{mh}{d} \right) \right] \right\}$$

$$+ \frac{d}{\rho \sqrt{2v_v^t f}} \left[\frac{\partial \tau_y}{\partial x} \left(1 - e^{-mh/d} \cos \frac{mh}{d} \right) \right.$$

$$+ \frac{\partial \tau_y}{\partial y} e^{-mh/d} \sin \frac{mh}{d} + \frac{\partial \tau_x}{\partial x} e^{-mh/d} \sin \frac{mh}{d}$$

$$\left. - \frac{\partial \tau_x}{\partial y} \left(1 - e^{-mh/d} \cos \frac{mh}{d} \right) \right]$$

After substituting (9.78) and (9.80) for the C terms, there results an elliptic equation for the pressure,

$$\nabla^2 p + a(x, y) \frac{\partial p}{\partial x} + b(x, y) \frac{\partial p}{\partial y} = c(x, y) \tag{9.83}$$

The boundary conditions for (9.83) come from the vertically integrated velocities

$$\bar{u}_x = \frac{1}{d} \int_{-d}^{0} u_x dz = h_1 \frac{\partial p}{\partial x} + h_4 \frac{\partial p}{\partial y} + h_2 \tau_y + h_3 \tau_x$$

$$\bar{u}_y = \frac{1}{d} \int_{-d}^{0} u_y dz = -h_4 \frac{\partial p}{\partial x} + h_1 \frac{\partial p}{\partial y} + h_3 \tau_y - h_2 \tau_x \tag{9.84}$$

in which h_1, h_2, h_3, and h_4 are functions of x and y. The vertically averaged velocities are set to zero at the lateral boundaries. Again, vertical lateral boundaries are excluded because it is the *average* velocities that are set to zero and not the velocity at all points in the depth. If a vertical wall were specified, the average velocity could be zero but a calculated transport across the wall over a part of the depth could occur with an equal transport in the opposite direction on another part.

The boundary conditions on the pressure are awkward. It is easier to introduce a stream function for the vertically integrated velocities

$$\bar{u}_x = -\frac{1}{d}\frac{\partial \Psi}{\partial y} \qquad \bar{u}_y = \frac{1}{d}\frac{\partial \Psi}{\partial x} \tag{9.85}$$

After additional algebra, an equation for the stream function is found that is of the form of (9.83) and has the boundary condition that Ψ is a constant on the lateral boundaries. Numerical methods must be employed to find the two-dimensional pressure distribution. Except for the lengthy algebra, the pressure equation is easy to solve accurately and thus the total system can be solved cheaply.

Figure 9.20 shows the solution in a rectangular lake with a wind stress along the axis. The Coriolis effect is clearly seen in the deep part where the surface velocity has a pronounced right turn. In the shallow water the balance in the equations is between the surface shear and the pressure gradient with the Coriolis term playing a smaller role. Because the basin is closed, the northern and southern transports must balance. The transport near the surface is in the wind direction and to the right of the wind whereas the return current is near the bottom.

Had the vertically integrated equations been used to compute the velocities, this detail would not have been observed. In fact, in a rectangular lake of constant depth, the average velocity would appear to be zero at any point (Fig. 9.21). All applications require a computation of the three-dimensional velocity.

9.6.3 Unsteady Circulation

A number of techniques are available to solve the problem of unsteady lake circulation, but all require numerical methods at some stage. By removing the time derivative with a Laplace transform, the analytical solution can be carried about as far as it was in the steady state problem. The inversion of the transform constitutes an additional step. As a general rule, analytical methods limit flexibility but are economical and tend to give more insight to the process than numerical methods. Thus there is merit in carrying the analysis as far as possible.

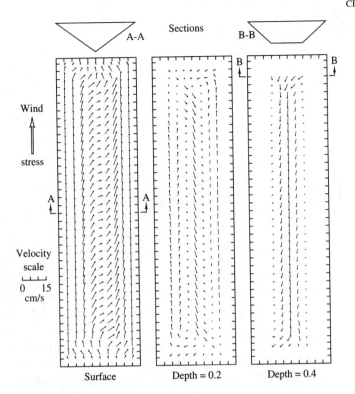

FIGURE 9.20
Velocities at three depths in a rectangular lake with triangular cross section (after J. A. Liggett and C. Hadjitheodorou, "Circulation in shallow, homogeneous lakes," *Journal of the Hydraulics Division*, ASCE, vol. 95, no. HY2, March 1969, p. 616, reproduced by permission of ASCE).

The following dimensionless variables are defined:

$$x_* = \frac{x}{L} \qquad y_* = \frac{y}{L} \qquad z_* = \frac{z}{d} \qquad t^* = ft$$

$$u_x^* = \frac{fL}{gd}u_x \qquad u_y^* = \frac{fL}{gd}u_y \qquad u_z^* = \frac{fL^2}{gd^2}u_z \qquad (9.86)$$

$$p_* = \frac{p}{\rho gd} + \frac{z}{d} \qquad \tau_x^* = \frac{fL\tau_x}{gv_v^t} \qquad \tau_y^* = \frac{Fl\tau_y}{gv_v^t}$$

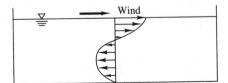

FIGURE 9.21
The velocity distribution in a lake of constant depth. Note that the vertically averaged velocity is zero.

and (9.68) and (9.69) become

$$\frac{\partial u_x^*}{\partial x_*} + \frac{\partial u_y^*}{\partial y_*} + \frac{\partial u_z^*}{\partial z_*} = 0 \qquad \frac{\partial u_x^*}{\partial t_*} - u_y^* = -\frac{\partial p_*}{\partial x_*} + \frac{1}{2m^2}\frac{\partial^2 u_x^*}{\partial z_*^2}$$

$$\frac{\partial u_y^*}{\partial t_*} + u_x^* = -\frac{\partial p_*}{\partial y_*} + \frac{1}{2m^2}\frac{\partial^2 u_y^*}{\partial z_*^2} \qquad \frac{\partial p_*}{\partial z_*} = 0 \tag{9.87}$$

with boundary conditions

$$u_z^* = u_y^* = u_z^* = 0 \quad \text{at} \quad z_* = -1$$

$$\frac{\partial u_x^*}{\partial z_*} = \tau_x^* \quad \text{and} \quad \frac{\partial u_y^*}{\partial z_*} = \tau_y^* \quad \text{at } z_* = 0 \tag{9.88}$$

These conditions imply the rigid lid approximation—the free surface does not move. That assumption was of no consequence in the steady flow problem since the coordinate system could be taken in the free surface, whatever its shape. In unsteady flow, however, it removes long surface waves. (Remember, short waves were removed by the shallow water approximation.) A disturbance in one part of the lake would now travel instantly to another part instead of traveling with the long-wave speed. The primary consequence in lakes is that the rigid lid approximation eliminates the *seiche* motion, the sloshing of lakes from end to end and side to side. Seiches can be important as the wind stress causes a buildup of water on the downwind shores. That buildup will often exceed the equilibrium value and rebound, thus causing a long wave to travel back and forth in the lake. It is dangerous to translate the pressure as calculated from a rigid lid model—either steady or unsteady—to water elevation for the purpose of predicting inundation. Generally the criterion for the rigid lid approximation is that

$$\frac{f^2 L^2}{(2\pi)^2 g D} \ll 1 \tag{9.89}$$

in which L is a wave length and D is a characteristic depth. The quantity is the ratio of the seiche period of a wave of length L to the inertial period. The maximum wave length should be taken as double the length of the basin since a seiche of one-half wave length can occur.

The Laplace transform is defined as

$$\breve{u} = \int_0^\infty u e^{-st} dt \tag{9.90}$$

All the terms of (9.87) are multiplied by e^{-st} and integrated with respect to t

$$\frac{\partial \check{u}_x^*}{\partial x_*} + \frac{\partial \check{u}_y^*}{\partial y_*} + \frac{\partial \check{u}_z^*}{\partial z_*} = 0 \qquad s\check{u}_x^* - \check{u}_y^* = \frac{\partial \check{p}_*}{\partial x_*} + \frac{1}{2m^2} \frac{\partial^2 \check{u}_x^*}{\partial z_*^2}$$

$$s\check{u}_y^* + \check{u}_x^* = \frac{\partial \check{p}_*}{\partial y_*} + \frac{1}{2m^2} \frac{\partial^2 \check{u}_y^*}{\partial z_*^2} \qquad \frac{\partial \check{p}_*}{\partial z_*} = 0 \tag{9.91}$$

The boundary conditions are

$$\check{u}_x^* = \check{u}_y^* = \check{u}_z^* = 0 \quad \text{at} \quad z_* = -1$$

$$\frac{\partial \check{u}_x^*}{\partial z_*} = \check{\tau}_x^* \quad \text{and} \quad \frac{\partial \check{u}_y^*}{\partial z_*} = \check{\tau}_y^* \quad \text{at} \quad z_* = 0 \tag{9.92}$$

Equations (9.91) are written for the initial conditions that $\vec{u}^*(x_*, y_*, z_*, 0) = 0$; otherwise, they would contain additional terms, $u_x^*(x_*, y_*, z_*, 0)$ and $u_y^*(x_*, y_*, z_*, 0)$, that stem from the evaluation of the Laplace transform at the lower limit.

Since $\check{p}_*$ is not a function of z_* (or, more importantly, its horizontal derivatives are not functions of z_*), the equation for the horizontal velocities can be integrated with respect to z_*

$$\check{u}_x^* = -\frac{1}{s^2 + 1} \left(s \frac{\partial \check{p}_*}{\partial x_*} + \frac{\partial \check{p}_*}{\partial y_*} \right) + \cos N z_* \left(k_2 e^{M z_*} - k_4 e^{-M z_*} \right)$$

$$- \sin N z_* \left(k_1 e^{M z_*} - k_3 e^{-M z_*} \right)$$

$$\check{u}_y^* = -\frac{1}{s^2 + 1} \left(-\frac{\partial \check{p}_*}{\partial x_*} + s \frac{\partial \check{p}_*}{\partial y_*} \right) + \cos N z_* \left(k_1 e^{M z_*} + k_3 e^{-M z_*} \right)$$

$$+ \sin N z_* \left(k_2 e^{M z_*} + k_4 e^{-M z_*} \right) \tag{9.93}$$

in which

$$M = m(s^2 + 1)^{1/4} \sin \left(\frac{\tan^{-1} \frac{1}{s}}{2} \right) \qquad N = \sqrt{2(1 + s^2)}\, m \sin \left(\frac{\tan^{-1} \frac{1}{s}}{2} \right) \tag{9.94}$$

The constants k_1, k_2, k_3, and k_4 can be determined through the boundary conditions. The process is algebraically intensive and is left to the references (Young and Liggett, 1977). The remainder of the process follows the steady state analysis. A stream function is defined and there results an equation of the type

$$\frac{\partial^2 \breve{\Psi}_*}{\partial x_*^2} + \frac{\partial^2 \breve{\Psi}_*}{\partial y_*^2} + A(x_*, y_*; s)\frac{\partial \breve{\Psi}_*}{\partial x_*} + B(x_*, y_*; s)\frac{\partial \breve{\Psi}_*}{\partial y_*} + C(x_*, y_*; s) = 0$$

(9.95)

Equation (9.95) must be solved numerically for the stream function.

After (9.95) is solved, the matter of inverting the Laplace transform to find the physical variables still remains. There are many, many techniques that can be used for the inversion and the best method is problem dependent. The method of Schapery (1962) has been used for this problem. If the boundary conditions are nasty functions of time, none of the methods is likely to work well. If the technique does work, it is much more efficient than time stepping through a numerical algorithm. The unsteady problem becomes "a steady state problem with a parameter (s)." It must be solved for a range of the s values, but ordinarily only 8 to 12 values are necessary, which is much less that the usual number of time steps.

9.7 SEICHE

Long-wave oscillations in lakes have often been observed. They are usually caused by set-up of the wind stress on the surface (Fig. 9.19) but can also be caused in very large lakes by the passage of a concentrated low pressure area over one portion of the lake. The latter was apparently the cause of a famous occurrence in Lake Michigan in June 1954. The water in the southern part of the lake, near Chicago, receded. Many people walked onto the newly formed beach to the water's edge and several were drowned when the seiche reversed. Harbor oscillations are a type of seiche.

In large lakes the Coriolis acceleration has a significant effect on the motion. The waves are long—up to twice the length of the lake—and thus the shallow water equations apply. Taking the shallow water equations without friction and without the nonlinear inertial terms gives

$$\frac{\partial \eta}{\partial t} + \frac{\partial}{\partial x}(d\bar{u}_x) + \frac{\partial}{\partial y}(d\bar{u}_y) = 0$$

(9.96)

$$\frac{\partial}{\partial t}(d\bar{u}_x) - f d\bar{u}_y = -c^2\frac{\partial \eta}{\partial x} \qquad \frac{\partial}{\partial t}(d\bar{u}_y) + f d\bar{u}_x = -c^2\frac{\partial \eta}{\partial y}$$

in which $c^2 = gd$ is the wave speed and d is taken as constant. Differentiating the second of (9.96) with respect to x, the third with respect to y, the first with respect to t, and combining produces

$$\frac{\partial^2 \eta}{\partial t^2} + f\left[\frac{\partial}{\partial x}(d\bar{u}_y) - \frac{\partial}{\partial y}(d\bar{u}_x)\right] - c^2\nabla^2\eta = 0$$

(9.97)

Differentiating the second of (9.96) with respect to y, the third with respect to x, and using the first gives

$$f\frac{\partial \eta}{\partial t} = \frac{\partial}{\partial t}\left[\frac{\partial}{\partial x}(d\bar{u}_y) - \frac{\partial}{\partial y}(d\bar{u}_x)\right] \quad \text{or} \quad f\eta = \frac{\partial}{\partial x}(d\bar{u}_y) - \frac{\partial}{\partial y}(d\bar{u}_x)$$

$$(9.98)$$

Now η is taken as harmonic in time, $\eta = \eta_0(x, y)e^{i\sigma t}$ where η_0 is an amplitude function. Substituting (9.98) into (9.97) produces a form of the wave equation

$$(\sigma^2 - f^2)\eta_0 + c^2\nabla^2\eta_0 = 0 \qquad (9.99)$$

Equation (9.99) forms the basis for an eigenvalue problem. In general the eigenvalues must be found numerically. When solved numerically, (9.99) will form a set of linear, homogeneous equations that has a nontrivial solution only for certain values of the frequency σ (Stoker, 1957). For a rectangular basin of constant depth and dimensions L_x by L_y, a solution is that $\eta_0 = a\cos(n\pi x/L_x)\cos(m\pi y/L_y)$, leading to

$$\sigma = \sqrt{f^2 + c^2\pi^2\left(\frac{n^2}{L_x^2} + \frac{m^2}{L_y^2}\right)} \qquad (9.100)$$

where n and m are integers.

If f is taken as zero and the basin is rectangular, the conditions that the averaged velocities normal to vertical walls at the end are zero are satisfied by $k_x = n_x\pi/L_x$ and $k_y = n_y\pi/L_y$ where n_x and n_y are integers and L_x and L_y are the length and width of the basin. (Note that as in some previous sections, it is the *averaged* velocities that are satisfied at the end walls whereas it should be the point velocities.) In a rotating basin, or one with variable bottom topography and variable shape, the solution is not easy and numerical techniques are used.

In a circular basin, an elementary solution is that $\eta = 0$ so that the surface is flat and the current moves in circles with $\sigma = f$; that is, the "wave frequency" is the inertial frequency.

Another elementary solution results from setting $\bar{u}_y \equiv 0$, which could represent the solution in a very long channel. Then the *Kelvin wave* is

$$\eta = \eta_0 e^{fy/c}e^{ik_x(x-ct)} \qquad \bar{u}_x = -\frac{c^2}{df}\frac{\partial \eta}{\partial y} \qquad (9.101)$$

which indicates that the wave amplitude decreases exponentially with negative y; that is, the wave is trapped along the right bank (in the Northern Hemisphere).

Normally, seiche modes contribute little of the current velocity except, perhaps, in shallow water. The longest wave length is $L/2$ where L is length as in Fig. 9.19. The volume of displaced water (per unit of width) in the fundamental oscillation is

$$\text{Vol} = a \int_0^{L/4} \sin \frac{2\pi \xi}{L} d\xi = \frac{aL}{2\pi} \qquad (9.102)$$

in which a is the amplitude. The velocity created by the seiche is of order

$$\text{velocity} = \frac{\text{displaced volume}}{\text{area} \times \text{time}} = \frac{2aL}{\pi T d} \qquad T = \frac{L}{c} \qquad c = \sqrt{gd} \quad (9.103)$$

which is often small compared to the wind driven current. In narrow parts of the lake and inlets to bays and harbors, it can be a substantial velocity, just as tidal oscillations in the oceans can contribute to very significant velocities in harbors, estuaries, and bays. Also, internal seiche modes are often prominent in stratified lakes where there is an oscillation of the interface between the more dense bottom water and the lighter water near the surface. Since an internal seiche may be three orders of magnitude larger than the surface seiche, it can produce velocities of the order of the wind-driven velocities (or much higher in the lower portions of a lake where wind-driven velocities are small—see Sec. 9.9).

9.8   STRATIFICATION

Changes in density—due mostly to temperature and salinity—in oceans or lakes can have a very marked effect on circulation. These bodies of water can divide themselves into layers with little communication between the layers. The stratification affects turbulence that, in turn, alters the flow. Often very small density differences can change the current patterns so that any calculation based on homogeneity is completely invalid and misleading. In oceans and the atmosphere stratification causes changes in the transport of oxygen, nutrients, heat, and pollutants. Changes in oceanic circulation caused by stratification can alter the weather (for example, the El Niño phenomenon).

9.8.1   Lakes

In freshwater lakes stratification is due primarily to temperature. Small amounts of dissolved solids have an effect as does pressure. Fig. 9.22 shows the effect due to temperature. The *density anomaly* is defined as

$$\sigma = (\rho - 1) \times 10^3 \qquad (9.104)$$

where ρ is the density in gm/cm^3. The variation of the density is expressed in this manner so that the numerical significance is not lost with leading nines after the decimal. The coefficient of thermal expansion for water is approximated by

$$C_t = \frac{b_1 + b_2 T}{1 + b_3 T} \qquad b_1 = -66.321 \qquad b_2 = 16.6961 \qquad b_3 = 0.014544$$

(9.105)

where temperature is expressed in degrees Celsius. Then density is

$$\ln \rho = -\int_4^T \frac{b_1 + b_2 T}{1 + b_3 T} dT = \frac{1}{b_3} \left[\left(b_1 - \frac{b_2}{b_3} \right) \ln \frac{1 + 4T}{1 + b_3 T} - b_2 (T - 4) \right]$$

(9.106)

A more common approximation is to define the density anomaly due to temperature as

$$\sigma_T = 6.8 \times 10^{-6} (T - 4)^2$$

(9.107)

which is plotted in the right side of Fig. 9.22 and compared with the more accurate value.

Fresh water reaches a maximum density at 4°C (actually about 3.96°C) and is the only liquid that shows such a maximum. The fact that a maximum density occurs before freezing has a profound influence on the physics of lakes. Warming in the summer above 4°C occurs in the upper layers of a lake; the warmer surface water "floats" on the cooler bottom water, creating density stratification. An idealized picture is shown in Fig. 9.23. The upper (usually mixed) layer is called the *epilimnion*, the lower layer (often stratified but sometimes of nearly constant density) is called the

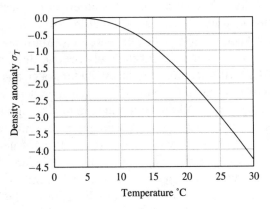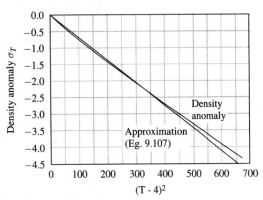

FIGURE 9.22
Density anomaly vs. temperature for fresh water.

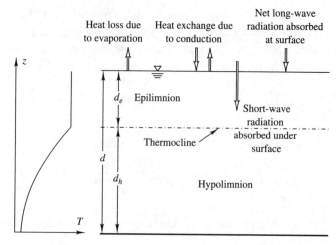

FIGURE 9.23
Idealized zones and heat exchange in a stratified lake.

hypolimnion, the dividing line is the *thermocline*, and the zone around the thermocline (not shown in the figure) is the *metalimnion*.

During a heating season, the thermocline is driven deeper into the lake, but in many temperate-zone lakes, stratification remains all season. When a lake begins to cool during the winter, cooler surface water sinks and in the process thoroughly mixes the epilimnion. Finally, when the epilimnion reaches a density close to the hypolimnion, the lake mixes entirely, usually due to a strong wind. The final mixing process is called the fall turnover. Summer is the stratified season and winter is the mixed season. The circulation and heat transfer processes are entirely different during these seasons.

The shape of the density-temperature curve has an effect on freezing of lakes. Because epilimnetic water becomes more dense at it cools, it is replaced by warmer water when it sinks. Thus, the entire lake must cool to maximum density before any part can cool substantially beyond that point. Once the lake reaches 4°C, further cooling produces lighter water and the surface can readily cool to the freezing point. Even then it is somewhat difficult to freeze a lake since wind will create mixing—a mixing that may involve the entire lake since it is then (almost) homogeneous. Only with shallow lakes or in harsh climates can the temperature of all the water be brought to the freezing point. Surface freezing occurs after a few calm, clear nights—clear to allow radiative cooling. Once a film of ice forms on the surface, the wind does not influence circulation (unless it can break the ice) and thus does not mix the lake. If the ice can become thicker and stronger during several calm days, it may last the duration of the winter.

Stratification also profoundly affects the turbulence in the lake and changes by orders of magnitude the eddy viscosities, the heat exchange, and the transport of all sorts of substances. In stable stratification—the metalimnion and hypolimnion during the heating season—the turbulence is damped. The other extreme is the mixing that occurs when turbulence is intensified due to thermal instability. A wind on the free surface can create turbulence, which propagates downward. When the turbulent front reaches the metalimnion, it tries to mix the more dense water with that of the epilimnion, but loses energy at the same time. Measurement often finds that the metalimnion does not have a smooth density distribution, but is stepped, the steps being remnants of turbulence penetration of past events.

The hypolimnion tends to be much less turbulent and is often nearly free of turbulence. Hypolimnetic turbulence is created by friction on the bottom and—more importantly—by breaking of internal waves. The result is often patchy turbulence. Measurements often show areas of turbulence surrounded by areas of turbulent-free water. The internal waves are gravity waves in the stratified medium and can be long or short. Much of the current in the hypolimnion results from internal seiches where part or all of the hypolimnion is in oscillation.

One-dimensional calculations of the stratification cycles are often accurate enough for many purposes. When the circulation is considered, the entire process is extremely complex and defies mathematical description. Any sort of calculation would be complex even if the turbulent process could be defined and determined, but we are a very long way from describing it. The result is that all calculation in stratified bodies of water must be regarded with suspicion.

9.8.2 Oceans

Stratification in the ocean and in brackish or salt lakes has the additional complexity of dissolved salt. Even pressure is a significant factor because of the great depths of the ocean. The density anomaly due solely to salinity is

$$\sigma_s = -0.093 + 0.1849\,S - 0.000482\,S^2 + 0.0000068\,S^3 \qquad (9.108)$$

in which S is the amount of salinity of water and is measured in parts per thousands, denoted by the symbol $^o/oo$. Figure 9.24 indicates how salinity affects water density. Note that with sufficient salinity the density of water does not reach a maximum above the freezing point. The critical value of salinity is 24.7$^o/oo$—standard sea water has a salinity of 35$^o/oo$—at which the freezing point is $-1.33°C$. At or above that value the entire water column must be cooled to the freezing point before ice can occur, making it more

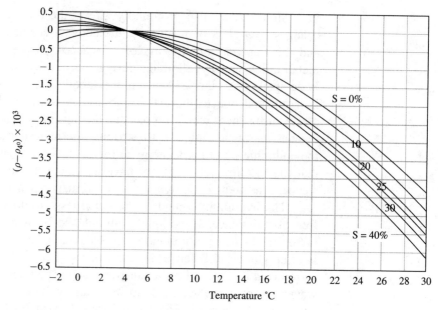

FIGURE 9.24
Density of water as a function of temperature and salinity.

difficult to freeze the ocean than to freeze freshwater lakes. Because ice excludes salt, once the surface freezes, the concentration of salt in the water increases, the density increases, the surface water sinks, and the freezing point of the remainder is lowered.

The effect of salinity on density can overwhelm the temperature effect. That fact is used in *solar ponds*. The arrangement is a stable stratification with dense, highly saline water on the bottom of a shallow pond. Short-wave solar radiation is absorbed below the surface, heating the dense saline water. The surface water, on the other hand, is cooled by the atmosphere and long-wave back radiation, but serves as an insulator between the atmosphere and the hot, salty water underneath. The insulation is effective because the stable stratification damps turbulence. Heat can be harvested from the lower layers for a variety of purposes.

The density as a function of temperature is important in scenarios of global warming. The rise in the level of the oceans for an increase of temperature of 2°C is approximately 0.3 m, a small but nontrivial amount due solely to the temperature rise and discounting the melting of land-based ice.

The compressibility of water becomes important not only in the stratification of the oceans but also to compute the speed of sound. The *coefficient of compressibility* is

$$K_p = -\frac{1}{V}\frac{\Delta V}{\Delta p} \qquad (9.109)$$

Ekman (1908) expressed the compressibility in terms of an empirical constant, k_p, that he called the *mean compressibility* such that

$$V_{sv} = V_{svo}(1 - k_p p) \qquad (9.110)$$

in which $V_{sv} = 1/\rho$ is the *specific volume* and V_{svo} is the specific volume at atmospheric pressure. Then

$$K_p = -\frac{1}{V}\frac{dV}{dp} = -\frac{1}{V_{sv}}\frac{dV_{sv}}{dp} = \frac{k_p + p\dfrac{dk_p}{dp}}{1 - k_p p} \qquad (9.111)$$

The value of k_p is plotted in Fig. 9.25. In the very deep parts of oceans, temperature tends to increase slightly with depth while salinity remains constant. The decrease in density due to the increase in temperature and the increase in density due to pressure tend to cancel. The correct measure in such cases is the *potential temperature*, which would be the temperature of a water sample if it is raised to the surface without gain or loss of heat. At a salinity of about 35°/oo, depth of 3000 m, and temperature of

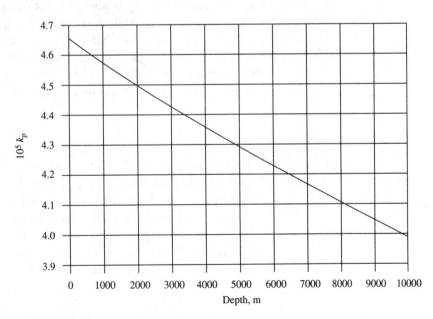

FIGURE 9.25
Pressure effect on density.

4°C, the potential temperature would be about 1.4°C lower than the actual temperature. That is, if a sample were raised to the surface without heat exchange, it would arrive at a temperature of 2.6°C.

The concept of potential temperature is more frequently used in atmospheric calculations where the potential temperature has a much larger effect on the stability of the atmosphere. Under conditions of neutral stability, a mass of air that is raised or lowered arrives at its new position with the same temperature as its new surroundings. If it arrives warmer when raised or cooler when lowered, the atmosphere is unstable and convective mixing will take place. If it arrives cooler when raised or warmer when lowered, the atmosphere is stably stratified, the situation of a common temperature inversion.

9.9 STRATIFIED (BAROCLINIC) CIRCULATION

Remember, the definition of baroclinicity—as opposed to barotropicity—is that the pressure gradient is not in the same direction as the density gradient, that there are density variations not caused by pressure. As shown in the previous section, the actual change in density in water from the effects of temperature and salinity is small. The equations of motion are usually taken with the *Boussinesq approximation*—not to be confused with the Boussinesq correction to the hydrostatic approximation (Sec. 8.9)—that the density is constant in the inertial terms of the equation of motion and in the equation for conservation of mass. That is, the total effect of the variation of density is the change it implies in the pressure gradient; the divergence of the velocity is still zero and the reference density multiplies the substantial derivative in the Navier-Stokes equations.

9.9.1 Lakes

Both stratification and rotation have a distinct effect on circulation. In a stratified lake, for example, the wind stress causes circulation in the epilimnion. The hypolimnion receives stress from the return circulation in the epilimnion, but the hypolimnetic circulation is an order of magnitude weaker than that of the epilimnion (Fig. 9.26). There is a water surface set-up in the direction of the wind and also a tilting of the thermocline, as shown in the figure. The slope of the thermocline may be of the order of three magnitudes greater than the slope of the water surface, or of the order of $\rho/\Delta\rho$. In the unsteady problem, the rigid lid approximation may be valid for the water surface but not for the thermocline. A steady state analysis similar to that of unstratified lakes can be found in Liggett and Lee (1971).

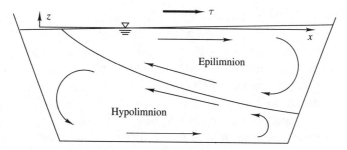

FIGURE 9.26
Case where the hypolimnion touches the surface.

The same type of analysis is made for the epilimnion and hypolimnion separately with the boundary condition of a continuous velocity on a sharp thermocline. For a wind stress in the x-direction, the slopes of the surface and the thermocline were found to be

$$\frac{\partial \eta_e}{\partial x} = \frac{\tau_x}{g(\rho_h - \rho_e)d_e} \frac{1 - \alpha}{(1 - \alpha)^2 + \alpha^2}$$

$$\frac{\partial \eta_e}{\partial y} = \frac{\tau_x}{g(\rho_h - \rho_e)d_e} \frac{\alpha}{(1 - \alpha)^2 + \alpha^2}$$

$$\frac{\partial \eta_h}{\partial x} = \frac{\tau_x}{g\rho_e d_e} \frac{(1 - \alpha)(d_h - 2\alpha d_e) + d_e}{(d_e + d_h)\left[(1 - \alpha)^2 + \alpha^2\right]}$$

$$\frac{\partial \eta_h}{\partial y} = \frac{\tau_x}{\rho_e g d_e} \frac{d_h}{d_e + d_h} \frac{\alpha}{(1 - \alpha)^2 + \alpha^2} \qquad (9.112)$$

$$\alpha = \frac{\kappa}{f \rho_h d_e} \frac{1}{1 + \dfrac{d_e}{d_h}}$$

$$\kappa = \sqrt{\frac{f v_e^t v_h^t}{2}} \frac{\rho_e \rho_h}{\rho_e \sqrt{v_e^t} + \rho_h \sqrt{v_h^t}}$$

if the epilimnion is sufficiently thick ($\alpha < 0.3$).

For a sufficiently large wind stress, the hypolimnion may reach the surface, a condition called *upwelling* (Fig. 9.26) (with downwelling taking place on the opposite side). The sort of analysis that was done for the unstratified case and (9.112) obviously does not apply to this case. Actually, the stratified circulation is much more complex than is implied by the assumptions. Additional complexity can be built into the models by adding more layers to the two that are considered here. The stratification itself is much more complex and the circulation and stratification interact in a nonlinear manner. The circulation breaks up into "cells" of more or less closed circulation. The number and configuration of the cells are highly dependent on the parameters such as the eddy viscosity in the epilimnion and hypolimnion, the densities and density distribution, the surface stress as a function of position, etc. These factors are poorly known at best and therefore, calculations based on such values may not even approximate reality. Stratified circulation remains largely an unsolved problem.

The variation of eddy viscosity is recognized to be a strong function of stratification. Most often the eddy viscosity is written as a function of the *Richardson number*, which can take on a number of different forms but is generally written as

$$Ri = \frac{gz^2 \dfrac{\partial \rho}{\partial z}}{\rho_0 U_0^2} \tag{9.113}$$

For example, the eddy viscosity might be expressed as

$$v_z^t = v_{0z}^t (1 - a\,Ri) \tag{9.114}$$

in which v_{0z}^t and a are constants that can be tuned to a particular situation. A general problem with this sort of calculation is that it includes an increasing number of semi-empirical constants as it grows in complexity. These constants can be tuned to a given situation to reproduce whatever—always inadequate—data that exist. But unless these constants have a good basis in physics, the models cannot be used to predict other events—even in the same geometry—with confidence. That doesn't mean that such models are useless but it does mean that their use is limited in that the details of the results are not dependable.

When used to determine the general aspects of circulation, the results can be qualitatively correct. For example, the crude results of Liggett and Lee (1971) indicate that the speed in a two-layer lake has a minimum in the epilimnion (Fig. 9.27). The same behavior has been measured in Lake Michigan (Verber, 1965).

9.9.2 Geostrophic Flow

In the matter of stratification, as in the other features of circulation, the differences between lakes and oceans are due to scale. Lakes, being smaller, are affected less by rotation, more by frictional effects, and more by lateral

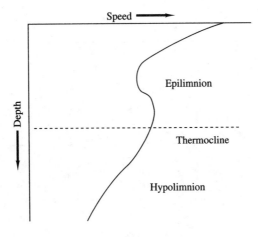

FIGURE 9.27

Speed as a function of depth in a two-layer lake (after Liggett and Lee, 1971).

boundaries. All of the features found in oceans exist to some degree in lakes.

In the case of geostrophic flow ($R_0 = 0$ and $E = 0$) the equations are

$$-fu_y = -\frac{1}{\rho}\frac{\partial p}{\partial x} \qquad fu_x = -\frac{1}{\rho}\frac{\partial p}{\partial y} \qquad 0 = -\frac{1}{\rho}\frac{\partial p}{\partial z} - g \qquad (9.115)$$

To simplify the algebra without loss in generality, take the y-axis as the direction of the current so that $u_x = 0$ and $\partial p/\partial y = 0$. Pressure is eliminated by differentiating the hydrostatic relationship with respect to x, the x-momentum equation with respect to z, and substituting for the cross derivatives,

$$\frac{\partial u_y}{\partial z} = -\frac{g}{\rho f}\frac{\partial \rho}{\partial x} - \frac{u_y}{\rho}\frac{\partial \rho}{\partial z} \qquad (9.116)$$

The last term is often small and can be neglected. Typical values are

$$\frac{\partial \rho}{\partial z} \approx 10^3 \frac{\partial \rho}{\partial x} \qquad u_y \approx 100 \text{ cm/s}$$

$$g \approx 10^3 \text{cm/s}^2 \qquad f \approx 10^{-4}/\text{s} \qquad (9.117)$$

which leads to the last term being two orders of magnitude smaller than the term after the equals sign. The approximation is

$$\frac{\partial u_y}{\partial z} = -\frac{g}{\rho f}\frac{\partial \rho}{\partial x} \qquad (9.118)$$

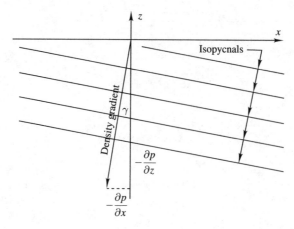

FIGURE 9.28
Stratification.

Consider the situation in Fig. 9.28. The *isopycnals*—lines of equal density—(or isopycnic surfaces) are slanted so that there is a horizontal component to the density gradient. Then

$$\tan \gamma = \frac{\dfrac{\partial \rho}{\partial x}}{\dfrac{\partial \rho}{\partial z}} \qquad \frac{\partial u_y}{\partial z} = -\frac{g}{\rho f}\frac{\partial \rho}{\partial z}\tan \gamma \qquad (9.119)$$

Meteorologists know this situation as the *thermal wind*. If the stratification is stable, $\partial \rho / \partial z < 0$, then the velocity increases upward, $\partial u_y / \partial z > 0$ and $\gamma > 0$ (i.e., as shown in the figure). If the isopycnals slant upward to the right (more dense water on the right), then the velocity increases downward and γ is opposite that shown in the figure. Typical numbers might be

$$\frac{\partial \rho}{\partial z} \approx 10^{-8} \text{ gm/cm}^4 \qquad f \approx 10^{-4}/\text{s} \qquad g \approx 1000 \text{ cm/s}^2$$

$$\tan \gamma \approx 10^{-3} \qquad \frac{\partial u_y}{\partial z} \approx \frac{10 \text{ cm/s}}{1000 \text{ m}} = 10^{-4}/\text{s}$$

$$(9.120)$$

The angle γ, although small, is measurable in the ocean and thus the vertical distribution of velocity can be determined by measurable quantities; however, to determine the absolute value, we need to know the velocity at some point. Also, the equations do not answer the question of whether the current causes the density gradient or the density gradient causes the current.

The configuration for the Gulf Stream is shown in Fig. 9.29. Oceanic gyres with geostrophic flow in a stratified system are shown in Fig. 9.30.

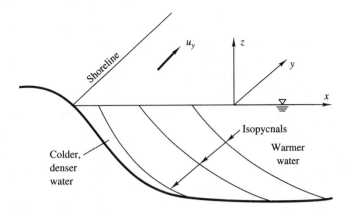

FIGURE 9.29
The Gulf Stream.

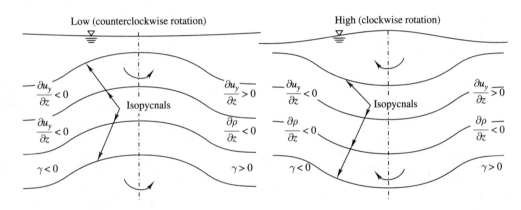

FIGURE 9.30
Circulation about highs and lows in a stratified ocean.

PROBLEMS

9.1. Compute the surface slope of a lake that is 15 m deep subject to a wind velocity of 10 m/s.

9.2. Assume that the area-depth relationship of a lake is parabolic and is given by $A = Kz^2$ where z is the distance from the bottom.
  (a) Using a two-layer approximation, what is the thermocline depth when the lake has maximum stability?
  (b) If the depths of the epilimnion and hypolimnion are equal, how much energy would be required to destratify the lake?
  (c) The specific weight of fresh water is given by the table. For the conditions of part (b), what is the specific weight of the mixed lake if the epilimnion was originally at 30°C and the hypolimnion was originally at 5°C?
  (d) What is the temperature of the lake after mixing?
  (e) A power plant that would use the lake water for cooling is to be built on the lake. The heat rejection to the lake will be 1.76×10^6 kW. The surface area of the lake is 1.7×10^6 m^2 and the maxi-

mum depth is 30 m. How long would it take the power plant to supply to the lake the energy computed in part (b) with epilimnion of 30°C and hypolimnion of 5°C? How long would it take to change the thermal energy in the lake from a homogeneous 5°C to the unstratified condition of part (c)?

Table for Problem 9.2

Temp (°C)	Sp. wt. kN/m^3
0	9.805
5	9.807
10	9.804
15	9.798
20	9.789
25	9.777
30	9.764
40	9.730

9.3. Answer the previous question if the area-depth equation is $A = K\sqrt{z}$.

9.4. The table gives density measurements at two oceanic stations, A and B. Complete the remainder of the table (i.e., compute the relative velocity at the stations, last column). The stations are 50 km apart. Use a Coriolis factor of $f = 10^{-4}$/s.

Table for Problem 9.4

Depth meters	Density A	B	Pressure $p_A - p_B$ g	$h_A - h_B$	$u_A - u_B$
0	1.020	1.030	0	0	0
100	1.022	1.031	5	4.9	
200	1.024	1.032	20	19.5	
300	1.026	1.033	45	44.3	
400	1.028	1.034	80		
500	1.030	1.035	125		
600	1.032	1.036			
700	1.034	1.037	245	238	233
800	1.036	1.038			
900	1.038	1.039	405		
1000	1.040	1.040	500	484	474

9.5. The table gives density measurements at two oceanic stations, A and B. Compute the relative velocity at the stations. They are 50 km apart. Use a Coriolis factor of $f = 10^{-4}$/s.

Table for Problem 9.5

Depth meters	Density A	B	Pressure $p_A - p_B$ g	$h_A - h_B$	$u_A - u_B$
0	1.020	1.022			
100	1.024	1.022			
200	1.028	1.022			
300	1.032	1.024			
400	1.036	1.030			
500	1.040	1.036			
600	1.040	1.040			
700	1.040	1.040			
800	1.040	1.040			
900	1.040	1.040			
1000	1.040	1.040			

9.6. Fig. 9.31 shows a "perpetual motion machine." A thin tube is lowered into the ocean a depth H and extends above the free surface a small distance h. The ocean is stratified both by temperature and salinity.

(a) Explain what causes the flow. There may have to be some mechanical effort to begin the flow, but once it starts it should continue without further mechanical input.

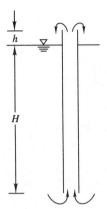

FIGURE 9.31

(*b*) What is the source of the energy?
(*c*) What equations apply to this problem to calculate the flow rate?

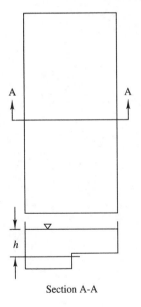

Section A-A

FIGURE 9.32

9.7. A wind blows along the long axis of a basin as shown in Fig. 9.32. The lake has two depths

as indicated in the cross-section. Reply to the following parts for the cases of (*i*) a large lake on the rotating earth in the northern hemisphere and (*ii*) a lake on a nonrotating earth (or a lake at the equator).

(*a*) Sketch the velocity vectors on the surface.
(*b*) Sketch the velocity vectors at the depth *h*.
(*c*) Sketch the transport vectors (the vertically integrated depth) in the lake.
(*d*) Locate the positions of maximum and minimum water elevation.

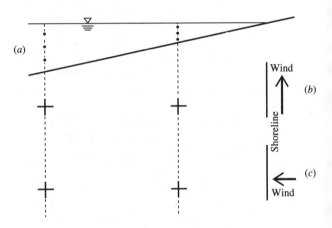

FIGURE 9.33

9.8. The top part (*a*) of Fig. 9.33 represents a cross-section of a beach on a large ocean. Plan views (*b*) and (*c*) are below the cross-section, one in which the wind is along the shore and the other in which the wind is normal to the shore. One of the sections is deep (where the depth is greater than the Ekman depth of influence but the beach effect is still present) and the other section is shallow (the depth is less than the Ekman depth of influence). From each + mark, sketch (without calculation) four vectors for (*a*) the surface velocity, (*b*) the velocity at one-fourth of the depth, (*c*) the velocity at one-half of the depth, (*d*) the velocity at three-fourths of the depth, and (*e*) the transport in the section. Assume steady flow.

9.9.

(*a*) A uniform wind blows over a lake as shown in Fig. 9.34. The cross-section of the lake is triangular. Sketch the surface velocity vectors in the left half of the diagram and the velocity vectors at one-half of the depth in the right half of the diagram.

(*b*) Sketch the streamlines of the mass transport (vertically integrated velocity).

(*c*) What is the net volume transport crossing Section A-A? Why?

(*d*) What part of the basin would you expect the surface to have the highest elevation? Why? What part of the basin would you expect the surface to have the lowest elevation?

(*e*) On the diagram of cross-section A-A sketch the velocity vectors (that is, show the velocity vectors in the *yz* plane).

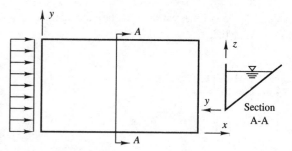

FIGURE 9.34

9.10. Compute the surface slope in a lake 15 m deep subject to a wind of 20 m/s. The stress is approximately

$$\tau = 3.2 \times 10^{-3} V^2$$

where τ is in units of Newtons/meter^2 and V is in meters/second.

9.11. Consider a column of seawater two km deep. If the water is initially at 10°C and warms to 12°C due to global warming, what is the increase in sea level?

CHAPTER 10

WAVES

The most obvious feature of free surfaces—in oceans and lakes especially, but also in rivers, channels, and on the fish pond—is the presence of waves. This feature was excluded from Chap. 4 because it was defined to be outside the realm of free surface flow problems. Wave problems include much greater complexity than can be covered in this chapter. In fact, waves are an area of active research in fluid mechanics. Linear wave theory has been around for many decades, but that is restricted to problems in which the amplitude is small compared to the wave length. Although linear theory is useful, many engineering problems need to be solved for extreme conditions, exactly those conditions where the wave amplitude is not small. These problems include the computation of forces on offshore structures, runup of storm waves on shore, beach erosion, the nasty conditions that plague ships in a storm, and destruction of shore facilities by tsunamis. Many of these problems are not yet solved, or at least not yet solved well.

10.1 THE LINEARIZED SERIES OF FREE SURFACE EQUATIONS

Most analytical solutions for wave problems are based on the linearized free surface boundary conditions. Actually, a series of equations can be derived for the free surface boundary conditions with the linear solution being the most elementary. Three fundamental nonlinearities exist: the kinematic condition, the dynamic condition, and the fact that the boundary conditions are applied on the actual (unknown) free surface. Failure to recognize the third nonlinearity does not affect the first-order solution but would lead to serious error in the higher order solutions.

The development follows a classical perturbation analysis and is patterned after Stoker (1957). Equation (4.9) is written with $p_a = 0$ as

$$F(x, y, z, t) = -\frac{\partial \Phi}{\partial t} + \frac{1}{2} \frac{\partial \Phi}{\partial x_i} \frac{\partial \Phi}{\partial x_i} + g\eta = 0 \qquad \text{on } z = \eta \qquad (10.1)$$

The function, F, is expanded in a Taylor series about $z = 0$ for evaluation at $z = \eta$

$$F(x, y, \eta, t) = \{F(x, y, 0, t)\} + \eta \left\{ \frac{\partial F}{\partial z} \right\} + \frac{\eta^2}{2!} \left\{ \frac{\partial^2 F}{\partial z^2} \right\} + \frac{\eta^3}{3!} \left\{ \frac{\partial^3 F}{\partial z^3} \right\} + \cdots$$
$$(10.2)$$

where the derivatives are to be evaluated at $z = 0$. The terms are in braces for identification in the more complex equations below.

Substituting (10.1) into (10.2)

$$\left\{ -\frac{\partial \Phi}{\partial t} + \frac{1}{2} \left(\frac{\partial \Phi}{\partial x} \right)^2 + \frac{1}{2} \left(\frac{\partial \Phi}{\partial y} \right)^2 + g\eta \right\}$$

$$+ \eta \left\{ \frac{\partial}{\partial z} \left[-\frac{\partial \Phi}{\partial t} + \frac{1}{2} \left(\frac{\partial \Phi}{\partial x} \right)^2 + \frac{1}{2} \left(\frac{\partial \Phi}{\partial y} \right)^2 + g\eta \right] \right\} \qquad (10.3)$$

$$+ \frac{\eta^2}{2} \left\{ \frac{\partial^2}{\partial z^2} \left[-\frac{\partial \Phi}{\partial t} + \frac{1}{2} \left(\frac{\partial \Phi}{\partial x} \right)^2 + \frac{1}{2} \left(\frac{\partial \Phi}{\partial y} \right)^2 + g\eta \right] \right\} + \cdots = 0$$

The kinematic equation is treated in a similar manner. Equation (4.5) is

$$f(x, y, z, t) = -\frac{\partial \eta}{\partial t} + \frac{\partial \Phi}{\partial x} \frac{\partial \eta}{\partial x} + \frac{\partial \Phi}{\partial y} \frac{\partial \eta}{\partial y} - \frac{\partial \Phi}{\partial z} = 0 \qquad (10.4)$$

The series (10.2) is written in f instead of F and using (10.4)

$$\left\{ \frac{\partial \eta}{\partial t} + \frac{\partial \Phi}{\partial z} - \frac{\partial \eta}{\partial x} \frac{\partial \Phi}{\partial x} - \frac{\partial \eta}{\partial y} \frac{\partial \Phi}{\partial y} \right\}$$

$$+ \eta \left\{ \frac{\partial}{\partial z} \left(\frac{\partial \eta}{\partial t} + \frac{\partial \Phi}{\partial z} - \frac{\partial \eta}{\partial x} \frac{\partial \Phi}{\partial x} - \frac{\partial \eta}{\partial y} \frac{\partial \Phi}{\partial y} \right) \right\} \qquad (10.5)$$

$$+ \frac{\eta^2}{2} \left\{ \frac{\partial^2}{\partial z^2} \left(\frac{\partial \eta}{\partial t} + \frac{\partial \Phi}{\partial z} - \frac{\partial \eta}{\partial x} \frac{\partial \Phi}{\partial x} - \frac{\partial \eta}{\partial y} \frac{\partial \Phi}{\partial y} \right) \right\} + \cdots = 0$$

Now the potential, Φ, and free surface elevation, η, are expanded in a perturbation series

$$\Phi = \epsilon\Phi^a + \epsilon^2\Phi^b + \epsilon^3\Phi^c + \cdots \tag{10.6}$$

$$\eta = \epsilon\eta^a + \epsilon^2\eta^b + \epsilon^3\eta^c + \cdots \tag{10.7}$$

where Φ^a, Φ^b, ..., η^a, η^b, ... are functions of the independent variables. The parameter ϵ is assumed to be small. In wave problems ϵ is the amplitude divided by the wave length and the resulting theory will be valid for waves in which the amplitude is small compared to the wave length. Since there is no imposed velocity, the series in Φ begins with a small term. Also the equilibrium free surface is taken at $z = 0$. Since Φ is a solution to Laplace's equation, all the Φ^a, Φ^b, ... satisfy Laplace's equation

$$\nabla^2\Phi^a = \nabla^2\Phi^b = \cdots = 0 \tag{10.8}$$

On the solid boundaries

$$\frac{\partial\Phi^a}{\partial n} = \frac{\partial\Phi^b}{\partial n} = \cdots = 0 \tag{10.9}$$

We now substitute the series for Φ and η into the free surface boundary conditions (10.3) and (10.5). First, using the dynamic condition (10.3)

$$\left\{ -\epsilon\frac{\partial\Phi^a}{\partial t} - \epsilon^2\frac{\partial\Phi^b}{\partial t} - \epsilon^3\frac{\partial\Phi^c}{\partial t} + \cdots + \frac{1}{2}\left(\epsilon\frac{\partial\Phi^a}{\partial x} + \epsilon^2\frac{\partial\Phi^b}{\partial x} + \cdots \right)^2 \right.$$

$$+ \frac{1}{2}\left(\epsilon\frac{\partial\Phi^a}{\partial y} + \epsilon^2\frac{\partial\Phi^b}{\partial y} + \cdots \right)^2 + g\left(\epsilon\eta^a + \epsilon^2\eta^b + \epsilon^3\eta^c + \cdots \right) \right\}$$

$$+ \left(\epsilon\eta^a + \epsilon^2\eta^b + \cdots \right)\left\{ \frac{\partial}{\partial z}\left[-\epsilon\frac{\partial\Phi^a}{\partial t} - \epsilon^2\frac{\partial\Phi^b}{\partial t} \right.\right.$$

$$+ \frac{1}{2}\left(\epsilon\frac{\partial\Phi^a}{\partial x} + \epsilon^2\frac{\partial\Phi^b}{\partial x} + \cdots \right)^2 + \frac{1}{2}\left(\epsilon\frac{\partial\Phi^a}{\partial y} + \epsilon^2\frac{\partial\Phi^b}{\partial y} + \cdots \right)^2$$

$$\left.\left. + \text{ terms in } \eta \right]\right\}$$

$$+ \frac{(\epsilon\eta^a + \cdots)^2}{2}\left\{ \frac{\partial^2}{\partial z^2}\left[-\epsilon\frac{\partial\Phi^a}{\partial t} + \cdots + \text{ terms in } \eta \right]\right\} = 0 \tag{10.10}$$

The omitted terms in η are to be differentiated with respect to z, which results in zero since $\eta = \eta(x, y, t)$. The series are carried out sufficiently to produce terms in ϵ^3. The dynamic boundary conditions for each of the terms come from equating like powers of ϵ from (10.10). Using first terms in ϵ, then terms in ϵ^2, and finally terms in ϵ^3 gives

$$\frac{\partial \Phi^a}{\partial t} = g\eta^a \quad \text{on } z = 0 \tag{10.11}$$

$$\frac{\partial \Phi^b}{\partial t} + \eta^a \frac{\partial^2 \Phi^a}{\partial z \partial t} - g\eta^b - \frac{1}{2}\left(\frac{\partial \Phi^a}{\partial x_i}\frac{\partial \Phi^a}{\partial x_i}\right) = 0 \quad \text{on } z = 0 \tag{10.12}$$

$$\frac{\partial \Phi^c}{\partial t} - \frac{\partial \Phi^a}{\partial x_i}\frac{\partial \Phi^b}{\partial x_i} - g\eta^c + \eta^b \frac{\partial \Phi^a}{\partial z \partial t} + \eta^a \frac{\partial \Phi^b}{\partial z \partial t}$$

$$+ \frac{\eta^a}{2}\frac{\partial \Phi^a}{\partial x_i}\frac{\partial \Phi^a}{\partial x_i} + \frac{(\eta^a)^2}{2}\frac{\partial^3 \Phi^a}{\partial z^2 \partial t} = 0 \quad \text{on } z = 0 \tag{10.13}$$

The kinematic boundary conditions are treated in a similar manner. Using (10.6) and (10.7) in (10.5)

$$\left\{ \frac{\partial}{\partial t}\left(\epsilon \eta^a + \epsilon^2 \eta^b + \epsilon^3 \eta^c + \cdots\right) + \frac{\partial}{\partial z}\left(\epsilon \Phi^a + \epsilon^2 \Phi^b + \epsilon^3 \Phi^c + \cdots\right) \right.$$

$$- \frac{\partial}{\partial x}\left(\epsilon \eta^a + \epsilon^2 \eta^b + \cdots\right)\frac{\partial}{\partial x}\left(\epsilon \Phi^a + \epsilon^2 \Phi^b + \cdots\right)$$

$$\left. - \frac{\partial}{\partial y}\left(\epsilon \eta^a + \epsilon^2 \eta^b + \cdots\right)\frac{\partial}{\partial y}\left(\epsilon \Phi^a + \epsilon^2 \Phi^b + \cdots\right) \right\}$$

$$+ \left(\epsilon \eta^a + \epsilon^2 \eta^b + \cdots\right)\left\{\frac{\partial}{\partial z}\left(\epsilon \frac{\partial \Phi^a}{\partial z} + \epsilon^2 \frac{\partial \Phi^b}{\partial z} + \cdots + \text{terms in } \eta\right)\right\}$$

$$+ \frac{\left(\epsilon \eta^a + \epsilon^2 \eta^b + \cdots\right)^2}{2}\left\{\frac{\partial^2}{\partial z^2}\left(\epsilon \frac{\partial \Phi^a}{\partial z} + \cdots + \text{terms in } \eta\right)\right\} = 0 \tag{10.14}$$

Equating terms in ϵ, ϵ^2, and ϵ^3

$$\frac{\partial \eta^a}{\partial t} + \frac{\partial \Phi^a}{\partial z} = 0 \quad \text{on } z = 0 \tag{10.15}$$

$$\frac{\partial \eta^b}{\partial t} + \frac{\partial \Phi^b}{\partial z} - \frac{\partial \eta^a}{\partial x}\frac{\partial \Phi^a}{\partial x} - \frac{\partial \eta^a}{\partial y}\frac{\partial \Phi^a}{\partial y} + \eta^a \frac{\partial^2 \Phi^a}{\partial z^2} = 0 \quad \text{on } z = 0 \tag{10.16}$$

$$\frac{\partial \eta^c}{\partial t} + \frac{\partial \Phi^c}{\partial z} - \frac{\partial \eta^a}{\partial x}\frac{\partial \Phi^b}{\partial x} - \frac{\partial \eta^a}{\partial y}\frac{\partial \Phi^b}{\partial y} - \frac{\partial \eta^b}{\partial x}\frac{\partial \Phi^a}{\partial x} - \frac{\partial \eta^b}{\partial y}\frac{\partial \Phi^a}{\partial y}$$

$$+ \eta^a \frac{\partial^2 \Phi^b}{\partial z^2} + \eta^b \frac{\partial^2 \Phi^a}{\partial z^2} + \frac{1}{2}\left(\eta^a\right)^2 \frac{\partial^3 \Phi^a}{\partial z^3} = 0 \quad \text{on } z = 0 \tag{10.17}$$

The first-order problem consists of solving Laplace's equation for Φ^a under the free surface boundary conditions (10.11) and (10.15). The second-order problem uses (10.12) and (10.16) to solve for Φ^b where Φ^a is already known from the first-order problem. The third-order problem would follow in sequence with both Φ^a and Φ^b known and where Φ^c is the unknown. The boundary conditions are always linear in the unknown quantities (and linear in the sense that they are to be applied at $z = 0$) but contain increasingly complex terms in the previously solved values of Φ and η.

10.2 LINEAR WAVES

10.2.1 Progressive Waves

The surface elevation, η, can easily be eliminated from (10.11) and (10.15) to obtain

$$\frac{\partial^2 \Phi^2}{\partial t^2} + g \frac{\partial \Phi^a}{\partial z} = 0 \quad \text{on } z = 0 \tag{10.18}$$

Linearizing Bernoulli's equation for any point in the fluid gives an equation for pressure in the linear problem. For the boundary condition on a horizontal bottom

$$\frac{\partial \Phi^a}{\partial n} = -\frac{\partial \Phi^a}{\partial z} = 0 \quad \text{on } z = -d \tag{10.19}$$

where d is the equilibrium water depth. A particular solution in infinitely deep water is

$$\Phi = A e^{kz} \cos(kx - \sigma t) \tag{10.20}$$

in which the potential goes to zero as $z \to -\infty$, A is an amplitude (an arbitrary number), σ is the *wave frequency*, and k is a *wave number*. The condition that (10.18) is satisfied is the *dispersion relationship*

$$\sigma = \sqrt{gk} \tag{10.21}$$

The free surface elevation is

$$\eta = \frac{1}{g} \frac{\partial \Phi}{\partial t} = \frac{A\sigma}{g} \sin(kx - \sigma t) = a \sin(kx - \sigma t) \tag{10.22}$$

where $a = A\sigma/g$ is the amplitude of the wave (the maximum height above mean water level and maximum trough depth so that the wave height—trough to crest distance—is $2a$ for linear waves only). The wave length is the distance between crests

$$L = \frac{2\pi}{k} \tag{10.23}$$

and the wave period (the time between passing crests at a fixed point) is

$$T = \frac{2\pi}{\sigma} \tag{10.24}$$

The wave celerity (speed at which the waves travel through the coordinate system) is

$$c = \frac{\sigma}{k} = \frac{L}{T} \tag{10.25}$$

In (10.20) and (10.22) the minus sign means that the wave is traveling in the positive x-direction; otherwise, the quantity in parentheses would be $kx + \sigma t$.

For finite depth with the wave traveling in the positive x-direction

$$\Phi = \frac{ag}{\sigma} \cosh k(z + d) \, \cos(kx - \sigma t)$$

$$\eta = a \cosh kd \, \sin(kx - \sigma t) \tag{10.26}$$

The vertical velocity is zero at $z = -d$, satisfying the condition that the normal velocity is zero on solid boundaries. The dispersion relationship is

$$\sigma = \sqrt{gk \, \tanh \, kd} \tag{10.27}$$

The wave speed depends on the depth and the wave length,

$$c = \sqrt{\frac{g}{k} \tanh kd} = \sqrt{\frac{gL}{2\pi} \tanh \frac{2\pi d}{L}} \tag{10.28}$$

10.2.2 Standing Waves

Although traveling waves are the ones seen at the beach—and those appreciated by surfers—a standing wave is one where the crests remain in the same position, alternately becoming crests and troughs. The solution is

$$\Phi = -\frac{ag}{\sigma} e^{kz} \cos \, \sigma t \, \cos kx \qquad \eta = a \, \sin \sigma t \, \cos \, kx \tag{10.29}$$

When $\sigma t = \pi/2, 5\pi/2, 9\pi/2, \ldots$, then $\sin \, \sigma t = 1$ and a crest is at the position $\cos \, kx = 1$; when $\sigma t = 3\pi/2, 7\pi/2, 11\pi/2, \ldots$, then $\sin \, \sigma t = -1$ and a trough is at the position $\cos \, kx = 1$; when $\sigma t = 0, \pi, 2\pi, \ldots$, then $\sin \sigma t = 0$ and the sea surface is temporarily flat. The solution is shown in Fig. 10.1 where the extreme positions show the crests and troughs of the wave and the x-axis represents the intermediate position.

A standing wave can be produced by two waves of equal amplitude traveling in opposite directions

$$\eta = a \, [\sin(kx - \sigma t) + \sin(kx + \sigma t)] = 2a \, \sin kx \, \cos \, \sigma t \tag{10.30}$$

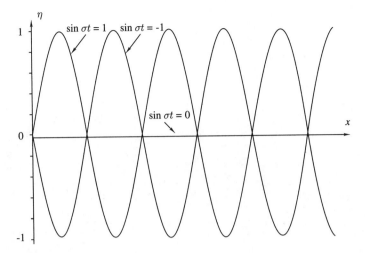

FIGURE 10.1
Standing wave.

The amplitude of the standing wave is twice the amplitude of the component traveling waves. This sort of wave is commonly formed from a traveling wave that reflects from a vertical wall. The wall becomes an antinode (the position of maximum amplitude of the standing wave).

10.2.3 Comparison of Linear Waves

Consider Table 10.1 in which three types of waves are compared: shallow water waves where the wave speed is a function of the depth, deep water waves in which the wave speed is a function of the wave length, and the intermediate case where the wave speed is a function of both depth and wave length. Deep water waves are special cases of the intermediate depth waves as $d \rightarrow \infty$. A shallow water wave, however, contains the approximation of hydrostatic pressure distribution; that is, the wave length is large compared to the depth. Thus, small amplitude wave theory and shallow water wave theory have entirely different bases.

Classically, deep water means that the depth is greater than one-half wave length. At that value $kd = \tanh \pi = 0.9963$, indicating that the intermediate depth wave speed differs by less than one-quarter percent from the deep water wave speed. On the shallow side, consider the intermediate case as kd becomes smaller. In that case $\tanh kh \rightarrow 1$ and $c \rightarrow \sqrt{gd}$. Defining the limit of the shallow water with the same accuracy as the limit for deep water means that

$$0.9963 \; kd = \tanh kd \quad \text{or} \quad kd = 0.1 \quad \text{or} \quad \frac{d}{L} = 0.015 \qquad (10.31)$$

TABLE 10.1
Comparison of linear waves

		Shallow water	Intermediate depth	Deep water
Surface elevation	$\eta =$	$a\sin(kx-\sigma t)$	$a\sin(kx-\sigma t)$	$a\sin(kx-\sigma t)$
Wave speed	$c =$	$\sqrt{gd}$	$\sqrt{\dfrac{gL}{2\pi}\tanh\dfrac{2\pi d}{L}}$	$\sqrt{\dfrac{gL}{2\pi}}$
Potential	$\Phi =$	$\dfrac{ac}{dk}\cos(kx-\sigma t)$	$ac\dfrac{\cosh k(z+d)}{\sinh kd}\cos(kx-\sigma t)$	$ac\,e^{kz}\cos(kx-\sigma t)$
Stream function	$\Psi =$	$-\dfrac{ac}{d}(z+d)\sin(kx-\sigma t)$	$-ac\dfrac{\sinh k(z+d)}{\sinh kd}\sin(kx-\sigma t)$	$-ac\,e^{kz}\sin(kx-\sigma t)$
x-velocity	$u_x =$	$\dfrac{ac}{d}\sin(kx-\sigma t)$	$ack\dfrac{\cosh k(z+d)}{\sinh kd}\sin(kx-\sigma t)$	$ack\,e^{kz}\sin(kx-\sigma t)$
z-velocity	$u_z =$	$-\dfrac{ack}{d}(z+d)\cos(kx-\sigma t)$	$-ack\dfrac{\sinh k(z+d)}{\sinh kd}\cos(kx-\sigma t)$	$-ack\,e^{kz}\sin(kx-\sigma t)$
Particle paths		tend to lines	$\dfrac{(\delta x)^2}{\ell_x^2}+\dfrac{(\delta z)^2}{\ell_y^2}=1$ (ellipse)	$(\delta x)^2+(\delta z)^2=R^2e^{2kz}$ (circle)
Pressure change	$\Delta p =$	$ag\rho\sin(kx-\sigma t)$	$ag\rho\dfrac{\cosh k(z+d)}{\cosh kd}\sin(kx-\sigma t)$	$ga\rho\,e^{kz}\sin(kx-\sigma t)$
Energy	$E =$			$\dfrac{\rho g a^2}{2}$
Power	$P =$	Ec	$\dfrac{Ec}{2}\left(1+\dfrac{2kd}{\sinh 2kh}\right)$	$\dfrac{Ec}{2}$
Group velocity	$U =$	c	$\dfrac{c}{2}\left(1+\dfrac{2kd}{\sinh 2kd}\right)$	$\dfrac{c}{2}$
Dispersion eq.	$\sigma =$	$k\sqrt{gd}$	$\sqrt{gk\tanh kd}$	$\sqrt{gk}$

Kinsman (1965) points out that these limits are much too severe for most calculations that contain error so that the accuracy is not justified. He defines "oceanographer's deep or shallow water" as

$$\text{deep water} \quad \frac{d}{L} > 0.277 \qquad \text{shallow water} \quad \frac{d}{L} < 0.048 \qquad (10.32)$$

which is accurate to within five percent.

10.2.4 Particle Paths

Taking the displacements of particles in the horizontal and vertical directions as δx and δz

$$
\begin{aligned}
u_x &= \frac{d}{dt}(\delta x) = Ak\frac{\cosh k(z+d)}{\sinh kd}\sin(kx - \sigma t) \\
u_z &= \frac{d}{dt}(\delta z) = -Ak\frac{\cosh k(z+d)}{\sinh kd}\cos(kx - \sigma t)
\end{aligned}
\qquad (10.33)
$$

Integrating (10.33) leads to parametric equations in the particle paths

$$
\begin{aligned}
\frac{\delta x}{\dfrac{Ak}{\sigma}\dfrac{\cosh k(z+D)}{\sinh kd}} &= \cos(kx - \sigma t) \\[2ex]
\frac{\delta z}{\dfrac{Ak}{\sigma}\dfrac{\sinh k(z+d)}{\sinh kd}} &= \sin(kx - \sigma t)
\end{aligned}
\qquad (10.34)
$$

Eliminating the sine and cosine functions by squaring and adding yields

$$
\frac{(\delta x)^2}{\left(\dfrac{Ak}{\sigma}\dfrac{\cosh k(z+d)}{\sinh kd}\right)^2} + \frac{(\delta z)^2}{\left(\dfrac{Ak}{\sigma}\dfrac{\sinh k(z+d)}{\sinh kd}\right)^2} = 1 \qquad (10.35)
$$

which shows that particles travel in closed paths. (When the nonlinear terms are included, the particle paths are open slightly, indicating a net current in the direction of travel of the waves.) The paths are elliptical with the major axis horizontal [$\sinh k(z+d) < \cosh k(z+d)$]. As $z \to -d$—the bottom—the minor axis of the ellipse becomes smaller; the ellipse tends to a line parallel to the bottom, thus satisfying the bottom boundary condition.

In deep water both the major and minor axes approach $Ak/\sigma \cdot e^{kz}$, signifying that the particle paths are circles with exponentially decreasing radius as depth increases. In the case of shallow water, the major axis approaches a constant since $\cosh k(z+d) \to 1$ (and shows no significant variation in depth for the shallow water approximation) and the minor axis becomes small as $\sinh k(z + d) \to 0$. Particle paths for shallow water are limiting cases where the entire motion is nearly horizontal.

10.2.5 Pressure

Since the flow is potential, the pressure is calculated by Bernoulli's equation. To be consistent with the linear theory, the velocity terms are dropped, leaving

$$p = \rho \left(\frac{\partial \Phi}{\partial t} - gz \right) \tag{10.36}$$

The pressure change due to the wave comes from subtracting the hydrostatic term and using the wave potential

$$\Delta p = a\rho g \frac{\cosh k(z + d)}{\cosh kd} \sin(kx - \sigma t) \tag{10.37}$$

For deep water the hyperbolic cosine term becomes the exponential, e^{kz}, so the pressure change due to the wave decreases exponentially with depth. Submarines can avoid a bumpy ride by submerging about one-half wave length where $(\Delta p)_{z=-L/2}/(\Delta p)_{z=0} \approx e^{-kL/2} = e^{-\pi} \approx 0.04$.

Experimentally a determination of the state of the sea surface from a pressure gage on the bottom is difficult because a small change in bottom pressure indicates a large wave. For very long waves—shallow water—the pressure is transmitted to the bottom undiminished, as indicated by the relationship

$$\Delta p = ag\rho \sin(kx - \sigma t) \tag{10.38}$$

which, of course, is the assumption of hydrostatic pressure. In a sea with many different wave lengths, a pressure gage on the bottom would see the long waves but the short waves would remain undetected.

10.2.6 Energy

Wave energy consists of two parts: potential energy due to elevation above the mean surface and kinetic energy due to the fluid motion. Potential energy over one wave length is the integral of the weight of water above sea level times the distance to the centroid (Fig. 10.2)

$$P.E. = \int_0^L \frac{\rho g \eta}{2} \eta \, dx = \frac{\rho g a^2 L}{4} \tag{10.39}$$

and the average potential energy per unit of surface area is $\rho g a^2/4$.

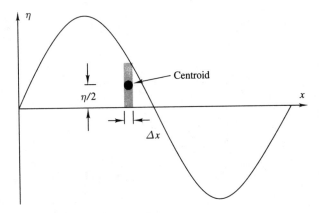

FIGURE 10.2
Element for energy integration.

To find the kinetic energy, we integrate over the wave length and over the depth

$$K.E. = \frac{\rho}{2} \int_0^L dx \int_{-d}^{\eta} \left(u_x^2 + u_z^2\right) dz \qquad (10.40)$$

In deep water $u_x^2 + u_z^2 = a^2 c^2 k^2 e^{2kz}$ and the integral is

$$K.E. = \frac{\rho g a^2 L}{4} \qquad (10.41)$$

The potential and kinetic energies are equal and the total energy per unit of surface area is their sum,

$$E = \frac{\rho g a^2}{2} \qquad (10.42)$$

Note that the energy is independent of wave period and wave length.

10.2.7 Power

Power is the transmission of energy. If one is interested in extracting energy from waves in the sea, it is the power arriving at the energy extractors that is relevant. We consider power from the work done by a wave. That work is pressure times velocity acting over an area (Fig. 10.3)

$$P = \frac{1}{T} \int_{\tau}^{\tau+T} dt \int_{-d}^{0} p u_x dz \qquad (10.43)$$

The hydrostatic part of the pressure when integrated over one wave period makes no contribution to the power. The term that contains the derivative

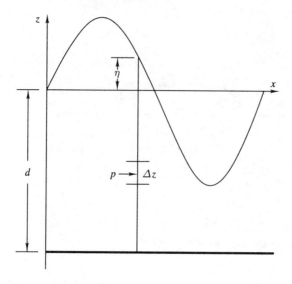

FIGURE 10.3
Integration for power.

of the potential with respect to time is

$$P = \frac{A^2 \sigma \rho k}{T \cosh kd \sinh kd} \int_{-d}^{0} \cosh^2 k(z+d)dz \int_{\tau}^{\tau+T} \sin^2(kx - \sigma t)dt$$

$$= \frac{A^2 \sigma \rho k}{2 \cosh kd \sinh kd} \int_{-d}^{0} \cosh^2 k(z+d)dz$$

$$= \frac{\rho g a^2}{2} \frac{c}{2} \left(1 + \frac{2kd}{\sinh 2kd} \right) = \frac{Ec}{2} \left(1 + \frac{2kd}{\sinh 2kd} \right)$$

$$(10.44)$$

In deep water $P = Ec/2$, whereas in shallow water $P = Ec$. Energy in deep water is transmitted at one-half of the wave speed, whereas in shallow water it is transmitted at the wave speed. The energy transmission speed is called the *group velocity*. For deep water that is the quantity

$$G = \frac{c}{2} \left(1 + \frac{2kd}{\sinh 2kd} \right) \qquad (10.45)$$

which appears prominently in the next section.

10.3 WAVES TRAINS

Up to this point we have considered only a single wave. The sea is actually composed of waves of a variety of wave lengths. Consider the sum of two waves of slightly different wave lengths and frequencies but the same amplitude

$$\eta_1 = a \sin(kx - \sigma t) \qquad \eta_2 = a \sin[(k + \Delta k)x - (\sigma + \Delta\sigma)t]$$
$$\eta = \eta_1 + \eta_2 \tag{10.46}$$

Using trigonometric identities

$$\eta = 2a \cos\left(\frac{\Delta k}{2}x - \frac{\Delta\sigma}{2}t\right) \sin\left[\left(k + \frac{\Delta k}{2}\right)x - \left(\sigma + \frac{\Delta\sigma}{2}\right)t\right] \tag{10.47}$$

The second trigonometric function, the sine, represents the shape of the free surface—the wave—but the first is a *modulating function*; when multiplied by $2a$ it is the amplitude of the wave, which changes slowly in time if $\Delta k/k$ and $\Delta\sigma/\sigma$ are small. The second trigonometric function is the *carrier wave*, modulated by the cosine. We say that the wave is "amplitude modulated." This principle is used in amplitude modulated (AM) radio. The length of the carrier wave is L and that of the modulating function is

$$L_{\text{mod}} = \frac{2\pi}{\frac{\Delta k}{2}} \tag{10.48}$$

We assume that $L_{\text{mod}} \gg L$. Figure 10.4 shows the result in which $L_{\text{mod}} \approx 32L$. The heavy line in Fig. 10.4 is the actual free surface. When the constituent waves reinforce each other, the amplitude is the sum of the components; when they cancel, the amplitude is the difference. An individual wave passes through the group, traveling faster than the group, growing and diminishing, depending on where it is in the pack. The observer sees several large waves followed by small waves, then a return to large waves. Every surfer knows that a large wave some distance away may lose amplitude by the time it reaches the position of the surfer. On the other hand a small wave may grow as it passes through the group.

From (10.47) the speed of the modulating function is $\Delta\sigma/\Delta k$. Using the dispersion equation $\sigma = \sqrt{gk \tanh kd}$, the speed of the modulating function is

$$c_{\text{mod}} = \frac{c}{2}\left(1 + \frac{2kd}{\sinh 2kd}\right) = G \tag{10.49}$$

showing that the wave group moves with the energy speed, one-half of the phase speed in deep water. A storm at sea generates waves of many lengths, frequencies, and speeds. As they travel across the sea, waves of nearly the same speed tend to travel together, separating the system into packs.

A similar phenomenon occurs with sound. Given two machines—generators in a power house, for example—running at slightly different speeds, the sound will be at slightly different frequencies and will alternately

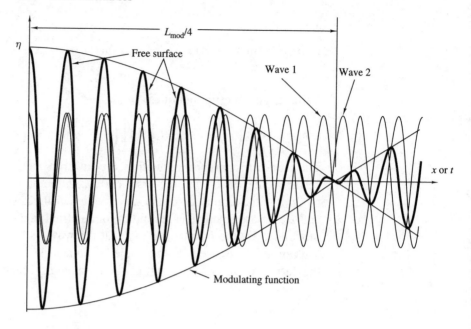

FIGURE 10.4
A wave packet.

reinforce and cancel. The listener will hear alternately a loud noise followed by a period of near silence. The result is called "beats." Beats are common in multi-engine, piston aircraft where the engine speeds are not perfectly synchronized. The pilot's job—or the job of an engine synchronizer—is to see that the engines run at the same speed. Sound waves, however, are not dispersive. If they were, a musical concert in a large hall would be impossible because a different sound would be heard at each position in the audience.

In general more than two waves make up a packet. Adding the potentials of all the waves within a narrow range of wave lengths gives

$$\Phi = \int_{k_0-\epsilon}^{k_0+\epsilon} A(k)e^{i(kx-\sigma t)}dk \tag{10.50}$$

in which k_0 is mean wave number, ϵ is the maximum departure from the average of wave numbers on each side of the mean, and A is the amplitude as a function of wave number. The wave is described by the real or the imaginary part of the function. Taking

$$kx - \sigma t = k_0 x - \sigma_0 t + (k - k_0)x - (\sigma - \sigma_0)t \tag{10.51}$$

(10.50) can be written

$$\Phi = \left\{ \int_{k_0-\epsilon}^{k_0+\epsilon} A(k) \exp\left\{ i\left[(k - k_0)\,x - (\sigma - \sigma_0)\,t \right] \right\} dk \right\} e^{i(k_0 x - \sigma_0 t)}$$

(10.52)

The integral in braces is the modulating term. The group speed is

$$\frac{\sigma - \sigma_0}{k - k_0} \approx \frac{d\sigma}{dk}$$

(10.53)

In most realistic analyses of the sea, the wave packet consists of a large number of components of different wave numbers

$$\eta(x, t) = \int_{-\infty}^{\infty} a(k) \exp\left\{ i\left[kx - \sigma(k)t \right] \right\} dk$$

(10.54)

where $a(k)$ is the *amplitude spectrum* and each wave length is associated with its own amplitude. As in any linear theory, we can choose the amplitude of each of the components, in this case the continuous amplitude spectrum as a function of wave number. For example, a Gaussian wave packet has the amplitude spectrum

$$a(k) = \frac{a_0}{\sqrt{2\pi}\,s} \exp\left\{ -\frac{(k - k_0)^2}{2s^2} \right\}$$

(10.55)

where a_0 is a constant (the amplitude of the wave with wave number k_0 and the maximum amplitude in the packet), s is the standard deviation of the Gaussian function, and k_0 is the wave number of the maximum amplitude wave. Such a packet appears as in Fig. 10.5. The envelope could be very

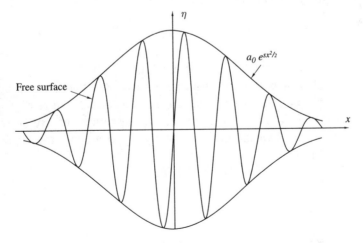

FIGURE 10.5
A Gaussian wave packet.

large and contain many waves or it could be more as shown with only a small number of waves. The modulating envelope can be stationary in time, in space, in both, or in neither.

10.4 FOURIER TRANSFORM

The Fourier integral is

$$f(x) = \frac{1}{\sqrt{2\pi}} \int_{-\infty}^{\infty} F(y)e^{iyx}dy \tag{10.56}$$

The inverse is

$$F(y) = \frac{1}{\sqrt{2\pi}} \int_{-\infty}^{\infty} f(x)e^{-iyx}dx \tag{10.57}$$

where the only restriction is that $f(x)$ is sufficiently integrable. Consider the wave packet

$$\eta(x, t) = \frac{1}{2\pi} \int_{-\infty}^{\infty} a(k)e^{i(kx-\sigma t)}dx \tag{10.58}$$

At $t = 0$

$$\eta(x, 0) = \frac{1}{2\pi} \int_{-\infty}^{\infty} a(k)e^{ikx}dk \tag{10.59}$$

The inverse Fourier transform gives the amplitude spectrum in terms of the initial conditions

$$a(k) = \int_{-\infty}^{\infty} \eta(x, 0)e^{-ikx}dx \tag{10.60}$$

If the initial conditions are known, (10.58) will give the sea surface for all x and t. The foregoing assumes (1) that there is no phase shift in the initial disturbance and (2) that the initial velocity of the surface is zero, $\partial\eta(x, 0)/\partial t = 0$. Although the assumptions are restrictive, they are no worse than the problem of trying to assume the shape of the initial free surface.

The real part of (10.58) is

$$\eta(x, t) = \frac{1}{2\pi} \int_{-\infty}^{\infty} a(k) \cos(kx - \sigma t)dk \tag{10.61}$$

Remember that σ is a function of the wave number through the dispersion equation. Assuming that $a(k)$ is known, we need to evaluate the integral.

Let

$$S(k) = kx - \sigma(k)t \qquad \frac{dS}{dk} = x - \frac{d\sigma}{dk}t \qquad \frac{d^2S}{dk^2} = -\frac{d^2\sigma}{dk^2}t \qquad (10.62)$$

Assuming deep water

$$\sigma = \sqrt{gk} \qquad \frac{d\sigma}{dk} = \frac{g}{2\sqrt{gk}} \qquad \frac{d^2\sigma}{dk^2} = -\frac{g^2}{4(gk)^{3/2}} \qquad (10.63)$$

We now set dS/dk to zero and solve for k, which we call k_0,

$$k_0 = g\left(\frac{t}{2x}\right)^2 \qquad (10.64)$$

which is the wave number at the minimum of $S(k)$. (Note that $d^2S/dk^2 > 0$, so the extremum must be a minimum.) The wave and $S(k)$ are shown in Fig. 10.6. The main contribution to the integral comes near k_0 since the positive and negative areas far from k_0 tend to cancel. Thus, we need only evaluate the integral near this value. The technique is called *Kelvin's method of stationary phase.*

First, expand $S(k)$ is a Taylor series about k_0

$$S(k) = S(k_0) + \frac{dS}{dk}(k - k_0) + \frac{1}{2}\frac{d^2S}{dk^2}(k - k_0)^2 + \cdots \qquad (10.65)$$

Using (10.62) and (10.63)

$$S(k) = k_0x - \sigma_0t - \frac{t}{2}\frac{d^2\sigma}{dk^2}(k - k_0)^2 + \cdots \qquad (10.66)$$

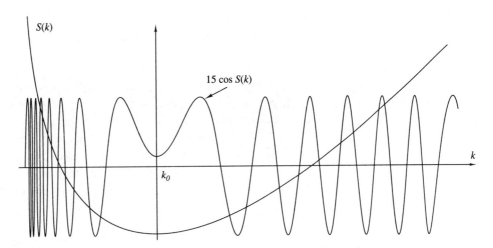

FIGURE 10.6
The method of stationary phase.

Taking the real part of (10.54) and using only these terms of (10.66)

$$\eta(x, t) = \frac{a(k_0)}{2\pi} \int_{-\infty}^{\infty} \cos\left[k_0 x - \sigma(k_0) t - \frac{t}{2} \frac{d^2\sigma}{dk^2}\bigg|_{k=k_0} (k - k_0)^2\right] dk$$
(10.67)

in which $a(k_0)$ was assumed to be slowly varying relative to $kx - \sigma t$ and is effectively constant over the critical range. The result of the integration is

$$\eta(x, t) \approx \frac{a(k_0)}{\sqrt{2\pi t \frac{dG}{dk}}} \cos\left(k_0 x - \sigma_0 \mp \frac{\pi}{4}\right)$$
(10.68)

in which the group velocity has been used as $G = d\sigma/dk$. From (10.68) we conclude:

1. Wave lengths increase toward the front of a wave train. For a given time—say t_1—and a given distance—say x_1—the group velocity is $G = x_1/t_1$ and wave number is k_1. For the same time and a lesser distance—say x_2—the group velocity is $G = x_2/t_1 < x_1/t_1$. In deep water $G = c/2$ and for all depths G increases as c. Therefore, $c_2 < c_1$, but c is a function of wave length—$c = \sqrt{gL/2\pi}$ in deep water—so $L_2 < L_1$. Thus, the wave length increases with distance.

2. The amplitude peak occurs in the neighborhood of $x = G(k_m) t$ where k_m maximizes $a(k)$. If the peak amplitude occurs at $x_1 = G(k_m) t_1$ at one point, then according to $x - Gt = 0$ it will continue to occur at $x = G(k_m) t$.

3. The amplitude decays as $1/\sqrt{t}$.

4. The amplitude variation, wave length, and wave period move with the group velocity.

The foregoing leads to a technique called *ray tracing*. Consider Fig. 10.7 where the origin represents the source of the wave energy. A given wave number (wave length) travels with the group velocity, which is $G = x/t$ and is constant along any ray. Thus, the wave length, wave period, and frequency are also constant along a ray. The amplitude is not constant but varies as $t^{-1/2}$. Since the wave energy is proportional to the square of the amplitude, it changes as t^{-1} and because the distance between rays is proportional to t, the energy between rays is constant (assuming a two-dimensional disturbance and no dissipation). Since the rays are further apart for larger time, the energy is spread over a greater distance.

An individual wave travels with twice the group velocity in deep water, $c = 2x/t$. Let x_w be the coordinate of a wave. Then the wave path

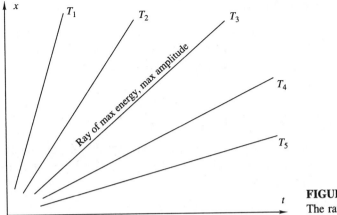

FIGURE 10.7
The ray diagram.

is described as

$$x_w = \text{constant } t^2 \tag{10.69}$$

As a wave outruns the group, it passes from the area of maximum energy and loses amplitude.

Actually the wave packet is made up of waves of a variety of lengths. The longest of these lengths is large compared to the depth so that shallow water theory applies. The situation is shown in Fig. 10.8. The longest wave—the shallow water wave—is the fastest; the short waves make up the end of the packet.

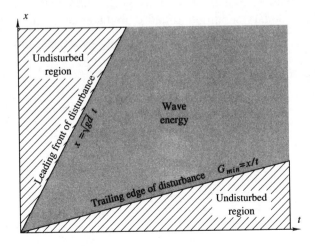

FIGURE 10.8
Spreading of wave energy from shallow water waves to deep water waves.

10.5   CAPILLARY WAVES

For the shortest waves surface tension can have a major effect. Consider two fluids of different densities separated by an interface (Fig. 10.9). For both fluids the Bernoulli equation applies. The linearized Bernoulli equation is

$$\frac{p_1}{\rho_1} = \frac{\partial \Phi_1}{\partial t} - g\eta \qquad \frac{p_2}{\rho_2} = \frac{\partial \Phi_2}{\partial t} - g\eta \tag{10.70}$$

in which we have used the fact that the displacement of the interface is the same in both fluids, $\eta_1 = \eta_2 = \eta$. The kinematic equation also applies to each fluid

$$\frac{\partial \eta}{\partial t} = -\frac{\partial \Phi}{\partial z} \tag{10.71}$$

The solution takes the same form as previously

$$\Phi_1 = Ae^{-kz}\cos(kx - \sigma t) \qquad \Phi_2 = Be^{kz}\cos(kx - \sigma t) \tag{10.72}$$

where the form assumes that both fluids are semi-infinite. The interface displacement is

$$\eta = \eta_1 = \eta_2 = -\frac{Ak}{\sigma}\sin(kx - \sigma t) \tag{10.73}$$

Due to the surface tension a jump in pressure occurs across the interface. Consider Fig. 10.10, which shows static equilibrium for a small portion of the interface. The balance of forces in the vertical is

$$T_s \sin\beta + p_2 \cos\frac{\alpha + \beta}{2}\delta x = T_s \sin\alpha + p_1 \cos\frac{\alpha + \beta}{2}\delta x \tag{10.74}$$

in which T_s is the surface tension. In conformity with the linearization of the boundary condition, α and β are taken as small so that (10.74) becomes

$$T_s\beta + p_2\delta x = T_s\alpha + p_1\delta x \tag{10.75}$$

Approximately,

$$\beta = \alpha + \frac{\partial \alpha}{\partial x}\delta x \qquad \alpha = \frac{\partial \eta}{\partial x} \tag{10.76}$$

ρ_1

z

ρ_2

FIGURE 10.9
Interface between two fluids.

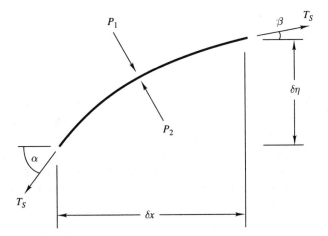

FIGURE 10.10
Forces on the interface.

so that the relationship between the surface tension and the pressure jump
is

$$\frac{p_1 - p_2}{T_s} = \frac{\partial^2 \eta}{\partial x^2} \tag{10.77}$$

From Bernoulli's equation

$$p_1 = \rho_1 \left(\frac{\partial \Phi_1}{\partial t} - g\eta \right)$$

$$= \rho_1 \left[A\sigma e^{-kz} \sin(kx - \sigma t) + \frac{gAk}{\sigma} \sin(kx - \sigma t) \right]$$

$$p_2 = \rho_2 \left(\frac{\partial \Phi_2}{\partial t} - g\eta \right) \tag{10.78}$$

$$= \rho_2 \left[-A\sigma e^{kz} \sin(kx - \sigma t) + \frac{gAk}{\sigma} \sin(kx - \sigma t) \right]$$

Applying the boundary conditions at $z = 0$

$$p_1 - p_2 = A\sigma \left(\rho_1 + \rho_2 \right) \sin(kx - \sigma t) - \frac{gAk}{\sigma} \left(\rho_2 - \rho_1 \right) \sin(kx - \sigma t) \tag{10.79}$$

Using (10.73), (10.77), and (10.79) gives the dispersion equation

$$\sigma^2 = \frac{1}{\rho_1 + \rho_2} \left[gk \left(\rho_2 - \rho_1 \right) + T_s k^3 \right] \tag{10.80}$$

in which the first term is the gravity wave contribution—if $T_s = 0$ and
$\rho_1 = 0$ (10.80) is (10.21)—and the second term is the capillary wave

contribution. The wave speed is

$$c_c = \frac{\sigma}{k} = \sqrt{\frac{g}{k}\frac{\rho_2 - \rho_1}{\rho_2 + \rho_1} + \frac{T_s k}{\rho_2 + \rho_1}} = \sqrt{c^2 + \tilde{c}^2} \qquad (10.81)$$

where $\tilde{c}$ is the speed of a capillary wave in the absence of gravity. For gravity waves the speed increases as the square root of the wave length whereas the wave speed due to the surface tension varies as the reciprocal of the square root of the wave length.

For water of infinite depth, the speed and group velocity of gravity waves considering the density of the second fluid (neglecting surface tension) are

$$c = \sqrt{\frac{g}{k}\frac{\rho_2 - \rho_1}{\rho_2 + \rho_1}} \qquad G = \frac{c}{2} \qquad (10.82)$$

For surface tension in the absence of gravity

$$\tilde{c} = \sqrt{\frac{T_s k}{\rho_2 + \rho_1}} \qquad \tilde{G} = \frac{d\sigma}{dk} = \frac{3}{2}\tilde{c} \qquad (10.83)$$

Unlike gravity waves, the group velocity of capillary waves is greater than that of the individual waves. Figure 10.11 shows the relationships. The wave length for minimum speed is

$$L_{\min} = 2\pi \sqrt{\frac{T_s}{g(\rho_2 - \rho_1)}} \qquad (10.84)$$

For an air-water interface, the surface tension can vary considerably depending on local conditions but using $T_s = 0.07$ N/m gives $L_{\min} = 1.7$ cm and a combined wave speed, c_c, of 0.23 m/s.

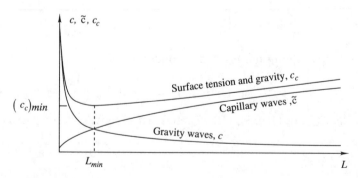

FIGURE 10.11
Wave speeds as a function of wave length.

The combined wave is stable only if the frequency is a real number

$$\frac{\sigma^2}{k^2} = \frac{gL}{2\pi}\frac{\rho_2 - \rho_1}{\rho_2 + \rho_1} + \frac{2\pi T s}{L(\rho_2 + \rho_1)} > 0 \qquad (10.85)$$

The stability condition is

$$\rho_2 - \rho_1 > -\frac{4\pi^2 T_s}{gL^2} \qquad (10.86)$$

Thus, stable waves are possible even if the heavy fluid is on the bottom, but the wave length must be sufficiently short. The only real application is for keeping liquids in a container with a very narrow mouth—such as a pipet—against gravity. Atmospheric pressure holds the fluid in the container, but the stability must be such that waves on the interface do not break and admit air. Even for widemouthed containers, the liquid could be contained if there is no disturbance, but the elimation of all disturbances is usually impossible. In narrowmouthed containers the maximum wave length is of the order of four times the diameter of the mouth. For air and water the critical wave length is about 1.7 cm so the mouth of the container should be less than 4 mm.

Consider the case of Fig. 10.9 in which there is a velocity in the two fluids, shown in Fig. 10.12. (See also Sec. 7.1.1, which treats the same problem in the absence of surface tension.) Potentials in each of the fluids are defined as

$$\Phi_1 = U_1 x + \phi_1 \qquad \Phi_2 = U_2 x + \phi_2$$

$$\frac{\partial \Phi_1}{\partial x} = U_1 + \frac{\partial \phi_1}{\partial x} \qquad \frac{\partial \Phi_2}{\partial x} = U_2 + \frac{\partial \phi_2}{\partial z} \qquad (10.87)$$

in which ϕ is the perturbation that represents the wave. The linearized kinematic and dynamic boundary conditions (assuming ϕ and η small and neglecting products of small quantities) are

$$U\frac{\partial \eta}{\partial x} + \frac{\partial \phi}{\partial z} + \frac{\partial \eta}{\partial t} = 0 \qquad \frac{p}{\rho} = \frac{\partial \phi}{\partial t} + U\frac{\partial \phi}{\partial x} - g\eta \qquad (10.88)$$

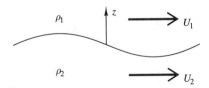

FIGURE 10.12
The interface with velocity in the fluids.

The kinematic condition applies to both fluids (two equations), but the dynamic condition sets the difference in pressures at the interface using (10.77)

$$T_s \frac{\partial^2 \eta}{\partial x^2} = \rho_1 \left(\frac{\partial \phi_1}{\partial t} + U_1 \frac{\partial \phi_1}{\partial x} + g\eta_1 \right) - \rho_2 \left(\frac{\partial \phi_2}{\partial t} + U_2 \frac{\partial \phi_2}{\partial x} + g\eta_2 \right)$$

(10.89)

which is (7.8) with surface tension term added.

The wave solution is written as

$$\eta = a e^{i(kx - \sigma t)} \qquad \phi_1 = A_1 e^{-kz} e^{i(kx - \sigma t)} \qquad \phi_2 = A_2 e^{kz} e^{i(kx - \sigma t)}$$

(10.90)

in which σ is complex. Using this solution in (10.88) and (10.89) gives

$$A_1 (i\rho_1\sigma - i\rho_1 U_1 k) + A_2 (-i\rho_2\sigma + i\rho_2 U_2 k) + a \left(\rho_2 g - \rho_1 g + T_s k^2 \right) = 0$$

$$A_1(k) \qquad\qquad\qquad\qquad\qquad + a (i\sigma - iU_1 k) \qquad\qquad = 0$$

$$\qquad\qquad A_2(-k) \qquad\qquad\qquad + a (i\sigma - iU_2 k) \qquad\qquad = 0$$

(10.91)

Considering (10.91) as a set of three equations in A_1, A_2, and a, a nontrivial solution will result only if the determinant of the coefficient matrix is zero. Setting the determinant to zero and solving for the wave speed gives

$$\frac{\sigma}{k} = \frac{\rho_1 U_1 + \rho_2 U_2}{\rho_2 + \rho_1} \pm \sqrt{\frac{g}{k}\frac{\rho_2 - \rho_1}{\rho_2 + \rho_1} - \frac{\rho_2 \rho_1 (U_2 - U_1)^2}{(\rho_2 + \rho_1)^2} + \frac{kT_s}{\rho_2 + \rho_1}}$$

(10.92)

which is the same as (7.16) except for the surface tension term. The stability condition is that the quantity under the square root is positive,

$$(U_2 - U_1)^2 \le \frac{gL}{2\pi} \frac{\rho_2^2 - \rho_1^2}{\rho_2\rho_1} + \frac{\rho_2 + \rho_1}{\rho_2\rho_1} \frac{2\pi T_s}{L}$$

(10.93)

In the case of (7.17) we argued that the wave length, L, could be very short so that for any difference in speeds there is an instability with sufficiently short waves. With the surface tension added, stability exists for finite velocity differences. Equation (10.93) can be stated as

$$(U_2 - U_1)^2 \le \frac{1 + \left(\dfrac{\rho_1}{\rho_2} \right)^2}{\dfrac{\rho_1}{\rho_2}} (c_c)_{min}^2$$

(10.94)

The "critical speed" of wind over water is about 6.6 m/s for the surface tension specified above. This value is probably near an upper limit. For greater wind speeds the water will appear ruffled; that is, reflected light will be scattered. This value also neglects the effect of viscosity, which is somewhat destabilizing. A practical value of critical speed has been elusive; investigators have proposed values from an order of magnitude less than 6.6 m/s to almost twice as much. The real value depends on exactly what is meant by "critical speed."

10.6 REFRACTION

Consider waves that approach a beach as shown in Fig. 10.13. Far from the beach the crests make an angle of α_d with the direction of the beach. Since the wave speed slows as the water becomes shallower, the part of the crest in shallower water travels more slowly than the part in deep water, creating a curved crest. Crests are conserved, however, so that an observer at any point will see the same number of crests in a given time as any other observer. Even though a wave travels more slowly in some parts, the frequency and the period are the same at all points. The wave speed is $c = L/T$ and is given by (10.28). Taking T as a constant gives an equation for the wave length in terms of the depth and deep water wave length, L_d, as

$$L = L_d \tanh \frac{2\pi d}{L} \tag{10.95}$$

which is transcendental in L.

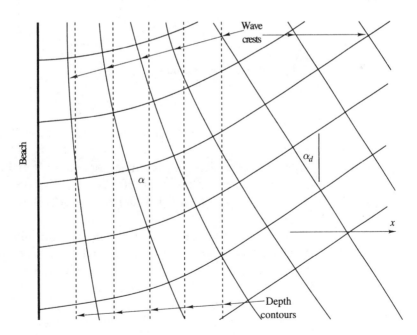

FIGURE 10.13
Refraction of waves in shoaling water.

The angle the crest makes with the beach is governed by *Snell's law*

$$\frac{\sin \alpha}{\sin \alpha_d} = \frac{c}{c_d} \tag{10.96}$$

which, together with (10.95) and (10.28), gives the crest angle as a function of depth. Using Snell's law, the entire wave pattern can be drawn for arbitrary bottom contours. The wave height can be computed if we assume that the energy between the orthogonals is constant and that linear theory is everywhere valid. Obviously, linear theory fails as the wave height grows and even the higher order theories fail when the wave breaks. In fact, it is the breaking wave, and the consequent velocity gradients that are created, that dissipate energy on the beach. The following equations apply to shoaling water but not near the beach.

The energy transmission (power) per unit length of crest is given by (10.44). Taking the power as constant between orthogonals

$$a^2 Gs = \text{constant} = a_d^2 G_d s_d \tag{10.97}$$

in which s is the distance between orthogonals. The amplitude is

$$a = a_d \sqrt{\frac{G_d s_d}{Gs}} \tag{10.98}$$

The ratio of group velocities is

$$\frac{G_d}{G} = \frac{L_d}{L} \frac{1}{1 + \dfrac{2kd}{\sinh 2kd}} \tag{10.99}$$

and the *refraction coefficient* is defined as

$$C_{ref} = \sqrt{\frac{s_d}{s}} \tag{10.100}$$

Since the space between the orthogonals becomes larger, the refraction coefficient becomes smaller and decreases the wave amplitude. The ratio of group velocities is a more complex function—it involves wave length as a function of depth as given by (10.95)—and is shown in Fig. 10.14. Note that the curve falls below 1.0 and then approaches unity asymptotically from below.

Orthogonals to the wave crests, unlike streamlines and equipotential lines, can cross. Since energy was assumed constant between orthogonals, a finite amount of energy is focused at the point of crossing and from (10.98) the wave amplitude becomes infinite. Of course, the assumptions are violated—and the wave becomes nonlinear—long before infinite waves are produced. Wave refraction over complex bottom topography can cause complicated wave patterns. These patterns can often be used to advantage. For example, a near-shore submarine canyon makes a good boat anchorage

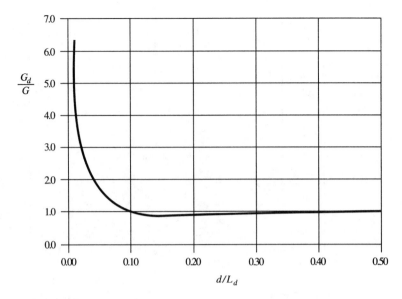

FIGURE 10.14
Ratio of group velocity versus depth over wave length.

because the orthogonals tend to diverge in the deep water, thus decreasing the wave amplitude.

Linear waves behave very much like optics and exhibit similar patterns of refraction and diffraction. Both phenomena satisfy similar equations. Waves can be focused or diffused by bottom topography just as light can be focused by a magnifying glass.

10.7 DIFFRACTION

Consider waves approaching an obstacle such as a breakwater (Fig. 10.15). The breakwater shields the water behind it from the waves, but the waves tend to bend around the end. *Huygens' principle* states that each point on a crest (or trough or anywhere else) is a disturbance that generates waves in all directions. The crest a short time later is the sum of these disturbances. For parallel waves traveling in the same direction, the envelope of these disturbances forms another crest; that is, the crest simply advances as expected. As the wave comes to the breakwater, part is cut off—actually reflected—and the other part passes the breakwater. Those points on the end of the wave generate disturbances that travel not only in the direction of the wave but also into the quiet zone. Since all of the energy is not directed in the original direction of travel, the diffracted wave is weaker than the original.

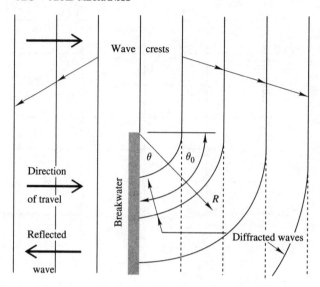

FIGURE 10.15
Wave diffraction around a breakwater.

The analysis makes use of cylindrical coordinates (Fig. 10.15). The potential is written

$$\Phi(R, \theta, t) = \phi(R, \theta) \cosh k(z + d)e^{i\sigma t} \qquad (10.101)$$

Using Laplace's equation for Φ to obtain an equation for ϕ results in

$$\frac{\partial^2 \phi}{\partial R^2} + \frac{1}{R}\frac{\partial \phi}{\partial R} + \frac{1}{R^2}\frac{\partial^2 \phi}{\partial \theta^2} + k^2\phi = 0 \qquad (10.102)$$

In Fig. 10.15 the wave approaches the breakwater at an angle θ_0 (shown as 90°). A solution that satisfies the boundary conditions is (Putnam and Authur, 1948; Ippen, 1966)

$$\phi(R, \theta) = \frac{1}{\sqrt{2}}e^{i[\pi/4 - kR\cos(\theta_0 - \theta)]}\left\{\int_{-\infty}^{u_1} e^{-i\pi u^2/2}du + \int_{-\infty}^{u_2} e^{-i\pi u^2/2}du\right\}$$

$$u_1 = 4\sqrt{\frac{kR}{\pi}}\sin\left(\frac{\theta_0 - \theta}{2}\right) \qquad u_2 = -\sqrt{4\frac{kR}{\pi}}\sin(\theta_0 + \theta)$$

$$(10.103)$$

The equation for the free surface is

$$\eta = a|F(R, \theta)|\cosh kd\ e^{i[(\pi/2)kct + \arg F(R,\theta)]} \qquad (10.104)$$

Eq. (10.103) can be solved numerically for a variety of problems.

Many real wave problems involve refraction, diffraction, and reflection. Obviously, the problems can be complex. Such practical problems as design of groins, breakwaters, piers, outfalls, harbors, and water intakes as well as the protection of beaches, wetlands, offshore oil fields, and so forth all depend on wave computations.

10.8 INTERNAL WAVES

Up to this point the development has neglected the density of the fluid (air) above the free surface. The consequence is that the pressure is taken as constant on the free surface even though the depth varies in the upper fluid and even though a velocity is induced in the upper fluid (although the latter factor is neglected in the linear equations). Since the oceans and other bodies of water are stratified, the upper fluid may be of significant density in the case of internal waves. An even more complex case is that where the fluid is continuously stratified. Each of these cases is treated briefly in this section.

10.8.1 Discrete Layers

The geometry of a two-fluid system is shown in Fig. 10.16. Waves can appear on the interface as well as the free surface. Although the interface cannot be perfectly sharp, we assume that is the case. (The velocities on each side of the interface are in the opposite direction, leading to an instability that thickens the transition from one fluid to the other. Nevertheless, the transition zone can be rather narrow.) The surface elevations are assumed as

$$\eta_1 = a_1 \cos(kx - \sigma t) + d_1 \qquad \eta_2 = a_2 \cos(kx - \sigma t) \qquad (10.105)$$

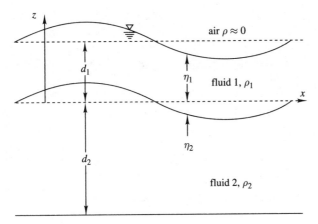

FIGURE 10.16
Waves in a two-layer fluid.

with the potentials

$$\Phi_1 = A_1 e^{kz} \sin(kx - \sigma t) + B_1 e^{-kz} \sin(kx - \sigma t)$$
$$\Phi_2 = A_2 e^{kz} \sin(kx - \sigma t) + B_2 e^{-kz} \sin(kx - \sigma t)$$

(10.106)

The linearized kinematic equations (10.15) in each fluid remain the same. The linearized Bernoulli equations take into account the different densities

$$p_1 = \rho_1 \left[\frac{\partial \Phi_1}{\partial t} - g(z - d_1) \right] \qquad p_2 = \rho_2 \left[\frac{\partial \Phi_2}{\partial t} - g\left(z - \frac{\rho_1}{\rho_2} d_1 \right) \right]$$

(10.107)

Setting the pressure to zero at the free surface and making it equivalent in each of the fluids at the interface leads to

$$A_1 \sigma e^{kd_1} + B_1 \sigma e^{-kd_1} + g a_1 = 0$$
$$\rho_2 (A_2 \sigma + B_2 \sigma) + \rho_2 g a_2 - \rho_1 (A_1 \sigma + B_1 \sigma) - \rho_1 g a_2 = 0$$

(10.108)

The kinematic conditions at the free surface and the interface are

$$A_1 k e^{kd_1} - B_1 k e^{-kd_1} + \sigma a_1 = 0 \qquad A_1 k - B_1 k + \sigma a_2 = 0$$
$$A_2 k - B_2 k + \sigma a_2 = 0$$

(10.109)

On the horizontal bottom the velocity is zero

$$A_2 k e^{-kd_2} - B_2 k e^{kd_2} = 0$$

(10.110)

As in the single fluid case, these equations—(10.108) through (10.110)—can be considered as equations in the arbitrary constants, A_1, B_1, A_2, B_2, a_1, a_2. Since these equations are homogeneous, the determinant of the coefficient matrix must be zero. That determinant is written

$$\begin{vmatrix} e^{kd_1} & e^{-kd_1} & 0 & 0 & \dfrac{g}{\sigma} & 0 \\[2mm] -\dfrac{\rho_1}{\rho_2} & -\dfrac{\rho_1}{\rho_2} & 1 & 1 & 0 & \dfrac{\rho_2 - \rho_1}{\rho_2} \dfrac{g}{\sigma} \\[2mm] e^{kd_1} & -e^{-kd_1} & 0 & 0 & \dfrac{\sigma}{k} & 0 \\[2mm] 1 & -1 & 0 & 0 & 0 & \dfrac{\sigma}{k} \\[2mm] 0 & 0 & 1 & -1 & 0 & \dfrac{\sigma}{k} \\[2mm] 0 & 0 & e^{-kd_2} & -e^{kd_2} & 0 & 0 \end{vmatrix}$$

(10.111)

Evaluating the determinant and making hyperbolic functions out of the exponentials gives

$$\sigma^4 [\rho_2 \coth kd_1 \coth kd_2 + \rho_1] - \sigma^2 gk [\rho_2 (\coth kd_1 + \coth kd_2)]$$

$$+ (\rho_2 - \rho_1) g^2 k^2 = 0 \tag{10.112}$$

Considered as an equation for σ^2,

$$\sigma^2 = \frac{gkq}{2} \pm \sqrt{\frac{g^2 k^2 q^2}{4} - \frac{g^2 k^2 (\rho_2 - \rho_1)}{\rho_2 \coth kd_1 \coth kd_2 + \rho_1}} \tag{10.113}$$

$$q = \frac{\rho_2 (\coth kd_1 + \coth kd_2)}{\rho_2 \coth kd_1 \coth kd_2 + \rho_1}$$

Taking both d_1 and d_2 as large so that the hyperbolic cotangent functions are approximately unity

$$\sigma^2 = \frac{gk}{\rho_2 + \rho_1} (\rho_2 \pm \rho_1) \tag{10.114}$$

Clearly, taking the plus sign, (10.114) is the deep water dispersion equation. Using this result, the wave amplitudes are related according to

$$a_2 = a_1 e^{-kd_1} \tag{10.115}$$

which indicates that the interfacial wave is small. Note that the function of (10.115) is the same as that of the particle velocities. If the interface is at a reasonable depth, its presence does not affect the waves on the free surface.

If the minus sign is taken in (10.114), the result is quite different. Then

$$\sigma^2 = gk \frac{\rho_2 - \rho_1}{\rho_2 + \rho_1} \tag{10.116}$$

For ρ_1 small, it is again the single fluid result, but in the usual case of a fluid stratified by temperature or salt, the two densities are nearly the same and the frequency is small (long period). In terms of the surface amplitude, the amplitude of the interface is

$$a_2 = -\frac{\rho_1}{\rho_2 - \rho_1} e^{kd_1} a_1 \tag{10.117}$$

Not only is the interface amplitude amplified by depth, it is large because of the small number in the denominator. Further, the phase is reversed (by the minus sign) so that a crest on the free surface is above a trough on the interface. However, it is the interface that is affected by a wave on the surface; the surface wave is only slightly dependent on a wave at the interface. The general solution for waves in a stratified fluid includes the superposition of both the above cases.

Taking the lower fluid as infinite but the upper fluid as finite gives the dispersion equation as

$$\sigma^2 = gk \frac{\rho_2 + \rho_2 \coth kd_1 \pm (\rho_2 - \rho_2 \coth kd_1 - 2\rho_1)}{2(\rho_1 + \rho_2 \coth kd_1)} \qquad (10.118)$$

The two roots are

$$\sigma^2 = gk \qquad \sigma^2 = gk \frac{\rho_2 - \rho_1}{\rho_2 \coth kd_1 + \rho_1} \qquad (10.119)$$

The amplitude functions are given by (10.115) and (10.117).

10.8.2 Continuous Stratification

Taking the density as a function of depth may be a more realistic assumption than that of distinct layers. The result is that a free surface exists (the air-water interface), but there is no sharp interface between liquids of different densities. We will consider only internal waves; that is, neglect movement of the free surface.

Following Lamb (1945) the density and pressure are taken as perturbations about a mean value

$$\rho = \rho_0 + \rho' \qquad p = p_0 + p' \qquad (10.120)$$

in which the primed quantities are small compared to the average quantities. The linearized equations of motion in the horizontal and vertical directions are

$$\rho_0 \frac{\partial u_x}{\partial t} = -\frac{\partial p'}{\partial x} \qquad \rho_0 \frac{\partial u_z}{\partial t} = -\frac{\partial p'}{\partial z} - \rho' g \qquad (10.121)$$

The pressure gradients cannot be discarded without eliminating the problem. Equation (10.121) indicates that the velocity components are of the order of the perturbation quantities.

The equation of conservation of mass is unchanged from the constant density case,

$$\frac{\partial u_x}{\partial x} + \frac{\partial u_z}{\partial z} = 0 \qquad (10.122)$$

We also assume that a fluid particle retains its density wherever it goes (density does not diffuse, or at least the time scale of diffusion is small compared to the wave period) so that

$$\frac{D}{Dt}(\rho_0 + \rho') = 0 \qquad (10.123)$$

Neglecting products of small quantities, including velocity, and taking $\partial \rho_0 / \partial x = 0$ gives

$$\frac{\partial \rho'}{\partial t} + u_z \frac{d\rho_0}{dz} = 0 \qquad (10.124)$$

which makes explicit the restriction that the reference density is a function of the vertical coordinate only.

Lamb now defines a stream function, eliminating p' and ρ' to find the equation

$$\nabla^2 \left(\frac{\partial^2 \Psi}{\partial t^2} \right) + \frac{1}{\rho_0} \frac{d\rho_0}{dz} \left(\frac{\partial}{\partial z} \frac{\partial^2 \Psi}{\partial t^2} - g \frac{\partial^2 \Psi}{\partial x^2} \right) = 0 \qquad (10.125)$$

Using (10.24)

$$\frac{\partial p'}{\partial t} = \rho_0 g \frac{\partial \Psi}{\partial x} \qquad (10.126)$$

and from the dynamic condition

$$\frac{\partial}{\partial z} \frac{\partial^2 \Psi}{\partial t^2} = g \frac{\partial^2 \Psi}{\partial x} \qquad (10.127)$$

The stream function is taken as

$$\Psi = \psi(z) e^{i(kx - \sigma t)} \qquad (10.128)$$

Equations (10.125) and (10.127) now become

$$\frac{\partial^2 \psi}{\partial z^2} - k^2 \psi + \frac{1}{\rho_0} \frac{d\rho_0}{dz} \left(\frac{\partial \psi}{\partial z} - \frac{gk^2}{\sigma^2} \psi \right) = 0 \qquad \frac{\partial \psi}{\partial z} - \frac{gk^2}{\sigma^2} \psi = 0 \quad (10.129)$$

If $\psi(z)$ is proportional to e^{kz}, (10.129) is satisfied if $\sigma^2 = gk$; that is, the dispersion equation is unchanged from the homogeneous case independently of the distribution of density.

The latter two equations can be solved for $\psi(z)$, but in general the solution must be done numerically for an arbitrary vertical distribution of density. If the density varies exponentially (Lamb, 1945), an analytical solution is possible. The density variation is

$$\rho_0 = C e^{-\beta z} \qquad (10.130)$$

Then,

$$\frac{\partial^2 \psi}{\partial z^2} - k^2 \psi - \beta \left(\frac{\partial \psi}{\partial z} - \frac{gk^2}{\sigma^2} \right) \psi = 0 \qquad (10.131)$$

The solution for Ψ is

$$\Psi = \left(A e^{\lambda_1 z} + B e^{\lambda_2 z} \right) e^{i(kx - \sigma t)} \qquad (10.132)$$

where λ satisfies the equation

$$\lambda^2 - \beta \lambda + \left(\frac{g\beta}{\sigma^2} - 1 \right) k^2 = 0 \qquad (10.133)$$

If the roots of (10.133) are real, the stratification has a small effect on the surface waves. Yih (1977) indicates that a condition for internal waves is

$$\sigma^2 \le -\frac{g}{\rho_0} \frac{d\rho_0}{dz} \tag{10.134}$$

Lamb goes on to assume that λ is complex

$$\lambda = \frac{\beta}{2} \pm im = \frac{\beta}{2} \pm i\sqrt{\left(\frac{g\beta}{\sigma^2} - 1\right)k^2 - \frac{\beta^2}{4}} \tag{10.135}$$

Then

$$\psi = Ce^{\beta(z+d)/2} \sin m(z+d) \tag{10.136}$$

Using the second of (10.129) gives the dispersion equation

$$\frac{\beta}{2} \sin md + m \cos md = \frac{gk^2}{\sigma^2} \sin md \tag{10.137}$$

σ can be eliminated by the definition of m from (10.135), giving

$$\tan md = \beta d \frac{md}{m^2 d^2 + k^2 d^2 - \dfrac{\beta^2 d^2}{4}} \tag{10.138}$$

which is a transcendental equation for m given the depth and the density distribution, β. There are multiple solutions to this equation that give the frequencies and wave lengths of the internal waves. These are the waves that result from stratification due to temperature or salt, the internal motion in lakes and oceans.

PROBLEMS

10.1. The captain of a submarine would like to know conditions on the surface of the ocean before surfacing. Only two gages are available to indicate the outside pressure, one gage at each end of the submarine. The submarine is oriented normal to the crests of the surface wave by maximizing the time between maximum (or minimum) pressure readings on each end. The following data are taken: (*i*) each pressure gage reads 13 m (of water) at the minimum and 16 m at the maximum; (*ii*) the time between maxima on either gage is 6 seconds; and (*iii*) the gage at the rear lags the one at the front by 5 s. The constants of the problem are: specific weight of sea water = 1024 kg/m^3; length of the submarine = 80 m. Assuming linear waves, find

(*a*) the wave period,
(*b*) the wave length,
(*c*) the speed of the waves,
(*d*) the speed of the submarine,
(*e*) the depth of the submarine below mean sea level, and
(*f*) the wave height.
(*g*) Comment on the accuracy of the above quantities.

10.2. Gerstner's wave is an example of a finite amplitude wave. Paradoxically its derivation begins by assuming that the particle paths (in infinitely deep water) are circles. The free surface is not sinusoidal, except in the small amplitude approximation. The particle paths are

$$\xi = x_0 - ae^{kz_0} \sin(kx_0 - \sigma t)$$

$$\zeta = z_0 + ae^{kz_0} \cos(kx_0 - \sigma t)$$

in which x_0 and z_0 are coordinates of the mean particle positions.

(a) Show that the particle paths are circles by the elimination of time.

(b) Find the particle velocities, u_ξ and u_ζ.

(c) On the free surface $z_0 = 0$ so

$$\xi = x_0 - a \sin(kx_0 - \sigma t)$$

$$\eta = \zeta = a \cos(kx_0 - \sigma t)$$

Show that the pressure condition on the free surface is satisfied if $\sigma^2 = gk$. View the problem as steady flow by subtracting the wave speed $c = \sigma/k$ from the horizontal particle velocity, then use Bernoulli's equation applied at the free surface.

(d) Show that the kinematic condition is satisfied. Showing that the stream function exists for the steady flow formulation is sufficient. Use x_0 and z_0 as coordinates.

(e) Show that a potential does not exist; Gerstner's wave is not irrotational.

10.3. The captain of a luxury cruise ship, the U. S. S. Smudgepot, is interested in maintaining a smooth voyage. Noticing that the waves are catching up to the ship, the captain reasons that if the ship can travel as fast as the crests it will sail more smoothly and will cure the passengers' seasickness. Upon reaching a speed of 8.2 m/s the ride becomes smooth although the wave crests are still going faster than the ship. The wave crests are at right angles to the direction of the ship.

(a) Compute the wave speed (not 8.2 m/s) and the wave length.

(b) A frightened passenger has just heard on the radio that an earthquake has oc-

curred at a point 1500 km distant at 2 pm and has sent out a very large tsunami. The passenger, of course, has a mental image of the ship lying on the ocean floor 300 m below. What will happen when the tsunami reaches the position of the ship? When will it happen?

10.4. A trochoid is a better approximation for a finite amplitude wave than a sine curve. The equations for the water surface of a trochoidal wave are

$$x = \frac{L}{2\pi}\alpha + \frac{H}{2}\sin\alpha \qquad z = \frac{H}{2}\cos\alpha$$

where L is the wave length, H is the wave height, x is measured horizontally, z is measured vertically, and α is a parameter. Compute the average potential energy per unit of surface area for the trochoidal wave.

10.5. For a line source where waves are generated by an instantaneous burst of energy, the wave amplitude varies as $t^{-1/2}$.

(a) How would the amplitude vary for waves generated by an instantaneous burst of energy at a point source?

(b) If the wave energy is continually being generated at the point source, how does the wave amplitude vary with distance from the source?

10.6. A ship 250 m long is traveling at right angle to the wave crests. A bored passenger notices that the wave crests are overtaking the ship each 10 s and a crest takes 10 s to travel the length of the ship.

(a) Find the wave length.

(b) Find the wave speed.

(c) Find the speed of the ship.

10.7. Consider a standing wave of amplitude a and length L in water of mean depth η_0. The standing wave has been created by a traveling wave being reflected from a vertical wall. Using linear wave theory, write an equation for the wave force on the wall. Express the equation in terms of wave length and period, T.

10.8. Many have made proposals to harness wave energy of the oceans. Suppose a plant of 30 percent efficiency could be constructed and suppose the average wave height (not wave amplitude) is 2 m with a wave length of 80 m.

(*a*) For what linear distance along the coast would energy have to be collected to produce 1000 megawatts (about the same as a small nuclear plant)?

(*b*) Where does this wave energy originate and how does it end up in waves?

(*c*) In the absence of the plant to extract energy from the waves, what happens to it?

10.9. For the wave shown in Fig. 10.17 calculate the potential energy per unit area of surface.

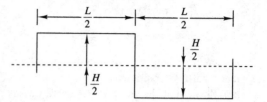

FIGURE 10.17

CHAPTER
11

TRANSPORT

Fluid mechanics is the study of the transport of mass, momentum, and energy. In this chapter we wish to study the transport of a substance by fluid, a study that may or may not be considered a part of fluid mechanics. The objective of many fluid flow calculations, however, is to form the basis for transport calculations that define the movement of natural substances such as water vapor as well as a host of manmade pollutants. In general the transport equations are more difficult to solve accurately than the corresponding fluid flow problem, which is not surprising when one considers that the fluid flow problem must be solved *very* accurately—meaning that some of the common approximations are not suitable—if the solution to the transport problem is to be acceptable. Most practical solutions to the advection-diffusion equation are numerical, but some of the elementary solutions and the behavior of the equation are contained in this chapter.

At the outset we wish to make a central assumption: the transported substance has no effect on the density, viscosity, or other properties of the fluid; nor does it affect the flow of the fluid. Thus, the transport calculation is independent of, and can be applied after, the flow calculation. In solving the equations of transport, we assume that the velocity of the fluid is given at every point. The assumptions omit a great many transport problems, namely those that treat two or more miscible fluids in which neither can be uniquely identified as the substrate and those problems of stratified flow where the

425

contaminant has an effect on density. Common problems of pollution and the transport of chemicals in small concentrations are included within the assumptions.

The amount of substance in the fluid is specified by the *concentration*, designated by the symbol $\mathbb{C}$, which is the mass of the substance per unit volume (in three dimensions) of fluid.

11.1 EQUATIONS

We consider only two methods of transport, *advection* and *diffusion*. In addition there is *dispersion*, which is really not a separate process but results from other assumptions or approximations of the flow.

11.1.1 Advection

Equation (1.20) applies to the transport of the fluid mass; it could just as well apply to the transport of another substance,

$$\frac{\partial \mathbb{C}}{\partial t} + \vec{\nabla} \cdot (\mathbb{C}\vec{u}) = 0 \qquad \text{or} \qquad \frac{\partial \mathbb{C}}{\partial t} + \frac{\partial}{\partial x_i}(\mathbb{C}u_i) = 0 \qquad (11.1)$$

The assumption of constant fluid density ($\vec{\nabla} \cdot \vec{u} = 0$) gives

$$\frac{D\mathbb{C}}{Dt} = \frac{\partial \mathbb{C}}{\partial t} + u_i \frac{\partial \mathbb{C}}{\partial x_i} = 0 \qquad (11.2)$$

which indicates that the concentration is constant for the fluid particle, that is, the substance is *conservative*; it does not decay, is not adsorbed, is not absorbed, and it does not undergo chemical, biological, or nuclear transformation.

In the transport literature the particle path lines are often called *characteristics*. Indeed, (11.2) is a hyperbolic equation that possesses a single characteristic and has the unusual property that any change in $\mathbb{C}$ is felt only along that characteristic. The characteristic is exactly the path line of the fluid particle, given by $d\vec{x}_p/dt = \vec{u}$.

11.1.2 Diffusion

Fick's law of diffusion states that the transport of concentration of a substance in a motionless fluid is proportional to the negative of the concentration gradient

$$q_i = -Ð\frac{\partial \mathbb{C}}{\partial x_i} \qquad (11.3)$$

in which D is the *diffusion coefficient* (or *dispersion coefficient*; dispersion is defined in a subsequent section). Fick's law is based on molecular transport and states that a substance tends to equalize its distribution in a substrate; that is, it flows from a zone of high concentration to a zone of low concentration. For example, Fick's law applies to heat flow and indicates that heat moves from a relatively high temperature zone to one of relatively low temperature, thus equalizing throughout the fluid. The diffusion coefficient is a constant of proportionality that depends on both the fluid and the contaminant.

If the contaminant is a conservative substance, the equation of continuity applies

$$\frac{\partial C}{\partial t} + \frac{\partial q_i}{\partial x_i} = \frac{\partial C}{\partial t} - \frac{\partial}{\partial x_i}\left(D\frac{\partial C}{\partial x_i}\right) = 0 \tag{11.4}$$

Equation (11.4) comes directly from the integral relationships or by applying the flow of the contaminant to an elementary volume (Fig. 11.1). Summing the net inflow and equating it to the rate of change of concentration in the volume produces (11.4). Steady state, homogeneous diffusion is governed by the Laplace equation, (11.4) without the time derivative and with D independent of the space variables.

11.1.3 Advection-Diffusion

Applying the diffusion equation to a fluid particle gives

$$\frac{DC}{Dt} = \frac{\partial}{\partial x_i}\left(D\frac{\partial C}{\partial x_i}\right) \tag{11.5}$$

Or, using the definition of the substantial derivative

$$\frac{\partial C}{\partial t} + u_i\frac{\partial C}{\partial x_i} = \frac{\partial}{\partial x_i}\left(D\frac{\partial C}{\partial x_i}\right) \tag{11.6}$$

A reaction and a source term are often added to the equation

$$\frac{\partial C}{\partial t} + u_i\frac{\partial C}{\partial x_i} = \frac{\partial}{\partial x_i}\left(D\frac{\partial C}{\partial x_i}\right) + RC + S \tag{11.7}$$

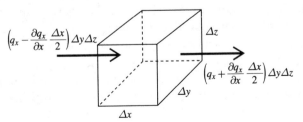

FIGURE 11.1
Flow in and out of an elementary volume.

in which $\mathbb{R}$ is the reaction rate and S is the source term, which may be a function of all the dependent variables and of $\mathbb{C}$.

The diffusion coefficient has been taken as a scalar. In anisotropic media—most porous media—diffusion is dependent on direction and the coefficient becomes a tensor

$$\frac{\partial \mathbb{C}}{\partial t} + u_i \frac{\partial \mathbb{C}}{\partial x_i} = \frac{\partial}{\partial x_i}\left(Ð_{ij}\frac{\partial \mathbb{C}}{\partial x_j}\right) + \mathbb{R}\mathbb{C} + S \tag{11.8}$$

Since the diffusion tensor is symmetric, $Ð_{ij} = Ð_{ji}$, it possesses principal axes and can be defined by three diagonal terms. For molecular diffusion, $Ð_{ij}$ is not a function of position or direction and (11.8) becomes

$$\frac{\partial \mathbb{C}}{\partial t} + u_i \frac{\partial \mathbb{C}}{\partial x_i} = Ð\frac{\partial^2 \mathbb{C}}{\partial x_i \partial x_i} + \mathbb{R}\mathbb{C} + S \tag{11.9}$$

Defining dimensionless variables

$$x_i^* = \frac{x_i}{L} \qquad t_* = \frac{U}{L}t \qquad u_i^* = \frac{u_i}{U} \qquad \mathbb{C}_* = \frac{\mathbb{C}}{\mathbb{C}_0} \tag{11.10}$$

the advection-diffusion equation becomes

$$\frac{\partial \mathbb{C}_*}{\partial t_*} + u_i^* \frac{\partial \mathbb{C}_*}{\partial x_i^*} = \frac{ÐL}{U}\frac{\partial^2 \mathbb{C}_*}{\partial x_i^* \partial x_i^*} \tag{11.11}$$

and the *Péclet number* is defined as

$$Pe = \frac{UL}{Ð} \tag{11.12}$$

The Péclet number is the ratio of diffusion to advection over the characteristic length, L. A small Péclet number indicates that diffusion dominates the transport of a contaminant.

11.2 SOLUTIONS FOR CONSTANT DIFFUSION COEFFICIENT

A number of methods are available for solving the diffusion and advection-diffusion equations (Crank, 1975). We shall treat these lightly, especially since most applications are complex enough to use numerical methods. Aside from applications to simple cases, the analytical solutions can serve to show properties of the equations and, secondly, to serve as a qualitative and quantitative check on numerical methods. The latter is important because some of the numerical solutions are notoriously inaccurate, especially for high Péclet numbers.

11.2.1 One-dimensional Diffusion

As illustrated in Sec. 8.2, the diffusion equation is parabolic. Thus, it requires two boundary conditions—say $\mathbb{C}$ specified at $x = 0$ and $x = L$—and one initial condition. Any solution, numerical or analytical, depends on the initial and boundary conditions.

First, consider the case where $\mathbb{C}(x, 0) = \mathbb{C}(\infty, t) = 0$ and where a given amount of the contaminant is injected at $x = 0$ at time $t = 0$. The solution of the diffusion equation is

$$\mathbb{C} = \frac{M_0}{2\sqrt{\pi Ðt}} e^{-x^2/4Ðt} \tag{11.13}$$

in which M_0 is the constant of integration and represents the mass of the substance that is initially at $x = 0$. The solution is shown in Fig. 11.2. For all times $\mathbb{C} \to 0$ as $x \to \infty$, so the substance spreads with an infinite velocity but the quantity away from the source is small initially and grows as time passes. The total quantity of the substance is conserved, so the concentration at the source must decrease as the substance spreads.

Second, consider the situation in which an infinite reservoir of the contaminant with concentration $\mathbb{C}_0$ is located on the negative x-axis and on the positive x-axis the concentration is initially zero. The boundary and initial conditions are

$$\mathbb{C}(x, t) = \mathbb{C}_0 \quad \text{for } x \le 0 \qquad \mathbb{C}(x, 0) = 0 \quad \text{for } x > 0 \qquad \mathbb{C}(\infty, t) = 0 \tag{11.14}$$

The solution is

$$\mathbb{C}(x, t) = \frac{\mathbb{C}_0}{2} \text{erfc} \frac{x}{2\sqrt{Ðt}} \tag{11.15}$$

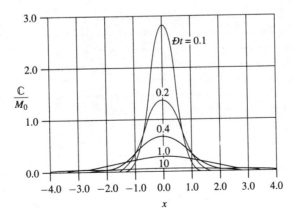

FIGURE 11.2
Solution of point diffusion.

where the function is the complimentary error function, defined as

$$\text{erfc}(\xi) = 1 - \text{erf}(\xi) = 1 - \frac{2}{\sqrt{\pi}} \int_0^\xi e^{-\xi^2} d\xi \qquad (11.16)$$

in which erf is an error function. Error functions and complimentary error functions are tabulated in handbooks and appear as computer subroutines. The solution is shown in Fig. 11.3. The full concentration eventually spreads to the entire positive x-axis.

11.2.2 Multi-dimensional Diffusion

In two dimensions with constant diffusion coefficient

$$\frac{\partial \mathbb{C}}{\partial t} = D \left(\frac{\partial^2 \mathbb{C}}{\partial x^2} + \frac{\partial^2 \mathbb{C}}{\partial y^2} \right) \qquad (11.17)$$

For a constant mass the solution is

$$\mathbb{C} = \frac{M_0}{4\pi Dt} e^{-R^2/4Dt} \qquad (11.18)$$

In three dimensions (Crank, 1975)

$$\mathbb{C} = \frac{M_0}{8(\pi Dt)^{3/2}} e^{-r^2/4Dt} \qquad (11.19)$$

There are, of course, many more types of solution. The method of separation of variables (e.g., Churchill, 1987) leads to Fourier series and Bessel function solutions. Most practical solutions, however, come from numerical calculations.

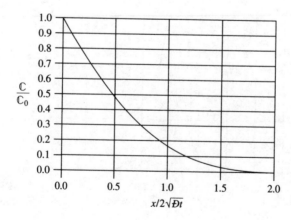

FIGURE 11.3
Solution for a fixed concentration.

11.2.3   Simple Advection-Diffusion

In pure advection the concentration of a substance stays with the particle. Adding diffusion, the concentration at the fluid particle changes according to the above formulas by simple translation. For example, (11.13) becomes

$$\mathbb{C} = \frac{M_0}{2\sqrt{\pi Dt}} e^{-(x-Ut)^2/4Dt} \tag{11.20}$$

in which U is the (constant) velocity of translation.

A major use of the analytic solutions for simple advection-diffusion is in checking numerical solutions to this problem. Typically, the numerical solution to diffusion-dominated problems is simple and accurate, but the solution to advection-dominated problems (small diffusion coefficient) gives unwanted numerical oscillations and "numerical diffusion." Since $\mathbb{C}$ is infinite at $t = 0$, no numerical method can accurately compute from $x = 0$ and $t = 0$ in (11.20); therefore, a numerical calculation should begin with an initial condition that is set at a later time. Equation (11.20) can be written

$$\mathbb{C} = \mathbb{C}_0 \sqrt{-\frac{t_0}{t-t_0}} \exp\left[-\frac{(x-Ut)^2}{4D(t-t_0)}\right] \tag{11.21}$$

in which x_0 and t_0 are the distance and time of the original source and the latter is assumed to be a negative number. The equation is normalized so that $\mathbb{C} = \mathbb{C}_0$ at $x = 0$ and $t = 0$. Then the concentration can be computed for all values of positive x and t and compared to a numerical calculation that uses the $x = 0$ and $t = 0$ curve as the initial condition.

A more interesting case is that of (11.15) in which there is flow in the positive x-direction, but the boundary condition on $\mathbb{C}$ remains at $x = 0$. The solution is (Carslaw and Jaeger, 1959)

$$\mathbb{C} = \frac{\mathbb{C}_0}{2}\left(\operatorname{erfc}\frac{x-Ut}{2\sqrt{Dt}} + e^{Ux/D}\operatorname{erfc}\frac{x+Ut}{2\sqrt{Dt}}\right) \tag{11.22}$$

which is pictured in Fig. 11.4. Both (11.21) and (11.22) are often used to check one-dimensional numerical programs.

Example 11.1.   Pollution from a continuous source.   Consider a source of a contaminant in an infinite flow field. The steady state advection-diffusion equation is [see (11.9)]

$$u_i \frac{\partial \mathbb{C}}{\partial x_i} = D \frac{\partial^2 \mathbb{C}}{\partial x_i \partial x_i}$$

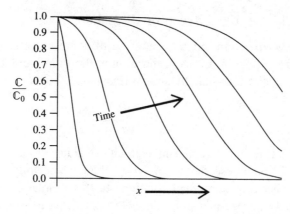

FIGURE 11.4
Advection-diffusion solution with an infinite reservoir on the negative x-axis.

Writing the Laplacian in cylindrical coordinates and taking the advective velocity as a constant in the x-direction gives

$$u_0 \frac{\partial \mathbb{C}}{\partial x} = \mathbb{D} \left[\frac{1}{R} \frac{\partial}{\partial R} \left(R \frac{\partial \mathbb{C}}{\partial R} \right) + \frac{\partial^2 \mathbb{C}}{\partial x^2} \right]$$

with the boundary conditions

$$\mathbb{C} = 0 \quad \text{as } r \to \infty \qquad \frac{\partial \mathbb{C}}{\partial r} = 0 \quad \text{at } r = 0$$

$$\text{and} \qquad -4\pi r^2 \mathbb{D} \frac{\partial \mathbb{C}}{\partial r} = Q \quad \text{as } r \to 0$$

where Q is the quantity of contaminant being discharged, $R = \sqrt{y^2 + z^2}$, and $r = \sqrt{x^2 + y^2 + z^2}$. The solution is

$$\mathbb{C} = \frac{Q}{4\pi \mathbb{D} r} \exp \left[-\frac{u_0}{2\mathbb{D}} (r - x) \right]$$

11.3 DISPERSION

11.3.1 Hydrodynamic Dispersion

Consider laminar flow through a pipe. The (uniform, no change in the x-direction) velocity distribution is a parabola of revolution (Fig. 11.5). At time $t = 0$ a substance is inserted uniformly across the pipe. The portion of the substance near the centerline is carried downstream with greater velocity than that portion near the pipe walls. After a time $t = t_1$ the substance appears as shown in the lower part of the figure. Dispersion has occurred as a result of the uneven velocity distribution and it would occur in the absence of molecular diffusion.

A correct two-dimensional solution of the advection-diffusion equation would find the concentration of the substance after the time t_1. If, however,

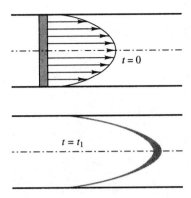

FIGURE 11.5
Dispersion in laminar pipe flow.

the flow is treated as one-dimensional with only the average velocity defined, and if samples are taken across the channel, a superficial look might indicate that the distribution is due to a *very* high value of molecular diffusion. Thus, the one-dimensional approximation to the advection-diffusion equation is to treat the dispersion as diffusion and to take a large value of Đ. Unlike the diffusion coefficient, the *dispersion coefficient* will depend on the state of flow and on the dimensional approximation; that is, solving a two-dimensional problem in one dimension or a three-dimensional problem in one or two dimensions leads to dispersion.

Three conditions must be considered: (1) molecular diffusion in which the diffusion coefficient is a property of the fluid and changes with the molecular activity (which is influenced by temperature) but not with the flow; (2) turbulent diffusion in which the diffusion coefficient reflects the mixing due to the turbulence of the flow and is analogous to the eddy viscosity; and (3) dispersion in which the "diffusion" coefficient (better called the dispersion coefficient) is made to account for the dimensional approximation.

11.3.2 Porous Media

The situation in porous media is more complicated because the "velocity" is the fluid transport velocity calculated by dividing the quantity of flow by the cross-sectional area of the media, including the area occupied by solids (i.e., the specific discharge). Even if the macroscopic flow is one-dimensional, the fluid must wind around the individual grains and thus the microscopic flow is three-dimensional; hence, the dispersion is three-dimensional regardless of the dimensionality of the averaged flow (Fig. 11.6). Flow that passes through a three-dimensional labyrinth of solid particles may carry a contaminant that is dispersed to all parts. Remember that in flow in porous

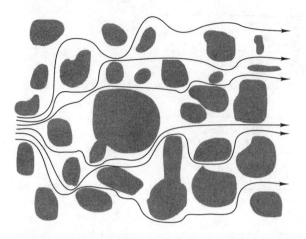

FIGURE 11.6
Flow in porous media.

media there are three scales:

$$\text{molecular scale} \ll \text{Darcy scale} \ll \text{problem scale}$$

The molecular diffusion takes place on the molecular scale, the dispersion takes place on the Darcy scale, and the solution is averaged on the problem scale. In general the dispersion is much larger than the diffusion, but if there is no velocity, there is no dispersion. Nevertheless, dispersion is commonly represented by Fick's law. In that case the diffusion equation is

$$\frac{\partial \mathbb{C}}{\partial t} + u_i \frac{\partial \mathbb{C}}{\partial x_i} = \left(\mathcal{D}_{ij}^d + \mathcal{D}_{ij}^m \right) \frac{\partial^2 \mathbb{C}}{\partial x_i \partial x_j} + \mathbb{R}\mathbb{C} + S \qquad (11.23)$$

in which $\mathcal{D}_{ij}^m$ is the molecular diffusion coefficient and $\mathcal{D}_{ij}^d$ is the hydrodynamic dispersion coefficient, both assumed constant. Again, $\mathcal{D}_{ij}^d$ is highly dependent on the state of the flow.

Measurements in porous media indicate that the dispersivity in the direction of the flow is larger than that in the transverse direction. It thus becomes a tensorial quantity, even in the case of an isotropic medium. Bear (1979) defines the dispersion coefficient as

$$\mathcal{D}_{ij}^d = \alpha_T \bar{u} \delta_{ij} + (\alpha_L - \alpha_T) \frac{\bar{u}_i \bar{u}_j}{\bar{u}} \qquad (11.24)$$

in which α_L is the *longitudinal dispersivity* (in the direction of flow), α_T is the *transverse dispersivity* (at right angles to the flow) and the velocities are the transport velocities with $\bar{u}$ the square root of the sum of the squares of the component velocities. The overscore indicates that the "velocities" are really specific discharges and not point velocities. The components of the dispersion tensor are

$$Đ_{xx} = \alpha_T \bar{u} + (\alpha_L - \alpha_T)\frac{\bar{u}_x^2}{\bar{u}} \qquad Đ_{xy} = Đ_{yx} = (\alpha_L - \alpha_T)\frac{\bar{u}_x \bar{u}_y}{\bar{u}}$$

$$Đ_{xz} = Đ_{zx} = (\alpha_L - \alpha_T)\frac{\bar{u}_x \bar{u}_z}{\bar{u}} \qquad Đ_{yy} = \alpha_T \bar{u} + (\alpha_L - \alpha_T)\frac{\bar{u}_y^2}{\bar{u}}$$

$$Đ_{yz} = Đ_{zy} = (\alpha_L - \alpha_T)\frac{\bar{u}_y \bar{u}_z}{\bar{u}} \qquad Đ_{zz} = \alpha_T \bar{u} + (\alpha_L - \alpha_T)\frac{\bar{u}_z^2}{\bar{u}}$$

$$(11.25)$$

Notice that even if $\bar{u}_y = \bar{u}_z = 0$, $Đ_{yy}$ and $Đ_{zz}$ are not zero—only the off-diagonal terms $Đ_{xy} = Đ_{xz} = Đ_{yz} = 0$—and dispersion takes place normal to the flow. Taking one of the coordinates in the direction of the flow places the dispersion tensor in its principal axes.

11.4 HEAT

The equation for thermal energy (1.99) is [see (1.50) for an explanation of terms]

$$\rho\frac{De}{Dt} = -\frac{\partial q_i}{\partial x_i} - p\frac{\partial u_i}{\partial x_i} - \tau_{ij}\frac{\partial u_j}{\partial x_i} \qquad (11.26)$$

Taking $v_s = 1/\rho$ as the specific volume, the change of internal energy can be written

$$\frac{De}{Dt} = \left(\frac{\partial e}{\partial v_s}\right)_T \frac{Dv_s}{Dt} + \left(\frac{\partial e}{\partial T}\right)_{v_s}\frac{DT}{Dt} = \left[-p + T\left(\frac{\partial p}{\partial T}\right)_{v_s}\right]\frac{Dv_s}{Dt} + c_v\frac{DT}{Dt} \qquad (11.27)$$

in which c_v is the specific heat capacity at constant volume. The derivative of the specific volume is

$$\frac{Dv_s}{Dt} = \frac{D}{Dt}\left(\frac{1}{\rho}\right) = -\frac{1}{\rho^2}\frac{D\rho}{Dt} = -\frac{1}{\rho}\frac{\partial u_i}{\partial x_i} \qquad (11.28)$$

Fick's law for temperature (now called *Fourier's law*) is

$$q_i = -\rho c_v Đ^h \frac{\partial T}{\partial x_i} \qquad (11.29)$$

in which $\rho c_v Đ^h$ is the thermal conductivity of the fluid. Using these relationships in (11.26), the energy equation can be written in terms of temperature

$$\rho c_v\frac{DT}{Dt} = \rho c_v Đ^h\frac{\partial^2 T}{\partial x_i \partial x_i} - T\left(\frac{\partial p}{\partial T}\right)_{v_s}\frac{\partial u_i}{\partial x_i} - \tau_{ij}\frac{\partial u_i}{\partial x_j} \qquad (11.30)$$

For most heat transfer problems the heat produced by the viscous effects is small. Neglecting those terms and specializing to a constant density fluid produces

$$\frac{\partial T}{\partial t} + u_i \frac{\partial T}{\partial x_i} = Ɖ^h \frac{\partial^2 T}{\partial x_i \partial x_i} \qquad (11.31)$$

Just as $Ɖ$ is dependent on the substance and the substrate, $Ɖ^h$ is not necessarily the same as $Ɖ$ and is dependent on the fluid. For air $\rho c_v Ɖ^h = 90.8 \times 10^{-6}$ cal/s cm °K; for water $\rho c_v Ɖ^h = 0.0015$ cal/s cm °K, but both of these quantities change with molecular activity, which is dependent on temperature.

For problems in *forced convection*—the impetus for the flow is not the heat but an outside driving mechanism—the equations of fluid mechanics are solved without regard to the transport of heat and then (11.31) is solved separately. For problems in *free convection*—the driving mechanism is buoyancy that results from density changes due to temperature—the energy equation (11.26) must be coupled with the equations for conservation of mass and momentum.

11.5 TURBULENCE

The concentration can be divided into average and fluctuating parts

$$ℂ = \overline{ℂ} + ℂ' \qquad (11.32)$$

in which the time average is

$$\overline{ℂ} = \frac{1}{2T} \int_{-T}^{T} ℂ \, dt \qquad (11.33)$$

and the average of the fluctuating terms is zero. The advection-diffusion equation becomes

$$\frac{\partial \overline{ℂ}}{\partial t} + \overline{u}_i \frac{\partial \overline{ℂ}}{\partial x_i} = \frac{\partial}{\partial x_i} \left(Ɖ \frac{\partial \overline{ℂ}}{\partial x_i} - \overline{u_i' ℂ'} \right) \qquad (11.34)$$

Empirical expressions for turbulent transport of a contaminant are written in analogy with turbulent flow. An "eddy diffusion coefficient," $Ɖ^t$, can replace $Ɖ$ in Fick's law (11.3) and the subsequent equations. All the disadvantages of the eddy viscosity apply: The eddy diffusion is dependent on the flow and not a property of the fluids. If its distribution for a given flow can be guessed with reasonable accuracy, the resulting calculations can be satisfactory. Other empirical formulas are:

Prandtl mixing length (see Sec. 7.3.2)

$$D^t = l^2 \left| \frac{d\bar{u}_x}{dy} \right| \tag{11.35}$$

Kármán similarity theory (see Sec. 7.3.3) on

$$D^t = -\kappa^2 \frac{\left| \left(\frac{d\bar{u}_x}{dy} \right)^3 \right|}{\left(\frac{d^2\bar{u}_x}{dy^2} \right)^2} \tag{11.36}$$

and Deissler's law of the wall (see Sec. 7.3.4)

$$D^t = -n^2\bar{u}_x y \left[1 - \exp\left(-\frac{\rho n^2 \bar{u}_x y}{\mu} \right) \right] \tag{11.37}$$

All of these relationships assume (if the same constants are used) that the turbulent mass transfer coefficient is the same as the turbulent momentum transfer coefficient; that is, that the *turbulent Schmidt number*

$$Sc^t = \frac{v^t}{D^t} \tag{11.38}$$

is unity.

11.6 SOLUTIONS

There are few exact solutions to the advection-diffusion equation. Cohen and Lewis (1967) developed the so-called ray method patterned after characteristic solutions to hyperbolic equations. The method has been expanded by Smith (1981) to obtain asymptotic solutions of the advection-diffusion equation. Smith's solutions appear in infinite series for large Péclet number. In some limiting cases, however, the method gives exact solutions in that the series contains only one term. Smith shows several examples.

11.6.1 Shear Flow Solution

Okubo and Karweit (1969) have found solutions for instantaneous and continuous sources in flows with nonuniform velocity distributions. They consider only those flows in which the velocity is in the x-direction but is given by $u_x = u_0 + \Omega_y y + \Omega_z z$ in which Ω_y and Ω_z represent the cross-stream shears. The following development closely follows their paper. The advection-diffusion equation is

$$\frac{\partial \mathbb{C}}{\partial t} + \left(u_0 + \Omega_y y + \Omega_z z \right) \frac{\partial \mathbb{C}}{\partial x} = D_{xx} \frac{\partial^2 \mathbb{C}}{\partial x^2} + D_{yy} \frac{\partial^2 \mathbb{C}}{\partial y^2} + D_{zz} \frac{\partial^2 \mathbb{C}}{\partial z^2} \tag{11.39}$$

The solution for a unit release of contaminant from an instantaneous source in the origin (Carter and Okubo, 1965) is

$$
\mathbb{C}_I = \frac{\exp\left\{-\dfrac{\left[x - \int_0^t u_0(t')dt' - (\Omega_y y + \Omega_z z)\dfrac{t}{2}\right]}{4 D_{xx} t (1 + \tau^2 t^2)} - \dfrac{y^2}{4 D_{yy} t} - \dfrac{z^2}{4 D_{zz} t}\right\}}{8(\pi t)^{3/2}\sqrt{D_{xx} D_{yy} D_{zz}(1 + \tau^2 t^2)}}
$$

$$
\tau^2 = \frac{1}{12}\left(\Omega_y^2 \frac{D_{yy}}{D_{xx}} + \Omega_z^2 \frac{D_{zz}}{D_{xx}}\right) \tag{11.40}
$$

Okubo and Karweit interpret $1/\tau$ as the time at which the shear begins to significantly affect the flow. The concentration contours are ellipsoids in the y- or z-planes. The maximum concentration for time much larger than $1/\tau$ changes as $t^{-5/2}$.

If the source is continuous—as opposed to instantaneous—Okubo and Karweit obtain the solution by simply summing over time so that the steady state solution is

$$
\mathbb{C}(x, y, z) = Q_s \int_0^\infty \mathbb{C}_I(x - u_0, t, y, z, t)dt \tag{11.41}
$$

in which Q_s is the quantity of contaminant emitted by the source per unit of time. Although (11.41) cannot be integrated explicitly, simple numerical techniques will give accurate results.

11.6.2 Ray Method

The results of Cohen and Lewis (1967) and of Smith (1981) are presented without detailed derivation. Those solutions that are exact have the form

$$
\mathbb{C}(x, y, z, t) = A(t)e^{-\Psi(x,y,z,t)} \tag{11.42}
$$

in which A and Ψ must be determined for any specific case. Consider a velocity that is similar to the case of the shear flows of the last section

$$
\begin{Bmatrix} u_x \\ u_y \\ u_z \end{Bmatrix} = [\Omega]\begin{Bmatrix} x \\ y \\ z \end{Bmatrix} + \begin{Bmatrix} u_{0x} \\ u_{0y} \\ u_{0z} \end{Bmatrix} \tag{11.43}
$$

in which Ω is a "shear matrix" and $\vec{u}_0$ represents the uniform flow. Although more general solutions are available, we restrict the shear matrix to the form

$$
[\Omega] = \begin{bmatrix} 0 & \omega_{12} & 0 \\ \omega_{21} & 0 & \omega_{23} \\ 0 & \omega_{32} & 0 \end{bmatrix} \tag{11.44}
$$

which has nonzero eigenvalues of

$$\lambda = \pm\sqrt{\omega_{12}\omega_{21} + \omega_{23}\omega_{32}} \tag{11.45}$$

The value of λ can be either real or imaginary. If it is real, the trajectories of the fluid particles form hyperbolas; if it is imaginary, the trajectories form ellipses. These trajectories are given by

$$\begin{Bmatrix} x \\ y \\ z \end{Bmatrix} = \begin{Bmatrix} \omega_{23} \\ 0 \\ -\omega_{21} \end{Bmatrix} \alpha t + \begin{Bmatrix} \omega_{12}\left(\beta\dfrac{e^{\lambda t}+e^{-\lambda t}}{2} + \gamma\dfrac{e^{\lambda t}-e^{-\lambda t}}{2\lambda}\right) \\[2ex] \gamma\dfrac{e^{\lambda t}+e^{-\lambda t}}{2} + \beta\lambda^2\dfrac{e^{\lambda t}-e^{-\lambda t}}{2\lambda} \\[2ex] \omega_{32}\left(\beta\dfrac{e^{\lambda t}+e^{-\lambda t}}{2} + \gamma\dfrac{e^{\lambda t}-e^{-\lambda t}}{2\lambda}\right) \end{Bmatrix}$$

$$+ \begin{Bmatrix} x_0 - \omega_{12}\beta \\ y_0 - \gamma \\ z_0 - \omega_{32}\beta \end{Bmatrix} \tag{11.46}$$

in which

$$\alpha = \frac{\omega_{32}u_{0x} - \omega_{12}u_{0z}}{\lambda^2} \qquad \beta = \frac{\omega_{21}x_0 + \omega_{23}z_0 + u_{0y}}{\lambda^2}$$

$$\gamma = y_0 + \frac{\omega_{23}u_{0z} + \omega_{21}u_{0x}}{\lambda^2}$$

Note that if λ is imaginary, the exponential terms in (11.46) become sine and cosine and the flow field is rotational with angular velocity of $|\lambda|$. Then (11.46) at time $t = 2\pi/|\lambda|$ becomes

$$\begin{Bmatrix} x \\ y \\ z \end{Bmatrix} = \begin{Bmatrix} x_0 + \dfrac{2\pi}{|\lambda|}\omega_{23}\dfrac{\omega_{32}u_{0x} - \omega_{12}u_{0z}}{\lambda^2} \\[2ex] y_0 \\[2ex] z_0 - \dfrac{2\pi}{|\lambda|}\omega_{21}\dfrac{\omega_{32}u_{0x} - \omega_{12}u_{0z}}{\lambda^2} \end{Bmatrix} \qquad \text{at } t = \frac{2\pi}{|\lambda|} \tag{11.47}$$

Thus, the flow field is periodic—a particle returns to its original position—if $u_{0x} = 0$ and $u_{0z} = 0$ for any value of u_{0y}. The effect of u_{0y} is to change the center of rotation.

With these assumptions, the values of the functions in (11.42) are

$$A(t) = \frac{1}{(4\pi)^{n/2}}\sqrt{\frac{\det[M(0)]\det[M(t)]}{\det[S(t)]}}$$

$$\Psi(x, y, z, t) = \frac{1}{4} \left[[M]^T \left\{ \begin{array}{c} x \\ y \\ z \end{array} \right\} - \int_0^t [M]^T \left\{ \begin{array}{c} u_{0x} \\ u_{0y} \\ u_{0z} \end{array} \right\} d\tau \right]^T$$

$$\times [S]^{-1} \left[[M]^T \left\{ \begin{array}{c} x \\ y \\ z \end{array} \right\} - \int_0^t [M]^T \left\{ \begin{array}{c} u_{0x} \\ u_{0y} \\ u_{0z} \end{array} \right\} d\tau \right]$$

$$(11.48)$$

where n is the dimension of the problem (1, 2, or 3) and we have introduced two additional matrices. $[M]$ is any fundamental solution of the matrix equation

$$\frac{d}{dt} M = -[\Omega]^T [M] \tag{11.49}$$

and

$$[S(t)] = \int_0^t [M]^T [D][M] d\tau \tag{11.50}$$

For $[\Omega]$ given by (11.44)

$$[M(t)] = \frac{1}{\lambda^2} \left(\frac{e^{\lambda t} + e^{-\lambda t}}{2} - 1 \right) [\Omega]^T [\Omega]^T - \frac{e^{\lambda t} - e^{-\lambda t}}{2\lambda} [\Omega]^T + [I] \tag{11.51}$$

in which $[I]$ is the unity matrix.

These equations complete the solution for which calculations can be done with a variety of flow fields. The solutions in this form have no reference to boundaries. The ray method can be used for flows contained in simple boundaries by reflecting the rays from the boundaries.

Example 11.2. Analytic solutions are often used as tests for numerical programs. Park and Liggett (1991) used the following three-dimensional "rotating Gaussian hill" to test the Taylor-Least Squares method. The Gaussian hill is simply a mound of contaminant such that the concentration takes the shape of a Gaussian curve with distance from the point of maximum concentration

$$\mathbb{C} = \exp \left[-\frac{(r - r_0)^2}{2\sigma^2} \right]$$

in which r_0 is the center of the hill and σ is a constant. This "blob" of contamination is made to circulate around the solution region in some sort of three-dimensional curve. At the end of the test period, the analytical solution is compared with the numerical solution.

The current example uses a flow field where

$$\Omega = \frac{\pi\sqrt{2}}{2000} \begin{bmatrix} 0 & 1 & 0 \\ -1 & 0 & 1 \\ 0 & -1 & 0 \end{bmatrix}$$

and the velocity vector is

$$\left\{ u_x \quad u_y \quad u_z \right\} = \left\{ 0.5 \quad \pi\frac{\sqrt{2}}{2} \quad 0.5 \right\}$$

The ray method assumes a point source initially at the origin. The numerical method cannot resolve a point source so the initial conditions for the numerical method are taken at $t = 4000$, after the source has diffused somewhat. For this particular example the diffusion vector is taken as

$$[Đ] = \begin{bmatrix} 5 & 0 & 0 \\ 0 & 5 & 0 \\ 0 & 0 & 5 \end{bmatrix}$$

which is isotropic.

PROBLEMS

11.1. Eq. (11.13) is the solution for the diffusion of a point source of a substance in the origin. Show by integrating (11.13) over a distributed source that the solution for an initially constant concentration of the substance between $x = -h$ and $x = h$ is given by

$$\mathbb{C}(x, t) = \frac{\mathbb{C}_0}{2} \left[\text{erf}\left(\frac{h - x}{2\sqrt{Đt}} \right) + \text{erf}\left(\frac{h + x}{2\sqrt{Đt}} \right) \right]$$

11.2. Consider the steady state problem of a smokestack discharging into the atmosphere. If the stack discharges at a height H above level ground into a wind with velocity u_0, write the equation for the concentration. (Use Example 11.1 and satisfy the condition at the ground surface by taking an image point.)

11.3. Use (11.21) to plot the concentration at times $t = 0$ and $t = 5$. Use $U = 4$, $Đ = 1$ and $t_0 = -5$.

11.4. Although (11.21) is used to check one-dimensional numerical codes, it can also be used for two- and three-dimensional programs when these are run so that nothing changes in the y- or z-directions. A better check, however, is to use the analytic solution to two-dimensional advection-diffusion. Begin with the diffusion equation (11.18) and derive the analog of (11.21).

11.5. A common case for checking two-dimensional numerical solutions is the "rotating Gaussian hill" with initial concentration

$$\mathbb{C}(x, y, 0)$$
$$= \mathbb{C}_0 \exp \left\{ -\frac{1}{2\sigma^2} \left[(x - x_0)^2 + (y - y_0)^2 \right] \right\}$$

An example is shown in Fig. 11.7, where a hill of concentration is placed on a two-dimensional grid. The velocity of the fluid is given so that after a period of time the center of the concentration comes back to its initial position. The velocity field is

$$u_x = -\omega y \qquad u_y = \omega x$$

Find the concentration at the peak when the hill has rotated once about the center of the grid. Hint: Use the ray method to find the Green's function for the advection-diffusion equation with a source point located at ξ, η.

11.6. For the problem given in Example 11.2 plot the shape of the Gaussian hill (i.e., plot the concentration vs. distance from the center of the hill) for $t = 4000$ and $t = 6000$. Normalize the hill at $t = 4000$ so that the concentration at the center is unity. Locate the center of the hill at these times.

11.7. You wish to find the speed of the flow and the diffusivity (or dispersivity) in a one-dimensional aquifer. For this purpose you inject salt into a well at position $x = 0$ m and observe the concentration of salt in the injection well and in observation wells at distances of 100 m, 200 m and 300 m downstream. The data are shown in the table. Using (11.20), find the velocity and the diffusion coefficient for the data at 60 days.

11.8. Solve Prob. 11.7 using the data from 30 days and 60 days but not from 120 days and not from the injection well (at $x = 0$). (A nonlinear equation solver is required.)

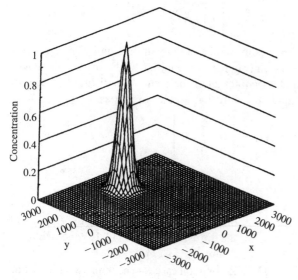

FIGURE 11.7

Table of observed concentrations for Problem 11.7

	Observation wells			
Time	1 (at 0 m)	2 (at 100 m)	3 (at 200 m)	4 (at 300 m)
30 days	0.2570	0.5913	0.0003	0.0000
60 days	0.0405	0.7492	0.2147	0.0010
120 days	0.0014	0.0747	0.4874	0.3958

The student may believe that vectors and tensors are instruments invented to make a subject such as fluid mechanics more difficult—to bedevil and harass those trying to learn one complex subject by imbedding another within it. The opposite is demonstrated in Chap. 1 in which we develop the equations of fluid mechanics. These equations would be much longer and much more difficult to read if they could not be written in vectors and tensors. Tensors provide a feel for the meaning of the terms; a feel that is hard to obtain without them, with much longer equations. Equally important is the fact that much of the literature is written in tensor notation and thus an elementary knowledge of Cartesian tensors is necessary to have access to that literature.

A.1 VECTORS IN A CARTESIAN COORDINATE SYSTEM

A vector is ordinarily represented in three-dimensional space by a directed line segment or arrow. We symbolize a vector by the arrow overbar such as $\vec{u}$. A vector is defined as a quantity that has all of the properties to be developed herein; a more fundamental definition in other terms appears to

443

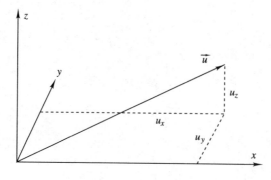

FIGURE A.1
A vector in Cartesian coordinates.

be inadequate. This point is clearly expressed by Block, et al. (1964), who state, "We take the position that a *vector* is a *vector*."

The vector $\vec{u}$ can also be represented by its components parallel to a given set of coordinate axes. The components are denoted by a subscript as shown in Fig. A.1 and are written as $\lfloor u_x\ u_y\ u_z \rfloor = \{u_x\ u_y\ u_z\}^T$ in which $\{\cdot\}$ indicates a column vector with $\{\cdot\}^T$ its transpose and $\lfloor\cdot\rfloor$ is a row vector. The x, y, z components in the array must always appear in that order; that is, the array $\lfloor 1\ 2\ 4 \rfloor$ means that $u_x = 1, u_y = 2, u_z = 4$ and is not equal to the array written in any other order. We now have two ways of representing a vector

$$\vec{u} = \lfloor u_x\ u_y\ u_z \rfloor \tag{A.1}$$

A vector can also be expressed in terms of *unit vectors*. These are written as vectors since they *are* really vectors in every sense. Using the unit vectors, $\vec{u}$ is written as

$$\vec{u} = \vec{e}_x u_x + \vec{e}_y u_y + \vec{e}_z u_z \tag{A.2}$$

A.2 VECTOR ALGEBRA

Vector addition is accomplished by simply adding the components of two vectors to form a new vector. Geometrically, it is represented by the parallelogram construction shown in Fig. A.2. Then

$$\vec{a} + \vec{b} = \lfloor a_x + b_x \quad a_y + b_y \quad a_z + b_z \rfloor \tag{A.3}$$

Subtraction is performed by reversing the sign (i.e., reversing the direction) of the vector being subtracted and adding. Vector addition and subtraction obey the following laws:

$$\vec{a} + \vec{b} = \vec{b} + \vec{a} \qquad \text{commutative} \tag{A.4}$$

$$\left(\vec{a} + \vec{b}\right) + \vec{c} = \vec{a} + \left(\vec{b} + \vec{c}\right) \qquad \text{associative} \tag{A.5}$$

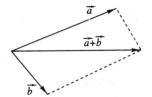

FIGURE A.2
Addition of vectors.

When a scalar is multiplied into a vector, the magnitude (length) of the vector is changed but its direction is not; hence

$$a\vec{u} = \lfloor au_x \quad au_y \quad au_z \rfloor \tag{A.6}$$

The following laws apply to multiplication:

$$a\vec{u} = \vec{u}a \qquad \text{commutative} \tag{A.7}$$

$$b(a\vec{u}) = (ba)\vec{u} \qquad \text{associative} \tag{A.8}$$

$$(a + b)\vec{u} = a\vec{u} + b\vec{u} \qquad \text{distributive} \tag{A.9}$$

Two vectors may be multiplied in different ways. The first is called the *scalar product* or *dot product* and is defined by

$$\vec{a} \cdot \vec{b} = ab \, \cos\left(\vec{a}, \vec{b}\right) \tag{A.10}$$

in which $\cos\left(\vec{a}, \vec{b}\right)$ is the cosine of the angle between the two vectors (always less than 180 degrees). The result of this type of multiplication is a scalar. The following laws apply to the scalar product:

$$\vec{a} \cdot \vec{b} = \vec{b} \cdot \vec{a} \qquad \text{commutative} \tag{A.11}$$

$$\left(\vec{a} \cdot \vec{b}\right)\vec{c} \neq \vec{a}\left(\vec{b} \cdot \vec{c}\right) \qquad \textit{not} \text{ associative} \tag{A.12}$$

$$\vec{a}\left(\vec{b} + \vec{c}\right) = \vec{a} \cdot \vec{b} + \vec{a} \cdot \vec{c} \qquad \text{distributive} \tag{A.13}$$

The other type of vector multiplication is called the *vector product* or *cross product* and is defined by

$$\vec{a} \times \vec{b} = \left[ab \, \sin\left(\vec{a}, \vec{b}\right)\right]\vec{n} \tag{A.14}$$

in which $\vec{n}$ is a unit vector normal to the plane containing $\vec{a}$ and $\vec{b}$. The normal $\vec{n}$ points in the direction that a right-handed screw would move if turned from $\vec{a}$ to $\vec{b}$ by the shortest route. Geometrically $\vec{a} \times \vec{b}$ has the magnitude of the area of the parallelogram containing $\vec{a}$ and $\vec{b}$ and it points in the direction normal to the plane of that parallelogram. The following laws apply to the vector product:

$$\vec{a} \times \vec{b} = -\vec{b} \times \vec{a} \qquad \textit{not} \text{ commutative} \tag{A.15}$$

$$\vec{a} \times \left(\vec{b} \times \vec{c}\right) \neq \left(\vec{a} \times \vec{b}\right) \times \vec{c} \qquad \textit{not} \text{ associative} \qquad \text{(A.16)}$$

$$\vec{a} \times \left(\vec{b} + \vec{c}\right) = \vec{a} \times \vec{b} + \vec{a} \times \vec{c} \qquad \text{distributive} \qquad \text{(A.17)}$$

In addition to the above operations, vectors may form multiple sums and products; however, all combinations are not possible. For example, $\vec{u} + a$ and $\vec{u} \times (\vec{a} \cdot \vec{b})$ would be nonsense since a vector cannot be added to a scalar in the first instance and a vector product must involve two vectors in the second example.

A.3   THE SUMMATION CONVENTION

Usually numbers are used in place of letters for the subscript that indicates the direction of a vector; 1 is associated with x, 2 with y, and 3 with z. Then $u_1 \equiv u_x, u_2 \equiv u_y$, and $u_3 \equiv u_z$. From (A.2)

$$\vec{u} = \vec{e}_1 u_1 + \vec{e}_2 u_2 + \vec{e}_3 u_3 = \sum_{i=1}^{3} \vec{e}_i u_i \qquad \text{(A.18)}$$

Thus, the summation sign gives another method of writing a vector and leads to a shorthand notation that carries the name *Einstein convention*. By use of this shorthand notation we can avoid writing the summation sign when performing vector and tensor operations. The convention simply states that if the same index occurs twice in any one term (i.e., on symbols unseparated by a plus sign, minus sign, or equal sign), the index is to be summed from 1 to 3 (or from 1 to 2 in two dimensions) even though the summation sign does not appear. Thus, (A.18) is written

$$\vec{u} = \vec{e}_i u_i \qquad \text{(A.19)}$$

and the summation is understood. The utility of this convention will be apparent shortly.

Two new quantities are introduced for use with the summation convention. The *Kronecker delta* is defined as

$$\delta_{ij} = 1 \quad \text{if } i = j$$
$$\delta_{ij} = 0 \quad \text{if } i \neq j \qquad \text{(A.20)}$$

The *alternating unit tensor* is defined by

$$\epsilon_{ijk} = 1 \qquad \text{if } ijk = 123, \ 231, \ \text{or } 312$$
$$\epsilon_{ijk} = -1 \quad \text{if } ijk = 321, \ 213, \ \text{or } 132 \qquad \text{(A.21)}$$
$$\epsilon_{ijk} = 0 \qquad \text{if } i = j, \ j = k, \ \text{or } i = k$$

In connection with these quantities there are several important identities. The first four are

$$\vec{e}_i \cdot \vec{e}_j = \delta_{ij} \tag{A.22}$$

$$\vec{e}_i \times \vec{e}_j = \epsilon_{ijk}\vec{e}_k \tag{A.23}$$

$$\epsilon_{ijk}\epsilon_{hjk} = 2\delta_{ih} \tag{A.24}$$

$$\epsilon_{ijk}\epsilon_{mnk} = \delta_{im}\delta_{nj} - \delta_{in}\delta_{jm} \tag{A.25}$$

To illustrate the summation convention, the validity of equations (A.23) and (A.24) is demonstrated. Expanding the right side of (A.23)

$$\epsilon_{ijk}\vec{e}_k = \epsilon_{ij1}\vec{e}_1 + \epsilon_{ij2}\vec{e}_2 + \epsilon_{ij3}\vec{e}_3 \tag{A.26}$$

Now i and j must be different or the right side of (A.26) is zero from the definition of the alternating unit tensor, equations (A.21). The nonzero possibilities are that i, j is 1, 2; 1, 3; 2, 3; 2, 1; 3, 1; or 3, 2. For these combinations the right side of (A.26) is, respectively, $\vec{e}_3, -\vec{e}_2, \vec{e}_1, -\vec{e}_3, \vec{e}_2$, or $-\vec{e}_1$, all of which conform to (A.14).

The proof of (A.24) follows in the same vein. Expanding the left side

$$\epsilon_{ijk}\epsilon_{hjk} = \epsilon_{i11}\epsilon_{h11} + \epsilon_{i12}\epsilon_{h12} + \epsilon_{i13}\epsilon_{h13} + \epsilon_{i21}\epsilon_{h21} + \epsilon_{i22}\epsilon_{h22} + \epsilon_{i23}\epsilon_{h23}$$

$$+ \epsilon_{i31}\epsilon_{h31} + \epsilon_{i32}\epsilon_{h32} + \epsilon_{i33}\epsilon_{h33}$$

$$= \epsilon_{i12}\epsilon_{h12} + \epsilon_{i13}\epsilon_{h13} + \epsilon_{i21}\epsilon_{h21} + \epsilon_{i23}\epsilon_{h23} + \epsilon_{i31}\epsilon_{h31} + \epsilon_{i32}\epsilon_{h32} \tag{A.27}$$

If $i \neq h$ one of the two terms of each pair is zero, making the right side zero; if $i = h$, two terms are equal to unity, which add to make the right side equal to 2, thus conforming with the definition of δ_{ih}.

Let's review vector algebra using this new notation. Addition is

$$\vec{a} + \vec{b} = \vec{e}_i a_i + \vec{e}_i b_i = \vec{e}_i (a_i + b_i) \tag{A.28}$$

Multiplication by a scalar is

$$a\vec{u} = \vec{e}_i (au_i) = a\vec{e}_i u_i \tag{A.29}$$

The scalar product becomes

$$\vec{a} \cdot \vec{b} = (\vec{e}_i a_i) \cdot (\vec{e}_j b_j) = \vec{e}_i \cdot \vec{e}_j \, a_i b_j = \delta_{ij} a_i b_j = a_i b_i \tag{A.30}$$

Also notice that the scalar product is simply the matrix product of the components when written in an array

$$\lfloor a_1 \ a_2 \ a_3 \rfloor \begin{Bmatrix} b_1 \\ b_2 \\ b_3 \end{Bmatrix} = a_i b_i \tag{A.31}$$

The vector product is

$$\vec{a} \times \vec{b} = (\vec{e}_i a_i) \times (\vec{e}_j b_j) = \vec{e}_i \times \vec{e}_j \, a_i b_j = \epsilon_{ijk} \vec{e}_k a_i b_j = \begin{vmatrix} \vec{e}_1 & \vec{e}_2 & \vec{e}_3 \\ a_1 & a_2 & a_3 \\ b_1 & b_2 & b_3 \end{vmatrix} \quad (A.32)$$

Multiple operations may easily be worked out using the foregoing expressions. For example,

$$\vec{a} \cdot \left(\vec{b} \times \vec{c}\right) = a_i \left(\vec{b} \times \vec{c}\right)_i = \epsilon_{ijk} a_i b_j c_k = \begin{vmatrix} a_1 & a_2 & a_3 \\ b_1 & b_2 & b_3 \\ c_1 & c_2 & c_3 \end{vmatrix} \quad (A.33)$$

A.4   VECTOR TRANSFORMATION BETWEEN RECTANGULAR COORDINATE SYSTEMS

The transformation properties of vectors and tensors are the keys to defining and using these quantities. In general there are three levels of difficulty: (1) Cartesian (rectangular) coordinates, (2) curvilinear coordinates (*e.g.*, cylindrical, spherical, hyperbolic, etc.), and (3) nonorthogonal coordinates. For the purposes of most Newtonian fluid mechanics problems, a Cartesian coordinate system is sufficient, although a large number of problems are solved in cylindrical or spherical coordinates. We will consider only the transformation properties of vectors and tensors in Cartesian coordinates with a brief introduction to curvilinear coordinates.

Consider the vector $\vec{u}$ in the x, y coordinate system of Fig. A.3. The quantities u_x and u_y represent the components in the x and y directions, respectively. We now ask: What are the components $u_{x'}$ and $u_{y'}$ in the primed coordinate system x', y'? From the geometry of the figure

$$u_{x'} = u_x \cos(x', x) + u_y \cos(x', y)$$
$$u_{y'} = u_x \cos(y', x) + u_y \cos(y', y) \quad (A.34)$$

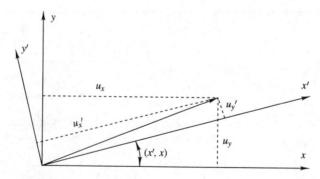

FIGURE A.3
Transformation of coordinates.

Extending these equations to three dimensions and using numbers for subscripts

$$u_{1'} = u_1 \cos\left(x_1', x_1\right) + u_2 \cos\left(x_1', x_2\right) + u_3 \cos\left(x_1', x_3\right) = u_i \cos\left(x_1', x_i\right)$$

$$u_{2'} = u_1 \cos\left(x_2', x_1\right) + u_2 \cos\left(x_2', x_2\right) + u_3 \cos\left(x_2', x_3\right) = u_i \cos\left(x_2', x_i\right)$$

$$u_{3'} = u_1 \cos\left(x_3', x_1\right) + u_2 \cos\left(x_3', x_2\right) + u_3 \cos\left(x_3', x_3\right) = u_i \cos\left(x_3', x_i\right)$$

$$(A.35)$$

To keep from having to write the cosine every time, the *angle* cosine is defined as

$$a_{ij} = \cos\left(x_i, x_j\right) \tag{A.36}$$

Then (A.35) can be written more economically

$$u_{1'} = a_{1'i} u_i \qquad u_{2'} = a_{2'i} u_i \qquad u_{3'} = a_{3'i} u_i \tag{A.37}$$

Equations (A.37) can be shortened still further by adopting the second part of the Einstein convention, which states: If an index is unrepeated in each term of the equation, it is understood that there are *three* (*two* in two dimensions) equations, one each for the unrepeated index being 1, 2, and 3. Thus, (A.37) become

$$u_{i'} = a_{i'j} u_j \tag{A.38}$$

in which the sum on j (the repeated index) is understood and where three separate equations are written for each value of i' (the unrepeated index). This method is truly economical because all of the information in (A.35) are written into (A.38).

Notice that under the summation convention the repeated index is a "dummy" index; the two equations

$$u_i = a_{ij} u_j \qquad \text{and} \qquad u_i = a_{ik} u_k \tag{A.39}$$

are exactly the same when expanded. In manipulating equations using the Einstein convention, the dummy index is changed so that it does not take on the same symbol as one of the unrepeated indices; that is, more than two of any one index should never appear in a term. Occasionally, the convention does not work properly, in which case the equations can be written in their full form, or a side notation is made to indicate the proper operation. The side notation often takes the form "no summation on i" where i is a repeated index, thus noting an exception to the rule.

A.5 CARTESIAN TENSORS

Having considered the scalar and vector products of two vectors, we now consider the *dyadic product* (or *outer product* or *direct product*), which is written without any sign between the two vectors

$$\mathbf{T} = \vec{a}\vec{b} = \begin{bmatrix} a_1b_1 & a_1b_2 & a_1b_3 \\ a_2b_1 & a_2b_2 & a_2b_3 \\ a_3b_1 & a_3b_2 & a_3b_3 \end{bmatrix} \qquad (A.40)$$

The dyadic product has nine components that are arranged in a tensor. It is the following matrix product of the component arrays

$$\begin{Bmatrix} a_1 \\ a_2 \\ a_3 \end{Bmatrix} \lfloor b_1 \ b_2 \ b_3 \rfloor = \vec{a}\vec{b} = \mathbf{T} \qquad (A.41)$$

Higher order tensor products such as $\vec{a}\vec{b}\vec{c}$ with 27 components or $\vec{a}\vec{b}\vec{c}\vec{d}$ with 81 components can be formed. The $\vec{a}\vec{b}\vec{c}$ components can be written with a three-dimensional matrix, but the components of higher order tensorial products are difficult to write efficiently. The following table shows the relationship between scalars, vectors, and tensors:

	Order of tensor	Number of components in 3-D space
Scalar	0	$3^0 = 1$
Vector	1	$3^1 = 3$
Dyad	2	$3^2 = 9$
3rd order tensor	3	$3^3 = 27$
4th order tensor	4	$3^4 = 81$
nth order tensor	n	3^n

A quantity related to the previously defined *unit vectors* is the unit tensor

$$\mathbf{e} = \begin{bmatrix} 1 & 0 & 0 \\ 0 & 1 & 0 \\ 0 & 0 & 1 \end{bmatrix} \qquad (A.42)$$

The components of the unit tensor are δ_{ij}, the Kronecker delta. Also, *unit dyads* are defined as

$$\vec{e}_1\vec{e}_1 = \begin{bmatrix} 1 & 0 & 0 \\ 0 & 0 & 0 \\ 0 & 0 & 0 \end{bmatrix} \qquad \vec{e}_1\vec{e}_2 = \begin{bmatrix} 0 & 1 & 0 \\ 0 & 0 & 0 \\ 0 & 0 & 0 \end{bmatrix}$$

$$\vec{e}_1\vec{e}_3 = \begin{bmatrix} 0 & 0 & 1 \\ 0 & 0 & 0 \\ 0 & 0 & 0 \end{bmatrix} \qquad \vec{e}_2\vec{e}_1 = \begin{bmatrix} 0 & 0 & 0 \\ 1 & 0 & 0 \\ 0 & 0 & 0 \end{bmatrix} \text{ etc.} \qquad (A.43)$$

Using unit dyads, dyadic equations can be written using the summation convention in the same manner as vector equations

$$\tau = \vec{e}_i \vec{e}_j \tau_{ij} \tag{A.44}$$

A.6 TENSOR ALGEBRA

The addition of two tensors is performed by adding their components

$$\tau + \sigma = \vec{e}_i \vec{e}_j \left(\tau_{ij} + \sigma_{ij} \right) \tag{A.45}$$

The multiplication of a tensor by a scalar is accomplished by multiplying each component by the scalar

$$a\tau = a\vec{e}_i \vec{e}_j \tau_{ij} \tag{A.46}$$

Tensors are multiplied in much the same way as vectors. The *scalar product* of two tensors is indicated by two dots

$$\tau : \sigma = \left(\vec{e}_i \vec{e}_j \tau_{ij} \right) : \left(\vec{e}_m \vec{e}_n \sigma_{mn} \right) = \vec{e}_i \vec{e}_j : \vec{e}_m \vec{e}_n \tau_{ij} \sigma_{mn} = \delta_{in} \delta_{jm} \tau_{ij} \sigma_{mn} = \tau_{ij} \sigma_{ji} \tag{A.47}$$

Note that the scalar products are taken by the inner quantities (the second and third) and then the outer quantities (the first and fourth). Incorrect scalar products would be δ_{im} and δ_{jn}. The *tensor product* of two tensors is indicated by a single dot

$$\begin{aligned}
\tau \cdot \sigma &= \left(\vec{e}_i \vec{e}_j \tau_{ij} \right) \cdot \left(\vec{e}_m \vec{e}_n \sigma_{mn} \right) = \vec{e}_i \vec{e}_j \cdot \vec{e}_m \vec{e}_n \tau_{ij} \sigma_{mn} \\
&= \delta_{jm} \vec{e}_i \vec{e}_n \tau_{ij} \sigma_{mn} = \vec{e}_i \vec{e}_n \tau_{ij} \sigma_{jn}
\end{aligned} \tag{A.48}$$

Again, the order of the indices is important. The latter operation is a simple matrix multiplication

$$\tau \cdot \sigma = \begin{bmatrix} \tau_{11} & \tau_{12} & \tau_{13} \\ \tau_{21} & \tau_{22} & \tau_{23} \\ \tau_{31} & \tau_{32} & \tau_{33} \end{bmatrix} \begin{bmatrix} \sigma_{11} & \sigma_{12} & \sigma_{13} \\ \sigma_{21} & \sigma_{22} & \sigma_{23} \\ \sigma_{31} & \sigma_{32} & \sigma_{33} \end{bmatrix} \tag{A.49}$$

A *vector product* is formed by multiplication of a dyad and a vector

$$\tau \cdot \vec{v} = \left(\vec{e}_i \vec{e}_j \tau_{ij} \right) \cdot \left(\vec{e}_k v_k \right) = \vec{e}_i \vec{e}_j \cdot \vec{e}_k \tau_{ij} v_k = \vec{e}_i \delta_{jk} v_k \tau_{ij} = \vec{e}_i \tau_{ij} v_j \tag{A.50}$$

This operation is also a matrix multiplication

$$\tau \cdot \vec{v} = \begin{bmatrix} \tau_{11} & \tau_{12} & \tau_{13} \\ \tau_{21} & \tau_{22} & \tau_{23} \\ \tau_{31} & \tau_{32} & \tau_{33} \end{bmatrix} \begin{Bmatrix} v_1 \\ v_2 \\ v_3 \end{Bmatrix} \tag{A.51}$$

The following table indicates the order of the results of the various multiplications:

	Symbol	Number to subtract from sum of orders
Dot	·	2
Cross	×	1
Double dot	:	4
None		0

A.7   PRINCIPAL AXES OF A SYMMETRIC TENSOR

In many physical problems the significant tensors are *symmetric*, which means

$$\tau_{ij} = \tau_{ji} \tag{A.52}$$

Symmetric tensors have special properties that can be useful. One such property is that they can be transformed into a coordinate system where all of the off-diagonal terms are zero; that is,

$$\tau_{ij} = 0 \quad \text{for } i \neq j \tag{A.53}$$

Such a coordinate system is called the *principal axes* of the tensor. The following derivation indicates how tensors can be transformed.

First, the vector $\vec{A}$ is defined such that

$$\tau \cdot \vec{A} = \lambda \vec{A} \tag{A.54}$$

where τ is a symmetric tensor and λ is a scalar. At this point both $\vec{A}$ and λ are unknown quantities. Then

$$\tau_{ij} A_j = \lambda A_i \tag{A.55}$$

Equations (A.55) written out in full are

$$A_1 (\tau_{11} - \lambda) + A_2 \tau_{12} + A_3 \tau_{13} = 0$$

$$A_1 \tau_{21} + A_2 (\tau_{22} - \lambda) + A_3 \tau_{23} = 0 \tag{A.56}$$

$$A_1 \tau_{31} + A_2 \tau_{32} + A_3 (\tau_{33} - \lambda) = 0$$

Equations (A.56) can be considered as three simultaneous equations in the three components of $\vec{A}$. In order that the A_i are not all zero (the trivial solution) the coefficient determinant must be zero

$$\begin{vmatrix} \tau_{11} - \lambda & \tau_{12} & \tau_{13} \\ \tau_{21} & \tau_{22} - \lambda & \tau_{23} \\ \tau_{31} & \tau_{32} & \tau_{33} - \lambda \end{vmatrix} = 0 \tag{A.57}$$

Equation (A.57) is a cubic equation in λ. In general this equation has three roots, λ^I, λ^{II}, and λ^{III}. Working with λ^I, the first two equations of (A.56) give

$$\tau_{11} - \lambda^I + \frac{A_2^I}{A_1^I}\tau_{12} + \frac{A_3^I}{A_1^I}\tau_{13} = 0$$

$$\tau_{21} + \frac{A_2^I}{A_1^I}\left(\tau_{22} - \lambda^I\right) + \frac{A_3^I}{A_1^I}\tau_{23} = 0$$

(A.58)

where A_i^I is the ith component of the vector corresponding to λ^I. Equations (A.58) can be solved for A_2^I/A_1^I and A_3^I/A_1^I. Using the unit vectors

$$\vec{A}^I = A_1^I\vec{e}_1 + A_2^I\vec{e}_2 + A_3^I\vec{e}_3 \tag{A.59}$$

or

$$\frac{\vec{A}^I}{A_1^I} = \vec{e}_1 + \frac{A_2^I}{A_1^I}\vec{e}_2 + \frac{A_3^I}{A_1^I}\vec{e}_3 \tag{A.60}$$

Equation (A.60) gives the direction of $\vec{A}^I$ but not its magnitude. The directions of $\vec{A}^{II}$ and $\vec{A}^{III}$ are found in a similar manner using the other roots of (A.57), λ^{II} and λ^{III}. For the purpose of defining the principal axes, only the directions of the three $\vec{A}$ are needed; the magnitudes do not matter.

The three vectors $\vec{A}^I$, $\vec{A}^{II}$, and $\vec{A}^{III}$ are mutually perpendicular, a fact that is demonstrated in the following development. From (A.55)

$$\tau_{ij}A_j^I = \lambda^I A_i^I \qquad \tau_{ij}A_j^{II} = \lambda^{II}A_i^{II} \qquad \tau_{ij}A_j^{III} = \lambda^{III}A_i^{III} \tag{A.61}$$

We now take the scalar products

$$\vec{A}^I \cdot \left(\tau \cdot \vec{A}^{II}\right) = A_i^I \tau_{ij} A_j^{II} \tag{A.62}$$

$$\vec{A}^{II} \cdot \left(\tau \cdot \vec{A}^I\right) = A_i^{II} \tau_{ij} A_j^I \tag{A.63}$$

Subtracting (A.63) from (A.62) gives

$$\vec{A}^I \cdot \left(\tau \cdot \vec{A}^{II}\right) - \vec{A}^{II} \cdot \left(\tau \cdot \vec{A}^I\right) = A_i^I \tau_{ij} A_j^{II} - A_i^{II} \tau_{ij} A_j^I \tag{A.64}$$

Reversing the dummy indices in the last term and using the fact that τ is symmetric yield

$$\vec{A}^I \cdot \left(\tau \cdot \vec{A}^{II}\right) - \vec{A}^{II} \cdot \left(\tau \cdot \vec{A}^I\right) = A_i^I A_j^{II}\left(\tau_{ij} - \tau_{ji}\right) = 0 \tag{A.65}$$

Using (A.54) to rewrite the left side of (A.65) gives

$$\vec{A}^I \cdot \left(\tau \cdot \vec{A}^{II}\right) - \vec{A}^{II} \cdot \left(\tau \cdot \vec{A}^I\right) = A^I \cdot \lambda^{II} \vec{A}^{II} - \vec{A}^{II} \cdot \lambda^I \vec{A}^I$$
$$= \vec{A}^I \cdot \vec{A}^{II} \left(\lambda^{II} - \lambda^I\right) = 0 \tag{A.66}$$

If the λ are distinct (that is, if $\lambda^I \neq \lambda^{II}$), then $\vec{A}^I$ must be normal to $\vec{A}^{II}$ in order that (A.66) is zero. The demonstration that $\vec{A}^{III}$ is normal to $\vec{A}^I$ if $\lambda^{III} \neq \lambda^I$ is done similarly. Thus, the three $\vec{A}^I$ are mutually perpendicular.

We can now define a coordinate system so that the unit vectors $\vec{e}_{1'}, \vec{e}_{2'}$, and $\vec{e}_{3'}$ are in the $\vec{A}^I, \vec{A}^{II}$, and $\vec{A}^{III}$ directions, respectively. Dividing (A.61) by $|\vec{A}|$ gives

$$\tau_{ij} a_{j1'} = \lambda^I a_{i1'} \qquad \tau_{ij} a_{j2'} = \lambda^{II} a_{i2'} \qquad \tau_{ij} a_{j3'} = \lambda^{III} a_{i3} \tag{A.67}$$

where $a_{ji'}$ is the projection of $\vec{e}_{i'}$ on $\vec{e}_j$. The tensors in the two coordinate systems are related by the transformation

$$\tau_{ij} = a_{ii'} a_{jj'} \tau_{i'j'} \tag{A.68}$$

Using (A.68) in the first of (A.67) yields

$$a_{ii'} a_{jj'} a_{j1'} \tau_{i'j'} = \lambda^I a_{i1'} \tag{A.69}$$

But

$$a_{jj'} a_{j1'} = \frac{\partial x_j}{\partial x_{j'}} \frac{\partial x_j}{\partial x_{1'}} = \delta_{j'1'} \tag{A.70}$$

Therefore,

$$a_{ii'} \tau_{i'1'} = \lambda^I a_{i1'} \tag{A.71}$$

We now multiply (A.71) through by $a_{i1'}$ and use the relationships $a_{ii'} a_{i1'} = \delta_{i'1'}$ and $a_{i1'} a_{i1'} = 1$ to obtain

$$\tau_{1'1'} = \lambda^I \tag{A.72}$$

An off-diagonal element in the primed coordinate system is obtained by multiplying (A.71) through by $a_{i2'}$ and using $a_{ii'} a_{i2'} = \delta_{i'2'}$ and $a_{i1'} a_{i2'} = \delta_{1'2'} = 0$ to give

$$\tau_{2'1'} = 0 \tag{A.73}$$

Similarly,

$$\tau_{2'2'} = \lambda^{II} \qquad \tau_{3'3'} = \lambda^{III} \qquad \tau_{i'j'} = 0 \qquad \text{for } i' \neq j' \tag{A.74}$$

The conditions for the principal axes have been satisfied; that is, the off-diagonal terms of τ are zero. Also, we see that the diagonal terms of τ are the λ. Written in matrix form

$$\tau = \begin{bmatrix} \tau_{1'1'} & 0 & 0 \\ 0 & \tau_{2'2'} & 0 \\ 0 & 0 & \tau_{3'3'} \end{bmatrix} = \begin{bmatrix} \lambda^I & 0 & 0 \\ 0 & \lambda^{II} & 0 \\ 0 & 0 & \lambda^{III} \end{bmatrix} \tag{A.75}$$

A.8 DIFFERENTIAL OPERATIONS

Just as there are various vector products, vectors and tensors may be differentiated in various ways. Consider a particle moving through space along curve C of Fig. A.4. The position of this particle is given by the vector $\vec{r}$, which is a function of time. The derivative of $\vec{r}$ with respect to time is the velocity of the particle, $\vec{u} = d\vec{r}/dt$. Written in its components, the velocity is

$$\frac{d\vec{r}}{dt} = \vec{e}_1 \frac{dr_1}{dt} + \vec{e}_2 \frac{dr_2}{dt} + \vec{e}_3 \frac{dr_3}{dt} = \vec{e}_i \frac{dr_i}{dt} \tag{A.76}$$

Higher derivatives such as acceleration are also defined

$$\frac{d\vec{u}}{dt} = \frac{d^2\vec{r}}{dt^2} = \vec{e}_i \frac{d^2 r_i}{dt^2} \tag{A.77}$$

Differentiation of vectors follows the rules

$$\frac{d}{dt}(a\vec{u}) = a\frac{d\vec{u}}{dt} + \vec{u}\frac{da}{dt} \tag{A.78}$$

$$\frac{d}{dt}(\vec{a} \cdot \vec{b}) = \vec{a} \cdot \frac{d\vec{b}}{dt} + \frac{d\vec{a}}{dt} \cdot \vec{b} \tag{A.79}$$

$$\frac{d}{dt}(\vec{a} \times \vec{b}) = \vec{a} \times \frac{d\vec{b}}{dt} + \frac{d\vec{a}}{dt} \times \vec{b} \tag{A.80}$$

In addition to the time derivative the *vector operator* has many important uses. The differential vector operator (using the so-called Gibbs notation) is known as *del* or *nabla* and is defined (in rectangular coordinates) as

$$\vec{\nabla} = \vec{e}_1 \frac{\partial}{\partial x_1} + \vec{e}_2 \frac{\partial}{\partial x_2} + \vec{e}_3 \frac{\partial}{\partial x_3} = \vec{e}_i \frac{\partial}{\partial x_i} \tag{A.81}$$

This vector operator means nothing when standing alone but must "operate" on a scalar, vector, or tensor. Such operations are defined below.

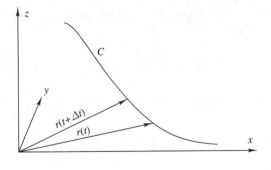

FIGURE A.4
A particle moving along a curve.

The *gradient of a scalar* is defined as

$$\vec{\nabla} a = \vec{e}_1 \frac{\partial a}{\partial x_i} \tag{A.82}$$

and is itself a vector. The following laws apply to the gradient:

$$\vec{\nabla} a \neq a\vec{\nabla} \qquad\qquad \text{\textit{not} commutative} \tag{A.83}$$

$$(\vec{\nabla} a)b \neq \vec{\nabla}(ab) \qquad\qquad \text{\textit{not} associative} \tag{A.84}$$

$$\vec{\nabla}(a+b) = \vec{\nabla} a + \vec{\nabla} b \qquad\qquad \text{distributive} \tag{A.85}$$

The scalar product of $\vec{\nabla}$ and a vector is known as the *divergence of a vector field* or simply the *divergence* and is defined

$$\vec{\nabla} \cdot \vec{u} = \left(\vec{e}_i \frac{\partial}{\partial x_i}\right) \cdot (\vec{e}_j u_j) = (\vec{e}_i \cdot \vec{e}_j)\frac{\partial u_j}{\partial x_i} = \frac{\partial u_i}{\partial x_i} \tag{A.86}$$

The following laws apply to the divergence:

$$\vec{\nabla} \cdot \vec{u} \neq \vec{u} \cdot \vec{\nabla} \qquad\qquad \text{\textit{not} commutative} \tag{A.87}$$

$$\vec{\nabla} \cdot a\vec{u} \neq \vec{\nabla} a \cdot \vec{u} \qquad\qquad \text{\textit{not} associative} \tag{A.88}$$

$$\vec{\nabla} \cdot \left(\vec{a}+\vec{b}\right) = \vec{\nabla} \cdot \vec{a} + \vec{\nabla} \cdot \vec{b} \qquad\qquad \text{distributive} \tag{A.89}$$

The vector product of $\vec{\nabla}$ and a vector is the *curl of a vector field* and is written

$$\vec{\nabla} \times \vec{u} = \left(\vec{e}_i \frac{\partial}{\partial x_i}\right) \times (\vec{e}_j u_j) = \epsilon_{ijk} \vec{e}_k \frac{\partial u_j}{\partial x_i} = \begin{vmatrix} \vec{e}_1 & \vec{e}_2 & \vec{e}_3 \\ \dfrac{\partial}{\partial x_1} & \dfrac{\partial}{\partial x_2} & \dfrac{\partial}{\partial x_1} \\ u_1 & u_2 & u_3 \end{vmatrix} \tag{A.90}$$

The same laws apply to the curl as to the divergence. In fluid mechanics the divergence of the velocity of an incompressible fluid indicates the amount of fluid that is being created at a point; the curl of the velocity is a measure of the rotation of a fluid particle. More precise physical interpretations of these quantities appear in Chap. 1.

Some combinations of vector operators are useful. The *Laplacian of a scalar* is defined as

$$\nabla^2 a = \vec{\nabla} \cdot \vec{\nabla} a = \left(\vec{e}_i \frac{\partial}{\partial x_i}\right) \cdot \left(\vec{e}_j \frac{\partial}{\partial x_j}\right) a = \frac{\partial^2 a}{\partial x_i \partial x_i} \tag{A.91}$$

In rectangular coordinates the *Laplacian operator* is defined

$$\nabla^2 = \frac{\partial^2}{\partial x_1^2} + \frac{\partial^2}{\partial x_2^2} + \frac{\partial^2}{\partial x_3^2} = \frac{\partial^2}{\partial x_i \partial x_i} \tag{A.92}$$

Often ∇^2 is written as Δ. Similarly, the Laplacian of a vector is

$$\nabla^2\vec{u} = \vec{\nabla}(\vec{\nabla} \cdot \vec{u}) - \vec{\nabla} \times (\vec{\nabla} \times \vec{u}) \tag{A.93}$$

In a rectangular coordinate system (A.93) reduces to

$$\nabla^2\vec{u} = \nabla^2\,(\vec{e}_i u_i) = \vec{e}_i \nabla^2 u_i \tag{A.94}$$

however, (A.94) does not transform properly into a curvilinear coordinate system.

The *divergence of a tensor* is defined as

$$\vec{\nabla} \cdot \tau = \left(\vec{e}_i \frac{\partial}{\partial x_i}\right) \cdot \left(\vec{e}_j \vec{e}_k \tau_{jk}\right) = \delta_{ij}\vec{e}_k \frac{\partial \tau_{jk}}{\partial x_i} = \vec{e}_k \frac{\partial \tau_{ik}}{\partial x_i} \tag{A.95}$$

Other combinations of scalars, vectors, tensors, and differential operators can be worked out by similar methods. Note, however, that the operations of gradient, divergence, and curl are defined above and not derived; the symbolism using the summation convention shows only that these definitions follow a consistent pattern.

A.9 TRANSFORMATIONS

In most of the study of fluid mechanics and in most of the literature in fluid mechanics it is sufficient to deal with Cartesian tensors. Some problems, however, are better solved in curvilinear coordinate systems. A number of such systems exist, but the ones in primary use are cylindrical and spherical coordinates. We deal briefly with these systems.

The transformation of vectors and tensors into alternate coordinate systems is an important topic. In fact the transformation properties are often used in the definition of what is and what is not a tensor; that is, for a quantity to be a tensor it must transform like a tensor. For a more complete exposition of that subject the reader is referred to one of the many books on vector and tensor analysis (e.g., Lass, 1950).

Consider the arc length in a rectangular coordinate system

$$ds^2 = dx^2 + dy^2 + dz^2 \tag{A.96}$$

For cylindrical coordinates we apply the transformation

$$x = r\cos\theta \qquad y = r\sin\theta \tag{A.97}$$

with the inverse transformation

$$r = \sqrt{x^2 + y^2} \qquad \theta = \arctan\frac{y}{x} \tag{A.98}$$

Now

$$dx = \frac{\partial x}{\partial r}dr + \frac{\partial x}{\partial \theta}d\theta = \cos\theta dr - r\sin\theta d\theta$$

$$dy = \frac{\partial y}{\partial r}dr + \frac{\partial y}{\partial \theta}d\theta = \sin\theta dr + r\cos\theta d\theta$$

(A.99)

Using (A.99) in (A.96)

$$ds^2 = dr^2 + r^2 d\theta^2 + dz^2 \qquad (A.100)$$

For spherical coordinates we apply the transformation

$$x = r\sin\theta\cos\phi \qquad y = r\sin\theta\sin\phi \qquad z = r\cos\theta \qquad (A.101)$$

with the inverse transformation

$$r = \sqrt{x^2 + y^2 + z^2} \qquad \theta = \arccos\frac{z}{\sqrt{x^2 + y^2 + z^2}} \qquad \phi = \arctan\frac{y}{x}$$

(A.102)

which gives an arc length as

$$ds^2 = dr^2 + r^2 d\theta^2 + r^2\sin^2\theta d\phi^2 \qquad (A.103)$$

In general the arc lengths can be expressed as

$$ds^2 = h_1^2 dv_1^2 + h_2^2 dv_2^2 + h_3^2 dv_3^2 \qquad (A.104)$$

in which the v_i are the coordinates of the transformed system and the h_i are *scale factors*. Scale factor have a number of uses. For example a differential volume is

$$dV = dxdydz = h_1 h_2 h_3 dv_1 dv_2 dv_3 \qquad (A.105)$$

Without derivation, some of the differential vector operations are stated below for a function $f(v_1, v_2, v_3)$ or a vector $\vec{u}$. The gradient is

$$\vec{\nabla}f = \frac{\partial f}{\partial v_1}\vec{\nabla}v_1 + \frac{\partial f}{\partial v_2}\vec{\nabla}v_2 + \frac{\partial f}{\partial v_3}\vec{\nabla}v_3$$

$$= \frac{1}{h_1}\frac{\partial f}{\partial v_1}\vec{e}_1 + \frac{1}{h_2}\frac{\partial f}{\partial v_2}\vec{e}_2 + \frac{1}{h_3}\frac{\partial f}{\partial v_3}\vec{e}_3$$

(A.106)

in which the $\vec{e}_i$ now represents the unit vectors in the v_i directions. The divergence is

$$\vec{\nabla} \cdot \vec{u} = \frac{1}{h_1 h_2 h_3}\left[\frac{\partial(h_2 h_3 u_1)}{\partial v_1} + \frac{\partial(h_3 h_1 u_2)}{\partial v_2} + \frac{\partial(h_1 h_2 u_3)}{\partial v_3}\right] \qquad (A.107)$$

The curl is

$$\vec{\nabla} \times \vec{u} = \frac{\vec{e}_1}{h_2 h_3} \left[\frac{\partial (h_3 u_3)}{\partial v_2} - \frac{\partial (h_2 u_2)}{\partial v_3} \right] + \frac{\vec{e}_2}{h_3 h_1} \left[\frac{\partial (h_1 u_1)}{\partial v_3} - \frac{\partial (h_3 u_3)}{\partial v_1} \right]$$
$$+ \frac{\vec{e}_3}{h_1 h_2} \left[\frac{\partial (h_2 u_2)}{\partial v_1} - \frac{\partial (h_1 u_1)}{\partial v_2} \right]$$

$$(A.108)$$

The Laplacian is

$$\nabla^2 f$$
$$= \frac{1}{h_1 h_2 h_3} \left[\frac{\partial}{\partial v_1} \left(\frac{h_2 h_3}{h_1} \frac{\partial f}{\partial v_1} \right) + \frac{\partial}{\partial v_2} \left(\frac{h_3 h_1}{h_2} \frac{\partial f}{\partial v_2} \right) + \frac{\partial}{\partial v_3} \left(\frac{h_1 h_2}{h_3} \frac{\partial f}{\partial v_3} \right) \right]$$

$$(A.109)$$

Other quantities may be worked out in a similar manner. Section A.11 gives some of the vector and tensor operations in rectangular, cylindrical, and spherical coordinates.

A.10 NOTATION

It seems that there is almost no limit to the variations of notation in mathematics; some consider the mark of originality the invention of a new notation. In vector calculus, however, the situation is not so bad; there are only a few different methods of specifying the same quantities. Since we will be dealing with Cartesian tensors, the material in this Appendix is sufficient for the development and the solution of the equations of fluid mechanics. There is, however, considerably more to the analysis of tensors and thus those who deal with fluid mechanics sometimes adopt a notation that is useful in the theory of relativity or when dealing with non-Newtonian fluids.

A common notation is the comma that represents differentiation. For example the gradient of a scalar is

$$a_{,i} = \frac{\partial a}{\partial x_i} \qquad (A.110)$$

The divergence is

$$u_{i,i} = \vec{\nabla} \cdot \vec{u} \qquad \text{and} \qquad u^i{}_{,i} = \vec{\nabla} \cdot \vec{u} \qquad (A.111)$$

In the second example of (A.111) the summation convention still holds.

Often the index is written as a superscript. The tensor, τ, may be written τ^{ij}, τ_{ij}, or τ_j^i. In a Cartesian coordinate system these three types of notation all have the same meaning; in curvilinear coordinate systems they are different. (They are called, respectively, a *contravariant tensor*, a *covariant tensor* and a *mixed tensor*.) The reason for the difference is that they have different transformation properties. For the purposes of interpreting the equations, they can all be read as Cartesian tensors. Examples of differentiation are

Laplacian of a scalar $\quad a_{,ii} = \nabla^2 a$ (A.112)

Laplacian of a vector $\quad u_{i,jj} \quad$ or $\quad u^i{}_{,jj}$ (A.113)

Divergence of a tensor $\quad \tau_{ij,i} \quad$ or $\quad \tau^{ij}{}_{,i} \quad$ or $\quad \tau^i_j{}_{,i}$ (A.114)

Curl of a vector (a second-order tensor) $\quad u_{i,j} - u_{j,i}$
$\quad$ or $\quad u^i{}_{,j} - u^j{}_{,i}$ (A.115)

Although one finds some variations on this notation (such as the exact placement of the comma and the indices), the equations should be easily read in terms of Cartesian tensors. The physical interpretation of the operations is enhanced by the economy of the notation.

Finally, some use the great compromise in which the partial derivative is fitted with a subscript to indicate the variable that should appear in the denominator of the derivative, for example

$$\partial_x = \frac{\partial}{\partial x}$$

or even

$$\partial_i = \frac{\partial}{\partial x_i}$$

In this text the subscript of a dependent variable always indicates a component (u_x is the component of $\vec{u}$ in the x-direction) whereas such notation is often used in the literature to indicate partial differentiation ($u_x = \partial u/\partial x$). Obviously, mathematical notation is not mathematically precise.

A.11 SUMMARY OF DIFFERENTIAL OPERATIONS

In the following expressions, operations involving τ may be valid only for a symmetric tensor.

A.11.1 Rectangular Coordinates

$$\vec{\nabla} \cdot \vec{u} = \frac{\partial u_i}{\partial x_i} \quad \nabla^2 a = \frac{\partial^2 a}{\partial x_i \partial x_i} \quad \left(\vec{\nabla} \cdot \tau\right)_i = \frac{\partial \tau_{ij}}{\partial x_j} \quad \left(\vec{u} \cdot \vec{\nabla}\vec{u}\right)_i = u_j \frac{\partial u_i}{\partial x_j}$$

$$\left(\vec{\nabla} \times \vec{u}\right)_x = \frac{\partial u_z}{\partial y} - \frac{\partial u_y}{\partial z} \quad \left(\vec{\nabla} \times \vec{u}\right)_y = \frac{\partial u_x}{\partial z} - \frac{\partial u_z}{\partial x} \quad \left(\vec{\nabla} \times \vec{u}\right)_z = \frac{\partial u_y}{\partial x} - \frac{\partial u_x}{\partial y}$$

$$\tau : \vec{\nabla}\vec{u} = \frac{1}{2}\tau_{ij}\left(\frac{\partial u_i}{\partial x_j} - \frac{\partial u_j}{\partial x_i}\right) \quad \text{if} \quad \tau_{ij} = \tau_{ji}$$

A.11.2 Cylindrical Coordinates

$$R = \sqrt{y^2 + z^2} \quad \theta = \arctan \frac{z}{y}$$

$$\vec{\nabla} \cdot \vec{u} = \frac{1}{R} \frac{\partial}{\partial R} (R u_R) + \frac{1}{R} \frac{\partial u_\theta}{\partial \theta} + \frac{\partial u_x}{\partial x} \quad \nabla^2 a = \frac{1}{R} \frac{\partial}{\partial R} \left(R \frac{\partial a}{\partial R} \right) + \frac{1}{R^2} \frac{\partial^2 a}{\partial \theta^2} + \frac{\partial^2 a}{\partial x^2}$$

$$\boldsymbol{\tau} : \vec{\nabla} \vec{u} = \tau_{RR} \frac{\partial u_R}{\partial R} + \tau_{\theta\theta} \left(\frac{1}{R} \frac{\partial u_\theta}{\partial \theta} + \frac{u_R}{R} \right) + \tau_{xx} \frac{\partial u_x}{\partial x}$$

$$+ \tau_{R\theta} \left[R \frac{\partial}{\partial R} \left(\frac{u_\theta}{R} \right) + \frac{1}{R} \frac{\partial u_R}{\partial \theta} \right] + \tau_{\theta x} \left(\frac{1}{R} \frac{\partial u_x}{\partial \theta} + \frac{\partial u_\theta}{\partial x} \right) + \tau_{Rx} \left(\frac{\partial u_x}{\partial R} + \frac{\partial u_R}{\partial x} \right)$$

$$(\vec{\nabla} s)_R = \frac{\partial s}{\partial R} \quad (\vec{\nabla} s)_\theta = \frac{1}{R} \frac{\partial s}{\partial \theta} \quad (\vec{\nabla} s)_x = \frac{\partial s}{\partial x}$$

$$(\vec{\nabla} \times \vec{u})_R = \frac{1}{R} \frac{\partial u_x}{\partial \theta} - \frac{\partial u_\theta}{\partial x} \quad (\vec{\nabla} \times \vec{u})_\theta = \frac{\partial u_R}{\partial x} - \frac{\partial u_x}{\partial R}$$

$$(\vec{\nabla} \times \vec{u})_x = \frac{1}{R} \frac{\partial}{\partial R} (R u_\theta) - \frac{1}{R} \frac{\partial u_R}{\partial \theta}$$

$$(\vec{\nabla} \cdot \boldsymbol{\tau})_R = \frac{1}{R} \frac{\partial}{\partial R} (R \tau_{RR}) + \frac{1}{R} \frac{\partial \tau_{R\theta}}{\partial \theta} - \frac{\tau_{\theta\theta}}{R} + \frac{\partial \tau_{Rx}}{\partial x}$$

$$(\vec{\nabla} \cdot \boldsymbol{\tau})_\theta = \frac{1}{R} \frac{\partial \tau_{\theta\theta}}{\partial \theta} + \frac{\partial \tau_{R\theta}}{\partial R} + \frac{2\tau_{R\theta}}{R} + \frac{\partial \tau_{\theta x}}{\partial x}$$

$$(\vec{\nabla} \cdot \boldsymbol{\tau})_x = \frac{1}{R} \frac{\partial}{\partial R} (R \tau_{Rx}) + \frac{1}{R} \frac{\partial \tau_{\theta x}}{\partial \theta} + \frac{\partial \tau_{xx}}{\partial x}$$

$$\left(\vec{u} \cdot \vec{\nabla} \vec{u} \right)_x = u_R \frac{\partial u_x}{\partial R} + \frac{u_\theta}{R} \frac{\partial u_x}{\partial \theta} + u_x \frac{\partial u_x}{\partial x}$$

$$\left(\vec{u} \cdot \vec{\nabla} \vec{u} \right)_R = u_R \frac{\partial u_R}{\partial R} + \frac{u_\theta}{R} \frac{\partial u_R}{\partial \theta} - \frac{u_\theta^2}{R} + u_x \frac{\partial u_R}{\partial x}$$

$$\left(\vec{u} \cdot \vec{\nabla} \vec{u} \right)_\theta = u_R \frac{\partial u_\theta}{\partial R} + \frac{u_\theta}{R} \frac{\partial u_\theta}{\partial \theta} + \frac{u_R u_\theta}{R} + u_x \frac{\partial u_\theta}{\partial x}$$

A.11.3 Spherical Coordinates

$$r = \sqrt{x^2 + y^2 + z^2} \quad \theta = \arctan \frac{\sqrt{x^2 + y^2}}{z} \quad \phi = \arctan \frac{y}{x}$$

$$\vec{\nabla} \cdot \vec{u} = \frac{1}{r^2} \frac{\partial}{\partial r}\left(r^2 u_r\right) + \frac{1}{r \sin\theta} \frac{\partial}{\partial \theta}\left(u_\theta \sin\theta\right) + \frac{1}{r \sin\theta} \frac{\partial u_\phi}{\partial \phi}$$

$$\nabla^2 a = \frac{1}{r^2}\left(r^2 \frac{\partial a}{\partial r}\right) + \frac{1}{r^2 \sin\theta} \frac{\partial}{\partial \theta}\left(\sin\theta \frac{\partial a}{\partial \theta}\right) + \frac{1}{r^2 \sin\theta} \frac{\partial^2 a}{\partial \phi^2}$$

$$\tau : \vec{\nabla}\vec{u} = \tau_{rr}\frac{\partial u_r}{\partial r} + \tau_{\theta\theta}\left(\frac{1}{r}\frac{\partial u_\theta}{\partial \theta} + \frac{u_r}{r}\right) + \tau_{\phi\phi}\left(\frac{1}{r \sin\theta}\frac{\partial u_\phi}{\partial \phi} + \frac{u_r}{r} + \cot\theta \frac{u_\theta}{r}\right)$$

$$+ \tau_{r\theta}\left(\frac{\partial u_\theta}{\partial r} + \frac{1}{r}\frac{\partial u_r}{\partial \theta} - \frac{u_\theta}{r}\right) + \tau_{r\phi}\left(\frac{\partial u_\phi}{\partial r} + \frac{1}{r \sin\theta}\frac{\partial u_r}{\partial \phi} - \frac{u_\phi}{r}\right)$$

$$+ \tau_{\theta\phi}\left(\frac{1}{r}\frac{\partial u_\phi}{\partial \theta} + \frac{1}{r \sin\theta}\frac{\partial u_\theta}{\partial \phi} - \frac{\cot\theta}{r}u_\phi\right)$$

$$(\vec{\nabla}s)_r = \frac{\partial s}{\partial r} \qquad (\vec{\nabla}s)_\theta = \frac{1}{r}\frac{\partial s}{\partial \theta} \qquad (\vec{\nabla}s)_\phi = \frac{1}{r \sin\theta}\frac{\partial s}{\partial \phi}$$

$$(\vec{\nabla} \times \vec{u})_r = \frac{1}{r \sin\theta}\frac{\partial}{\partial \theta}\left(u_\phi \sin\theta\right) - \frac{1}{r \sin\theta}\frac{\partial u_\theta}{\partial \phi}$$

$$(\vec{\nabla} \times \vec{u})_\theta = \frac{1}{r \sin\theta}\frac{\partial u_r}{\partial \phi} - \frac{1}{r}\frac{\partial}{\partial r}\left(u_r u_\phi\right)$$

$$(\vec{\nabla} \times \vec{u})_\phi = \frac{1}{r}\frac{\partial}{\partial r}\left(r u_\theta\right) - \frac{1}{r}\frac{\partial u_r}{\partial \theta}$$

$$\left(\vec{\nabla} \cdot \tau\right)_r = \frac{1}{r^2}\frac{\partial}{\partial r}\left(r^2 \tau_{rr}\right) + \frac{1}{r \sin\theta}\frac{\partial}{\partial \theta}\left(\tau_{r\theta} \sin\theta\right) + \frac{1}{r \sin\theta}\frac{\partial \tau_{r\phi}}{\partial \phi} - \frac{1}{r}\left(\tau_{\theta\theta} + \tau_{\phi\phi}\right)$$

$$\left(\vec{\nabla} \cdot \tau\right)_\theta = \frac{1}{r^2}\frac{\partial}{\partial r}\left(r^2 \tau_{r\theta}\right) + \frac{1}{r \sin\theta}\frac{\partial}{\partial \theta}\left(\tau_{\theta\theta} \sin\theta\right) + \frac{1}{r \sin\theta}\frac{\partial \tau_{\theta\phi}}{\partial \phi} - \frac{\tau_{r\theta}}{r} - \frac{\cot\theta}{r}\tau_{\phi\phi}$$

$$\left(\vec{\nabla} \cdot \tau\right)_\phi = \frac{1}{r^2}\frac{\partial}{\partial r}\left(r^2 \tau_{r\phi}\right) + \frac{1}{r}\frac{\partial \tau_{\theta\phi}}{\partial \theta} + \frac{1}{r \sin\theta}\frac{\partial \tau_{\phi\phi}}{\partial \phi} + \frac{\tau_{r\phi}}{r} + \frac{2\cot\theta}{r}\tau_{\theta\phi}$$

$$(\nabla^2\vec{u})_r = \nabla^2 u_r - \frac{2u_r}{r} - \frac{2}{r^2}\frac{\partial u_\theta}{\partial \theta} - \frac{2u_\theta}{r^2}\cot\theta - \frac{2}{r^2 \sin\theta}\frac{\partial u_\phi}{\partial \phi}$$

$$\left(\nabla^2 \vec{u}\right)_\theta = \nabla^2 u_\theta + \frac{2}{r^2}\frac{\partial u_r}{\partial \theta} - \frac{u_\theta}{r^2 \sin^2 \theta} - \frac{2\cos\theta}{r^2 \sin^2 \theta}\frac{\partial u_\phi}{\partial \phi}$$

$$\left(\nabla^2 \vec{u}\right)_\phi = \nabla^2 u_\phi - \frac{u_\phi}{r^2 \sin^2 \theta} + \frac{2}{r^2 \sin\theta}\frac{\partial u_r}{\partial \phi} + \frac{2\cos\theta}{r^2 \sin^2 \theta}\frac{\partial u_\theta}{\partial \phi}$$

$$\left(\vec{u} \cdot \nabla \vec{u}\right)_r = u_r\frac{\partial u_r}{\partial r} + \frac{u_\theta}{r}\frac{\partial u_r}{\partial \theta} + \frac{u_\phi}{r\sin\theta}\frac{\partial u_r}{\partial \phi} - \frac{u_\theta^2 + u_r^2}{r}$$

$$\left(\vec{u} \cdot \nabla \vec{u}\right)_\theta = u_r\frac{\partial u_\theta}{\partial r} + \frac{u_\theta}{r}\frac{\partial u_\theta}{\partial \theta} + \frac{u_\phi}{r\sin\theta}\frac{\partial u_\theta}{\partial \phi} + \frac{u_r u_\phi}{r} - \frac{u_\phi^2}{r}\cot\theta$$

$$\left(\vec{u} \cdot \nabla \vec{u}\right)_\phi = u_r\frac{\partial u_\phi}{\partial r} + \frac{u_\theta}{r}\frac{\partial u_\phi}{\partial \theta} + \frac{u_\phi}{r\sin\theta}\frac{\partial u_\phi}{\partial \phi} + \frac{u_r u_\phi}{r} + \frac{u_\theta u_\phi}{r}\cot\theta$$

PROBLEMS

A.1. Find the cosine and sine of the angle between $\vec{a}$ and $\vec{b}$ and give the vector and scalar products where

$$\vec{a} = 2\vec{e}_x - 3\vec{e}_y + \vec{e}_z \qquad \vec{b} = 3\vec{e}_x - \vec{e}_y - 2\vec{e}_z$$

A.2. Show that $\vec{a} - \vec{b}$ is parallel to the xy-plane and find its length where

$$\vec{a} = 2\vec{e}_x - \vec{e}_y - \vec{e}_z \qquad \vec{b} = -\vec{e}_x + \vec{e}_y - \vec{e}_z$$

A.3. Find the cosine and sine of the angle between $\vec{a}$ and $\vec{b}$ and give the vector and scalar products where

$$\vec{a} = \vec{e}_x - \vec{e}_y + \vec{e}_z \qquad \vec{b} = 2\vec{e}_x + 3\vec{e}_y - \vec{e}_z$$

A.4. Prove the following identities:

$$\vec{a} \cdot \left(\vec{b} \times \vec{c}\right) = \vec{b} \cdot \left(\vec{c} \times \vec{a}\right)$$

$$\left(\vec{a} \times \vec{b}\right) \cdot \left(\vec{c} \times \vec{d}\right) = \left(\vec{a} \cdot \vec{c}\right)\left(\vec{b} \cdot \vec{d}\right)$$
$$- \left(\vec{a} \cdot \vec{d}\right)\left(\vec{b} \cdot \vec{c}\right)$$

$$\left(\vec{a} \times \vec{b}\right) \times \left(\vec{c} \times \vec{d}\right) = \left[\left(\vec{a} \times \vec{b}\right) \cdot \vec{d}\right]\vec{c}$$
$$- \left[\left(\vec{a} \times \vec{b}\right) \cdot \vec{c}\right]\vec{d}$$

$$\vec{a} \times \left(\vec{b} \times \vec{c}\right) + \vec{b} \times \left(\vec{c} \times \vec{a}\right) + \vec{c} \times \left(\vec{a} \times \vec{b}\right) = 0$$

A.5. The rotation vector, $\vec{\omega}$, is a vector with the magnitude of angular velocity and points in the direction normal to the plane of rotation according to the right-hand screw rule. Show that the velocity of a particle about a point is given by $\vec{u} = \vec{\omega} \times \vec{r}$ and the acceleration vector (for $\vec{\omega}$ = constant) is $\vec{a} = (\vec{\omega} \cdot \vec{r})\vec{\omega} - (\vec{\omega} \cdot \vec{\omega})\vec{r}$

A.6. Show that for Cartesian coordinates the inverse of the transformation matrix is equal to its transpose (an *orthonormal* matrix). Taking δ as the unit tensor with the components δ_{ij}, we have the transformations

$$\delta_{ij} = a_{ii'}a_{jj'}\delta_{i'j'} = a_{ii'}a_{ji'}$$
$$\delta_{i'j'} = b_{i'i}b_{j'j}\delta_{ij} = b_{i'j}b_{j'j}$$

Next, consider the pair of inverse transformations

$$x_i = a_{ii'}x_{i'} \qquad x_{i'} = b_{i'j}x_j$$

Substitution of the second of these transformations into the first gives

$$x_i = a_{ii'}b_{i'j}x_j \quad \text{or} \quad a_{ii'}b_{i'j} = \delta_{ij} = a_{ii'}a_{ji'}$$

Thus, $b_{i'j} = a_{ji'}$ and the second of the transformations is $x_{i'} = a_{ji'}x_j$ leading to

$$a_{ji} = a_{ij}^{-1}$$

Notice that $\delta_{ij} = a_{ik}a_{jk}$ gives six equations for the a_{ij}, leaving us free to rotate the transformed coordinate system three ways to provide the other three equations. The three equations

$$a_{ji}a_{ji} = 1 \quad \text{no sum on } i$$

are called the *normalizing conditions* and the three equations

$$a_{1i}a_{2i} = 0 \qquad a_{2i}a_{3i} = 0 \qquad a_{1i}a_{3i} = 0$$

are called the *orthogonality conditions*. Write these six equations in full.

A.7. Give the transformation matrix for (*a*) a 90-degree rotation about the z-axis, (*b*) a 180-degree rotation about the x-axis, and (*c*) the combination of parts (*a*) and (*b*) [that is, the (*a*) rotation followed by a 180-degree rotation about the x-axis]. Is the matrix the same if the rotations are performed in reverse order?

A.8. The transformation of a vector to a primed coordinate system is given by $u_{i'} = a_{i'i}u_i$
(*a*) Find $a_{i'i}$ for a rotation of 45 degrees counterclockwise about the x_2-axis.
(*b*) An additional rotation is made into a double-primed coordinate system that is described as

$$a_{i''i'} = \begin{bmatrix} 1 & 0 & 0 \\ 0 & \dfrac{1}{2} & \dfrac{\sqrt{3}}{2} \\ 0 & -\dfrac{\sqrt{3}}{2} & \dfrac{1}{2} \end{bmatrix}$$

What is the transformation from the un-primed to the double-primed system?
(*c*) Would the result be the same if the rotation were made in reverse order? Use parts (*a*) and (*b*) to demonstrate your answer.

A.9. Show that the determinant of a 3×3 matrix is

$$\begin{vmatrix} a_{11} & a_{12} & a_{13} \\ a_{21} & a_{22} & a_{23} \\ a_{31} & a_{32} & a_{33} \end{vmatrix} = \epsilon_{ijk}a_{1i}a_{2j}a_{3k}$$

A.10. Prove the following identities:
(*a*) $\vec{a}\vec{b} \cdot \vec{c} = \vec{a}\left(\vec{b} \cdot \vec{c}\right)$
(*b*) $\vec{a}\vec{b} : \vec{c}\vec{d} = \vec{a}\vec{c} : \vec{b}\vec{d}$
(*c*) $\boldsymbol{\tau} : \vec{a}\vec{b} = (\boldsymbol{\tau} \cdot \vec{a}) \cdot \vec{b}$

A.11. Show in matrix form the components of the tensor

$$\boldsymbol{\tau} = \begin{bmatrix} \tau_{11} & \tau_{12} & \tau_{13} \\ \tau_{21} & \tau_{22} & \tau_{23} \\ \tau_{31} & \tau_{32} & \tau_{33} \end{bmatrix}$$

for the rotations of Prob. 7.

A.12. Prove the following identities for Cartesian coordinates:
(*a*) $\vec{\nabla} \times (a\vec{u}) = \vec{\nabla}a \times \vec{u} + a\vec{\nabla} \times \vec{u}$
(*b*) $\vec{\nabla} \times \vec{\nabla}a = 0$
(*c*) $\vec{\nabla} \cdot \left(\vec{\nabla} \times \vec{u}\right) = 0$
(*d*) $\vec{\nabla} \cdot (\vec{u} \times \vec{v}) = \left(\vec{\nabla} \times \vec{u}\right) \cdot \vec{v} - \left(\vec{\nabla} \times \vec{v}\right) \cdot \vec{u}$
(*e*) $\vec{\nabla} \times (\vec{u} \times \vec{v}) = \left(\vec{v} \cdot \vec{\nabla}\right)\vec{u} - \vec{v}\left(\vec{\nabla} \cdot \vec{u}\right)$ $+\vec{u}\left(\vec{\nabla} \cdot \vec{v}\right) - \left(\vec{u} \cdot \vec{\nabla}\right)\vec{v}$
(*f*) $\vec{\nabla} \times \left(\vec{\nabla} \times \vec{u}\right) = \vec{\nabla}\left(\vec{\nabla} \cdot \vec{u}\right) - \nabla^2\vec{u}$
(*g*) $\vec{\nabla}(\vec{u} \cdot \vec{v}) = \vec{u} \times \left(\vec{\nabla} \times \vec{v}\right) + \vec{v} \times \left(\vec{\nabla} \times \vec{u}\right)$ $+ \left(\vec{u} \cdot \vec{\nabla}\right)\vec{v} + \left(\vec{v} \cdot \vec{\nabla}\right)\vec{u}$
(*h**)$\left(\vec{u} \cdot \vec{\nabla}\right)\vec{r} = \vec{u}$
(*i**)$\vec{\nabla} \cdot \vec{r} = 3$
(*j**)$\vec{\nabla} \times \vec{r} = 0$
(*k*) $\boldsymbol{\tau} : \vec{\nabla}\vec{u} = \vec{\nabla} \cdot (\boldsymbol{\tau} \cdot \vec{u}) - \vec{u} \cdot \left(\vec{\nabla} \cdot \boldsymbol{\tau}\right)$ for τ_{ij} $= \tau_{ji}$
(*l*) $\vec{\nabla} \cdot a\mathbf{e} = \vec{\nabla}a$
(*m*) $\vec{\nabla} \cdot \vec{u}\vec{v} = \vec{u} \cdot \vec{\nabla}\vec{v} + \vec{v}\left(\vec{\nabla} \cdot \vec{u}\right)$
(*n*) $a\mathbf{e} : \vec{\nabla}\vec{u} = a\left(\vec{\nabla} \cdot \vec{u}\right)$
*In (*h*), (*i*), and (*j*), $\vec{u}$ is the velocity vector of a particle and $\vec{r}$ is the vector distance to the particle.

A.13. Use the divergence theorem

$$\int_V \vec{\nabla} \cdot \vec{u} \, dV = \int_{\partial V} \vec{u} \cdot \vec{n} \, dA$$

where ∂V is the surface enclosing the volume, V, and $\vec{n}$ is the unit outward normal from the volume to show that

$$\int_V \left[\vec{P} \cdot \left(\vec{\nabla} \times \vec{\nabla} \times \vec{Q} \right) - \vec{Q} \cdot \left(\vec{\nabla} \times \vec{\nabla} \times \vec{P} \right) \right] dV$$

$$= \int_{\partial V} \left(\vec{Q} \times \vec{\nabla} \times \vec{P} - \vec{P} \times \vec{\nabla} \times \vec{Q} \right) \cdot \vec{n} \, dA$$

A.14. Use the chain rule of differentiation to find $\vec{\nabla} \cdot \vec{u}$ in cylindrical coordinates.

POSTFACE

In the first part of the twentieth century knowledge in fluid mechanics was exploding. The rapid growth continued into the middle of the century, fueled by government spending in armaments. As we approach the end of the century, we can ask: What has been the progress in the last thirty years? What is likely to occur in the next thirty years?

In many aspects the last thirty years have been a period of consolidation with moderate progress, not at all comparable to the earlier advance. True, we have gone from solving one- and two-dimensional problems to two- and three-dimensional, unsteady problems, but those developments owe more to progress in numerical methods and computation than to fluid mechanics. And in some respects even the numerical solutions have been disappointing. We are not able to predict the weather six months in advance; we cannot accurately compute the drag on an automobile without reference to experiment; tracing the path of pollutants in the air, water, and ground is still difficult; prediction of floods is much too inaccurate; ocean circulation remains a mystery (El Niño, for example); calculations of sediment transport in air and water are wildly inaccurate. Most of the unsolved problems contain the difficulty of *scale*. The difficult problems are those that mix the small scales (the shortest eddy lengths of turbulence, grains of sediment) with problem lengths that are many orders of magnitude greater. Although we will make progress in computing over a range of

466

scales—massive computer power will help— nothing on the horizon indicates that such problems will be fully, or even satisfactorily, resolved. Thus, a very large set of important problems remains for the talented researcher to resolve.

Computational fluid mechanics has changed the relative importance of some subjects. Thirty years ago much of a fluid mechanics course was devoted to instrluction in how to solve problems. Perturbation methods, complex mappings, and transforms were important components. Although these are still valuable tools, the flexibility and power of numerical methods have made them the primary method of engineering calculation. The serious student of fluid mechanics must have a course in computational fluid mechanics.

REFERENCES

Baltzer, R. A., and C. Lai: "Computer Simulation of Unsteady Flows in Waterways," *Journal of the Hydraulics Division*, ASCE, vol. 94, no. HY4, pp. 1083–1117, July 1968.

Bear, Jacob: *Hydraulics of Groundwater*, McGraw-Hill, 1979.

—— and Yehuda Bachmat: *Introduction to Modeling of Transport Phenomena in Porous Media*, Kluwer, 1990.

Betts, P. L., and T. T. Mohamad: "A Boundary Approach to Water Waves," in *Finite Element Analysis*, T. Kawai, ed., University of Tokyo Press, pp. 923–929, 1982.

Binder, R. C.: *Advanced Fluid Mechanics*, vols. I, II, and III, Prentice-Hall, 1958.

Bird, R. B., W. E. Stewart, and E. N. Lightfoot: *Transport Phenomena*, Wiley, 1960.

—— R. C. Armstrong, and Ole Hassager: *Dynamics of Polymeric Liquids*, vol. 1, *Fluid Mechanics*, Wiley, 1977.

Blasius, H.: "Grenzschichten in Flüssigkeiten mit kleiner Reibung," *Z. Math. u Phys.*, vol. 56, 1 (English translation in NACA Tech. Memo. no. 1256), 1908.

Block, H. D., E. T. Cranch, P. J. Hilton, and R. J. Walker: *Engineering Mathematics*, Cornell, 1964.

Bradshaw, P.: "The Understanding and Prediction of Turbulent Flow," *Aeronautical Journal*, vol. 6, 1972.

Brebbia, C. A., and J. Dominguez: *Boundary Elements—An Introductory Course*, McGraw-Hill, 1992.

Bromley, D. Allan: "Physics: Natural Philosophy and Invention," *American Scientist*, vol. 74, no. 6, pp. 622–639, November–December 1986.

Brutsaert, Wilfried: "On the Anisotropy of the Eddy Diffusivity," *Journal of the Meteorological Society of Japan*, vol. XXXXVIII, no. 5, pp. 411–416, August 28, 1970.

Carslaw, H. S., and J. C. Jaeger: *Conduction of Heat in Soils*, 2nd ed., Clarendon, Oxford, 1959.

Carter, H. H., and Akira Okubo: "A Study of the Physical Process of Movement and Dispersion in the Cape Kennedy Area," Final Report, U. S. Atomic Energy Commission, Contract no. AT(30-1)-2973, 1965.

Cebeci, T., and A. M. O. Smith: *Analysis of Turbulent Boundary Layers*, Academic Press, 1974.

Chow, Ven Te: *Open-Channel Hydraulics*, McGraw-Hill, 1959.

Churchill, Ruel V., *Complex Variables*, McGraw-Hill, 1948.

——— *Fourier Series and Boundary Value Problems*, 4th ed., McGraw-Hill, 1987.

Cohen, J. K., and R. M. Lewis: "A Ray Method for the Asymptotic Solution of the Diffusion Equation," *Journal of the Institute of Mathematical Applications*, vol. 3, pp. 266–290, 1967.

Crank, J.: *The Mathematics of Diffusion*, Oxford University Press, 1975.

Cunge, J. A., F. M. Holly, Jr., and A. Verwey: *Practical Aspects of Computational Hydraulics*, Pitman, 1980.

Daily, James and Jordaan: "Effects of Unsteadiness on Resistance and Energy Dissipation," *Hydrodynamics Laboratory*, MIT, 1956.

Davies, J. T.: *Turbulence Phenomena*, Academic Press, 1972.

Deissler, R. G.: NACA Report 1210, 1955.

Dracos, T. A., and B. Glenne: "Stability Criteria for Open-channel Flow," *Journal of the Hydraulics Division*, ASCE, vol. 93, no. HY6, pp. 79–101, November 1967.

Dryden, H. L., F. P. Murnaghan, and H. Bateman: *Hydrodynamics*, Dover, 1956.

Einstein, H. A., and Huon Li: "Secondary Currents in Straight Channels," *Transactions, American Geophysical Union*, vol. 39, no. 6, pp. 1085–1088, December 1958.

Ekman, V. W.: "Die Zusammendrückbarkeit des Meerwassers," *Conseil Perm. Intern. p. l'Expl. de la Mer*, Pub. de Circonstance, no. 43, 1908.

Escoffier, F. F., and M. B. Boyd: "Stability Aspects of Flow in Open Channels," *Journal of the Hydraulics Division*, ASCE, vol. 88, no. HY6, November 1962.

Garabedian, P. R.: "Calculation of Axially Symmetric Cavities and Jets," *Pacific Journal of Mathematics*, vol. 6, pp. 611–684, 1956.

Gresho, P. M.: "Incompressible Fluid Dynamics: Some Fundamental Formulation Issues," *Annual Reviews of Fluid Mechanics*, vol. 23, January 1991.

Hadley, George: *Linear Algebra*, Addison-Wesley, 1961.

Hinze, J. O.: *Turbulence*, McGraw-Hill, 1959.

Holmboe, J.: "On the Behavior of Symmetric Waves in Stratified Shear Layers," *Geofys. Publ.*, vol. 24, pp. 39–62, 1962.

Holstein, H., and T. Bohlen: "Ein einfaches Verfahren zur Berechnung laminarer Reibungsschichten, die dem Nährungsansatz von K. Pohlhausen genügen," Lilienthall-Bericht S 10, pp. 5–16, 1940.

Howarth, L.: "On the Solution of the Laminar Boundary Layer Equations," *Proceedings of the Royal Society*, vol. A 164, p. 547, 1938.

Hunt, B.: "Numerical Solution of an Integral Equation for Flow from a Circular Orifice," *Journal of Fluid Mechanics*, vol. 31, part 2, pp. 361–377, 1968.

Ippen, Arthur T. ed.: *Estuary and Coastline Hydrodynamics*, McGraw-Hill, 1966.

Jones, Bradley: *Elements of Practical Aerodynamics*, John Wiley and Sons, 1939.

Kinsman, Blair: *Wind Waves*, Prentice-Hall, 1965.

Kline, S. J., M. V. Morkovin, G. Sovran, and D. S. Cockrell, eds.: *Computation of Turbulent Boundary Layers—1968*, AFOSR-IFP Stanford Conference, Stanford University Press, Stanford, California, 1968.

Kolmogoroff, A. N.: "On Degeneration of Isotropic Turbulence in an Incompressible Viscous Liquid," *Comp. Rend. Acad. Sci. U. S. S. R.*, vol. 31, 1941.

Lamb, Horace: *Hydrodynamics*, Dover, 1945.

Lanchester, F. W.: *Aerial Flight: Aerodynamics*, Van Nostrand, 1908.

Landsford, W. M.: "Discharge Coefficients for Pipe Orifices," *Civil Engineering*, vol. 4, pp. 245–247, 1934.

Lass, Harry: *Vector and Tensor Analysis*, McGraw-Hill, 1950.

LeMehaute, B.: "Progressive Wave Absorber," *Journal of Hydraulic Research*, vol. 10, no. 2, pp. 153–169, 1972.

Levi, Enzo: *El Agua Según la Sciencia*, Consejo Nacional de Ciencia y Technología and Ediciones Castell, Mexico, 1989.

Liggett, J. A.: "Unsteady Circulation in Shallow, Homogeneous Lakes," *Journal of the Hydraulics Division*, ASCE, vol. 95, no. HY4, pp. 1273–1288, July, 1969.

———— "Stability," Chap. 9 of *Unsteady Flow in Open Channels*. Edited by K. Mahmood, and V. Yevjevich, Water Resources Publications, P. O. Box 303, Fort Collins, Colo., 1975.

———— and W. H. Graf: "Steady and Unsteady Effects on Discharge in a River Connecting Two Reservoirs," pub. no. 15, Great Lakes Research Division, The University of Michigan, pp. 249–258, 1966.

———— and C. Hadjitheodorou: "Circulation in Shallow Homogeneous Lakes," *Journal of the Hydraulics Division*, ASCE, vol. 95, no. HY2, pp. 609–620, March 1969.

———— and K. K. Lee: "Properties of Circulation in Stratified Lakes," *Journal of the Hydraulics Division*, ASCE, vol. 97, no. HY1, pp. 15–29, January 1971.

———— and P. L-F. Liu: *The Boundary Integral Equation Method for Porous Media Flow*, George Allen and Unwin, London, 1983.

Lin, C. C.: *The Theory of Hydrodynamic Stability*, Cambridge University Press, 1955.

Long, R. R.: *Mechanics of Solids and Fluids*, Prentice-Hall, 1961.

Mahmood, K., and V. Yevjevich: *Unsteady Flow in Open Channels*, Water Resources Publications, P. O. Box 303, Fort Collins, Colo., 1975.

Mayer, P. G. W.: "A Study of Roll Waves and Slug Flows in Inclined Open Channels," Ph. D. Thesis, Cornell University, 1957.

Mei, Chiang C.: *The Applied Dynamics of Ocean Surface Waves*, World Scientific, 1989.

Montouri, Carlo: Discussion to "Stability Aspects of Flow in Open Channels" (by Escoffier and Boyd, 1962), *Journal of the Hydraulics Division*, ASCE, vol. 89, no. HY4, July 1963.

Moody, L. F.: "Friction Factors for Pipe Flow," *Transactions of the ASME*, vol. 66, 1944.

Milne-Thomson, L. M.: *Theoretical Hydrodynamics*, Macmillan, 1960.

Neumann, Gerhard, and W. J. Pierson, Jr.: *Principles of Physical Oceanography*, Prentice-Hall, 1966.

Nikuradse, J.: "Strömungsgesetze in rauhen Rohren," *VDI-Forschungsheft*, (English translation in NACA Tech. Memo no. 1292), 1933.

Okubo, Akira, and M. J. Karweit: "Diffusion from a Continuous Source in a Uniform Shear Flow," *Limnology and Oceanography*, vol. 14, no. 4, pp. 514–520, July 1969.

Orr, W. M. F.: "The Stability or Instability of the Steady Motions of a Perfect Liquid and of a Viscous Liquid," *Proceeding of the Royal Irish Academy*, vol. 27, pp. 9–138, 1907.

Park, N. S., and J. A. Liggett: "Application of Taylor-Least Squares Finite Element to Three-Dimensional Advection-Diffusion Equation," *International Journal for Numerical Methods in Fluids*, vol. 13, pp. 759–773, 1991.

Panton, Ronald L.: *Incompressible Flow*, Wiley, 1984.

Pohlhausen, K.: "Zur näherungsweisen Integration der Differentialgleichung der laminaren Grenzschicht," *Zeitschrift für angewandte Mathematik und Mechanik (ZAMM)*, vol. 1, pp. 252–268, 1921.

Polubarinova-Kochina, P. Y.: *Theory of Groundwater Movement*, Gostekhizdat, Moscow, 1952; translated into English by Roger J. M. DeWiest, Princeton University Press, 1962.

Potter, M. C., and J. F. Foss: *Fluid Mechanics*, Great Lakes Press, Inc., Okemos Mich., 1982.

Powell, Ralph W.: "Vedernikov's Criterion for Ultra-rapid Flow," *Transactions of the American Geophysical Union*, vol. 29, no. 6, 1948.

Prandtl, Ludwig: "Über die ausgebildete Turbulenz," *Zeitschrift für angewandte Mathematik und Mechanik (ZAMM)*, vol. 5, pp. 136–139, 1925.

———— *Essentials of Fluid Dynamics with Applications to Hydraulics, Aeronautics, Meteorology, and Other Subjects*, Hafner Publishing Company, New York, 1952.

Preiswerk, E.: "Application of the Methods of Gas Dynamics to Water Flows with Free Surface: I. Flows with no Energy Dissipation; II. Flows with Momentum Discontinuities (Hydraulic Jumps)," NACA Tech. Memo 934 and 935, 1940.

Proudman, I., and J. R. A. Pearson: "Expansions at Small Reynolds Number for the Flow Past a Sphere and a Circular Cylinder," *Journal of Fluid Mechanics*, vol. 2, pp. 237–262, 1957.

Putnam, J. A., and R. S. Authur: "Diffraction of Water Waves by Breakwaters," *Transactions of the American Geophysical Union*, vol. 29, no. 4, 1948.

Richards, G. J.: *Phil. Trans. Roy. Soc. London, ser. A.*, vol. 223, 1935.

Rouse, Hunter, and Simon Ince: *History of Hydraulics*, Iowa Institute of Hydraulic Research, State University of Iowa, 1957.

Saph, A. V., and E. W. Schoder: *An Experimental Study of the Resistances to the Flow of Water in Pipes*, Ph. D. Thesis, Cornell University, 1902.

Schapery, R. A.: "Approximate Methods of Transform Inversion for Viscoelastic Stress Analysis," *Proceedings of the Fourth U. S. National Congress on Applied Mechanics*, vol. 2, pp. 1075–1085, 1962.

Schlichting, Hermann: *Boundary Layer Theory*, 4th ed., McGraw-Hill, 1960.

Schoder, Ernest W., and Francis M. Dawson, *Hydraulics*, McGraw-Hill, 1934.

Shen, S. F.: "Calculated Amplified Oscillations in Plane Poiseuille and Blasius Flow," *Journal of the Aeronautical Sciences*, vol. 21, pp. 62–64, 1954.

Simmons, T. J.: *Circulation Models of Lakes and Inland Seas*, Canadian Bulletin of Fisheries and Aquatic Sciences, Bulletin 203, Ottawa, 1980.

Smith, R.: "The Early Stages of Contaminant Dispersion in Shear Flows," *Journal of Fluid Mechanics*, vol. 111, pp. 107–122, 1981.

Smyth, W. D., and W. R. Peltier: "Instability and Transition in Finite-amplitude Kelvin-Helmholtz and Holmboe Waves," *Journal of Fluid Mechanics*, vol. 228, pp. 387–415, 1991.

Sommerfeld, A.: "Ein Beitrag zur hydrodynamischen Erklärung der turbulenten Flüssigkeitsbewegungen", *Atti del IV Congresso Internazionale dei Matematici*, Rome, pp. 116–124, 1908.

Stern, M. E.: *Ocean Circulation Physics*, Academic Press, 1975.

Stoker, J. J.: *Water Waves*, Interscience, 1957.

Strack, O. D. L.: *Groundwater Mechanics*, Prentice-Hall, 1989.

Streeter, V. L.: *Fluid Dynamics*, McGraw-Hill, 1948.

Tennekes, H., and J. L. Lumley: *A First Course in Turbulence*, MIT Press, 1972.

Townsend, A. A.: *The Structure of Turbulent Shear Flow*, Cambridge University Press, 1956.

Trefftz, E.: "Uber die kontraktion kreisformiger flussigkeitsstrahlen," *Z. Math. Phys.*, vol. 64, 34, pp. 34–61, 1916.

Vedernikov, V. V.: "Characteristic Features of a Liquid Flow in an Open Channel," C. R. (Doklady), USSR Acad. Sci., vol. 52, 1946.

Vedernikov, V. V.: "Conditions at the Front of a Translatory Wave Distributing a Steady Motion of a Real Fluid," C. R. (Doklady), USSR Acad. Sci., vol. 49, no. 4, 1945.

Verber, J. L.: "Current Profiles to Depth in Lake Michigan," pub. no. 13, Great Lakes Research Division, University of Michigan, 1965.

Von Kármán, T.: "Uber laminare und Zeitschrift für angewante Mathematik und Mechanik turbulente Reibung," *(ZAMM)*, vol. 1 (English translation in NACA Tech. Memo no. 1092), 1921.

———— "Mechanische Ähnlichkeit und Turbulenz," *Proceedings of the Third International Congress of Applied Mechanics*, Stockholm, 1930.

———— "Eine praktische Anwendung der Analogie zwischen Überschallströmung in gasen und über kritischer strömung in offenen Gerinnen," *Zeitschrift für angewandte Mathematik und Mechanik (ZAMM)*, 1938.

Whitehead, A. N.: "Second Approximations to Viscous Fluid Motion," *Quart. Jour. Math.*, vol. 23, pp. 143–152, 1889.

Woo, Dah-Cheng, and E. F. Brater: "Spatially Varied Flow from Controlled Rainfall," *Journal of the Hydraulics Division*, ASCE, vol. 88, no. HY6, pp. 31–56, November 1962.

Wu, Jin: "Wind Stress on a Water Surface," *Journal of Geophysical Research*, vol. 74, no. 4, January 1969.

Yih, Chia-Shun: *Fluid Mechanics*, West River Press, Ann Arbor, 1977.

Young, A. D.: *Boundary Layers*, AIAA Education Series, American Institute of Aeronautics and Astronautics, Inc., Washington, D. C. and BSP Professional Books, London, 1989.

Young, D. L., J. A. Liggett, and R. H. Gallagher: "Unsteady Stratified Circulation in a Cavity," *Journal of the Engineering Mechanics Division*, ASCE, vol. 102, no. EM6, pp. 1009–1023, December 1976.

Young, D. L., and J. A. Liggett: "Transient Finite Element Shallow Lake Circulation," *Journal of the Hydraulics Division*, ASCE, vol. 103, no. HY2, pp. 109–121, February 1977.

A FLUID
MECHANICS
BIBLIOGRAPHY

In compiling a bibliography of fluid mechanics, the question quickly arises as to the boundaries of the field. Aerodynamics, boundary layer theory, and computational fluid mechanics are part of the field, but even there one asks how deeply should the list go into the specialization. Should magnetohydrodynamics, meteorology, wave mechanics, and others be included? What about non-Newtonian flow? The result has been a solid equivocation; these subjects are neither excluded nor comprehensively included.

The list is confined to the English language except for some historically significant books.

Abbott, I. H., and A. E. von Doenhoff: *Theory of Wing Sections*, Dover, 1959.

Abbott, Michael B.: *Computational Hydraulics*, Pitman, 1979.

——— and D.R. Basco: *Computational Fluid Dynamics: An Introduction for Engineers*, Wiley, 1989.

Abramovich, G. N.: *The Theory of Turbulent Jets*, MIT Press, 1963.

Acheson, D. J.: *Elementary Fluid Dynamics*, Clarendon, 1990.

Ahmed, Nazeer: *Fluid Mechanics*, Engineering Press, 1987.

Airy, G. B.: "Tides and Waves," *Encyclopedia Metropolitana*, London, 1845.

Albertson, M. L., J. R. Barton, and D. B. Simons: *Fluid Mechanics for Engineers*, Prentice-Hall, 1960.

Aldama, A. A.: *Filtering Techniques for Turbulent Flow Simulation*, Springer-Verlag, 1990.

Allaire, Paul E.: *Basics of the Finite Element Method*, W. C. Brown, 1985.

473

Allen, Theodore, and Richard L. Ditsworth: *Fluid Mechanics*, McGraw-Hill, 1972.

Anderson, J. D.: *Modern Compressible Flow with Historical Perspective*, McGraw-Hill, 1982.

Aris, Rutherford: *Vectors, Tensors, and the Basic Equations of Fluid Mechanics*, Dover, 1989; Prentice-Hall, 1962.

Ashley, Holt, and Marten Landahl: *Aerodynamics of Wings and Bodies*, Addison-Wesley, 1965.

Astarita, Giovanni, and G. Marrucci: *Principles of Non-Newtonian Fluid Mechanics*, McGrawHill, 1974.

Azbel, David, and Nicholas P. Cheremisinoff: *Fluid Mechanics and Unit Operations*, Ann Arbor Science, 1983.

Baker, A. J.: *Finite Element Computational Fluid Mechanics*, Hemisphere Pub. Corp., 1983.

Bakhmeteff, B. A.: *Hydraulics of Open Channels*, McGraw-Hill, 1932.

———— *The Mechanics of Turbulent Flow*, Princeton University Press, 1941.

Barlow, Peter: *Hydrodynamics*, reprinted from the *Encyclopedia Metropolitana*, vol. 3, 1848.

Barna, P. S.: *Fluid Mechanics for Engineers*, 3rd ed., Butterworths, 1969.

Barton, Edwin Henry: *An Introduction to the Mechanics of Fluids*, Longmans, Green and Co., 1915.

Basset, Alfred Barnard: *A Treatise on Hydrodynamics with Numerous Examples*, Deighton Bell, 1888.

———— *An Elementary Treatise on Hydrodynamics and Sound*, Deighton Bell, 1890.

Batchelor, George Keith: *The Theory of Homogeneous Turbulence*, Cambridge University Press, 1953.

———— *An Introduction to Fluid Dynamics*, Cambridge, University Press, 1967.

Bayley, Frederick John: *An Introduction to Fluid Dynamics*, Allen and Unwin, Interscience, 1958.

Benque, J. P., A. Haugel, and P. L. Viollet: *Numerical Models in Environmental Fluid Mechanics*, Pitman Advanced Pub. Program, 1982.

Bergel, D. H., ed.: *Cardiovascular Fluid Dynamics*, Academic Press, 1972.

Berger, Stanley A.: *Laminar Wakes*, Elsevier, 1971.

Bernoulli, Daniel: *Hydrodynamica, Sive de Viribus et Motibus Fluidorum Commentarii*, Strasbourg, 1738.

———— *Hydrodynamics*, Dover, 1968.

Bertin, J. J., and M. L. Smith: *Aerodynamics*, Prentice-Hall, 1979.

Bertin, John L.: *Engineering Fluid Mechanics*, 2nd ed., Prentice-Hall, 1987.

Besant, William Henry: *Treatise on Hydrostatics and Hydrodynamics*, Deighton Bell, 1859.

Binder, Raymond Charles: *Advanced Fluid Dynamics and Fluid Machinery*, Prentice-Hall, 1951.

———— *Advanced Fluid Mechanics*, 2 vols., Prentice-Hall, 1958.

———— *Fluid Mechanics*, 5th ed., Prentice-Hall, 1973.

Bird, R. B., W. E. Stewart, and E. N. Lightfoot: *Transport Phenomena*, Wiley, 1960.

Birkhoff, Garrett: *Hydrodynamics*, 2nd ed., Princeton University Press, 1960.

———— *Hydrodynamics*, Princeton University Press, 1960; Greenwood Press, Inc., 1978.

———— and E. H. Zarantonello: *Jets, Wakes, and Cavities*, Academic Press, 1957.

Blevins, Robert D.: *Applied Fluid Dynamics Handbook*, Van Nostrand Reinhold Co., 1984.

Bober, William, and Richard A. Kenyon: *Fluid Mechanics*, Wiley, 1980.

Boon, Jean-Pierre, and Sidney Yip: *Molecular Hydrodynamics*, McGraw-Hill, 1980.

Bose, Tarit Kumar: *Computational Fluid Dynamics*, Wiley, 1988.

Bossut, C.: *Traité théorique et expérimental d'hydrodynamique*, 3rd ed., Imprimerie Royale, 1786–87.

Boxer, G.: *Work Out Fluid Mechanics*, Macmillan, 1988.

Boyle, W. P.: *Applied Fluid Mechanics*, McGraw-Hill Ryerson, 1986.

Bradshaw, Peter: *Experimental Fluid Mechanics*, 2nd ed., Pergamon Press, 1970.

———— *An Introduction to Turbulence and Its Measurement*, Pergamon Press, 1971.

———— ed.: *Turbulence*, Springer-Verlag, 1978.

————, Tuncer Cebeci, and James H. Whitelaw: *Engineering Calculation Methods for Turbulent Flow*, Academic Press, 1984.

Brenkert, Karl: *Elementary Theoretical Fluid Mechanics*, Wiley, 1960.

Brown, Robert Alan: *Fluid Mechanics of the Atmosphere*, Academic Press, 1991.

Brutsaert, Wilfried: *Evaporation into the Atmosphere*, Reidel, 1982.

Campbell, Robert Gordon: *Foundations of Fluid Flow Theory*, Addison-Wesley, 1973.

Cebeci, Tuncer, and Peter Bradshaw: *Momentum Transfer in Boundary Layers*, Hemisphere, 1977.

———— and A. M. O. Smith: *Analysis of Turbulent Boundary Layers*, Academic Press, 1974.

Cermak, Jack E.: *Applications of Fluid Mechanics to Wind Engineering*, ASME, 1974.

Chadwick, A. J., and J. C. Morfett: *Hydraulics in Civil and Environmental Engineering*, 2nd ed., E & F. N. Spon, Ltd. 1993.

Chang, P. A.: *Separation of Flow*, Pergamon Press, 1970.

Chang, P. K.: *Control of Flow Separation*, McGraw-Hill, 1976.

Chang, Tien-Sheng: *Intermediate Fluid Mechanics*, Edwards Brothers, 1962.

Chaudhry, M. Hanif: *Open Channel Flow*, Prentice-Hall, 1993.

Chen, Peter J.: *Selected Topics in Wave Propagation*, Noordhoff International Publishing, 1976.

Cheremisinoff, Nicholas P.: *Fundamentals of Wind Energy*, Ann Arbor Science, 1978.

———— *Applied Fluid Flow Measurement: Fundamentals and Technology*, Dekker, 1979.

———— *Fluid Flow: Pumps, Pipes, and Channels*, Ann Arbor Science, 1981.

———— *Instrumentation for Complex Fluid Flows*, Technomic Pub. Co., 1986.

———— *Flow Measurement for Engineers and Scientists*, Dekker, 1988.

———— *Practical Fluid Mechanics for Engineers and Scientists*, Technomic Pub. Co., 1990.

———— and Paul N. Cheremisinoff: *Hydrodynamics of Gas-Solids Fluidization*, Gulf Pub. Co., 1984.

Cheremisinoff, Nicholas P., ed.: *Encyclopedia of Fluid Mechanics*, Gulf Pub. Co., 1986.

———— and Ramesh Gupta, eds.: *Handbook of Fluids in Motion*, Ann Arbor Science, 1983.

Chevray, René, and Jean Mathieu: *Topics in Fluid Mechanics*, Cambridge University Press, 1993.

Chirgwin, Brian H., and C. Plumpton: *Elementary Classical Hydrodynamics*, Pergamon Press, 1967.

Chorin, Alexandre Joel: *Lectures on Turbulence Theory*, Publish or Perish, 1975.

———— *Numerical Methods in Statistical Hydrodynamics*, Presses de l'Université de Montréal, 1977.

———— *Computational Fluid Mechanics*, Academic Press, 1989.

———— and J. E. Marsden: *A Mathematical Introduction to Fluid Mechanics*, 3rd ed., Springer-Verlag, 1992.

Chorlton, Frank: *Textbook of Fluid Dynamics*, Van Nostrand, 1967.

Chow, Chuen-Yen: *An Introduction to Computational Fluid Mechanics*, Wiley, 1979.

Chow, Ven Te: *Open-Channel Hydraulics*, McGraw-Hill, 1959.

Chung, T. J.: *Finite Element Analysis in Fluid Dynamics*, McGraw-Hill, 1978.

Church, Irving Porter: *Hydraulics: or Mechanics of Fluids, part IV of the Mechanics of Engineering*, Cornell University, 1886.

——— *Mechanics of Engineering (Fluids): A Treatise on Hydraulics and Pneumatics for Use in Technical Schools*, Wiley, 1889.

Churchill, Stuart W.: *Viscous Flows: The Practical Use of Theory*, Butterworths, 1988.

Cole, G. H. A.: *Fluid Dynamics, An Introductory Account of Certain Theoretical Aspects Involving Low Velocities and Small Amplitudes*, Wiley, 1962.

Courant, Richard, and K. O. Friedrichs: *Supersonic Flow and Shock Waves*, Springer-Verlag, 1977.

Cox, Glen Nelson, and F. J. Ermano: *Fluid Mechanics*, Van Nostrand, 1941.

Coxon, W. F.: *Flow Measurement and Control*, Macmillan, 1959.

Csanady, G. T.: *Theory of Turbomachines*, McGraw-Hill, 1964.

——— *Turbulent Diffusion in the Environment*, Reidel, 1973.

——— *Circulation in the Coastal Ocean*, Reidel, 1982.

Curle, N., and H. J. Davies: *Modern Fluid Dynamics*, 2 vols., Van Nostrand, 1968.

Currie, Iain G.: *Fundamental Mechanics of Fluids*, McGraw-Hill, 1974.

D' Alembert, J.: *Traité de dynamique*, Gauthier-Villars, 1921.

——— *Oeuvres de d'Alembert*, vol. 1, Slatkine Reprints, 1967.

Dahlberg, Eric Charles: *Applied Hydrodynamics in Petroleum Exploration*, Springer-Verlag, 1982.

Daily, James, W., and R. F. Harleman: *Fluid Dynamics*, Addison-Wesley, 1966.

Daneshyar, H.: *One-dimensional Compressible Flow*, Pergamon Press, 1976.

Darcy, H.: *Recherches expérimentales relatives au mouvement de l'eau dans les tuyaux*, Paris, 1857.

——— and H. Bazin: *Recherches hydrauliques: 1re partie, recherches expérimentales sur l'écoulement de l'eau dans les canaux découverts; 2e partie, recherches expérimentales relatives aux remous et à la propagation des ondes*, Paris, 1865.

——— and H. Bazin: *Recherches hydrauliques entreprises par H. Darcy, continuées par H. Bazin*, Imprimerie Impériale, 1865.

Daugherty, Robert Long: *Hydraulics, A Text on Practical Fluid Mechanics*, McGraw-Hill, 1937.

——— and A. C. Ingersoll: *Fluid Mechanics with Engineering Applications*, 5th ed., McGraw-Hill, 1954.

——— and Joseph B. Franzini: *Fluid Mechanics with Engineering Applications*, 7th ed., McGraw-Hill, 1977.

——— Joseph B. Franzini, and E. John Finnemore: *Fluid Mechanics, with Engineering Applications*, 8th ed., McGraw-Hill, 1985.

Davies, J. T.: *Turbulence Phenomena*, Academic Press, 1972.

Debler, Walter R.: *Fluid Mechanics Fundamentals*, Prentice-Hall, 1990.

DeGroot, A.: *Fluid Mechanics*, International Correspondence Schools, 1975.

DeMarchi, G.: *Guglielmini*, Brescia, 1947.

DeNevers, Noel: *Fluid Mechanics for Chemical Engineers*, McGraw-Hill, 1991.

DeVillamil, Richard: *ABC of Hydrodynamics*, E. & F. N. Spon, Ltd., 1912.

——— *Motion of Liquids*, E. & F. N. Spon, Ltd., 1914.

Dinnar, Uri: *Cardiovascular Fluid Dynamics*, CRC Press, 1981.

Dixon, S. L.: *Fluid Mechanics, Thermodynamics of Turbomachinery*, 3rd ed., Pergamon Press, 1978.

Dommasch, D. O., S. S. Sherby, and T. F. Connally: *Airplane Aerodynamics*, 4th ed., Pitman, 1967.

Donkin, C. T. B.: *Elementary Practical Hydraulics of Flow in Pipelines*, Oxford University Press, 1959.

Douglas, John F.: *Solving Problems In Fluid Mechanics*, Longman Scientific & Technical, 1986.

—— J. M. Gasiorek, and J. A. Swaffield: *Fluid Mechanics*, 2nd ed., Pitman, 1985.

Drazin, P. G., and W. H. Reid: *Hydrodynamic Stability*, Cambridge University Press, 1981.

Dryden, H. L., F. P. Murnaghan, and H. Bateman: *Hydrodynamics*, Dover, 1956.

Duckworth, Roger Alan: *Mechanics of Fluids*, Longman, 1977.

Duncan, William Jolly, A. S. Thom, and A. D. Young: *Mechanics of Fluids*, 2nd ed., E. Arnold, 1970.

Dupuit, A. J.: *Traité théorique et pratique de la conduite et de la distribution des eaux*, Paris, 1865.

Durant, W. F., ed.: *Aerodynamic Theory*, Springer, 1934.

Eckart, Carl Henry: *Hydrodynamics of Oceans and Atmospheres*, Pergamon Press, 1960.

Einstein, Hans Albert: *Progress Report of the Analogy Between Surface Shock Waves on Liquids and Shocks in Compressible Gases*, Hydrodynamics Laboratory, California Institute of Technology, 1946.

Elevatorski, E. A.: *Hydraulic Energy Dissipators*, McGraw-Hill, 1959.

Emrich, Raymond J., ed.: *Fluid Dynamics*, Academic Press, 1981.

Eskinazi, Salamon: *Vector Mechanics of Fluids and Magnetofluids*, Academic Press, 1967.

—— *Principles of Fluid Mechanics*, 2nd ed., Allyn and Bacon, 1968.

—— *Fluid Mechanics and Thermodynamics of Our Environment*, Academic Press, 1975.

—— *Fluid Mechanics*, Sigma Pub. Co., 1985.

Evett, Jack B., and Cheng Liu: *Fundamentals of Fluid Mechanics*, McGraw-Hill, 1987.

—— and Cheng Liu: *2500 Solved Problems in Fluid Mechanics and Hydraulics*, McGraw-Hill, 1989.

Ewald, Peter Paul: *The Physics of Solids and Fluids, with Recent Developments*, 2nd ed., Blackie, 1936.

Fanning, John Thomas: *A Practical Treatise on Hydraulic and Water-Supply Engineering*, 11th ed., Van Nostrand, 1893; 3rd ed., 1882.

Farrar, John: *An Elementary Treatise on Mechanics*, MIT University Press, 1825.

Flamant, A.: *Hydraulique*, Baudry, Paris, 1891; 2nd ed., 1900.

Fletcher, C. A. J.: *Computational Techniques for Fluid Dynamics*, 2nd ed., Springer-Verlag, 1991.

Forchheimer, P.: *Hydraulik*, 3rd ed., Leipzig, 1901; T. G. Teubner, 1930.

Foster, Keith, and G. A. Parker: *Fluidics, Components, and Circuits*, Wiley-Interscience, 1970.

Fox, J. A.: *An Introduction to Engineering Fluid Mechanics*, McGraw-Hill, 1975.

Fox, Robert W., and Alan T. McDonald: *Introduction to Fluid Mechanics*, 4th ed., Wiley, 1992.

Francis, John Robert Dark: *Problems in Hydraulics and Fluid Mechanics for Engineering Students*, E. Arnold, 1964.

—— *A Textbook of Fluid Mechanics for Engineering Students*, E. Arnold, 4th ed., 1975.

—— and P. Minton: *Civil Engineering Hydraulics*, 5th ed., E. Arnold, 1984.

Frederick, D., and T. S. Chang: *Continuum Mechanics*, Allyn and Bacon, 1965.

French, Richard H.: *Open-Channel Hydraulics*, McGraw-Hill, 1985.

Friedlander, Susan: *Introduction to the Mathematical Theory of Geophysical Fluid Dynamics*, Elsevier North-Holland, 1980.

Friedrichs, Kurt Otto: *Fluid Dynamics and Applications*, New York University, 1941.

—— *Selected Topics in Fluid Mechanics*, Gordon and Breach, 1966.

Frost, Walter, and Trevor H. Moulden: *Handbook of Turbulence*, Plenum Press, 1977.

Gasiorek, Janusz Maria, and W. G. Carter: *Mechanics of Fluids for Mechanical Engineers*, Hart Pub. Co., 1968.

Gatski, T. B., S. Sarkar, and C. G. Speziale, eds.: *Studies in Turbulence*, Springer-Verlag, 1992.

Gebhart, Benjamin, Yogesh Jaluria, Roop L. Mahajan, and Bahgat Sammakia: *Buoyancy-Induced Flows and Transport*, Hemisphere Pub. Corp., 1988.

Gentzsch, Wolfgang, and K. W. Neves: *Computational Fluid Dynamics: Algorithms and Supercomputers*, distributed by National Aeronautics and Space Administration, 1988.

Gerhart, Philip M., Richard Gross, and John I. Hochstein: *Fundamentals of Fluid Mechanics*, 2nd edtion, Addison-Wesley, 1992.

Ghil, Michael, and S. Childress: *Topics in Geophysical Fluid Dynamics: Atmospheric Dynamics, Dynamo Theory, and Climate Dynamics*, Springer-Verlag, 1986.

Gibson, Arnold Hartley: *Osborne Reynolds and His Work in Hydraulics and Hydrodynamics*, Longmans, Green and Co., 1946.

Gilbrech, Donald A.: *Fluid Mechanics*, Wadsworth Pub. Co., 1965.

Giles, Ranald V.: *Theory and Problems of Fluid Mechanics and Hydraulics*, McGraw-Hill, 1962.

Ginzburg, Isaak P.: *Applied Fluid Dynamics*, Office of Technical Services, U.S. Dept. of Commerce, Washington, 1963.

Glauert, H.: *Airfoil and Airscrew Theory*, Cambridge University Press, 1932.

Goldstein, Richard J., ed.: *Fluid Mechanics Measurements*, Hemisphere Pub. Corp., 1983.

Goldstein, Sydney: *Lectures on Fluid Mechanics*, Interscience, 1960.

Goldstein, Sydney, ed.: *Modern Developments in Fluid Dynamics*, Clarendon, 1938; Dover, 1965.

Goodbody, A. M.: *Cartesian Tensors with Applications to Mechanics, Fluid Mechanics and Elasticity*, E. Horwood, 1982.

Granet, Irving: *Fluid Mechanics for Engineering Technology*, 3rd ed., Prentice-Hall, 1989.

Granger, Robert Alan: *Incompressible Fluid Dynamics*, U. S. Naval Academy, 1975.

Greenspan, H. P.: *The Theory of Rotating Fluids*, Cambridge University Press, 1968.

Greig, Dorothy Margaret, and T. H. Wise: *Hydrodynamics and Vector Field Theory*, Van Nostrand, 1962.

Grimson, John: *Mechanics and Thermodynamics of Fluids*, McGraw-Hill, 1970.

—— *Advanced Fluid Dynamics and Heat Transfer*, McGraw-Hill, 1971.

Gunzburger, Max D.: *Finite Element Methods for Viscous Incompressible Flows: A Guide to Theory, Practice, and Algorithms*, Academic Press, 1989.

Gupta, Ram S.: *Hydrology and Hydraulic Systems*, Prentice-Hall, 1989.

Hadley, G. A.: *Linear Algebra*, 1961.

Hansen, Arthur G.: *Fluid Mechanics*, Wiley, 1967.

Happel, John, and Howard Brenner: *Low Reynolds Number Hydrodynamics, With Special Applications to Particulate Media*, 2nd ed., M. Nijhoff, 1973.

Heath, T. L.: *The Works of Archimedes*, Cambridge, 1897.

Henderson, F. M.: *Open Channel Flow*, Macmillan, 1966.

Henke, Russell W.: *Introduction to Fluid Mechanics*, Addison-Wesley, 1966.

Hill, M. N.: *The Sea*, vol. I, *Physical Oceanography*, Interscience, 1962.

Hinze, J. O.: *Turbulence*, 2nd ed., McGraw-Hill, 1975.

Hoerner, S. F.: *Fluid-Dynamic Drag*, Hoerner Fluid Dynamics, Brick Town, N. J., 1964.

Hoffman, Klaus A.: *Computational Fluid Mechanics for Engineers*, Engineering Education System, Austin, Tex., 1989.

Holt, Maurice: *Numerical Methods in Fluid Dynamics*, Springer-Verlag, 1988.

Hughes, Thomas J. R., and Jerrold E. Marsden: *A Short Course in Fluid Mechanics*, Publish or Perish, 1976.

Hughes, William F.: *An Introduction to Viscous Flow*, Hemisphere Pub. Corp., 1979.

Hunsaker, J. C., and B. G. Rightmire: *Engineering Applications of Fluid Mechanics*, McGraw-Hill, 1947.

Hunt, J. N.: *Incompressible Fluid Dynamics*, Elsevier, 1964.

Hwang, Ned H. C., and Carlos E. Hita: *Fundamentals of Hydraulic Engineering Systems*, Prentice-Hall, 1987.

Ippen, Arthur T., ed.: *Estuary and Coastline Hydrodynamics*, McGraw-Hill, 1966.

Ireland, Jack William: *Mechanics of Fluids*, Butterworths, 1971.

Jaeger, Charles: *Engineering Fluid Mechanics*, Blackie, 1956.

——— *Fluid Transients in Hydro-Electric Engineering Practice*, Blackie, 1977.

Jameson, Alexander Hope: *An Introduction to Fluid Mechanics*, 2nd ed., Longmans, Green and Co., 1959.

Janna, William S.: *Introduction To Fluid Mechanics*, 3rd ed., PWS Engineering, 1993.

John, James E. A., and William L. Haberman: *Introduction to Fluid Mechanics*, 3rd ed., Prentice-Hall, 1988.

Jones, A. Clement, and C. H. Blomfield: *Elementary Mechanics of Solids and Fluids*, Longmans, Green and Co., 1909.

Joseph, Daniel D.: *Fluid Dynamics of Viscoelastic Liquids*, Springer-Verlag, 1990.

Kaplun, Saul: *Fluid Mechanics and Singular Perturbations*, Academic Press, 1967.

Karamcheti, K.: *Principles of Ideal Fluid Aerodynamics*, Wiley, 1966.

Kaufmann, Walther: *Fluid Mechanics*, McGraw-Hill, 1963.

Kay, J. M., and R. M. Nedderman: *An Introduction To Fluid Mechanics and Heat Transfer*, 3rd ed., Cambridge University Press, 1974.

Kendall, P. C., and Charles Plumpton: *Magnetohydrodynamics*, Pergamon Press, 1964.

Kenyon, Richard A.: *Principles of Fluid Mechanics*, Ronald, 1960.

Khan, Irfan A.: *Fluid Mechanics*, Holt, Rinehart & Winston, 1987.

King, Horace H., and Chester O. Wisler: *Hydraulics*, Wiley, 1927.

Kinsky, Roger: *Applied Fluid Mechanics*, McGraw-Hill, 1982.

Kinsman, Blair: *Wind Waves, Their Generation and Propagation on the Ocean Surface*, Prentice-Hall, 1965.

Kitagawa, K.: *Boundary Element Analysis of Viscous Flow*, Springer-Verlag, 1990.

Knapp, R. T., J. W. Daily, and F. G. Hammitt: *Cavitation*, McGraw-Hill, 1970.

Knudsen, James George, and Donald L. Katz: *Fluid Dynamics and Heat Transfer*, McGraw-Hill, 1958.

Kochin, N. E., I. A. Kibel, and N. V. Roze: *Theoretical Hydrodynamics*, translated from the 5th Russian edition by D. Boyanovitch, ed. J. R. M. Radok, Interscience, 1964.

Kolupaila, S.: *Bibliography of Hydrometry*, University of Notre Dame Press, 1961.

Kraut, G. P.: *Fluid Mechanics for Technicians*, Maxwell Macmillan International, 1992.

Kreider, Jan F.: *Principles of Fluid Mechanics*, Allyn and Bacon, 1985.

Kuethe, A. M., and J. D. Schetzer: *Foundations of Aerodynamics*, Wiley, 1959.

—— and Chuen-Yen Chow: *Foundations of Aerodynamics*, 4th ed., Wiley, 1986.

Kundu, Pijush K.: *Fluid Mechanics*, Academic Press, 1990.

Ladyzhenskaia, O. A.: *The Mathematical Theory of Viscous Incompressible Flow*, Gordon and Breach, 1969.

Laikhtman, David L.: *Physics of the Boundary Layer of the Atmosphere*, Office of Technical Services, U. S. Dept. of Commerce, Washington, 1964.

Lamb, Horace, Sir: *Hydrodynamics*, 6th ed., Cambridge University Press, 1975.

Lanchester, F. W.: *Aerial Flight: Aerodynamics*, Van Nostrand, 1908.

Landahl, Marten: *Unsteady Transonic Flow*, Pergamon Press, 1961.

—— and E. Mollo-Christensen: *Turbulence and Random Processes in Fluid Mechanics*, 2nd ed., Cambridge University Press, 1992.

Landau, L. D., and E. M. Lifshitz: *Fluid Mechanics*, 2nd ed., Pergamon Press, 1987.

Langlois, W. E.: *Slow Viscous Flow*, Macmillan, 1964.

Lapple, C. E.: *Fluid and Particle Mechanics*, University of Delaware, 1951.

Lea, F. C.: *Hydraulics*, E. Arnold, 1907.

LeBlond, P. H., and L. A. Mysak: *Waves in the Oceans*, Elsevier, 1978.

Leliavsky, S.: *Irrigation and Hydraulic Design*, 3 vols., Chapman and Hall, 1965.

—— *An Introduction to Fluvial Hydraulics*, 2nd ed., Dover, 1966.

LeMehaute, Bernard: *An Introduction to Hydrodynamics and Water Waves*, Springer-Verlag, 1976.

Lesieur, Marcel: *Turbulence in Fluids*, 2nd ed., Kluwer, 1990.

Li, Wen-Hsiung, and Sau-Hai Lam: *Principles of Fluid Mechanics*, Addison-Wesley, 1964.

—— *Fluid Mechanics in Water Resources Engineering*, Allyn and Bacon, 1983.

Lichnerowicz, Andre: *Relativistic Hydrodynamics and Magnetohydrodynamics*, W. A. Benjamin, 1967.

Lighthill, James: *An Informal Introduction to Theoretical Fluid Mechanics*, Clarendon, 1986.

—— *Physiological Fluid Mechanics*, Springer-Verlag, 1972.

—— *Waves in Fluids*, Cambridge University Press, 1978.

Lin, C. C.: *The Theory of Hydrodynamic Stability*, Cambridge University Press, 1955.

—— *Laminar Flow and Transition to Turbulence*, Princeton University Press, 1959.

Lliboutry, Luis: *Very Slow Flows of Solids: Basics of Modeling in Geodynamics and Glaciology*, Kluwer, 1987.

Long, Robert R.: *Mechanics of Solids and Fluids*, Prentice-Hall, 1961.

Longwell, Paul A.: *Mechanics of Fluid Flow*, McGraw-Hill, 1966.

Lovejoy, S., and Daniel Schertzer: *Multifractals and Turbulence: Fundamentals and Applications*, World Scientific, 1993.

Lu, Pau-Chang: *Introduction to the Mechanics of Viscous Fluids*, Hemisphere Pub. Corp., 1977.

Macagno, Enzo O.: *Leonardian Fluid Mechanics*, Iowa Institute of Hydraulics Research, University of Iowa, 1989.

Madill, William: *Fluid Mechanics Level III*, Macdonald and Evans, 1983.

Manohar, Madhev, and P. Krishnamachar: *Fluid Mechanics*, 3rd ed., Vikas Pub. House, 1983.

Massel, Stanislaw R.: *Hydrodynamics of Coastal Zones*, Elsevier, 1989.

Massey, B. S.: *Mechanics of Fluids*, 6th ed., Van Nostrand Reinhold, 1989.

McComb, W. D.: *The Physics of Fluid Turbulence*, Clarendon, 1992.

McCormack, Percival D., and Lawrence Crane: *Physical Fluid Dynamics*, Academic Press, 1973.

McDowell, D. M., and J. D. Jackson, eds.: *Osborne Reynolds and the Engineering Science Today*, Manchester University Press, 1970.

McLeod, Edward B.: *Introduction to Fluid Dynamics*, Pergamon Press, 1963.

Medaugh, F. W.: *Elementary Hydraulics*, Van Nostrand, 1924.

Mei, Chiang C.: *The Applied Dynamics of Ocean Surface Waves*, World Scientific, 1989.

Meyer, Richard E.: *Introduction to Mathematical Fluid Dynamics*, Dover, 1971.

Milne-Thomson, L. M.: *Theoretical Hydrodynamics*, 5th ed., Macmillan, 1968.

Mironer, Alan: *Engineering Fluid Mechanics*, McGraw-Hill, 1979.

Mohanty, A. K.: *Fluid Mechanics*, Prentice-Hall of India, 1986.

Monin, A. S.: *Fundamentals of Geophysical Fluid Dynamics*, Kluwer, 1990.

—— *Theoretical Geophysical Fluid Dynamics*, Kluwer Academic Publishers, 1990.

—— and R. V. Ozmidov: *Turbulence in the Ocean*, Reidel, 1985.

—— and A. M. Yaglom: *Statistical Fluid Mechanics; Mechanics of Turbulence*, MIT Press, 1975.

Moore, F. K., ed.: *Theory of Laminar Flows*, Princeton University Press, 1964.

Morris, Henry M.: *Applied Hydraulics in Engineering*, Ronald, 1963.

Moseley, Henry: *A Treatise on Hydrostatics and Hydrodynamics*, Cambridge, 1830.

Mott, Robert L.: *Applied Fluid Mechanics*, 4th ed., Merrill, 1994.

Munk, Max M.: *Fundamentals of Fluid Dynamics for Aircraft Designers*, Ronald, 1929.

Munson, Bruce Roy, Donald F. Young, and Theodore H. Okiishi: *Fundamentals of Fluid Mechanics*, Wiley, 1990.

Murdock, J. W.: *Fluids Mechanics and Its Applications*, Houghton-Mifflin, 1976.

—— *Fluids Mechanics for the Practicing Engineer*, Dekker, 1993.

Murphy, Glenn: *Mechanics of Fluids*, 2nd ed., International Textbook Co., 1952.

Nash, John F., and Virendra C. Patel: *Three-dimensional Turbulent Boundary Layers*, Scientific and Business Consultants, 1972.

Neumann, Gerhard, and Willard J. Pierson, Jr.: *Principles of Physical Oceanography*, Prentice-Hall, 1966.

Newman, John Nicholas: *Marine Hydrodynamics*, MIT Press, 1977.

Newton, Isaac: *Mathematical Principles of Natural Philosophy*, Printed for B. Moote, London, 1729.

Norrie, D. H.: *An Introduction to Incompressible Flow Machines*, Elsevier, 1963.

Nunn, R. H.: *Intermediate Fluid Mechanics*, Hemisphere Pub. Corp., 1989.

Obremski, H. J., M. V. Morkovin, and M. Landahl: *A Portfolio of Stability Characteristics of Incompressible Boundary Layers*, NATO Advisory Group for Aerospace Research and Development, 1969.

O'Brien, Morrough Parker, and George H. Hickox: *Elements of Fluid Mechanics with Applications to Hydraulics*, University of California, 1934.

—— and George H. Hickox: *Applied Fluid Mechanics*, McGraw-Hill, 1937.

Oke, T. R.: *Boundary Layer Climates*, 2nd ed., Methuen, 1987.

Olson, A. T., and K. A. Shelstad: *Introduction to Fluid Flow and the Transfer of Heat and Mass*, Prentice-Hall, 1987.

Olson, Reuben M.: *Essentials of Engineering Fluid Mechanics*, 4th ed., Harper & Row, 1980.

—— and Steven J. Wright: *Essentials of Engineering Fluid Mechanics*, 5th ed., Harper & Row, 1990.

O'Neill, M. E., and F. Chorlton: *Ideal and Incompressible Fluid Dynamics*, Halsted Press, 1986.

———— and F. Chorlton: *Viscous and Compressible Fluid Dynamics*, Halsted Press, 1989.

Owczarek, Jerzy A.: *Introduction to Fluid Mechanics*, International Textbook Co., 1968.

———— *Fundamentals of Gas Dynamics*, International Textbook Co., 1986.

Pai, Shih I.: *Fluid Dynamics of Jets*, Van Nostrand, 1954.

———— *Viscous Flow Theory*, Van Nostrand, 1956.

———— *Introduction to the Theory of Compressible Flow*, Van Nostrand, 1959.

———— *Magnetogasdynamics and Plasma Dynamics*, Prentice-Hall, 1962.

———— *Two-phase Flows*, ed. K. Oswatitsch, Vieweg, 1977.

———— *Modern Fluid Mechanics*, Van Nostrand Reinhold, 1981.

———— and Shijun Luo: *Theoretical and Computational Dynamics of a Compressible Flow*, Van Nostrand Reinhold, 1991.

Panchev, S.: *Random Functions and Turbulence*, Pergamon Press, 1971.

———— *Dynamic Meteorology*, Reidel, 1985.

Panton, Ronald L.: *Incompressible Flow*, Wiley, 1984.

Pao, Richard Hsien-Feng: *Fluid Mechanics*, Wiley, 1967.

Papanastasiou, Tasos C.: *Applied Fluid Mechanics*, Prentice-Hall, 1993.

Parker, Jerald D., James H. Boggs, and Edward F. Blick: *Introduction to Fluid Mechanics and Heat Transfer*, Addison-Wesley, 1969.

Parker, Jerald D., and Faye C. McQuiston: *Introduction to Fluid Mechanics and Heat Transfer*, 2nd ed., Kendall/Hunt Pub. Co., 1988.

Parker, Sybil P., ed.: *Fluid Mechanics Source Book*, McGraw-Hill, 1988.

Patankar, S. V.: *Numerical Heat Transfer and Fluid Flow*, Hemisphere Pub. Corp., 1980.

Paterson, Andrew Robert: *A First Course in Fluid Dynamics*, Cambridge University Press, 1983.

Pedley, T. J.: *The Fluid Mechanics of Large Blood Vessels*, Cambridge University Press, 1980.

Pedlosky, Joseph: *Geophysical Fluid Dynamics*, 2nd ed., Springer-Verlag, 1987.

Peerless, S. J.: *Basic Fluid Mechanics*, Pergamon Press, 1967.

Peterson, Aldor Cornelius: *Applied Mechanics: Fluids*, Allyn and Bacon, 1971.

Philips, O. M.: *Dynamics of the Upper Ocean*, 2nd ed., Cambridge University Press, 1977.

Plapp, John E.: *Engineering Fluid Mechanics*, Prentice-Hall, 1968.

Plint, M. A., and L. Boswirth: *Fluid Mechanics, a Laboratory Course*, Griffin, 1978.

Pnueli, David, and Chaim Gutfinger: *Fluid Mechanics*, Cambridge University Press, 1992.

Potter, Merle C., and John F. Foss: *Fluid Mechanics*, Great Lakes Press, 1982.

Potter, Merle C., and D. Wiggert: *Mechanics of Fluids*, Prentice-Hall, 1991.

Powell, Ralph Waterbury: *Mechanics of Liquids*, Macmillan, 1940.

———— *An Elementary Text in Hydraulics and Fluid Mechanics*, Macmillan, 1951.

Prager, W.: *Introduction to Mechanics of Continua*, Ginn, 1961.

Prandtl, Ludwig: *Essentials of Fluid Dynamics with Applications to Hydraulics, Aeronautics, Meteorology, and Other Subjects*, Hafner Publishing Company, New York, 1952.

———— and O. G. Tietjens: *Fundamentals of Hydro- and Aeromechanics*, Dover, 1957; McGraw-Hill, 1934.

———— and O. G. Tietjens: *Applied Hydro- and Aeromechanics*, Dover, 1957; McGraw-Hill, 1934.

Prasuhn, Alan L.: *Fundamentals of Fluid Mechanics*, Prentice-Hall, 1980.

Prier, I. Larry: *Theory and Applications of Hydrodynamics*, LUS Ltd., 1979.

Purday, H. F. P: *Streamline Flow*, Constable, 1949.

Putterman, Seth J.: *Superfluid Hydrodynamics*, North-Holland Pub. Co., 1974.

Rahman, Matiur: *The Hydrodynamics of Waves and Tides, with Applications*, Computational Mechanics, Southampton, 1988.

Raudkivi, A. J.: *Advanced Fluid Mechanics*, Wiley, 1975.

———— *Loose Boundary Hydraulics*, 3rd ed., Pergamon Press, 1990.

Reynolds, A. J.: *Turbulent Flows in Engineering*, Wiley, 1974.

Rich, George R.: *Hydraulic Transients*, 2nd ed., Dover, 1963.

Richardson, Edward Gick: *Dynamics of Real Fluids*, 2nd ed., E. Arnold, 1961.

Richardson, Stephen M.: *Fluid Mechanics*, Hemisphere Pub. Corp., 1989.

Roache, Patrick J.: *Computational Fluid Dynamics*, Hermosa Publishers, 1985.

Roberson, John A., and Clayton T. Crowe: *Engineering Fluid Mechanics*, 5th ed., Houghton-Mifflin, 1993.

———— John J. Cassidy, and M. Hanif Chaudhry: *Hydraulic Engineering*, Houghton-Mifflin, 1988.

Robertson, James M.: *Hydrodynamics in Theory and Application*, Prentice-Hall, 1965.

Robinson, J. Lister: *Basic Fluid Mechanics*, McGraw-Hill, 1963.

Rodi, Wolfgang: *Turbulence Models and their Application in Hydraulics*, 2nd ed., International Association for Hydraulic Research, 1979.

Rogers, Ruth H.: *Fluid Mechanics*, Routledge & Kegan Paul, 1978.

Rosenhead, Louis: *Laminar Boundary Layers*, Clarendon, 1963.

Round, G. F., and V. K. Garg: *Applications of Fluid Dynamics*, E. Arnold, 1986.

Rouse, Hunter: *Fluid Mechanics for Hydraulic Engineers*, Dover, 1961; McGraw-Hill, 1938.

———— *Elementary Mechanics of Fluids*, Wiley, 1946.

———— *Hydraulics, Mechanics of Fluids, Engineering Education: Selected Writings of Hunter Rouse*, Dover, 1971.

———— and J. W. Howe: *Basic Mechanics of Fluids*, Wiley, 1953.

———— and Simon Ince: *History of Hydraulics*, Iowa Institute of Hydraulic Research, State University of Iowa, 1957.

Rouse, Hunter, ed.: *Advanced Fluid Mechanics*, Wiley, 1959.

Roy, D. N.: *Applied Fluid Mechanics*, Ellis Horwood, 1988.

Rutherford, Daniel Edwin: *Fluid Dynamics*, Interscience, 1959.

Sabersky, Rolf H., Alan J. Acosta, and Edward G. Hauptman: *Fluid Flow*, 3rd ed., Macmillan, 1989.

Sand, Stig Erik: *Three-dimensional Deterministic Structure of Ocean Waves*, Institute of Hydrodynamics and Hydraulic Engineering, Technical University of Denmark, 1979.

Saunders, Harold Eugene: *Hydrodynamics in Ship Design*, 3 vols., Society of Naval Architects and Marine Engineers, 1957–1965.

Schippers, H.: *Multiple Grid Methods for Equations of the Second Kind, with Applications in Fluid Mechanics*, Mathematisch Centrum, 1983.

Schlichting, Hermann: *Boundary Layer Theory*, McGraw-Hill, 1955.

———— and E. Truckenbrodt: *Aerodynamics of the Airplane*, McGraw-Hill, 1979.

Schoder, Ernest W., and Francis M. Dawson, *Hydraulics*, McGraw-Hill, 1934

Schowalter, William Raymond: *Mechanics of Non-Newtonian Fluids*, Pergamon Press, 1979.

Sears, William Rees: *Theoretical Aerodynamics*, Ithaca, New York, 1970.

Selby, Arthur Laidlaw: *Elementary Mechanics of Solids and Fluids*, Clarendon, 1893.

Shames, Irving H.: *Mechanics of Fluids*, 3rd ed., McGraw-Hill, 1992.

Shapiro, Ascher H.: *The Dynamics and Thermodynamics of Compressible Fluid Flow*, Ronald, 1958.

———— *Shape and Flow: The Fluid Dynamics of Drag*, Heinemann, 1964.

────── *Fluid Dynamics*, MIT Center for Advanced Engineering Study, 1984.

Sharp, James J.: *Hydraulic Modeling*, Butterworths, 1981.

────── *Basic Fluid Mechanics*, Butterworths, 1988.

Shepherd, D. G.: *Elements of Fluid Mechanics*, Harcourt, Brace & World, 1965.

Sherman, Frederick S.: *Viscous Flow*, McGraw-Hill, 1990.

Shinbrot, Marvin: *Lectures on Fluid Mechanics*, Gordon and Breach, 1973.

Shivamoggi, Bhimsen K.: *Theoretical Fluid Dynamics*, M. Nijhoff, 1985.

Smith, Augustus William: *An Elementary Treatise on Mechanics*, Harper & Brothers, 1855.

Smith, Douglas Richard Leonard: *Fluid Mechanics Through Worked Examples*, Macmillan, 1965.

Sod, Gary A.: *Numerical Methods in Fluid Dynamics: Initial and Initial Boundary-value Problems*, Cambridge University Press, 1985.

Soo, Shao-lee: *Fluid Dynamics of Multiphase Systems*, Blaisdell Pub. Co., 1967.

Sparenberg, J. A.: *Elements of Hydrodynamic Propulsion*, Kluwer, 1984.

Stern, M. E.: *Ocean Circulation Physics*, Academic Press, 1975.

Stoker, J. J.: *Water Waves*, Interscience, 1957.

Straughan, Brian: *Instability, Nonexistence and Weighted Energy Methods in Fluid Dynamics and Related Theories*, Pitman, 1982.

Streeter, Victor L.: *Fluid Dynamics*, McGraw-Hill, 1948.

────── *Handbook of Fluid Dynamics*, McGraw-Hill, 1961.

────── and E. Benjamin Wylie: *Fluid Mechanics*, 8th ed., McGraw-Hill, 1985.

Sullivan, James A.: *Fundamentals of Fluid Mechanics*, Reston Pub. Co., 1978.

────── *Fluid Power: Theory and Applications*, Prentice-Hall, 1989.

Svendsen, I. A., and Ivar G. Jonsson: *Hydrodynamics of Coastal Regions*, Den Private Ingeniorfond, Technical University, 1980.

Swanson, W. M.: *Fluid Mechanics*, Holt, Rinehart, and Winston, 1970.

Tan, Wei-Yan: *Shallow Water Hydrodynamics: Mathematical Theory and Numerical Solution for a Two-Dimensional System of Shallow Water Equations*, Elsevier, 1992.

Telionis, Demetri P.: *Unsteady Viscous Flows*, Springer-Verlag, 1981.

Temple, George Frederick James: *An Introduction to Fluid Dynamics*, Clarendon, 1958.

Tennekes, H., and J. L. Lumley: *A First Course in Turbulence*, MIT Press, 1972.

Thwaites, Bryan, ed.: *Incompressible Aerodynamics; An Account of the Theory and Observation of the Steady Flow of Incompressible Fluid Past Aerofoils, Wings, and other Bodies*, Clarendon, 1960.

Timman, R., A. J. Hermans, and G. C. Hsiao: *Water Waves and Ship Hydrodynamics*, M. Nijhoff, 1985.

Townsend, A. A.: *The Structure of Turbulent Shear Flow*, 2nd ed., Cambridge University Press, 1976.

Tritton, D. J.: *Physical Fluid Dynamics*, 2nd ed., Clarendon, 1988.

Truesdell, C.: *The Kinematics of Vorticity*, Indiana University Press, 1954.

────── *Continuum Mechanics*, Springer-Verlag, 1966.

────── *The Mechanical Foundations of Elasticity and Fluid Dynamics*, Gordon and Breach, 1966.

────── *A First Course in Rational Continuum Mechanics*, 2nd ed., Academic Press, 1990.

Vallentine, H. R.: *Applied Hydrodynamics*, 2d ed., Plenum Press, 1967.

Van den Berg, Bernard: *Three-dimensional Turbulent Boundary Layers*, Vieweg, 1988.

Van den Ven, Theo G. M.: *Colloidal Hydrodynamics*, Academic Press, 1989.

Van Dyke, Milton: *Perturbation Methods in Fluid Mechanics*, Parabolic Press, 1975.

——— *An Album of Fluid Motion*, Parabolic Press, 1982.

Vennard, John K.: *Elementary Fluid Mechanics*, Wiley, 1940.

——— and Robert L. Street: *Elementary Fluid Mechanics*, 6th ed., Wiley, 1982.

Vincenti, W. G., and C. H. Kruger: *Introduction to Physical Gas Dynamics*, Kreiger, 1965.

Von Kármán, T.: *Aerodynamics*, McGraw-Hill, 1954, 1963.

——— *Collected Works*, Butterworths, 1956.

——— *The Wind and Beyond; Theordore von Kármán, Pioneer in Aviation and Pathfinder in Space*, with Lee Edson, Little Brown, 1967.

——— *The Collected Works of Theodore Von Kármán: 1952–1963*, Von Kármán Institute for Fluid Dynamics, 1975.

Von Mises, Richard: *Mathematical Theory of Compressible Fluid Flow*, Academic Press, 1958.

——— and K. O. Friedrichs: *Fluid Dynamics,* Springer-Verlag, 1971.

Von Schwind, Joseph J.: *Geophysical Fluid Dynamics for Oceanographers*, Prentice-Hall, 1980.

Wallis, B. B.: *One-dimensional Two-phase Flow*, McGraw-Hill, 1969.

Walshaw, Arthur Clifford, and D. A. Jobson: *Mechanics of Fluids*, 3rd ed., Longman, 1979.

Walton, William: *A Collection of Problems in Illustration of the Principles of Theoretical Hydrostatics and Hydrodynamics, Deightons*, Cambridge, 1847; W.P. Grant, 1842.

Ward-Smith, Alfred John: *Internal Fluid Flow: The Fluid Dynamics of Flow in Pipes and Ducts*, Clarendon, 1980.

Webb, Paul W.: *Hydrodynamics and Energetics of Fish Propulsion*, Canadian Department of the Environment Fisheries and Marine Service, 1975.

Webber, Norman Bruton: *Fluid Mechanics for Civil Engineers*, Chapman and Hall, 1971.

West, B. J.: *Deep Water Gravity Waves*, Springer-Verlag, 1981.

Whitaker, Stephen: *Introduction to Fluid Mechanics*, Krieger, 1984; Prentice-Hall, 1968.

White, Frank M.: *Fluid Mechanics*, 2nd ed., McGraw-Hill, 1986.

——— *Viscous Fluid Flow*, 2nd ed., McGraw-Hill, 1991.

Whitham, G. B.: *Linear and Nonlinear Waves*, Interscience, 1974.

Wiegel, R. L.: *Oceanographical Engineering*, Prentice-Hall, 1964.

Wilkinson, W. L.: *Non-Newtonian Fluids*, Pergamon Press, 1960.

Williams, Jerome, and Samuel A. Elder: *Fluid Physics for Oceanographers and Physicists, An Introduction to Incompressible Flow*, Pergamon Press, 1989.

Williams, John: *Fluid Mechanics*, Allen and Unwin, 1974.

Wilson, Derek Henry: *Hydrodynamics*, E. Arnold, 1959.

Wylie, E. Benjamin, and Victor L. Streeter: *Fluid Transients*, FEB Press, 1983.

Yang, Huijun: *Wave Packets and their Bifurcations in Geophysical Fluid Dynamics*, Springer-Verlag, 1990.

Yeh, Hsuan, and Joel I. Abrams: *Principles of Mechanics of Solids and Fluids*, McGraw-Hill, 1960.

Yih, Chia-shun: *Fluid Mechanics*, West River Press, 1977.

——— *Stratified Flows*, 2nd ed., Academic Press, 1980.

Young, A. D.: *Boundary Layers*, BSP Professional Books, Oxford, 1989.

Yuan, Shao Wen: *Foundations of Fluid Mechanics*, Prentice-Hall, 1967.

Zakharov, V. E., V. S. Lvov, and G. Falkovich: *Kolmogorov Spectra of Turbulence*, Springer-Verlag, 1992

Ziegler, Franz: *Mechanics of Solids and Fluids*, Springer-Verlag, 1991.

INDEX

Adsorption 427
Advection 426
Advection-diffusion 427, 431
  ray method 438
Airfoils, flow around 122
Alternating unit tensor 446
Ampere's Law 99
Amplitude spectrum 403
Analytic function of a complex variable
  117
Angle of attack 186
Anisotropic media 90
Apparent mass 114, 117
Aristotle 19
Axisymmetric
  flow 92, 105
  tube flow 157

Backwater 187
Baroclinic 35, 380
Barotropic 35
Beats 402
Bernoulli,
  Daniel 19, 31
  Jacob 31
  Johann 31
  Nikolaus 31
Bernoulli's equation 29, 30, 33, 88, 136,
  174
  linearized 393
Biharmonic equation 161
Biot-Savart 99
Bipotential equation 161
Blasius 122, 224, 232
  equation 195, 196
  theorem 124
Blasius, P. R. H. 253
Bohlen 202
Bond number 72
Borda, Jean Charles 94
Bore 291

Bottom layer 354
Boundary condition
  dynamic 136, 141, 217, 391
  far-field 142
  free surface 266
  kinematic 134, 139, 217, 392
  on Navier-Stokes equation 59
  radiation 143
  with damping 144
Boundary integral equation method 110
Boundary layer 58, 103
  Blasius 224
  control 226
  displacement thickness 192, 254, 255
  energy thickness 194
  equations 179
  fence 189
  momentum thickness 193, 254, 255
  theory 172
  thickness 178
  turbulent 254
Boussinesq
  approximation in stratified flow 380
  equations 333
  equations for shallow water 328
Boussinesq, Joseph 333
Breakwater 415
Buckingham pi-theorem 46
Buffer layer 249, 251
Bulk
  modulus 77
  viscosity 25
Butterfly effect 215

Capacity coefficient 76
Capillary
  number 72
  wave 408, 410
  wave, stability 411
Carrier wave 401
Cartesian

coordinate system 448
tensors 449
Cauchy 122
number 71
principle value 110
theorem 123
Cauchy, Augustin Louis 119
Cauchy-Riemann equations 118, 152
Cavitation 64, 165
number 64
Cells 382
Centered simple wave 287
Channel
circular 310
one dimensional flow 298
prismatic 299
rectangular 282, 308
trapezoid 309
triangular 310
Characteristics 273, 426
backward 274
forward 274
theory of 274
Circulation 93
Closure problem 230
Codeformational time derivative 14, 15
Coefficient of compressibility 379
Complex
potential 118
region 285
Complimentary error function 430
Compressibility
coefficient of 379
mean 379
Compressible flow 271
Conformal mapping 119
Conservation
form of the momentum equation 9, 10
of energy 11
of mass 7, 31, 228
of momentum 8, 32, 343
Conservative substance 427
Continuity equation 7, 228
Contravarient tensor 459
Control 303
surface 6
Control volume 3
energy equation 11, 12
general integral 6
mass 7
momentum equation 8, 32
Convected derivative 15
Convection

forced 436
free 436
Coordinates
cylindrical 457
inertial 339
spherical 458
transformation of 457
Coriolis
force 340
parameter 344, 349
Corner
flow near a sharp corner 121
Couette flow 156, 157, 168
Courant number 73
Covarient tensor 460
Creeping flow 160
Critical
depth 283, 289
flow 283
Cross product 445
Curl of a vector 456, 459
Cutoff wall 130
Cyclostrophic motion 350
Cylinder
flow around 103, 121
flow around rotating 130
flow around with circulation 124
Cylindrical coordinate system 457

Dam
break 285, 288, 289, 293, 296
flow through 150
Darcy
friction factor 158, 232
number 72
Darcy's law 90, 140
Darcy-Weisbach equation 158, 232
Dean number 71
Deborah number 75
Deep water 396
oceanographer's 397
Deissler's equation 238, 239, 437
Del 455
Density anomaly 374
Depth of influence 353
Derivative
codeformational 15
directional 275, 276
Jaumann 14
material 14
partial 276
substantial 15
time 13

total 92, 276
Desorption 427
Determinant 464
Diagonal jumps 326
Differentiation 455
Diffraction 415
Diffusion 426
    coefficient 73, 427, 433
    turbulent 433
Dimensional
    analysis 45
    approximation 264
    matrix 48, 49
Dimensionless equations of motion 51
Dimensionless numbers 67
    Bond 72
    Brinkman 72
    capacity coefficient 76
    capillary 72
    Cauchy 71
    cavitation 64
    Courant 73
    Darcy 72
    Dean 71
    Deborah 75
    drag coefficient 74, 185
    Eckert 72
    Ekman 75, 346, 347, 363
    Euler 74
    Fourier number 74
    Froude 52, 55, 57, 66, 301
    Görtler 76
    Graetz 71
    Grashof 72
    grid Peclet 73
    head coefficient 76
    Jacob 72
    Keulegan-Carpenter 74
    kinematic flow 71
    Knudsen 74
    lift coefficient 74, 185
    Mach 62, 63, 90, 226
    Nusselt 72
    Peclet 73, 428, 437
    power coefficient 76
    Prandtl 71, 219
    pressure coefficient 52
    Rayleigh 72
    Reynolds 52, 54, 58, 158, 160, 163,
        167, 168, 197
    Reynolds, eddy 262
    Reynolds, roughness 251
    Richardson 75, 219, 382

    Rossby 75, 346
    Ruark 74
    Schmidt 73, 219, 437
    Sherwood 73
    skin friction coefficient 198
    specific speed 76
    Stanton 73
    Stokes 76
    Strouhal 74, 187
    Taylor 75, 363
    turbulent Knudsen 74
    turbulent Reynolds 244
    Ursell 75
    Vedernikov 307–310, 315, 318
    Weber 56
Direct product 449
Direction cosine 449
Directional derivative 275
Dirichlet boundary conditions 88, 139
Dispersion 426, 432
    coefficient 427, 433, 434
    equation 219, 409, 419
    hydrodynamic 432
    in porous media 433
    porous media 433
    relationship 394
Dispersion relationship 393
Dispersivity 434
Displacement thickness 192, 254, 255
Distorted models 65
Divergence
    of a tensor 457
    of a vector 456, 458, 459
    of the velocity 25
    theorem 6, 7, 465
Domain of dependence 281
Dot product 445, 447
Doublet 101, 103
    axisymmetric 106
    in complex notation 119
    line, axisymmetric flow 109
Downwelling 381
Drag 243, 246
    coefficient 74, 185, 232
    total 58
    viscous 58, 60
    wave 58
Du Buat, Pierre Louis Georges 282
Dyad 449, 451
    unit 450
Dynamic boundary condition 136, 141,
    217, 391

Eckert number 72
Eddy 230, 262
   Reynolds number 262
   scale 260
   viscosity 234, 235, 237, 244, 245, 248,
     250
   wave length 260
Effective porosity 140
Einstein convention 446, 449
Ekman
   depth of influence 359, 363
   number 75, 346, 347, 363
   spiral 352
   surface layer 353
Ekman, Vagn Walfrid 357
Ellipse, flow around 122
Ellipsoids, apparent mass factors 114
Elliptic equations 277
Elliptical conduit 170
Emerged 19
Energy
   cascade 262
   correction factor 32
   equation 13, 15, 291, 435
   internal 435
   mechanical 230
   spectrum 260
   thickness 194
   turbulent 260, 262
   wave 398
Enthalpy 30
Entropy 88
Epilimnion 375
Equation
   characteristic 282
   Chezy 308
   elliptic 277
   energy 291
   heat 278
   hyperbolic 277
   Laplace's 277
   Manning's 303
   of state 26, 64
   parabolic 277
   Saint Venant 282
   wave 272, 278
Equilibrium range 261
Equipotential lines 93, 88
Error function 430
Euler
   equations 271
   number 74
Euler, Leonard 31, 39

Eulerian method 2, 6
Expansion flow 40

Falkner-Skan equation 200
Fall turnover 376
Fanning friction factor 232
Far-field boundary conditions 142
Fetch 364
Fick's law 426, 435, 436
Field approach to fluid problems 2
Flow nets 94
Flutter 189
   speed 189
Fourier
   integral 404
   number 74
   transform 404
Fourier's law 435
Free shear flows 204
Free surface 36, 57
   boundary conditions 266
   flow 264
Fresh-salt water interface 152
Friction
   factor, Darcy 158, 232
   factor, Fanning 232
   slope 282, 299, 300, 305, 319
   velocity 238
Froude number 52, 55, 57, 66, 301
Froude, William 53
Fundamental
   dimensions 45
   units 45

Galileo Galilei 19
Garabedian, Paul R. 148
Gauge pressure 56
Geostrophic
   flow 347
   velocity 347
Gerstner's wave 422
Gibbs notation 455
Girard, Pierre 26
Golf ball 185
Görtler
   number 76
   vortices 76
Gradient
   current 349, 350
   of a scalar 456, 459
Graetz number 71
Grashof number 72
Gravity 10, 343
Grid Peclet number 73

Group velocity 396, 400, 401, 403
capillary waves 410
Gulf Stream 356, 357, 384

Head
coefficient 76
loss 33, 158
Heat equation 278, 436
Hele-Shaw flow 165
Helmholtz
first theorem 98
second theorem 98
Hodograph transformation 145
Holmboe instability 219
Holomorphic 117
Holstein 202
Hunt 148
Huygens' principle 415
Hydraulic
bore 290, 291, 305
bore formation 311
conductivity 90, 167
radius 300, 308, 310, 312
Hydraulic jump 290, 292
diagonal 326
strong 327
weak 327
Hydraulic radius 256
Hyperbolic equations 277
Hypolimnion 376

Images, method of 104
Incompressible fluid 24, 25
Inertial
coordinate system 339
current 349
frequency 373
period 349
sublayer 249–251
Initial line 281
Interfacial stability 216
Intermittency 244
Internal waves 417
Irrotationality condition 28, 88
Isentropic, pressure-density equation 28, 62
Isopycnals 384
Isovels 259

Jacob number 72
Jacobian 39
Jaumann derivative 14
Jet 204, 239, 242, 248
axisymmetric 209, 248

circular 248
plane 207, 246, 248
velocity and width in laminar and turbulent flow 248
Joukowski transformation 122

Karman
constant 236, 251
momentum equation 200, 201, 254
similarity theory 235, 437
vortex street 187
Karman-Pohlhausen method 201
Kelvin wave 373
Kelvin's method of stationary phase 405
Kelvin-Helmholtz instability 216, 219
Keulegan-Carpenter number 74
Kinematic
boundary condition 134, 139, 217, 392
flow number 71
Kinetic energy, in unsteady flow 114
Klein, Felix 181
Knudsen number 74
Kolmogoroff micro scale 262
Kronecker delta 446, 450
Kuroshio current 356

Lagrange, Joseph Louis 39, 91, 94, 119
Lagrangian
conservation of mass 39
description of fluid flow 2
equation of motion 38
Laminar sublayer 234
Lanchester, Frederick William 181
Lansford 148
Laplace 94, 119
equation 29, 88, 217, 218, 275, 277
equation, cylindrical coordinates 105
transform 368, 370
Laplace, Pierre Simon 39, 91
Laplacian 459
of a scalar 456
of a vector 457
operator 456
Lapse rate 364
Law of the wall 238, 437
Leibnitz' theorem 267
Lift coefficient 74, 185
Line source 91
Lubrication 168

Mach
angle 320
lines 320
number 62, 63, 90, 226

Manning's equation 303
Mass transfer coefficient 74
Material derivative 14
  of a tensor 14
Matrix multiplication 451
Mean
  compressibility 379
  free path 74, 91
  pressure 25, 250
Meander of rivers 190
Mechanical energy
  conversion from thermal 26, 230
  equation of 26
  turbulence equation 263
Metalimnion 376
Method of stationary phase 405
Mixed tensor 460
Mixing length 235, 237, 437
Model 44
Modulating function 401
Molecular diffusivity 73
Momentum
  conservation of 343
  correction factor 32, 268
  equation 9, 32
  in a turbulent jet 247, 248
  thickness 193, 254, 255
  transport 10, 229
  vertically integrated equation 66
Moody, Lewis F. 253
Multiply connected 97

Nabla 455
Nansen, Fridtjof 357
Navier, Louis 26
Navier-Stokes equation 25, 59, 230
Neumann boundary conditions 88
Newton, Sir Isaac 19, 26, 39
Newton's law
  of gravitation 45
  of viscosity 19, 234, 250
Newtonian fluids 19
Nikuradse 197
No-slip boundary condition 34, 36, 174,
    177, 249, 251
Non-circular conduits 158, 255
Non-Newtonian fluid 19, 210
Normal stress 24
Normalizing conditions 464
Numerical
  boundary 142
  wave maker 145
Nusselt number 72

Orifice, flow from 145
Orr-Sommerfeld equation 220, 222
Orthogonality conditions 464
Orthonormal matrix 463
Oseen correction 164
Outer
  core 249
  product 449

Parabolic equations 277
Parallel flow 89
  axisymmetric 106
  in complex notation 119
Peclet number 73, 428, 437
Perfect
  fluid 27
  gas 24
Perturbation potential 89
Pi-theorem 46
Pipe
  flow 155, 157, 232, 238, 249, 253, 262
  friction factor 252
  Reynolds number 262
  rough 232
  roughness 232
  smooth 232, 238, 249
Plane, viscous flow along 170
Plate
  boundary layer on 195, 253
  flow around 149
  flow between two 156, 223, 236
  stagnation flow 173
  wake behind 205
Pohlhausen 201, 202
  method 202
Poiseuille
  flow 156, 157, 168, 171, 223, 236
  law 157
Poiseuille, Jean Louis 157
Poisson, Simeon Denis 119
Poisson's equation 159
Polytropic flow 271
Porous media 90, 139, 150
  dispersion in 433
Potential 28
  temperature 379
Power coefficient 76
Prandtl
  mixing length 235, 237, 261, 262, 437
  number 71
Prandtl number 219
Prandtl, Ludwig 181
Prandtl-Meyer function 322

Pressure 9
coefficient 52
dimensionless 54
gauge 56
mean 25, 250
thermodynamic 24, 250
turbulent 250
vapor 64
work 12
Principal axes 21, 22, 452
Prismatic channel 299
Progressive damping 144
Prototype 44

Quasi-linear equations 274

Radiation boundary conditions 143
Range of influence 281
Rank of a matrix 49
Rankine
body 102
half-body 92
Rate of strain 18, 22, 231, 232
Ray
method 437, 438
tracing 406
Rayleigh number 72
Rectangular channels 282
Refraction
coefficient 414
wave 413
Relative roughness 254
Reverse flow 182
Reynolds
apparatus 215
equation of lubrication 169
experiment 215
number 52, 54, 58, 158, 160, 163, 167, 168, 197
number, critical 223, 224
number, eddy 262
number, roughness 251
number, turbulent 244
stresses 67, 229, 232, 236, 241, 243, 247, 258
stresses, axisymmetric wake 246
Reynolds, Osborne 61, 253, 333
Richardson number 75, 219, 382
Riemann invariants 279, 283, 303
Rising water 315
River 187
meanders 190
Roll waves 317
Rossby number 75, 346

Rotating
cylinders 224
earth 339
Rotation 14, 93, 338, 380, 456
tensor 16
vector 463
Rough wall 252
Roughness
height 251
relative 254
Reynolds number 251
Ruark number 74

Saint Venant
equations 282
Jean-Claude Barre de 282
Salt-fresh water interface 152
Saph, Augustus V. 253
Scalar
gradient 456, 459
Laplacian 456
product 445, 447, 451
Scale
factors 51, 458
model 52
Schlicting 197, 202
Schmidt number 73, 219, 437
Schoder, Ernest W. 253
Schrodinger's equation 19
Schwarz-Christoffel transformation 125
Second coefficient of viscosity 25
Secondary currents
in curves 189
in non-circular conduits 259
in straight conduits 256
Seepage surface 141
Seiche 370, 372
internal 422
Self-preserving 240
Separation 182, 187
point of 182
zone of 182
Shallow water 396, 398, 400
equations 265
oceanographer's 397
theory 264
Shape
factor 254
parameter 202
Shear 9, 10, 254
matrix 438
Sherwood number 73
Shock wave 290

Sideral day 343
Similarity hypothesis 243, 437
Similitude 50, 199
   approximate 55–57, 63
   complete 52, 63
   incomplete 53
Simple wave 284, 286, 287, 289, 294, 295,
      303, 323
   centered 287, 295, 296
Simply connected 97
Sink 90
Skin friction coefficient 198
Slider bearing 168
Slope current 358
Slow flow 160
Snell's law 414
Solar ponds 378
Solution system
   constant mass 1
   constant volume 1
   Eulerian 2
   Lagrangian 2
Sound, speed of 88
Source 40, 110, 427
   asymmetric flow 106
   flow 90
   flow, complex notation 119
   line 91, 109
   line, axisymmetric flow 107
   surface 91
Specific
   discharge 140, 433
   heat 28
   speed 76
   volume 379
Specific discharge 90
Specific discharge vector 90
Spectral analysis 261
Speed of sound 271
Sphere 106
   flow around 162, 171
   unsteady flow around 111
Spherical
   cavity 165
   coordinate system 458
Stability 305
   condition for interfacial flow 219, 412
   of capillary waves 411
   of laminar flow 172, 215
Stagnation
   flow 173
   flow, axially symmetric 175
   point 92, 173, 174

   pressure 174
Stall 182, 185
Standing waves 393, 394
Stanton number 73
Stanton, Thomas E. 253
Stanton-Moody diagram 232
Stationary phase 405
Stephan problem 134
Stokes
   flow 160, 232
   law 163
   number 76
   paradox 155, 163, 164
   theorem 94, 95, 122
Stokes, George 26
Strain tensor 16
Stratification 374, 380, 382
   lakes 374
   oceans 377
Stratified fluid 35
Stream function 41, 92, 89–91, 106, 107,
      118, 161, 347,
      368, 371
Streamlines 92
Stress tensor 19, 25, 229
Strouhal number 74, 187
Subcritical flow 283
Sublayer
   inertial 249–251
   laminar 234
   turbulent 250
   viscous 234, 249–251
Substantial derivative 15, 25, 35
   of a vector 14
Suction 226
Summation convention 446
Supercritical flow 283
   two-dimensional 319
Superposition 92
Surface
   layer 353
   source 91, 109
   tension 55, 57
Symmetric tensor 452
Systems approach 2

Taylor
   column 347
   number 75, 363
Telegrapher's equation 335
Temperature 435
Tensor
   addition 451

alternating unit 446
Cartesian 449
contraction 24
contravarient 459
covarient 460
differentiation 460
divergence 457
identities 464
mixed 460
multiplication 451, 452
product 451
rotation 16
strain 16, 22, 40
stress 19, 25, 229
symmetric 452
unit 450
Theory of characteristics 274
Thermal
    conversion from mechanical energy 27,
        230
    energy equation 26, 435
    expansion coefficient 375
    wind 384
Thermocline 376
Thermodynamic pressure 24, 250
Thin airfoil theory 89
Tides 265
Tornado 129
Transformation
    matrix 279, 321
    of coordinate systems 457
Trefftz, E. 148
Triangular duct 171
Trochoidal wave 423
Tsunami 265, 271
Tube flow 155
    non-circular conduits 158
Turbomachinery 76
Turbulence 436
    definition 214
    mechanical energy equation 263
    self-preserving 240
Turbulent
    Knudsen number 74
    pressure 250
    Reynolds number 244
    sublayer 250

Uniform
    flow 40
    region 284
Unit
    dyad 450

    tensor 450
    vector 444
Unsteady flow 110
Upwelling 381
Ursell parameter 75

Vapor pressure 64
Vector
    acceleration 455, 463
    addition 444, 447
    algebra 444
    column 444
    curl 456, 459
    differential operations 458
    differentiation 455, 460
    divergence 456, 458, 459
    identities 463, 464
    multiplication 445, 447
    notation 459
    operator 455
    product 445, 448, 449, 451
    rotation 463
    subtraction 444
    transformation 448
    unit 444
    unit normal 445
    vorticity 18
Vedernikov number 307–310, 315, 318
Velocity 455
    average 31, 158
    correlation 229, 232
    defect 241, 244–246
    distribution near a wall 238
    distribution, logarithmic 250
    Friction 238, 250
    potential 28, 87
    time averaged 226
Velocity defect 205
Virtual mass 112, 114
Viscosity 19
    bulk 25
    eddy 234, 235, 237, 244, 245, 248, 250
    Newton's law 19, 234, 250
    second coefficient 25
Viscous sublayer 234, 249
von Kármán, Theodore 181, 201, 238
Vortex
    bound vorticity 100
    decay of velocity in 41
    filament 97
    flow 91, 96
    flow, complex notation 119
    forced 40

free vorticity 100
generator 185
line 40, 97
potential 40
shedding 187, 210
sheet 99, 100
stretching 230, 261
tube 98
Vorticity 33, 93, 95, 110, 232
equation 33
transport equation 34, 41, 231
transport in two dimensions 34
vector 18, 231, 236

Wake 104, 180, 182, 204, 239, 240, 243, 248
axisymmetric 245
circular 248
function 251
plane 248
velocity and width in laminar and turbulent flow 248
Wall
effect 105
rough 251, 252
smooth 249
turbulence 248–250
Water hammer 336
Wave
between fluids of different densities 219
breaking 315
capillary 408, 410
carrier 401
comparison of deep and shallow 396
diffraction 415

dispersion equation 393, 394, 396, 419
energy 396, 398
equation 272, 278
Gerstner's 422
group velocity 396, 400
in a nonhomogeneous fluid 420
in a two-layered fluid 417
internal 417
linear 396
linearized equations 153, 389
particle paths 396, 397
potential 396
power 396, 399
pressure 398
progressive 393
refraction 413
simple 323
small amplitude 391
speed 271, 272, 283, 333, 396, 410
speed (critical) 412
standing 394
stream function 396
tank 144
trains 400
trochoidal 423
Weber number 56
Weber,
Ernst Hienrich 56
Moritz 53, 56
Wilhelm Eduard 56
Whitehead paradox 164
Williams, G. S. 253
Wind over water 364, 412

Zone of determinacy 281

Physical Properties of Water

Temp. °C	Density kg/m^3	Viscosity N-s/m$^2 \times 10^3$	Vapor pressure kN/m^2	Surface tension N/m$\times 10^2$
0	999.8	1.80	0.61	7.56
5	1000.0	1.52	0.87	7.49
10	999.7	1.31	1.23	7.42
15	999.2	1.15	1.70	7.35
20	998.3	1.00	2.34	7.28
25	997.1	0.897	3.17	7.20
30	995.7	0.801	4.24	7.12
40	992.3	0.659	7.38	6.96
50	988.0	0.544	12.33	6.79
60	983.2	0.470	19.92	6.62
70	977.7	0.405	31.16	6.44
80	971.6	0.356	47.34	6.26
90	965.1	0.318	70.10	6.08
100	958.1	0.284	101.33	5.89

U.S. Standard Atmosphere

Altitude (m)	Temperature (°C)	Pressure (kPa)	Density (kg/m^3)	Viscosity N-s/m$^2 \times 10^5$
0	15.000	101.325	1.2250	1.7894
1,000	8.501	89.876	1.1117	1.7579
2,000	2.004	79.501	1.0066	1.7260
3,000	− 4.491	70.121	0.90925	1.6938
4,000	−10.984	61.660	0.81935	1.6612
5,000	−17.474	54.048	0.73643	1.6282
6,000	−23.963	47.218	0.66011	1.5949
7,000	−30.450	41.105	0.59002	1.5612
8,000	−36.935	35.652	0.52579	1.5271
9,000	−43.417	30.801	0.46706	1.4926
10,000	−49.898	26.450	0.41351	1.4577
11,000	−56.376	22.670	0.36480	1.4223
12,000	−56.500	19.399	0.31194	1.4216
13,000	−56.500	16.580	0.26660	1.4216
14,000	−56.500	14.170	0.22786	1.4216
15,000	−56.500	12.112	0.19475	1.4216
16,000	−56.500	10.353	0.16647	1.4216
17,000	−56.500	8.8497	0.14230	1.4216
18,000	−56.500	7.5652	0.12165	1.4216
19,000	−56.500	6.4675	0.10400	1.4216
20,000	−56.500	5.5293	0.08891	1.4216
22,000	−54.576	4.0475	0.06451	1.4322
24,000	−52.590	2.9717	0.04694	1.4430
26,000	−50.606	2.1884	0.03426	1.4538
28,000	−48.623	1.6162	0.02508	1.4646
30,000	−46.641	1.1970	0.01841	1.4753